Probabilistic
Robotics

Probabilistic Robotics

Sebastian Thrun
Wolfram Burgard
Dieter Fox

The MIT Press
Cambridge, Massachusetts
London, England

© 2006 Massachusetts Institute of Technology

All rights reserved. No part of this book may be reproduced in any form by any electronic or mechanical means (including photocopying, recording, or information storage and retrieval) without permission in writing from the publisher.

Typeset in 10/13 Lucida Bright by the authors using LATEX 2_ε.
Printed and bound in the United States of America.

Library of Congress Cataloging-in-Publication Data

Thrun, Sebastian, 1967–
    Probabilistic robotics / Sebastian Thrun, Wolfram Burgard, Dieter Fox.
       p. cm. – (Intelligent robotics and autonomous agents series)
    Includes bibliographical references and index.
    ISBN-13: 978-0-262-20162-9 (alk. paper)
  1. Robotics. 2. Probabilities. I. Burgard, Wolfram. II. Fox, Dieter. III. Title. IV. Intelligent robotics and autonomous agents.
    TJ211.T575 2005
    629.8'92–dc22

                                                                                                                 2005043346

Brief Contents

I Basics   1
1. Introduction   3
2. Recursive State Estimation   13
3. Gaussian Filters   39
4. Nonparametric Filters   85
5. Robot Motion   117
6. Robot Perception   149

II Localization   189
7. Mobile Robot Localization: Markov and Gaussian   191
8. Mobile Robot Localization: Grid And Monte Carlo   237

III Mapping   279
9. Occupancy Grid Mapping   281
10. Simultaneous Localization and Mapping   309
11. The GraphSLAM Algorithm   337
12. The Sparse Extended Information Filter   385
13. The FastSLAM Algorithm   437

IV Planning and Control   485
14. Markov Decision Processes   487
15. Partially Observable Markov Decision Processes   513

16 *Approximate POMDP Techniques*   547
17 *Exploration*   569

Contents

Preface   xvii

Acknowledgments   xix

I Basics   1

1 *Introduction*   3
- 1.1   Uncertainty in Robotics   3
- 1.2   Probabilistic Robotics   4
- 1.3   Implications   9
- 1.4   Road Map   10
- 1.5   Teaching Probabilistic Robotics   11
- 1.6   Bibliographical Remarks   11

2 *Recursive State Estimation*   13
- 2.1   Introduction   13
- 2.2   Basic Concepts in Probability   14
- 2.3   Robot Environment Interaction   19
  - 2.3.1   State   20
  - 2.3.2   Environment Interaction   22
  - 2.3.3   Probabilistic Generative Laws   24
  - 2.3.4   Belief Distributions   25
- 2.4   Bayes Filters   26
  - 2.4.1   The Bayes Filter Algorithm   26
  - 2.4.2   Example   28
  - 2.4.3   Mathematical Derivation of the Bayes Filter   31
  - 2.4.4   The Markov Assumption   33

- 2.5 Representation and Computation 34
- 2.6 Summary 35
- 2.7 Bibliographical Remarks 36
- 2.8 Exercises 36

3 Gaussian Filters 39

- 3.1 Introduction 39
- 3.2 The Kalman Filter 40
    - 3.2.1 Linear Gaussian Systems 40
    - 3.2.2 The Kalman Filter Algorithm 43
    - 3.2.3 Illustration 44
    - 3.2.4 Mathematical Derivation of the KF 45
- 3.3 The Extended Kalman Filter 54
    - 3.3.1 Why Linearize? 54
    - 3.3.2 Linearization Via Taylor Expansion 56
    - 3.3.3 The EKF Algorithm 59
    - 3.3.4 Mathematical Derivation of the EKF 59
    - 3.3.5 Practical Considerations 61
- 3.4 The Unscented Kalman Filter 65
    - 3.4.1 Linearization Via the Unscented Transform 65
    - 3.4.2 The UKF Algorithm 67
- 3.5 The Information Filter 71
    - 3.5.1 Canonical Parameterization 71
    - 3.5.2 The Information Filter Algorithm 73
    - 3.5.3 Mathematical Derivation of the Information Filter 74
    - 3.5.4 The Extended Information Filter Algorithm 75
    - 3.5.5 Mathematical Derivation of the Extended Information Filter 76
    - 3.5.6 Practical Considerations 77
- 3.6 Summary 79
- 3.7 Bibliographical Remarks 81
- 3.8 Exercises 81

4 Nonparametric Filters 85

- 4.1 The Histogram Filter 86
    - 4.1.1 The Discrete Bayes Filter Algorithm 86
    - 4.1.2 Continuous State 87
    - 4.1.3 Mathematical Derivation of the Histogram Approximation 89

		4.1.4	Decomposition Techniques 92

    4.2   Binary Bayes Filters with Static State   94
    4.3   The Particle Filter   96
            4.3.1   Basic Algorithm   96
            4.3.2   Importance Sampling   100
            4.3.3   Mathematical Derivation of the PF   103
            4.3.4   Practical Considerations and Properties of Particle Filters   104
    4.4   Summary   113
    4.5   Bibliographical Remarks   114
    4.6   Exercises   115

5 *Robot Motion*   117
    5.1   Introduction   117
    5.2   Preliminaries   118
            5.2.1   Kinematic Configuration   118
            5.2.2   Probabilistic Kinematics   119
    5.3   Velocity Motion Model   121
            5.3.1   Closed Form Calculation   121
            5.3.2   Sampling Algorithm   122
            5.3.3   Mathematical Derivation of the Velocity Motion Model   125
    5.4   Odometry Motion Model   132
            5.4.1   Closed Form Calculation   133
            5.4.2   Sampling Algorithm   137
            5.4.3   Mathematical Derivation of the Odometry Motion Model   137
    5.5   Motion and Maps   140
    5.6   Summary   143
    5.7   Bibliographical Remarks   145
    5.8   Exercises   145

6 *Robot Perception*   149
    6.1   Introduction   149
    6.2   Maps   152
    6.3   Beam Models of Range Finders   153
            6.3.1   The Basic Measurement Algorithm   153
            6.3.2   Adjusting the Intrinsic Model Parameters   158
            6.3.3   Mathematical Derivation of the Beam Model   162

        6.3.4   Practical Considerations   167
        6.3.5   Limitations of the Beam Model   168
6.4   Likelihood Fields for Range Finders   169
        6.4.1   Basic Algorithm   169
        6.4.2   Extensions   172
6.5   Correlation-Based Measurement Models   174
6.6   Feature-Based Measurement Models   176
        6.6.1   Feature Extraction   176
        6.6.2   Landmark Measurements   177
        6.6.3   Sensor Model with Known Correspondence   178
        6.6.4   Sampling Poses   179
        6.6.5   Further Considerations   180
6.7   Practical Considerations   182
6.8   Summary   183
6.9   Bibliographical Remarks   184
6.10 Exercises   185

II Localization   189

7 *Mobile Robot Localization: Markov and Gaussian*   191

7.1   A Taxonomy of Localization Problems   193
7.2   Markov Localization   197
7.3   Illustration of Markov Localization   200
7.4   EKF Localization   201
        7.4.1   Illustration   201
        7.4.2   The EKF Localization Algorithm   203
        7.4.3   Mathematical Derivation of EKF Localization   205
        7.4.4   Physical Implementation   210
7.5   Estimating Correspondences   215
        7.5.1   EKF Localization with Unknown Correspondences   215
        7.5.2   Mathematical Derivation of the ML Data Association   216
7.6   Multi-Hypothesis Tracking   218
7.7   UKF Localization   220
        7.7.1   Mathematical Derivation of UKF Localization   220
        7.7.2   Illustration   223
7.8   Practical Considerations   229

7.9 Summary 232
7.10 Bibliographical Remarks 233
7.11 Exercises 234

8 *Mobile Robot Localization: Grid And Monte Carlo* 237

8.1 Introduction 237
8.2 Grid Localization 238
    8.2.1 Basic Algorithm 238
    8.2.2 Grid Resolutions 239
    8.2.3 Computational Considerations 243
    8.2.4 Illustration 245
8.3 Monte Carlo Localization 250
    8.3.1 Illustration 250
    8.3.2 The MCL Algorithm 252
    8.3.3 Physical Implementations 253
    8.3.4 Properties of MCL 253
    8.3.5 Random Particle MCL: Recovery from Failures 256
    8.3.6 Modifying the Proposal Distribution 261
    8.3.7 KLD-Sampling: Adapting the Size of Sample Sets 263
8.4 Localization in Dynamic Environments 267
8.5 Practical Considerations 273
8.6 Summary 274
8.7 Bibliographical Remarks 275
8.8 Exercises 276

III Mapping 279

9 *Occupancy Grid Mapping* 281

9.1 Introduction 281
9.2 The Occupancy Grid Mapping Algorithm 284
    9.2.1 Multi-Sensor Fusion 293
9.3 Learning Inverse Measurement Models 294
    9.3.1 Inverting the Measurement Model 294
    9.3.2 Sampling from the Forward Model 295
    9.3.3 The Error Function 296
    9.3.4 Examples and Further Considerations 298
9.4 Maximum A Posteriori Occupancy Mapping 299
    9.4.1 The Case for Maintaining Dependencies 299

    9.4.2 Occupancy Grid Mapping with Forward Models 301
 9.5 Summary 304
 9.6 Bibliographical Remarks 305
 9.7 Exercises 307

10 *Simultaneous Localization and Mapping*   309

 10.1 Introduction 309
 10.2 SLAM with Extended Kalman Filters 312
    10.2.1 Setup and Assumptions 312
    10.2.2 SLAM with Known Correspondence 313
    10.2.3 Mathematical Derivation of EKF SLAM 317
 10.3 EKF SLAM with Unknown Correspondences 323
    10.3.1 The General EKF SLAM Algorithm 323
    10.3.2 Examples 324
    10.3.3 Feature Selection and Map Management 328
 10.4 Summary 330
 10.5 Bibliographical Remarks 332
 10.6 Exercises 334

11 *The GraphSLAM Algorithm*   337

 11.1 Introduction 337
 11.2 Intuitive Description 340
    11.2.1 Building Up the Graph 340
    11.2.2 Inference 343
 11.3 The GraphSLAM Algorithm 346
 11.4 Mathematical Derivation of GraphSLAM 353
    11.4.1 The Full SLAM Posterior 353
    11.4.2 The Negative Log Posterior 354
    11.4.3 Taylor Expansion 355
    11.4.4 Constructing the Information Form 357
    11.4.5 Reducing the Information Form 360
    11.4.6 Recovering the Path and the Map 361
 11.5 Data Association in GraphSLAM 362
    11.5.1 The GraphSLAM Algorithm with Unknown Correspondence 363
    11.5.2 Mathematical Derivation of the Correspondence Test 366
 11.6 Efficiency Consideration 368
 11.7 Empirical Implementation 370

11.8 Alternative Optimization Techniques   376
11.9 Summary   379
11.10 Bibliographical Remarks   381
11.11 Exercises   382

12 *The Sparse Extended Information Filter*   **385**

12.1 Introduction   385
12.2 Intuitive Description   388
12.3 The SEIF SLAM Algorithm   391
12.4 Mathematical Derivation of the SEIF   395
    12.4.1 Motion Update   395
    12.4.2 Measurement Updates   398
12.5 Sparsification   398
    12.5.1 General Idea   398
    12.5.2 Sparsification in SEIFs   400
    12.5.3 Mathematical Derivation of the Sparsification   401
12.6 Amortized Approximate Map Recovery   402
12.7 How Sparse Should SEIFs Be?   405
12.8 Incremental Data Association   409
    12.8.1 Computing Incremental Data Association Probabilities   409
    12.8.2 Practical Considerations   411
12.9 Branch-and-Bound Data Association   415
    12.9.1 Recursive Search   416
    12.9.2 Computing Arbitrary Data Association Probabilities   416
    12.9.3 Equivalence Constraints   419
12.10 Practical Considerations   420
12.11 Multi-Robot SLAM   424
    12.11.1 Integrating Maps   424
    12.11.2 Mathematical Derivation of Map Integration   427
    12.11.3 Establishing Correspondence   429
    12.11.4 Example   429
12.12 Summary   432
12.13 Bibliographical Remarks   434
12.14 Exercises   435

13 *The FastSLAM Algorithm*   **437**

13.1 The Basic Algorithm   439

13.2 Factoring the SLAM Posterior   439
    13.2.1 Mathematical Derivation of the Factored SLAM Posterior   442
13.3 FastSLAM with Known Data Association   444
13.4 Improving the Proposal Distribution   451
    13.4.1 Extending the Path Posterior by Sampling a New Pose   451
    13.4.2 Updating the Observed Feature Estimate   454
    13.4.3 Calculating Importance Factors   455
13.5 Unknown Data Association   457
13.6 Map Management   459
13.7 The FastSLAM Algorithms   460
13.8 Efficient Implementation   460
13.9 FastSLAM for Feature-Based Maps   468
    13.9.1 Empirical Insights   468
    13.9.2 Loop Closure   471
13.10 Grid-based FastSLAM   474
    13.10.1 The Algorithm   474
    13.10.2 Empirical Insights   475
13.11 Summary   479
13.12 Bibliographical Remarks   481
13.13 Exercises   482

IV Planning and Control   485

14 Markov Decision Processes   487

14.1 Motivation   487
14.2 Uncertainty in Action Selection   490
14.3 Value Iteration   495
    14.3.1 Goals and Payoff   495
    14.3.2 Finding Optimal Control Policies for the Fully Observable Case   499
    14.3.3 Computing the Value Function   501
14.4 Application to Robot Control   503
14.5 Summary   507
14.6 Bibliographical Remarks   509
14.7 Exercises   510

15 Partially Observable Markov Decision Processes 513
- 15.1 Motivation 513
- 15.2 An Illustrative Example 515
  - 15.2.1 Setup 515
  - 15.2.2 Control Choice 516
  - 15.2.3 Sensing 519
  - 15.2.4 Prediction 523
  - 15.2.5 Deep Horizons and Pruning 526
- 15.3 The Finite World POMDP Algorithm 527
- 15.4 Mathematical Derivation of POMDPs 531
  - 15.4.1 Value Iteration in Belief Space 531
  - 15.4.2 Value Function Representation 532
  - 15.4.3 Calculating the Value Function 533
- 15.5 Practical Considerations 536
- 15.6 Summary 541
- 15.7 Bibliographical Remarks 542
- 15.8 Exercises 544

16 Approximate POMDP Techniques 547
- 16.1 Motivation 547
- 16.2 QMDPs 549
- 16.3 Augmented Markov Decision Processes 550
  - 16.3.1 The Augmented State Space 550
  - 16.3.2 The AMDP Algorithm 551
  - 16.3.3 Mathematical Derivation of AMDPs 553
  - 16.3.4 Application to Mobile Robot Navigation 556
- 16.4 Monte Carlo POMDPs 559
  - 16.4.1 Using Particle Sets 559
  - 16.4.2 The MC-POMDP Algorithm 559
  - 16.4.3 Mathematical Derivation of MC-POMDPs 562
  - 16.4.4 Practical Considerations 563
- 16.5 Summary 565
- 16.6 Bibliographical Remarks 566
- 16.7 Exercises 566

17 Exploration 569
- 17.1 Introduction 569
- 17.2 Basic Exploration Algorithms 571
  - 17.2.1 Information Gain 571

       17.2.2   Greedy Techniques   572
       17.2.3   Monte Carlo Exploration   573
       17.2.4   Multi-Step Techniques   575
17.3   Active Localization   575
17.4   Exploration for Learning Occupancy Grid Maps   580
       17.4.1   Computing Information Gain   580
       17.4.2   Propagating Gain   585
       17.4.3   Extension to Multi-Robot Systems   587
17.5   Exploration for SLAM   593
       17.5.1   Entropy Decomposition in SLAM   593
       17.5.2   Exploration in FastSLAM   594
       17.5.3   Empirical Characterization   598
17.6   Summary   600
17.7   Bibliographical Remarks   602
17.8   Exercises   604

Bibliography   **607**

Index   **639**

Preface

This book provides a comprehensive introduction into the emerging field of probabilistic robotics. Probabilistic robotics is a subfield of robotics concerned with perception and control. It relies on statistical techniques for representing information and making decisions. By doing so, it accommodates the uncertainty that arises in most contemporary robotics applications. In recent years, probabilistic techniques have become one of the dominant paradigms for algorithm design in robotics. This monograph provides a first comprehensive introduction into some of the major techniques in this field.

This book has a strong focus on algorithms. All algorithms in this book are based on a single overarching mathematical foundation: Bayes rule, and its temporal extension known as Bayes filters. This unifying mathematical framework is the core commonality of probabilistic algorithms.

In writing this book, we have tried to be as complete as possible with regards to technical detail. Each chapter describes one or more major algorithms. For each algorithm, we provide the following four things: (1) an example implementation in pseudo code; (2) a complete mathematical derivation from first principles that makes the various assumptions behind each algorithm explicit; (3) empirical results insofar as they further the understanding of the algorithms presented in the book; and (4) a detailed discussion of the strengths and weaknesses of each algorithm—from a practitioner's perspective. Developing all this for many different algorithms proved to be a laborious task. The result might at times be a bit difficult to digest for the casual reader—although skipping the mathematical derivation sections is always an option! We hope that a careful reader emerges with a much deeper level of understanding than any superficial, non-mathematical exposition of this topic would have been able to convey.

This book is the result of more than a decade of research by us, the authors, our students, and many of our colleagues in the field. We began writing it in 1999, hoping that it would take not much more than a few months to complete this book. Five years have passed, and almost nothing from the original draft has survived. Through working on this book, we have learned much more about information and decision theory than we thought we ever would. We are happy to report that much of what we learned has made it into this book.

This monograph is written for students, researchers, and practitioners in robotics. We believe everybody building robots has to develop software. Hence the material in this book should be relevant to every roboticist. It should also be of interest to applied statisticians, and people concerned with real-world sensor data outside the realm of robotics. To serve a wide range of readers with varying technical backgrounds, we have attempted to make this book as self-contained as possible. Some prior knowledge of linear algebra and basic probability and statistics will be helpful, but we have included a primer for the basic laws of probability, and avoided the use of advanced mathematical techniques throughout this text.

This book is also written for classroom use. Each chapter offers a number of exercises and suggests hands-on projects. When used in the classroom, each chapter should be covered in one or two lectures. Chapters should be skipped or reordered quite arbitrarily; in fact, in our own teaching we usually start right in the middle of the book, with Chapter 7. We recommend that the study of the book be accompanied by practical, hands-on experimentation as directed by the exercises at the end of each chapter. Nothing more important in robotics than doing it yourself!

Despite our very best efforts, we believe there will still be technical errors left in this book. Many of these errors have been corrected in this third printing of the book. We continue to post corrections on the book's Web site, along with other materials relevant to this book:

www.probabilistic-robotics.org

We hope you enjoy this book!

Sebastian Thrun
Wolfram Burgard
Dieter Fox

Acknowledgments

This book would not have been possible without the help and support from so many friends, family members, students, and colleagues in the field. We cannot possibly list all of them here.

Much of the material in this book is the result of collaborations with our current and past students and post-docs. We specifically would like to acknowledge Rahul Biswas, Matthew Deans, Frank Dellaert, James Diebel, Brian Gerkey, Dirk Hähnel, Johnathan Ko, Cody Kwok, John Langford, Lin Liao, David Lieb, Benson Limketkai, Michael Littman, Yufeng Liu, Andrew Lookingbill, Dimitris Margaritis, Michael Montemerlo, Mark Moors, Mark Paskin, Joelle Pineau, Charles Rosenberg, Nicholas Roy, Aaron Shon, Jamie Schulte, Dirk Schulz, David Stavens, Cyrill Stachniss, and Chieh-Chih Wang, along with all other past and present members of our labs. Greg Armstrong, Grinnell More, Tyson Sawyer, and Walter Steiner were instrumental in keeping our robots running over the years.

Much of our research was conducted while we were with Carnegie Mellon University in Pittsburgh, PA, and we thank our former colleagues and friends at CMU for many inspirational discussions. We also would like to express our gratitude to Armin Cremers from the University of Bonn who brought the three of us together in his research group and who set the seed for our wonderful collaboration.

We are indebted to numerous colleagues whose comments and insights were instrumental during the course of our research. We would specifically like to thank Gary Bradski, Howie Choset, Henrik Christensen, Hugh Durrant-Whyte, Nando de Freitas, Zoubin Gharamani, Geoffrey Gordon, Steffen Gutmann, Andrew Howards, Leslie Kaelbling, Daphne Koller, Kurt Konolige Ben Kuipers, John Leonard, Tom Mitchell, Kevin Murphy, Eduardo Nebot, Paul Newman, Andrew Y. Ng, Reid Simmons, Satinder Singh, Gaurav Sukhatme, Juan Tardós, Ben Wegbreit, and Alex Zelinsky for their feedback over the years.

Anita Araneda, Gal Elidan, Udo Frese, Gabe Hoffmann, John Leonard, Benson Limketkai, Rudolph van der Merwe, Anna Petrovskaya, Bob Wang, and Stefan Williams gave us extensive comments on earlier drafts of this book, which we gratefully acknowledge. Chris Manning kindly provided the Latex macros for this book, and Bob Prior was instrumental in getting the book published.

A number of agencies and corporations have made this work possible through generous financial contributions and technical advice. We are particularly grateful for DARPA's support under a number of programs (TMR, MARS, LAGR, SDR, MICA, and CoABS). We also acknowledge funding from the National Science Foundation, via its CAREER, ITR, and various CISE grant programs; from the German Research Foundation; and from the European Commission. Generous financial support was also provided by a number of corporate sponsors and individual donors: Android, Bosch, DaimlerChrysler, Intel, Google, Microsoft, Mohr Davidow Ventures, Samsung, and Volkswagen of America. We specifically thank John Blitch, Doug Gage, Sharon Heise, James Hendler, Larry Jackel, Alex Krott, Wendell Sykes, and Ed van Reuth for providing us with ample challenges and guidance throughout the years. Naturally, the views and conclusions contained in this document are those of the authors, and should not be interpreted as necessarily representing policies or endorsements of any of our sponsors.

We owe our most important acknowledgments to our families, whose love and dedication carried us through this immense project. We specifically thank Petra Dierkes-Thrun, Anja Gross-Burgard and Carsten Burgard, and Luz, Sofia and Karla Fox for their love and support.

Part I

Basics

1 *Introduction*

1.1 Uncertainty in Robotics

Robotics is the science of perceiving and manipulating the physical world through computer-controlled devices. Examples of successful robotic systems include mobile platforms for planetary exploration, industrial robotics arms in assembly lines, cars that travel by themselves, and manipulators that assist surgeons. Robotics systems are situated in the physical world, perceive information on their environments through sensors, and manipulate through physical forces.

While much of robotics is still in its infancy, the idea of "intelligent" manipulating devices has an enormous potential to change society. Wouldn't it be great if all our cars were able to safely steer themselves, making car accidents a notion of the past? Wouldn't it be great if robots, and not people, would clean up nuclear disaster sites like Chernobyl? Wouldn't it be great if our homes were populated by intelligent assistants that take care of all domestic repair and maintenance tasks?

To do these tasks, robots have to be able to accommodate the enormous uncertainty that exists in the physical world. There is a number of factors that contribute to a robot's uncertainty.

First and foremost, *robot environments* are inherently unpredictable. While the degree of uncertainty in well-structured environments such as assembly lines is small, environments such as highways and private homes are highly dynamic and in many ways highly unpredictable. The uncertainty is particularly high for robots operating in the proximity of people.

Sensors are limited in what they can perceive. Limitations arise from several factors. The range and resolution of a sensor is subject to physical limitations. For example, cameras cannot see through walls, and the spatial resolution of a camera image is limited. Sensors are also subject to noise, which perturbs sensor measurements in unpredictable ways and hence limits the information that can be extracted. And finally, sensors can break. Detecting a faulty sensor can be extremely difficult.

Robot actuation involves motors that are, at least to some extent, unpredictable. Uncertainty arises from effects like control noise, wear-and-tear, and mechanical failure. Some actuators, such as heavy-duty industrial robot arms, are quite accurate and reliable. Others, like low-cost mobile robots, can be extremely flaky.

Some uncertainty is caused by the robot's software. All *internal models* of the world are approximate. Models are abstractions of the real world. As such, they only partially model the underlying physical processes of the robot and its environment. Model errors are a source of uncertainty that has often been ignored in robotics, despite the fact that most robotic models used in state-of-the-art robotics systems are rather crude.

Uncertainty is further created through *algorithmic approximations*. Robots are real-time systems. This limits the amount of computation that can be carried out. Many popular algorithms are approximate, achieving timely response through sacrificing accuracy.

The level of uncertainty depends on the application domain. In some robotic applications, such as assembly lines, humans can cleverly engineer the system so that uncertainty is only a marginal factor. In contrast, robots operating in residential homes or on other planets will have to cope with substantial uncertainty. Such robots are forced to act even though neither their sensors, nor their internal models, will provide it with sufficient information to make the right decisions with absolute certainty. As robotics is now moving into the open world, the issue of uncertainty has become a major stumbling block for the design of capable robot systems. Managing uncertainty is possibly the most important step towards robust real-world robot systems.

Hence this book.

1.2 Probabilistic Robotics

This book provides a comprehensive overview of *probabilistic robotics*. Probabilistic robotics is a relatively new approach to robotics that pays tribute to the uncertainty in robot perception and action. The key idea in probabilistic robotics is to represent uncertainty explicitly using the calculus of probability theory. Put differently, instead of relying on a single "best guess" as to what might be the case, probabilistic algorithms represent information by probability distributions over a whole space of guesses. By doing so, they can represent ambiguity and degree of belief in a mathematically sound way. Control choices can be made robust relative the uncertainty that remains, and probabilistic robotics can even actively chose to reduce their uncertainty when this appears to be the superior choice. Thus, probabilistic algorithms degrade gracefully in the face of uncertainty. As a result, they outperform alternative techniques in many real-world applications.

We shall illustrate probabilistic robotics with two motivating examples: one pertaining to robot perception, and another to planning and control.

MOBILE ROBOT LOCALIZATION

Our first example is *mobile robot localization*. Robot localization is the problem of estimating a robot's coordinates relative to an external reference frame. The robot is given a map of its environment, but to localize itself relative to this map it needs to consult its sensor data. Figure 1.1 illustrates such a situation. The environment is known to possess three indistinguishable doors. The task of the robot is to find out where it is, through sensing and motion.

This specific localization problem is known as *global localization*. In global localization, a robot is placed somewhere in a known environment and has to localize itself from scratch. The probabilistic paradigm represents the robot's momentary *belief* by a probability density function over the space of all locations. This is illustrated in diagram (a) in Figure 1.1. This diagram shows a uniform distribution over all locations. Now suppose the robot takes a first sensor measurement and observes that it is next to a door. Probabilistic techniques exploit this information to update the belief. The 'posterior' belief is shown in diagram (b) in Figure 1.1. It places an increased probability at places near doors, and lower probability near walls. Notice that this distribution possesses three peaks, each corresponding to one of the indistinguishable doors in the environment. Thus, by no means does the robot *know* where it is. Instead, it now has three, distinct hypotheses which are each equally plausible given the sensor data. We also note that the robot assigns positive probability to places *not* next to a door. This is the natural result of the inherent uncertainty in sensing: With a small, non-zero probability, the robot might have erred in its assessment of seeing a door. The ability to maintain low-probability hypotheses is essential for attaining robustness.

Now suppose the robot moves. Diagram (c) in Figure 1.1 shows the effect on a robot's belief. The belief has been shifted in the direction of motion. It also possesses a larger spread, which reflects the uncertainty that is intro-

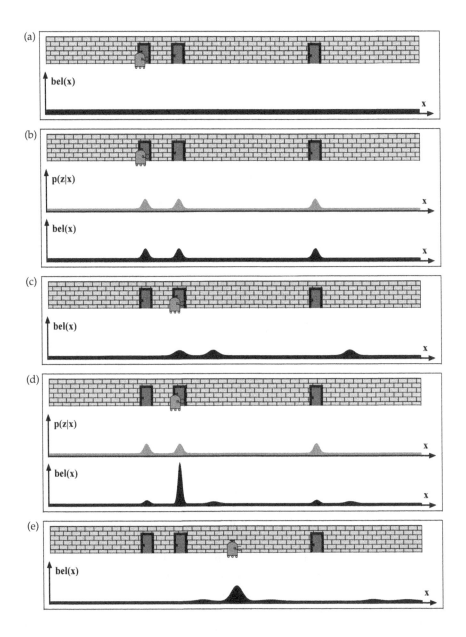

Figure 1.1 The basic idea of *Markov localization*: A mobile robot during global localization. Markov localization techniques will be investigated in Chapters 7 and 8.

(a)

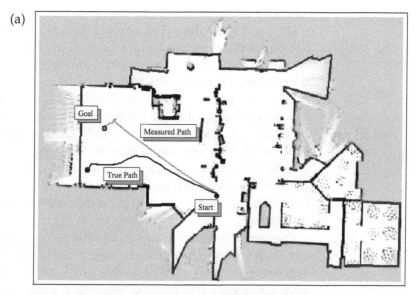

(b)

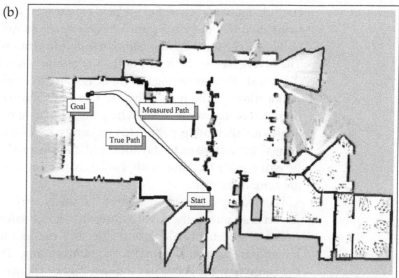

Figure 1.2 Top image: a robot navigating through open, featureless space may lose track of where it is. Bottom: This can be avoided by staying near known obstacles. These figures are results of an algorithm called *coastal navigation*, which will be discussed in Chapter 16. Images courtesy of Nicholas Roy, MIT.

duced by robot motion. Diagram (d) in Figure 1.1 depicts the belief after observing another door. This observation leads our algorithm to place most of the probability mass on a location near one of the doors, and the robot is now quite confident as to where it is. Finally, Diagram (e) shows a belief as the robot travels further down the corridor.

This example illustrates many aspects of the probabilistic paradigm. Stated probabilistically, the robot perception problem is a state estimation problem, and our localization example uses an algorithm known as *Bayes filter* for posterior estimation over the space of robot locations. The representation of information is a probability density function. The update of this function represents the information gained through sensor measurements, or the information lost through processes in the world that increase a robot's uncertainty.

BAYES FILTER

Our second example brings us into the realm of robotic planning and control. As just argued, probabilistic algorithms can compute a robot's momentary uncertainty. But they can also anticipate future uncertainty, and take such uncertainty into consideration when determining the right choice of control. One such algorithm is called *coastal navigation*. An example of coastal navigation is shown in Figure 1.2. This figure shows a 2-D map of an actual building. The top diagram compares an estimated path with an actual path: The divergence is the result of the uncertainty in robot motion that we just discussed. The interesting insight is: not all trajectories induce the same level of uncertainty. The path in Figure 1.2a leads through relatively open space, deprived of features that could help the robot to remain localized. Figure 1.2b shows an alternative path. This trajectory seeks a distinct corner, and then "hugs" the wall so as to stay localized. Not surprisingly, the uncertainty will be reduced for the latter path, hence chances of arriving at the goal location are actually higher.

COASTAL NAVIGATION

This example illustrates one of the many ways proper consideration of uncertainty affects the robot's controls. In our example, the anticipation of possible uncertainty along one trajectory makes the robot prefer a second, longer path, just so as to reduce the uncertainty. The new path is better, in the sense that the robot has a much higher chance of actually being at the goal when believing that it is. In fact, the second path is an example of active information gathering. The robot has, through its probabilistic consideration, determined that the best choice of action is to seek information along its path, in its pursuit to reach a target location. Probabilistic planning techniques anticipate uncertainty and can plan for information gathering, and probabilistic control techniques realize the results of such plans.

1.3 Implications

Probabilistic robotics seamlessly integrates models with sensor data, overcoming the limitations of both at the same time. These ideas are not just a matter of low-level control. They cut across all levels of robotic software, from the lowest to the highest.

In contrast with traditional programming techniques in robotics—such as model-based motion planning techniques or reactive behavior-based approaches—probabilistic approaches tend to be more robust in the face of sensor limitations and model limitations. This enables them to scale much better to complex real-world environments than previous paradigms, where uncertainty is of even greater importance. In fact, certain probabilistic algorithms are currently the only known working solutions to hard robotic estimation problems, such as the localization problem discussed a few pages ago, or the problem of building accurate maps of very large environments.

In comparison to traditional model-based robotic techniques, probabilistic algorithms have weaker requirements on the accuracy of the robot's models, thereby relieving the programmer from the insurmountable burden to come up with accurate models. Probabilistic algorithms have weaker requirements on the accuracy of robotic sensors than those made by many reactive techniques, whose sole control input is the momentary sensor input. Viewed probabilistically, the *robot learning problem* is a long-term estimation problem. Thus, probabilistic algorithms provide a sound methodology for many flavors of robot learning.

However, these advantages come at a price. The two most frequently cited limitations of probabilistic algorithms are *computational complexity*, and a *need to approximate*. Probabilistic algorithms are inherently less efficient than their non-probabilistic counterparts. This is due to the fact that they consider entire probability densities instead of a single guess. The need to approximate arises from the fact that most robot worlds are continuous. Computing exact posterior distributions tends to be computationally intractable. Sometimes, one is fortunate in that the uncertainty can be approximated tightly with a compact parametric model (e.g., Gaussians). In other cases, such approximations are too crude to be of use, and more complicated representations must be employed.

Recent developments in computer hardware has made an unprecedented number of FLOPS available at bargain prices. This development has certainly aided the field of probabilistic robotics. Further, recent research has successfully increased the computational efficiency of probabilistic algorithms, for a range of hard robotics problems—many of which are described in depth in this book. Nevertheless, computational challenges remain. We shall revisit this discussion at numerous places, where we investigate the strengths and weaknesses of specific probabilistic solutions.

1.4 Road Map

This book is organized in four major parts.

- Chapters 2 through 4 introduce the basic mathematical framework that underlies all of the algorithms described in this book, along with key algorithms. These chapters are the mathematical foundation of this book.

- Chapters 5 and 6 present probabilistic models of mobile robots. In many ways, these chapters are the probabilistic generalization of classical robotics models. They form the robotic foundation for the material that follows.

- The mobile robot localization problem is discussed in Chapters 7 and 8. These chapters combine the basic estimation algorithms with the probabilistic models discussed in the previous two chapters.

- Chapters 9 through 13 discuss the much richer problem of robotic mapping. As before, they are all based on the algorithms discussed in the foundational chapters, but many of them utilize tricks to accommodate the enormous complexity of this problem.

- Problems of probabilistic planning and control are discussed in Chapters 14 through 17. Here we begin by introducing a number of fundamental techniques, and then branch into practical algorithms for controlling a robot probabilistically. The final chapter, Chapter 17, discusses the problem of robot exploration from a probabilistic perspective.

The book is best read in order, from the beginning to the end. However, we have attempted to make each individual chapter self-explanatory. Frequent sections called *"Mathematical Derivation of ..."* can safely be skipped on first reading without compromising the coherence of the overall material in this book.

1.5 Teaching Probabilistic Robotics

When used in the classroom, we do *not* recommend to teach the chapters in order—unless the students have an unusually strong appreciation of abstract mathematical concepts. Particle filters are easier to teach than Gaussian filters, and students tend to get more excited by mobile robot localization problems than abstract filter algorithms. In our own teachings, we usually begin with Chapter 2, and move directly to Chapters 7 and 8. While teaching localization, we go back to the material in Chapters 3 through 6 as needed. We also teach Chapter 14 early, to expose students to the problems related to planning and control early on in a course.

As a teacher, feel free to use slides and animations from the book's Web site
$$\text{www.probabilistic-robotics.org}$$
to illustrate the various algorithms in this book. And feel free to send us, the authors, pointers to your class Web sites and any material that could help others in teaching Probabilistic Robotics.

The material in this book is best taught with hands-on implementation assignments. There is nothing more educational in robotics than programming an actual robot. And nobody can explain the pitfalls and challenges in robotics better than Nature!

1.6 Bibliographical Remarks

MODEL-BASED PARADIGM

The field of robotics has gone through a series of paradigms for software design. The first major paradigm emerged in the mid-1970s, and is known as the *model-based paradigm*. The model-based paradigm began with a number of studies showing the hardness of controlling a high-DOF robotic manipulator in continuous spaces (Reif 1979). It culminated in text like Schwartz et al.'s (1987) analysis of the complexity of robot motion, a first singly exponential general motion planning algorithm by Canny (1987), and Latombe's (1991) seminal introductory text into the field of model-based motion planning (additional milestone contributions will be discussed in Chapter 14). This early work largely ignored the problem of uncertainty—even though it extensively began using randomization as a technique for solving hard motion planning problems (Kavraki et al. 1996). Instead, the assumption was that a full and accurate model of the robot and the environment be given, and the robot be deterministic. The model had to be sufficiently accurate that the residual uncertainty was managed by a low-level motion controller. Most motion planning techniques simply produced a single reference trajectory for the control of a manipulator, although ideas such as *potential fields* (Khatib 1986) and *navigation functions* (Koditschek 1987) provided mechanisms for reacting to the unforeseen—as long as it could be sensed. Applications of these early techniques, if any, were confined to environments where every little bit of uncertainty could be engineered away, or sensed with sufficient accuracy.

BEHAVIOR-BASED ROBOTICS

The field took a radical shift in the mid-1980s, when the lack of sensory feedback became the focus of an entire community of researchers within robotics. With strong convictions, the field of *behavior-based robotics* rejected the idea of any internal model. Instead, it was the interaction with a physical environment of a *situated agent* (Kaelbling and Rosenschein 1991) that created the complexity in robot motion (a phenomena often called *emergent behavior* (Steels 1991)). Consequently, sensing played a paramount role, and internal models were rejected (Brooks 1990).

The enthusiasm in this field was fueled by some early successes that were far beyond the reach of traditional model-based motion planning algorithms. One of them was "Genghis," a hexapod robot developed by Brooks (1986). A relatively simple finite state automaton was able to control the gait of this robot even in rugged terrain. The key to success of such techniques lay in sensing: the control was entirely driven by environment interaction, as perceived through the robot's sensors. Some of the early work impressed by creating a seemingly complex robot through clever exploitation of environment feedback (Connell 1990). More recently, the paradigm enjoyed commercial success through a robotic vacuum cleaning robot (IRobots Inc. 2004), whose software follows the behavior-based paradigm.

Due to the lack of internal models and a focus on simple control mechanism, most robot systems were confined to relatively simple tasks, where the momentary sensor information was sufficient to determine the right choice of control. Recognizing this limitation, more recent work in this field embraced *hybrid control* architectures (Arkin 1998), in which behavior-based technique provided low-level control, whereas a model-based planner coordinated the robot's actions at a high, abstract level. Such hybrid architectures are commonplace in robotics today. They are not dissimilar to the seminal work on three-layered architectures by Gat (1998), which took its origins in "Shakey the Robot" (Nilsson 1984).

Modern probabilistic robotics has emerged since the mid-1990s, although its roots can be traced back to the invention of the Kalman filter (Kalman 1960). In many ways, probabilistic robotics falls in between model-based and behavior-based techniques. In probabilistic robotics, there are models, but they are assumed to be incomplete and insufficient for control. There are also sensor measurements, but they too are assumed to be incomplete and insufficient for control. Through the integration of both, models and sensor measurements, a control action can be devised. Statistics provides the mathematical glue to integrate models and sensor measurements.

Many of the key advances in the field of probabilistic robotics will be discussed in future chapters. Some of the cornerstones in this field include the advent of Kalman filter techniques for high-dimensional perception problems by Smith and Cheeseman (1986), the invention of occupancy grid maps by (Elfes 1987; Moravec 1988), and the re-introduction of partially observable planning techniques due to Kaelbling et al. (1998). The past decade has seen an explosion of techniques: Particle filters have become vastly popular (Dellaert et al. 1999), and researchers have developed new programming methodologies focused on Bayesian information processing (Thrun 2000b; Lebeltel et al. 2004; Park et al. 2005). This development went hand in hand with the deployment of physical robot systems driven by probabilistic algorithms, such as industrial machines for cargo handling by Durrant-Whyte (1996), entertainment robots in museums (Burgard et al. 1999a; Thrun et al. 2000a; Siegwart et al. 2003), and robots in nursing and health care (Pineau et al. 2003d). An open-source software package for mobile robot control that heavily utilizes probabilistic techniques is described in Montemerlo et al. (2003a).

The field of commercial robotics is also at a turning point. In its annual World Robotics Survey, the *United Nations and the International Federation of Robotics* 2004 finds a 19% annual increase in the size of the robotic market worldwide. Even more spectacular is the change of market segments, which indicates a solid transition from industrial applications to service robotics and consumer products.

2 Recursive State Estimation

2.1 Introduction

At the core of probabilistic robotics is the idea of estimating state from sensor data. State estimation addresses the problem of estimating quantities from sensor data that are not directly observable, but that can be inferred. In most robotic applications, determining what to do is relatively easy if one only knew *certain* quantities. For example, moving a mobile robot is relatively easy if the exact location of the robot and all nearby obstacles are known. Unfortunately, these variables are not directly measurable. Instead, a robot has to rely on its sensors to gather this information. Sensors carry only partial information about those quantities, and their measurements are corrupted by noise. State estimation seeks to recover state variables from the data. Probabilistic state estimation algorithms compute belief distributions over possible world states. An example of probabilistic state estimation was already encountered in the introduction to this book: mobile robot localization.

The goal of this chapter is to introduce the basic vocabulary and mathematical tools for estimating state from sensor data.

- Chapter 2.2 introduces basic probabilistic concepts used throughout the book.

- Chapter 2.3 describes our formal model of robot environment interaction, setting forth some of the key terminology used throughout the book.

- Chapter 2.4 introduces *Bayes filters*, the recursive algorithm for state estimation that forms the basis of virtually every technique presented in this book.

- Chapter 2.5 discusses representational and computational issues that arise when implementing Bayes filters.

2.2 Basic Concepts in Probability

This section familiarizes the reader with the basic notation and probabilistic facts used throughout the book. In probabilistic robotics, quantities such as sensor measurements, controls, and the states of a robot and its environment are all modeled as random variables. *Random variables* can take on multiple values, and they do so according to specific probabilistic laws. Probabilistic inference is the process of calculating these laws for random variables that are derived from other random variables and the observed data.

RANDOM VARIABLE

Let X denote a random variable and x denote a specific value that X might assume. A standard example of a random variable is a *coin flip*, where X can take on the values *heads* or *tails*. If the space of all values that X can take on is discrete, as is the case if X is the outcome of a coin flip, we write

$$(2.1) \quad p(X = x)$$

to denote the probability that the random variable X has value x. For example, a fair coin is characterized by $p(X = \text{head}) = p(X = \text{tail}) = \frac{1}{2}$. Discrete probabilities sum to one, that is,

$$(2.2) \quad \sum_x p(X = x) = 1$$

Probabilities are always non-negative, that is, $p(X = x) \geq 0$.

To simplify the notation, we will usually omit explicit mention of the random variable whenever possible, and instead use the common abbreviation $p(x)$ instead of writing $p(X = x)$.

Most techniques in this book address estimation and decision making in continuous spaces. Continuous spaces are characterized by random variables that can take on a continuum of values. Unless explicitly stated, we assume that all continuous random variables possess *probability density functions* (PDFs). A common density function is that of the one-dimensional *normal distribution* with mean μ and variance σ^2. The PDF of a normal distribution is given by the following *Gaussian* function:

PROBABILITY DENSITY FUNCTION
NORMAL DISTRIBUTION
GAUSSIAN

$$(2.3) \quad p(x) = \left(2\pi\sigma^2\right)^{-\frac{1}{2}} \exp\left\{-\frac{1}{2}\frac{(x-\mu)^2}{\sigma^2}\right\}$$

Normal distributions play a major role in this book. We will frequently abbreviate them as $\mathcal{N}(x; \mu, \sigma^2)$, which specifies the random variable, its mean, and its variance.

The Normal distribution (2.3) assumes that x is a scalar value. Often, x will be a multi-dimensional vector. Normal distributions over vectors are called *multivariate*. Multivariate normal distributions are characterized by density functions of the following form:

MULTIVARIATE DISTRIBUTION

$$(2.4) \quad p(x) = \det(2\pi\Sigma)^{-\frac{1}{2}} \exp\left\{-\tfrac{1}{2}(x-\mu)^T \Sigma^{-1}(x-\mu)\right\}$$

Here μ is the mean vector. Σ a *positive semidefinite* and *symmetric* matrix called the *covariance matrix*. The superscript T marks the transpose of a vector. The argument in the exponent in this PDF is quadratic in x, and the parameters of this quadratic function are μ and Σ.

COVARIANCE MATRIX

The reader should take a moment to realize that Equation (2.4) is a strict generalization of Equation (2.3); both definitions are equivalent if x is a scalar value and $\Sigma = \sigma^2$.

Equations (2.3) and (2.4) are examples of PDFs. Just as discrete probability distribution always sums up to 1, a PDF always integrates to 1:

$$(2.5) \quad \int p(x)\, dx = 1$$

However, unlike a discrete probability, the value of a PDF is not upper-bounded by 1. Throughout this book, we will use the terms *probability*, *probability density*, and *probability density function* interchangeably. We will silently assume that all continuous random variables are measurable, and we also assume that all continuous distributions actually possess densities.

JOINT DISTRIBUTION

The *joint distribution* of two random variables X and Y is given by

$$(2.6) \quad p(x, y) = p(X = x \text{ and } Y = y)$$

This expression describes the probability of the event that the random variable X takes on the value x *and* that Y takes on the value y. If X and Y are *independent*, we have

INDEPENDENCE

$$(2.7) \quad p(x, y) = p(x)\, p(y)$$

Often, random variables carry information about other random variables. Suppose we already know that Y's value is y, and we would like to know the probability that X's value is x conditioned on that fact. Such a probability will be denoted

$$(2.8) \quad p(x \mid y) = p(X = x \mid Y = y)$$

CONDITIONAL PROBABILITY and is called *conditional probability*. If $p(y) > 0$, then the conditional probability is defined as

$$(2.9) \quad p(x \mid y) = \frac{p(x,y)}{p(y)}$$

If X and Y are independent, we have

$$(2.10) \quad p(x \mid y) = \frac{p(x)\,p(y)}{p(y)} = p(x)$$

In other words, if X and Y are independent, Y tells us nothing about the value of X. There is no advantage of knowing the value of Y if we are interested in X. Independence, and its generalization known as conditional independence, plays a major role throughout this book.

An interesting fact, which follows from the definition of conditional probability and the axioms of probability measures, is often referred to as *Theorem of total probability*:

THEOREM OF TOTAL PROBABILITY

$$(2.11) \quad p(x) = \sum_y p(x \mid y)\,p(y) \quad \text{(discrete case)}$$

$$(2.12) \quad p(x) = \int p(x \mid y)\,p(y)\,dy \quad \text{(continuous case)}$$

If $p(x \mid y)$ or $p(y)$ are zero, we define the product $p(x \mid y)\,p(y)$ to be zero, regardless of the value of the remaining factor.

BAYES RULE Equally important is *Bayes rule*, which relates a conditional of the type $p(x \mid y)$ to its "inverse," $p(y \mid x)$. The rule, as stated here, requires $p(y) > 0$:

$$(2.13) \quad p(x \mid y) = \frac{p(y \mid x)\,p(x)}{p(y)} = \frac{p(y \mid x)\,p(x)}{\sum_{x'} p(y \mid x')\,p(x')} \quad \text{(discrete)}$$

$$(2.14) \quad p(x \mid y) = \frac{p(y \mid x)\,p(x)}{p(y)} = \frac{p(y \mid x)\,p(x)}{\int p(y \mid x')\,p(x')\,dx'} \quad \text{(continuous)}$$

Bayes rule plays a predominant role in probabilistic robotics (and probabilistic inference in general). If x is a quantity that we would like to infer from y, the probability $p(x)$ will be referred to as *prior probability distribution*, and y is called the *data* (e.g., a sensor measurement). The distribution $p(x)$ summarizes the knowledge we have regarding X prior to incorporating the data y. The probability $p(x \mid y)$ is called the *posterior probability distribution* over X. As (2.14) suggests, Bayes rule provides a convenient way to compute a posterior $p(x \mid y)$ using the "inverse" conditional probability $p(y \mid x)$ along with the prior probability $p(x)$. In other words, if we are interested in inferring

PRIOR PROBABILITY

POSTERIOR PROBABILITY

2.2 Basic Concepts in Probability

GENERATIVE MODEL

a quantity x from sensor data y, Bayes rule allows us to do so through the inverse probability, which specifies the probability of data y assuming that x was the case. In robotics, the probability $p(y \mid x)$ is often coined *generative model*, since it describes at some level of abstraction how state variables X *cause* sensor measurements Y.

An important observation is that the denominator of Bayes rule, $p(y)$, does not depend on x. Thus, the factor $p(y)^{-1}$ in Equations (2.13) and (2.14) will be the same for any value x in the posterior $p(x \mid y)$. For this reason, $p(y)^{-1}$ is often written as a *normalizer in Bayes rule* variable, and generically denoted η:

$$(2.15) \quad p(x \mid y) = \eta \, p(y \mid x) \, p(x)$$

The advantage of this notation lies in its brevity. Instead of explicitly providing the exact formula for a normalization constant—which can grow large very quickly in some of the mathematical derivations—we simply will use the normalization symbol η to indicate that the final result has to be normalized to 1. Throughout this book, normalizers of this type will be denoted η (or η', η'', ...). **Important:** We will freely use the same η in different equations to denote normalizers, even if their actual values differ.

We notice that it is perfectly fine to condition any of the rules discussed so far on arbitrary random variables, such as the variable Z. For example, conditioning Bayes rule on $Z = z$ gives us:

$$(2.16) \quad p(x \mid y, z) = \frac{p(y \mid x, z) \, p(x \mid z)}{p(y \mid z)}$$

as long as $p(y \mid z) > 0$.

Similarly, we can condition the rule for combining probabilities of independent random variables (2.7) on other variables z:

$$(2.17) \quad p(x, y \mid z) = p(x \mid z) \, p(y \mid z)$$

CONDITIONAL INDEPENDENCE

Such a relation is known as *conditional independence*. As the reader easily verifies, (2.17) is equivalent to

$$(2.18) \quad p(x \mid z) = p(x \mid z, y)$$
$$(2.19) \quad p(y \mid z) = p(y \mid z, x)$$

Conditional independence plays an important role in probabilistic robotics. It applies whenever a variable y carries no information about a variable x if another variable's value z is known. Conditional independence does not imply (absolute) independence, that is,

(2.20) $\quad p(x, y \mid z) = p(x \mid z)\, p(y \mid z) \;\not\Rightarrow\; p(x, y) = p(x)\, p(y)$

The converse is also in general untrue: absolute independence does not imply conditional independence:

(2.21) $\quad p(x, y) = p(x)\, p(y) \;\not\Rightarrow\; p(x, y \mid z) = p(x \mid z)\, p(y \mid z)$

In special cases, however, conditional and absolute independence may coincide.

A number of probabilistic algorithms require us to compute features, or statistics, of probability distributions. The *expectation* of a random variable X is given by

EXPECTATION OF A RV

(2.22) $\quad E[X] \;=\; \sum_x x\, p(x) \quad \text{(discrete)}$

(2.23) $\quad E[X] \;=\; \int x\, p(x)\, dx \quad \text{(continuous)}$

Not all random variables possess finite expectations; however, those that do not are of no relevance to the material presented in this book.

The expectation is a linear function of a random variable. In particular, we have

(2.24) $\quad E[aX + b] \;=\; a E[X] + b$

for arbitrary numerical values a and b. The covariance of X is obtained as follows

(2.25) $\quad \mathrm{Cov}[X] \;=\; E[X - E[X]]^2 \;=\; E[X^2] - E[X]^2$

The covariance measures the squared expected deviation from the mean. As stated above, the mean of a multivariate normal distribution $\mathcal{N}(x; \mu, \Sigma)$ is μ, and its covariance is Σ.

ENTROPY

A final concept of importance in this book is *entropy*. The entropy of a probability distribution is given by the following expression:

(2.26) $\quad H_p(x) \;=\; E[-\log_2 p(x)]$

which resolves to

(2.27) $\quad H_p(x) \;=\; -\sum_x p(x)\, \log_2 p(x) \quad \text{(discrete)}$

(2.28) $\quad H_p(x) \;=\; -\int p(x)\, \log_2 p(x)\, dx \quad \text{(continuous)}$

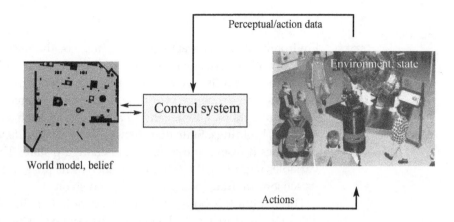

Figure 2.1 Robot environment interaction.

The concept of entropy originates in information theory. The entropy is the expected information that the value of x carries. In the discrete case, $-\log_2 p(x)$ is the number of bits required to encode x using an optimal encoding, assuming that $p(x)$ is the probability of observing x. In this book, entropy will be used in robotic information gathering, so as to express the information a robot may receive upon executing specific actions.

2.3 Robot Environment Interaction

ROBOT ENVIRONMENT

Figure 2.1 illustrates the interaction of a robot with its environment. The *environment*, or *world*, is a dynamical system that possesses internal state. The robot can acquire information about its environment using its sensors. However, sensors are noisy, and there are usually many things that cannot be sensed directly. As a consequence, the robot maintains an internal belief with regards to the state of its environment, depicted on the left in this figure.

The robot can also influence its environment through its actuators. The effect of doing so is often somewhat unpredictable. Hence, each control action affects both the environment state, and the robot's internal belief with regards to this state.

This interaction will now be described more formally.

2.3.1 State

STATE   Environments are characterized by *state*. For the material presented in this book, it will be convenient to think of the state as the collection of all aspects of the robot and its environment that can impact the future. Certain state variables tend to change over time, such as the whereabouts of people in the vicinity of a robot. Others tend to remain static, such as the location of walls in (most) buildings. State that changes will be called *dynamic state*, which distinguishes it from *static state*, or non-changing state. The state also includes variables regarding the robot itself, such as its pose, velocity, whether or not its sensors are functioning correctly, and so on.

Throughout this book, state will be denoted x; although the specific variables included in x will depend on the context. The state at time t will be denoted x_t. Typical state variables used throughout this book are:

POSE
- The robot *pose*, which comprises its location and orientation relative to a global coordinate frame. Rigid mobile robots possess six such state variables, three for their Cartesian coordinates, and three for their angular orientation (pitch, roll, and yaw). For rigid mobile robots confined to planar environments, the pose is usually given by three variables, its two location coordinates in the plane and its heading direction (yaw).

- In robot manipulation, the pose includes variables for the *configuration of the robot's actuators*. For example, they might include the joint angles of revolute joints. Each degree of freedom in a robot arm is characterized by a one-dimensional configuration at any point in time, which is part of the kinematic state of the robot. The robot configuration is often referred to as *kinematic state*.

- The *robot velocity and the velocities of its joints* are commonly referred to as *dynamic state*. A rigid robot moving through space is characterized by up to six velocity variables, one for each pose variables. Dynamic state will play only a minor role in this book.

- The *location and features of surrounding objects in the environment* are also state variables. An object may be a tree, a wall, or a pixel within a larger surface. Features of such objects may be their visual appearance (color, texture). Depending on the granularity of the state that is being modeled, robot environments possess between a few dozen and up to hundreds of billions of state variables (and more). Just imagine how many bits it will take to accurately describe your physical environment! For many of the

LANDMARK
problems studied in this book, the location of objects in the environment will be static. In some problems, objects will assume the form of *landmarks*, which are distinct, stationary features of the environment that can be recognized reliably.

- The *location and velocities of moving objects and people* are also potential state variables. Often, the robot is not the only moving actor in its environment. Other moving entities possess their own kinematic and dynamic state.

- There are many other state variables that may impact a robot's operation. For example, whether or not a sensor is broken can be a state variable, as can be the level of battery charge for a battery-powered robot. The list of potential state variables is endless!

COMPLETE STATE
A state x_t will be called *complete* if it is the best predictor of the future. Put differently, completeness entails that knowledge of past states, measurements, or controls carry no additional information that would help us predict the future more accurately. It is important to notice that our definition of completeness does not require the future to be a *deterministic* function of the state. The future may be stochastic, but no variables prior to x_t may influence the stochastic evolution of future states, unless this dependence is mediated through the state x_t. Temporal processes that meet these conditions are commonly known as *Markov chains*.

MARKOV CHAIN

The notion of state completeness is mostly of theoretical importance. In practice, it is impossible to specify a complete state for any realistic robot system. A complete state includes not just all aspects of the environment that may have an impact on the future, but also the robot itself, the content of its computer memory, the brain dumps of surrounding people, etc. Some of those are hard to obtain. Practical implementations therefore single out a small subset of all state variables, such as the ones listed above. Such a state is called *incomplete state*.

INCOMPLETE STATE

In most robotics applications, the state is continuous, meaning that x_t is defined over a continuum. A good example of a continuous state space is that of a robot pose, that is, its location and orientation relative to an external coordinate system. Sometimes, the state is discrete. An example of a discrete state space is the (binary) state variable that models whether or not a sensor is broken. State spaces that contain both continuous and discrete variables are called *hybrid* state spaces.

In most cases of interesting robotics problems, state changes over time. Time, throughout this book, will be discrete, that is, all interesting events will take place at discrete time steps $t = 0, 1, 2 \ldots$. If the robot starts its operation at a distinct point in time, we will denote this time as $t = 0$.

2.3.2 Environment Interaction

There are two fundamental types of interactions between a robot and its environment: The robot can influence the state of its environment through its actuators, and it can gather information about the state through its sensors. Both types of interactions may co-occur, but for didactic reasons we will separate them throughout this book. The interaction is illustrated in Figure 2.1.

MEASUREMENT

- **Environment sensor measurements.** Perception is the process by which the robot uses its sensors to obtain information about the state of its environment. For example, a robot might take a camera image, a range scan, or query its tactile sensors to receive information about the state of the environment. The result of such a perceptual interaction will be called a *measurement*, although we will sometimes also call it *observation* or *percept*. Typically, sensor measurements arrive with some delay. Hence they provide information about the state a few moments ago.

CONTROL ACTION

- **Control actions** change the state of the world. They do so by actively asserting forces on the robot's environment. Examples of *control actions* include robot motion and the manipulation of objects. Even if the robot does not perform any action itself, state usually changes. Thus, for consistency, we will assume that the robot *always* executes a control action, even if it chooses not to move any of its motors. In practice, the robot continuously executes controls and measurements are made concurrently.

Hypothetically, a robot may keep a record of all past sensor measurements and control actions. We will refer to such a collection as the *data* (regardless of whether they are being memorized or not). In accordance with the two types of environment interactions, the robot has access to two different data streams.

- **Environment measurement data** provides information about a momentary state of the environment. Examples of measurement data include camera images, range scans, and so on. For most parts, we will simply ignore small timing effects (e.g., most laser sensors scan environments sequentially at very high speeds, but we will simply assume the measurement corresponds to a specific point in time). The measurement data at time t will be denoted z_t.

Throughout most of this book, we simply assume that the robot takes exactly one measurement at a time. This assumption is mostly for notational convenience, as nearly all algorithms in this book can easily be extended to robots that can acquire variable numbers of measurements within a single time step. The notation

(2.29) $$z_{t_1:t_2} = z_{t_1}, z_{t_1+1}, z_{t_1+2}, \ldots, z_{t_2}$$

denotes the set of all measurements acquired from time t_1 to time t_2, for $t_1 \leq t_2$.

- **Control data** carry information about the *change of state* in the environment. In mobile robotics, a typical example of control data is the velocity of a robot. Setting the velocity to 10 cm per second for the duration of five seconds suggests that the robot's pose, after executing this motion command, is approximately 50 cm ahead of its pose before command execution. Thus, control conveys information regarding the change of state.

ODOMETER

An alternative source of control data are *odometers*. Odometers are sensors that measure the revolution of a robot's wheels. As such they convey information about the change of state. Even though odometers are sensors, we will treat odometry as control data, since they measure the effect of a control action.

Control data will be denoted u_t. The variable u_t will always correspond to the change of state in the time interval $(t-1; t]$. As before, we will denote sequences of control data by $u_{t_1:t_2}$, for $t_1 \leq t_2$:

(2.30) $$u_{t_1:t_2} = u_{t_1}, u_{t_1+1}, u_{t_1+2}, \ldots, u_{t_2}$$

Since the environment may change even if a robot does not execute a specific control action, the fact that time passed by constitutes, technically speaking, control information. We therefore assume that there is exactly one control data item per time step t, and include as legal action "*do-nothing*".

The distinction between measurement and control is a crucial one, as both types of data play fundamentally different roles in the material yet to come. Environment perception provides information about the environment's state, hence it tends to increase the robot's knowledge. Motion, on the other hand, tends to induce a loss of knowledge due to the inherent noise in robot actuation and the stochasticity of robot environments. By no means is our distinction intended to suggest that actions and perceptions are separated in time. Rather, perception and control takes place concurrently. Our separation is strictly for convenience.

2.3.3 Probabilistic Generative Laws

The evolution of state and measurements is governed by probabilistic laws. In general, the state x_t is generated stochastically from the state x_{t-1}. Thus, it makes sense to specify the probability distribution from which x_t is generated. At first glance, the emergence of state x_t might be conditioned on all past states, measurements, and controls. Hence, the probabilistic law characterizing the evolution of state might be given by a probability distribution of the following form: $p(x_t \mid x_{0:t-1}, z_{1:t-1}, u_{1:t})$. Notice that through no particular motivation we assume here that the robot executes a control action u_1 first, and then takes a measurement z_1.

An important insight is the following: If the state x is complete then it is a sufficient summary of all that happened in previous time steps. In particular, x_{t-1} is a sufficient statistic of all previous controls and measurements up to this point in time, that is, $u_{1:t-1}$ and $z_{1:t-1}$. From all the variables in the expression above, only the control u_t matters if we know the state x_{t-1}.

In probabilistic terms, this insight is expressed by the following equality:

$$(2.31) \quad p(x_t \mid x_{0:t-1}, z_{1:t-1}, u_{1:t}) = p(x_t \mid x_{t-1}, u_t)$$

The property expressed by this equality is an example of *conditional independence*. It states that certain variables are independent of others if one knows the values of a third group of variables, the conditioning variables. Conditional independence will be exploited pervasively in this book. It is the primary reason why many of the algorithms presented in the book are computationally tractable.

One might also want to model the process by which measurements are being generated. Again, if x_t is complete, we have an important conditional independence:

$$(2.32) \quad p(z_t \mid x_{0:t}, z_{1:t-1}, u_{1:t}) = p(z_t \mid x_t)$$

In other words, the state x_t is sufficient to predict the (potentially noisy) measurement z_t. Knowledge of any other variable, such as past measurements, controls, or even past states, is irrelevant if x_t is complete.

2.3 Robot Environment Interaction

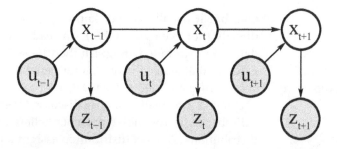

Figure 2.2 The dynamic Bayes network that characterizes the evolution of controls, states, and measurements.

STATE TRANSITION PROBABILITY

This discussion leaves open as to what the two resulting conditional probabilities are: $p(x_t \mid x_{t-1}, u_t)$ and $p(z_t \mid x_t)$. The probability $p(x_t \mid x_{t-1}, u_t)$ is the *state transition probability*. It specifies how environmental state evolves over time as a function of robot controls u_t. Robot environments are stochastic, which is reflected by the fact that $p(x_t \mid x_{t-1}, u_t)$ is a probability distribution, not a deterministic function. Sometimes the state transition distribution does not depend on the time index t, in which case we may write it as $p(x' \mid u, x)$, where x' is the successor and x the predecessor state.

MEASUREMENT PROBABILITY

The probability $p(z_t \mid x_t)$ is called the *measurement probability*. It also may not depend on the time index t, in which case it shall be written as $p(z \mid x)$. The measurement probability specifies the probabilistic law according to which measurements z are generated from the environment state x. It is appropriate to think of measurements as noisy projections of the state.

The state transition probability and the measurement probability together describe the dynamical stochastic system of the robot and its environment. Figure 2.2 illustrates the evolution of states and measurements, defined through those probabilities. The state at time t is stochastically dependent on the state at time $t-1$ and the control u_t. The measurement z_t depends stochastically on the state at time t. Such a temporal generative model is also known as *hidden Markov model* (HMM) or *dynamic Bayes network* (DBN).

2.3.4 Belief Distributions

BELIEF

Another key concept in probabilistic robotics is that of a *belief*. A belief reflects the robot's internal knowledge about the state of the environment. We already discussed that state cannot be measured directly. For example, a robot's pose might be $x_t = \langle 14.12, 12.7, 45° \rangle$ in some global coordinate system, but it usually cannot know its pose, since poses are not measurable directly (not even with GPS!). Instead, the robot must infer its pose from data. We therefore distinguish the true state from its internal *belief* with regards to that state. Synonyms for belief in the literature are the terms *state of knowledge* and *information state* (not to be confused with the information vector and information matrix discussed below).

INFORMATION STATE

Probabilistic robotics represents beliefs through conditional probability distributions. A belief distribution assigns a probability (or density value) to each possible hypothesis with regards to the true state. Belief distributions are posterior probabilities over state variables conditioned on the available data. We will denote belief over a state variable x_t by $bel(x_t)$, which is an abbreviation for the posterior

$$(2.33) \quad bel(x_t) = p(x_t \mid z_{1:t}, u_{1:t})$$

This posterior is the probability distribution over the state x_t at time t, conditioned on all past measurements $z_{1:t}$ and all past controls $u_{1:t}$.

The reader may notice that we silently assume that the belief is taken *after* incorporating the measurement z_t. Occasionally, it will prove useful to calculate a posterior *before* incorporating z_t, just after executing the control u_t. Such a posterior will be denoted as follows:

$$(2.34) \quad \overline{bel}(x_t) = p(x_t \mid z_{1:t-1}, u_{1:t})$$

PREDICTION

This probability distribution is often referred to as *prediction* in the context of probabilistic filtering. This terminology reflects the fact that $\overline{bel}(x_t)$ predicts the state at time t based on the previous state posterior, before incorporating the measurement at time t. Calculating $bel(x_t)$ from $\overline{bel}(x_t)$ is called *correction* or the *measurement update*.

2.4 Bayes Filters

2.4.1 The Bayes Filter Algorithm

BAYES FILTER

The most general algorithm for calculating beliefs is given by the *Bayes filter* algorithm. This algorithm calculates the belief distribution bel from measurement and control data. We will first state the basic algorithm and then will elucidate it with a numerical example. After that, we will derive it mathematically from the assumptions made so far.

Table 2.1 depicts the basic Bayes filter in pseudo-algorithmic form. The Bayes filter is recursive, that is, the belief $bel(x_t)$ at time t is calculated from

2.4 Bayes Filters

1:          **Algorithm Bayes_filter**($bel(x_{t-1}), u_t, z_t$):
2:          for all x_t do
3:              $\overline{bel}(x_t) = \int p(x_t \mid u_t, x_{t-1}) \, bel(x_{t-1}) \, dx_{t-1}$
4:              $bel(x_t) = \eta \, p(z_t \mid x_t) \, \overline{bel}(x_t)$
5:          endfor
6:          return $bel(x_t)$

Table 2.1 The general algorithm for Bayes filtering.

UPDATE RULE OF A BAYES FILTER

the belief $bel(x_{t-1})$ at time $t-1$. Its input is the belief bel at time $t-1$, along with the most recent control u_t and the most recent measurement z_t. Its output is the belief $bel(x_t)$ at time t. Table 2.1 only depicts a single iteration of the Bayes Filter algorithm: the *update rule*. This update rule is applied recursively, to calculate the belief $bel(x_t)$ from the belief $bel(x_{t-1})$, calculated previously.

The Bayes filter algorithm possesses two essential steps. In line 3, it processes the control u_t. It does so by calculating a belief over the state x_t based on the prior belief over state x_{t-1} and the control u_t. In particular, the belief $\overline{bel}(x_t)$ that the robot assigns to state x_t is obtained by the integral (sum) of the product of two distributions: the prior assigned to x_{t-1}, and the probability that control u_t induces a transition from x_{t-1} to x_t. The reader may recognize the similarity of this update step to Equation (2.12). As noted above, this update step is called the control update, or *prediction*.

The second step of the Bayes filter is called the *measurement update*. In line 4, the Bayes filter algorithm multiplies the belief $\overline{bel}(x_t)$ by the probability that the measurement z_t may have been observed. It does so for each hypothetical posterior state x_t. As will become apparent further below when actually deriving the basic filter equations, the resulting product is generally not a probability. It may not integrate to 1. Hence, the result is normalized, by virtue of the normalization constant η. This leads to the final belief $bel(x_t)$, which is returned in line 6 of the algorithm.

To compute the posterior belief recursively, the algorithm requires an initial belief $bel(x_0)$ at time $t = 0$ as boundary condition. If one knows the value of x_0 with certainty, $bel(x_0)$ should be initialized with a point mass distribution that centers all probability mass on the correct value of x_0, and

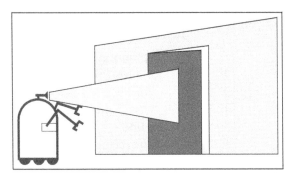

Figure 2.3 A mobile robot estimating the state of a door.

assigns zero probability anywhere else. If one is entirely ignorant about the initial value x_0, $bel(x_0)$ may be initialized using a uniform distribution over the domain of x_0 (or a related distribution from the Dirichlet family of distributions). Partial knowledge of the initial value x_0 can be expressed by non-uniform distributions; however, the two cases of full knowledge and full ignorance are the most common ones in practice.

The algorithm Bayes filter can only be implemented in the form stated here for very simple estimation problems. In particular, we either need to be able to carry out the integration in line 3 and the multiplication in line 4 in closed form, or we need to restrict ourselves to finite state spaces, so that the integral in line 3 becomes a (finite) sum.

2.4.2 Example

Our illustration of the Bayes filter algorithm is based on the scenario in Figure 2.3, which shows a robot estimating the state of a door using its camera. To make this problem simple, let us assume that the door can be in one of two possible states, open or closed, and that only the robot can change the state of the door. Let us furthermore assume that the robot does not know the state of the door initially. Instead, it assigns equal prior probability to the two possible door states:

$$bel(X_0 = \textbf{open}) \;\;=\;\; 0.5$$
$$bel(X_0 = \textbf{closed}) \;\;=\;\; 0.5$$

Let us now assume the robot's sensors are noisy. The noise is characterized by the following conditional probabilities:

$$p(Z_t = \textbf{sense\_open} \mid X_t = \textbf{is\_open}) \;\;=\;\; 0.6$$

2.4 Bayes Filters

$$p(Z_t = \text{sense\_closed} \mid X_t = \text{is\_open}) = 0.4$$

and

$$p(Z_t = \text{sense\_open} \mid X_t = \text{is\_closed}) = 0.2$$
$$p(Z_t = \text{sense\_closed} \mid X_t = \text{is\_closed}) = 0.8$$

These probabilities suggest that the robot's sensors are relatively reliable in detecting a *closed* door, in that the error probability is 0.2. However, when the door is open, it has a 0.4 probability of an erroneous measurement.

Finally, let us assume the robot uses its manipulator to push the door open. If the door is already open, it will remain open. If it is closed, the robot has a 0.8 chance that it will be open afterwards:

$$p(X_t = \text{is\_open} \mid U_t = \text{push}, X_{t\_1} = \text{is\_open}) = 1$$
$$p(X_t = \text{is\_closed} \mid U_t = \text{push}, X_{t\_1} = \text{is\_open}) = 0$$
$$p(X_t = \text{is\_open} \mid U_t = \text{push}, X_{t\_1} = \text{is\_closed}) = 0.8$$
$$p(X_t = \text{is\_closed} \mid U_t = \text{push}, X_{t\_1} = \text{is\_closed}) = 0.2$$

It can also choose not to use its manipulator, in which case the state of the world does not change. This is stated by the following conditional probabilities:

$$p(X_t = \text{is\_open} \mid U_t = \text{do\_nothing}, X_{t\_1} = \text{is\_open}) = 1$$
$$p(X_t = \text{is\_closed} \mid U_t = \text{do\_nothing}, X_{t\_1} = \text{is\_open}) = 0$$
$$p(X_t = \text{is\_open} \mid U_t = \text{do\_nothing}, X_{t\_1} = \text{is\_closed}) = 0$$
$$p(X_t = \text{is\_closed} \mid U_t = \text{do\_nothing}, X_{t\_1} = \text{is\_closed}) = 1$$

Suppose at time $t = 1$, the robot takes no control action but it senses an open door. The resulting posterior belief is calculated by the Bayes filter using the prior belief $bel(X_0)$, the control $u_1 = \text{do\_nothing}$, and the measurement **sense_open** as input. Since the state space is finite, the integral in line 3 turns into a finite sum:

$$\overline{bel}(x_1)$$
$$= \int p(x_1 \mid u_1, x_0) \, bel(x_0) \, dx_0$$
$$= \sum_{x_0} p(x_1 \mid u_1, x_0) \, bel(x_0)$$
$$= p(x_1 \mid U_1 = \text{do\_nothing}, X_0 = \text{is\_open}) \, bel(X_0 = \text{is\_open})$$
$$+ p(x_1 \mid U_1 = \text{do\_nothing}, X_0 = \text{is\_closed}) \, bel(X_0 = \text{is\_closed})$$

We can now substitute the two possible values for the state variable X_1. For the hypothesis $X_1 = \text{is\_open}$, we obtain

$\overline{bel}(X_1 = \text{is\_open})$
$= p(X_1 = \text{is\_open} \mid U_1 = \text{do\_nothing}, X_0 = \text{is\_open})$
$\quad bel(X_0 = \text{is\_open})$
$\quad + p(X_1 = \text{is\_open} \mid U_1 = \text{do\_nothing}, X_0 = \text{is\_closed})$
$\quad bel(X_0 = \text{is\_closed})$
$= 1 \cdot 0.5 + 0 \cdot 0.5 = 0.5$

Likewise, for $X_1 = \text{is\_closed}$ we get

$\overline{bel}(X_1 = \text{is\_closed})$
$= p(X_1 = \text{is\_closed} \mid U_1 = \text{do\_nothing}, X_0 = \text{is\_open})$
$\quad bel(X_0 = \text{is\_open})$
$\quad + p(X_1 = \text{is\_closed} \mid U_1 = \text{do\_nothing}, X_0 = \text{is\_closed})$
$\quad bel(X_0 = \text{is\_closed})$
$= 0 \cdot 0.5 + 1 \cdot 0.5 = 0.5$

The fact that the belief $\overline{bel}(x_1)$ equals our prior belief $bel(x_0)$ should not surprise, as the action **do_nothing** does not affect the state of the world; neither does the world change over time by itself in our example.

Incorporating the measurement, however, changes the belief. Line 4 of the Bayes filter algorithm implies

$bel(x_1) = \eta \, p(Z_1 = \text{sense\_open} \mid x_1) \, \overline{bel}(x_1)$

For the two possible cases, $X_1 = \text{is\_open}$ and $X_1 = \text{is\_closed}$, we get

$bel(X_1 = \text{is\_open})$
$= \eta \, p(Z_1 = \text{sense\_open} \mid X_1 = \text{is\_open}) \, \overline{bel}(X_1 = \text{is\_open})$
$= \eta \, 0.6 \cdot 0.5 = \eta \, 0.3$

and

$bel(X_1 = \text{is\_closed})$
$= \eta \, p(Z_1 = \text{sense\_open} \mid X_1 = \text{is\_closed}) \, \overline{bel}(X_1 = \text{is\_closed})$
$= \eta \, 0.2 \cdot 0.5 = \eta \, 0.1$

The normalizer η is now easily calculated:

$\eta = (0.3 + 0.1)^{-1} = 2.5$

2.4 Bayes Filters

Hence, we have

$$bel(X_1 = \text{is\_open}) = 0.75$$
$$bel(X_1 = \text{is\_closed}) = 0.25$$

This calculation is now easily iterated for the next time step. As the reader easily verifies, for $u_2 = \text{push}$ and $z_2 = \text{sense\_open}$ we get

$$\overline{bel}(X_2 = \text{is\_open}) = 1 \cdot 0.75 + 0.8 \cdot 0.25 = 0.95$$
$$\overline{bel}(X_2 = \text{is\_closed}) = 0 \cdot 0.75 + 0.2 \cdot 0.25 = 0.05$$

and

$$bel(X_2 = \text{is\_open}) = \eta\, 0.6 \cdot 0.95 \approx 0.983$$
$$bel(X_2 = \text{is\_closed}) = \eta\, 0.2 \cdot 0.05 \approx 0.017$$

At this point, the robot believes that with 0.983 probability the door is open.

At first glance, this probability may appear to be sufficiently high to simply accept this hypothesis as the world state and act accordingly. However, such an approach may result in unnecessarily high costs. If mistaking a closed door for an open one incurs costs (e.g., the robot crashes into a door), considering both hypotheses in the decision making process will be essential, as unlikely as one of them may be. Just imagine flying an aircraft on auto pilot with a perceived chance of 0.983 for not crashing!

2.4.3 Mathematical Derivation of the Bayes Filter

The correctness of the Bayes filter algorithm is shown by induction. To do so, we need to show that it correctly calculates the posterior distribution $p(x_t \mid z_{1:t}, u_{1:t})$ from the corresponding posterior one time step earlier, $p(x_{t-1} \mid z_{1:t-1}, u_{1:t-1})$. The correctness follows then by induction under the assumption that we correctly initialized the prior belief $bel(x_0)$ at time $t = 0$.

Our derivation requires that the state x_t is complete, as defined in Chapter 2.3.1, and it requires that controls are chosen at random. The first step of our derivation involves the application of Bayes rule (2.16) to the target posterior:

$$(2.35) \quad p(x_t \mid z_{1:t}, u_{1:t}) = \frac{p(z_t \mid x_t, z_{1:t-1}, u_{1:t})\, p(x_t \mid z_{1:t-1}, u_{1:t})}{p(z_t \mid z_{1:t-1}, u_{1:t})}$$
$$= \eta\, p(z_t \mid x_t, z_{1:t-1}, u_{1:t})\, p(x_t \mid z_{1:t-1}, u_{1:t})$$

We now exploit the assumption that our state is complete. In Chapter 2.3.1, we defined a state x_t to be complete if no variables prior to x_t may influence the stochastic evolution of future states. In particular, if we (hypothetically) knew the state x_t and were interested in predicting the measurement z_t, no past measurement or control would provide us additional information. In mathematical terms, this is expressed by the following conditional independence:

$$(2.36) \qquad p(z_t \mid x_t, z_{1:t-1}, u_{1:t}) \;=\; p(z_t \mid x_t)$$

Such a statement is another example of *conditional independence*. It allows us to simplify (2.35) as follows:

$$(2.37) \qquad p(x_t \mid z_{1:t}, u_{1:t}) \;=\; \eta \, p(z_t \mid x_t) \, p(x_t \mid z_{1:t-1}, u_{1:t})$$

and hence

$$(2.38) \qquad bel(x_t) \;=\; \eta \, p(z_t \mid x_t) \, \overline{bel}(x_t)$$

This equation is implemented in line 4 of the Bayes filter algorithm in Table 2.1.

Next, we expand the term $\overline{bel}(x_t)$, using (2.12):

$$(2.39) \qquad \overline{bel}(x_t) \;=\; p(x_t \mid z_{1:t-1}, u_{1:t})$$
$$= \int p(x_t \mid x_{t-1}, z_{1:t-1}, u_{1:t}) \, p(x_{t-1} \mid z_{1:t-1}, u_{1:t}) \, dx_{t-1}$$

Once again, we exploit the assumption that our state is complete. This implies if we know x_{t-1}, past measurements and controls convey no information regarding the state x_t. This gives us

$$(2.40) \qquad p(x_t \mid x_{t-1}, z_{1:t-1}, u_{1:t}) \;=\; p(x_t \mid x_{t-1}, u_t)$$

Here we retain the control variable u_t, since it does *not* predate the state x_{t-1}. In fact, the reader should quickly convince herself that $p(x_t \mid x_{t-1}, u_t) \neq p(x_t \mid x_{t-1})$.

Finally, we note that the control u_t can safely be omitted from the set of conditioning variables in $p(x_{t-1} \mid z_{1:t-1}, u_{1:t})$ for randomly chosen controls. This gives us the recursive update equation

$$(2.41) \qquad \overline{bel}(x_t) \;=\; \int p(x_t \mid x_{t-1}, u_t) \, p(x_{t-1} \mid z_{1:t-1}, u_{1:t-1}) \, dx_{t-1}$$

As the reader easily verifies, this equation is implemented by line 3 of the Bayes filter algorithm in Table 2.1.

To summarize, the Bayes filter algorithm calculates the posterior over the state x_t conditioned on the measurement and control data up to time t. The derivation assumes that the world is Markov, that is, the state is complete.

Any concrete implementation of this algorithm requires three probability distributions: The initial belief $p(x_0)$, the measurement probability $p(z_t \mid x_t)$, and the state transition probability $p(x_t \mid u_t, x_{t-1})$. We have not yet specified these densities for actual robot systems. But we will soon: Chapter 5 is entirely dedicated to $p(x_t \mid u_t, x_{t-1})$ and Chapter 6 to $p(z_t \mid x_t)$. We also need a representation for the belief $bel(x_t)$, which will be discussed in Chapters 3 and 4.

2.4.4 The Markov Assumption

MARKOV ASSUMPTION A word is in order on the *Markov assumption*, or the *complete state assumption*, since it plays such a fundamental role in the material presented in this book. The Markov assumption postulates that past and future data are independent if one knows the current state x_t. To see how severe an assumption this is, let us consider our example of mobile robot localization. In mobile robot localization, x_t is the robot's pose, and Bayes filters are applied to estimate the pose relative to a fixed map. The following factors may have a systematic effect on sensor readings. Thus, they induce violations of the Markov assumption:

- Unmodeled dynamics in the environment not included in x_t (e.g., moving people and their effects on sensor measurements in our localization example),

- inaccuracies in the probabilistic models $p(z_t \mid x_t)$ and $p(x_t \mid u_t, x_{t-1})$ (e.g., an error in the map for a localizing robot),

- approximation errors when using approximate representations of belief functions (e.g., grids or Gaussians, which will be discussed below), and

- software variables in the robot control software that influence multiple controls (e.g., the variable "target location" typically influences an entire sequence of control commands).

In principle, many of these variables can be included in state representations. However, incomplete state representations are often preferable to more complete ones to reduce the computational complexity of the Bayes filter algorithm. In practice, Bayes filters have been found to be surprisingly robust to such violations. As a general rule of thumb, however, one should exercise care when defining the state x_t, so that the effect of unmodeled state variables has close-to-random effects.

2.5 Representation and Computation

In probabilistic robotics, Bayes filters are implemented in several different ways. As we will see in the next two chapters, there exist quite a variety of techniques and algorithms that are all derived from the Bayes filter. Each such technique relies on different assumptions regarding the measurement and state transition probabilities and the initial belief. These assumptions then give rise to different types of posterior distributions, and the algorithms for computing them have different computational characteristics. As a general rule of thumb, exact techniques for calculating beliefs exist only for highly specialized cases; in general robotics problems, beliefs have to be approximated. The nature of the approximation has important ramifications on the complexity of the algorithm. Finding a suitable approximation is usually a challenging problem, with no unique best answer for all robotics problems.

When choosing an approximation, one has to trade off a range of properties:

1. **Computational efficiency.** Some approximations, such as linear Gaussian approximations that will be discussed further below, make it possible to calculate beliefs in time polynomial in the dimension of the state space. Others may require exponential time. Particle-based techniques, discussed further below, have an *any-time* characteristic, enabling them to trade off accuracy with computational efficiency.

2. **Accuracy of the approximation.** Some approximations can approximate a wider range of distributions more tightly than others. For example, linear Gaussian approximations are limited to unimodal distributions, whereas histogram representations can approximate multi-modal distributions, albeit with limited accuracy. Particle representations can approximate a wide array of distributions, but the number of particles needed to attain a desired accuracy can be large.

3. **Ease of implementation.** The difficulty of implementing probabilistic algorithms depends on a variety of factors, such as the form of the measurement probability $p(z_t \mid x_t)$ and the state transition probability $p(x_t \mid u_t, x_{t-1})$. Particle representations often yield surprisingly simple implementations for complex nonlinear systems—one of the reasons for their recent popularity.

The next two chapters will introduce concrete implementable algorithms, which fare quite differently relative to the criteria described above.

2.6 Summary

In this section, we introduced the basic idea of Bayes filters in robotics, as a means to estimate the state of an environment and the robot.

- The interaction of a robot and its environment is modeled as a coupled dynamical system, in which the robot manipulates its environment by choosing controls, and in which it can perceive the environment through its sensors.

- In probabilistic robotics, the dynamics of the robot and its environment are characterized in the form of two probabilistic laws: the state transition distribution, and the measurement distribution. The state transition distribution characterizes how state changes over time, possibly as the effect of robot controls. The measurement distribution characterizes how measurements are governed by states. Both laws are probabilistic, accounting for the inherent uncertainty in state evolution and sensing.

- The *belief* of a robot is the posterior distribution over the state of the environment (including the robot state) given all past sensor measurements and all past controls. The *Bayes filter* is the principal algorithm for calculating the belief in robotics. The Bayes filter is recursive; the belief at time t is calculated from the belief at time $t-1$.

- The Bayes filter makes a *Markov assumption* according to which the state is a complete summary of the past. This assumption implies the belief is sufficient to represent the past history of the robot. In robotics, the Markov assumption is usually only an approximation. We identified conditions under which it is violated.

- Since the Bayes filter is not a practical algorithm, in that it cannot be implemented on a digital computer, probabilistic algorithms use tractable approximations. Such approximations may be evaluated according to different criteria, relating to their accuracy, efficiency, and ease of implementation.

The next two chapters discuss two popular families of recursive state estimation techniques that are both derived from the Bayes filter.

2.7 Bibliographical Remarks

The basic statistical material in this chapter is covered in most introductory textbooks to probability and statistics. Some early classical texts by DeGroot (1975), Subrahmaniam (1979), and Thorp (1966) provide highly accessible introductions into this material. More advanced treatments can be found in (Feller 1968; Casella and Berger 1990; Tanner 1996), and in (Devroye et al. 1996; Duda et al. 2000). The robot environment interaction paradigm is common in robotics. It is discussed from the AI perspective by Russell and Norvig (2002).

2.8 Exercises

1. A robot uses a range sensor that can measure ranges from $0m$ and $3m$. For simplicity, assume that actual ranges are distributed uniformly in this interval. Unfortunately, the sensor can be faulty. When the sensor is faulty, it constantly outputs a range below $1m$, regardless of the actual range in the sensor's measurement cone. We know that the prior probability for a sensor to be faulty is $p = 0.01$.

   Suppose the robot queried its sensor N times, and every single time the measurement value is below $1m$. What is the posterior probability of a sensor fault, for $N = 1, 2, \ldots, 10$. Formulate the corresponding probabilistic model.

2. Suppose we live at a place where days are either sunny, cloudy, or rainy. The weather transition function is a Markov chain with the following transition table:

		tomorrow will be...		
		sunny	cloudy	rainy
	sunny	.8	.2	0
today it's...	cloudy	.4	.4	.2
	rainy	.2	.6	.2

   (a) Suppose Day 1 is a sunny day. What is the probability of the following sequence of days: Day2 = *cloudy*, Day3 = *cloudy*, Day4 = *rainy*?

   (b) Write a simulator that can randomly generate sequences of "weathers" from this state transition function.

   (c) Use your simulator to determine the stationary distribution of this Markov chain. The stationary distribution measures the probability that a random day will be sunny, cloudy, or rainy.

   (d) Can you devise a closed-form solution to calculating the stationary distribution based on the state transition matrix above?

2.8 Exercises

(e) What is the entropy of the stationary distribution?

(f) Using Bayes rule, compute the probability table of yesterday's weather given today's weather. (It is okay to provide the probabilities numerically, and it is also okay to rely on results from previous questions in this exercise.)

(g) Suppose we added seasons to our model. The state transition function above would only apply to the Summer, whereas different ones would apply to Winter, Spring, and Fall. Would this violate the Markov property of this process? Explain your answer.

3. Suppose that we cannot observe the weather directly, but instead rely on a sensor. The problem is that our sensor is noisy. Its measurements are governed by the following measurement model:

		our sensor tells us...		
		sunny	cloudy	rainy
	sunny	.6	.4	0
the actual weather is...	cloudy	.3	.7	0
	rainy	0	0	1

(a) Suppose Day 1 is sunny (this is known for a fact), and in the subsequent four days our sensor observes *cloudy, cloudy, rainy, sunny*. What is the probability that Day 5 is indeed sunny as predicted by our sensor?

(b) Once again, suppose Day 1 is known to be sunny. At Days 2 through 4, the sensor measures *sunny, sunny, rainy*. For each of the Days 2 through 4, what is the most likely weather on that day? Answer the question in two ways: one in which only the data available to the day in question is used, and one in hindsight, where data from future days is also available.

(c) Consider the same situation (Day 1 is sunny, the measurements for Days 2, 3, and 4 are *sunny, sunny, rainy*). What is the most likely sequence of weather for Days 2 through 4? What is the probability of this most likely sequence?

4. In this exercise we will apply Bayes rule to Gaussians. Suppose we are a mobile robot who lives on a long straight road. Our location x will simply be the position along this road. Now suppose that initially, we believe to be at location $x_{\text{init}} = 1,000m$, but we happen to know that this estimate

is uncertain. Based on this uncertainty, we model our initial belief by a Gaussian with variance $\sigma_{\text{init}}^2 = 900m^2$.

To find out more about our location, we query a GPS receiver. The GPS tells us our location is $z_{\text{GPS}} = 1,100m$. This GPS receiver is known to have an error variance of $\sigma_{\text{init}}^2 = 100m^2$.

(a) Write the probability density functions of the prior $p(x)$ and the measurement $p(z \mid x)$.

(b) Using Bayes rule, what is the posterior $p(x \mid z)$? Can you prove it to be Gaussian?

(c) How likely was the measurement $x_{\text{GPS}} = 1,100m$ given our prior, and knowledge of the error probability of our GPS receiver?

Hint: This is an exercise in manipulating quadratic expressions.

5. Derive Equations (2.18) and (2.19) from (2.17) and the laws of probability stated in the text.

6. Prove Equation (2.25). What are the implications of this equality?

3 Gaussian Filters

3.1 Introduction

This chapter describes an important family of recursive state estimators, collectively called *Gaussian filters*. Historically, Gaussian filters constitute the earliest tractable implementations of the Bayes filter for continuous spaces. They are also by far the most popular family of techniques to date—despite a number of shortcomings.

Gaussian techniques all share the basic idea that beliefs are represented by multivariate normal distributions. We already encountered a definition of the multivariate normal distribution in Equation (2.4), which is restated here for convenience:

$$(3.1) \quad p(x) = \det(2\pi\Sigma)^{-\frac{1}{2}} \exp\left\{-\tfrac{1}{2}(x-\mu)^T \Sigma^{-1}(x-\mu)\right\}$$

This density over the variable x is characterized by two sets of parameters: The mean μ and the covariance Σ. The mean μ is a vector that possesses the same dimensionality as the state x. The covariance is a quadratic matrix that is symmetric and positive-semidefinite. Its dimension is the dimensionality of the state x squared. Thus, the number of elements in the covariance matrix depends quadratically on the number of elements in the state vector.

The commitment to represent the posterior by a Gaussian has important ramifications. Most importantly, Gaussians are unimodal; they possess a single maximum. Such a posterior is characteristic of many tracking problems in robotics, in which the posterior is focused around the true state with a small margin of uncertainty. Gaussian posteriors are a poor match for many global estimation problems in which many distinct hypotheses exist, each of which forms its own mode in the posterior.

The parameterization of a Gaussian by its mean and covariance is called

MOMENTS PARAMETERIZATION

CANONICAL PARAMETERIZATION

the *moments parameterization*. This is because the mean and covariance are the first and second moments of a probability distribution; all other moments are zero for normal distributions. In this chapter, we will also discuss an alternative parameterization, called *canonical parameterization*, or sometimes *natural parameterization*. Both parameterizations, the moments and the canonical parameterizations, are functionally equivalent in that a bijective mapping exists that transforms one into the other. However, they lead to filter algorithms with somewhat different computational characteristics. As we shall see, the canonical and the natural parameterizations are best thought of as duals: what appears to be computationally easy in one parameterization is involved in the other, and vice versa.

This chapter introduces the two basic Gaussian filter algorithms.

- Chapter 3.2 describes the Kalman filter, which implements the Bayes filter using the moments parameterization for a restricted class of problems with linear dynamics and measurement functions.

- The Kalman filter is extended to nonlinear problems in Chapter 3.3, which describes the extended Kalman filter.

- Chapter 3.4 describes a different nonlinear Kalman filter, known as unscented Kalman filter.

- Chapter 3.5 describes the information filter, which is the dual of the Kalman filter using the canonical parameterization of Gaussians.

3.2 The Kalman Filter

3.2.1 Linear Gaussian Systems

Probably the best studied technique for implementing Bayes filters is the *Kalman filter*, or *(KF)*. The Kalman filter was invented by Swerling (1958) and Kalman (1960) as a technique for filtering and prediction in *linear Gaussian systems*, which will be defined in a moment. The Kalman filter implements belief computation for continuous states. It is not applicable to discrete or hybrid state spaces.

The Kalman filter represents beliefs by the moments parameterization: At time t, the belief is represented by the the mean μ_t and the covariance Σ_t.

GAUSSIAN POSTERIOR

Posteriors are *Gaussian* if the following three properties hold, in addition to the Markov assumptions of the Bayes filter.

1. The state transition probability $p(x_t \mid u_t, x_{t-1})$ must be a *linear* function in its arguments with added Gaussian noise. This is expressed by the following equation:

$$(3.2) \quad x_t = A_t x_{t-1} + B_t u_t + \varepsilon_t$$

Here x_t and x_{t-1} are state vectors, and u_t is the control vector at time t. In our notation, both of these vectors are vertical vectors. They are of the form

$$(3.3) \quad x_t = \begin{pmatrix} x_{1,t} \\ x_{2,t} \\ \vdots \\ x_{n,t} \end{pmatrix} \quad \text{and} \quad u_t = \begin{pmatrix} u_{1,t} \\ u_{2,t} \\ \vdots \\ u_{m,t} \end{pmatrix}$$

A_t and B_t are matrices. A_t is a square matrix of size $n \times n$, where n is the dimension of the state vector x_t. B_t is of size $n \times m$, with m being the dimension of the control vector u_t. By multiplying the state and control vector with the matrices A_t and B_t, respectively, the state transition function becomes *linear* in its arguments. Thus, Kalman filters assume linear system dynamics.

The random variable ε_t in (3.2) is a Gaussian random vector that models the uncertainty introduced by the state transition. It is of the same dimension as the state vector. Its mean is zero, and its covariance will be denoted R_t. A state transition probability of the form (3.2) is called a *linear Gaussian*, to reflect the fact that it is linear in its arguments with additive Gaussian noise. Technically, one may also include a constant additive term in (3.2), which is here omitted since it plays no role in the material to come.

Equation (3.2) defines the state transition probability $p(x_t \mid u_t, x_{t-1})$. This probability is obtained by plugging Equation (3.2) into the definition of the multivariate normal distribution (3.1). The mean of the posterior state is given by $A_t x_{t-1} + B_t u_t$ and the covariance by R_t:

$$(3.4) \quad p(x_t \mid u_t, x_{t-1}) = \det(2\pi R_t)^{-\frac{1}{2}} \\ \exp\left\{-\tfrac{1}{2}(x_t - A_t x_{t-1} - B_t u_t)^T R_t^{-1}(x_t - A_t x_{t-1} - B_t u_t)\right\}$$

1: **Algorithm Kalman_filter($\mu_{t-1}, \Sigma_{t-1}, u_t, z_t$):**
2: $\bar{\mu}_t = A_t \, \mu_{t-1} + B_t \, u_t$
3: $\bar{\Sigma}_t = A_t \, \Sigma_{t-1} \, A_t^T + R_t$
4: $K_t = \bar{\Sigma}_t \, C_t^T (C_t \, \bar{\Sigma}_t \, C_t^T + Q_t)^{-1}$
5: $\mu_t = \bar{\mu}_t + K_t (z_t - C_t \, \bar{\mu}_t)$
6: $\Sigma_t = (I - K_t \, C_t) \, \bar{\Sigma}_t$
7: return μ_t, Σ_t

Table 3.1 The Kalman filter algorithm for linear Gaussian state transitions and measurements.

2. The measurement probability $p(z_t \mid x_t)$ must also be *linear* in its arguments, with added Gaussian noise:

$$z_t = C_t x_t + \delta_t \tag{3.5}$$

Here C_t is a matrix of size $k \times n$, where k is the dimension of the measurement vector z_t. The vector δ_t describes the measurement noise. The distribution of δ_t is a multivariate Gaussian with zero mean and covariance Q_t. The measurement probability is thus given by the following multivariate normal distribution:

$$p(z_t \mid x_t) = \det(2\pi Q_t)^{-\frac{1}{2}} \exp\left\{-\tfrac{1}{2}(z_t - C_t x_t)^T Q_t^{-1} (z_t - C_t x_t)\right\} \tag{3.6}$$

3. Finally, the initial belief $bel(x_0)$ must be normally distributed. We will denote the mean of this belief by μ_0 and the covariance by Σ_0:

$$bel(x_0) = p(x_0) = \det(2\pi \Sigma_0)^{-\frac{1}{2}} \exp\left\{-\tfrac{1}{2}(x_0 - \mu_0)^T \Sigma_0^{-1} (x_0 - \mu_0)\right\} \tag{3.7}$$

These three assumptions are sufficient to ensure that the posterior $bel(x_t)$ is always a Gaussian, for any point in time t. The proof of this non-trivial result can be found below, in the mathematical derivation of the Kalman filter (Chapter 3.2.4).

3.2.2 The Kalman Filter Algorithm

The *Kalman filter algorithm* is depicted in Table 3.1. Kalman filters represent the belief $bel(x_t)$ at time t by the mean μ_t and the covariance Σ_t. The input of the Kalman filter is the belief at time $t-1$, represented by μ_{t-1} and Σ_{t-1}. To update these parameters, Kalman filters require the control u_t and the measurement z_t. The output is the belief at time t, represented by μ_t and Σ_t.

In lines 2 and 3, the predicted belief $\bar{\mu}$ and $\bar{\Sigma}$ is calculated representing the belief $\overline{bel}(x_t)$ one time step later, but before incorporating the measurement z_t. This belief is obtained by incorporating the control u_t. The mean is updated using the deterministic version of the state transition function (3.2), with the mean μ_{t-1} substituted for the state x_{t-1}. The update of the covariance considers the fact that states depend on previous states through the linear matrix A_t. This matrix is multiplied twice into the covariance, since the covariance is a quadratic matrix.

KALMAN GAIN

The belief $\overline{bel}(x_t)$ is subsequently transformed into the desired belief $bel(x_t)$ in lines 4 through 6, by incorporating the measurement z_t. The variable K_t, computed in line 4 is called *Kalman gain*. It specifies the degree to which the measurement is incorporated into the new state estimate, in a way that will become clearer in Chapter 3.2.4. Line 5 manipulates the mean, by adjusting it in proportion to the Kalman gain K_t and the deviation of the actual measurement, z_t, and the measurement predicted according to the measurement probability (3.5). The key concept here is the *innovation*, which is the difference between the actual measurement z_t and the expected measurement $C_t\,\bar{\mu}_t$ in line 5. Finally, the new covariance of the posterior belief is calculated in line 6, adjusting for the information gain resulting from the measurement.

INNOVATION

The Kalman filter is computationally quite efficient. For today's best algorithms, the complexity of matrix inversion is approximately $O(d^{2.4})$ for a matrix of size $d \times d$. Each iteration of the Kalman filter algorithm, as stated here, is lower bounded by (approximately) $O(k^{2.4})$, where k is the dimension of the measurement vector z_t. This (approximate) cubic complexity stems from the matrix inversion in line 4. Even for certain sparse updates discussed in future chapters, it is also at least in $O(n^2)$, where n is the dimension of the state space, due to the multiplication in line 6 (the matrix $K_t C_t$ may be sparse). In many applications—such as the robot mapping applications discussed in later chapters—the measurement space is much lower dimensional than the state space, and the update is dominated by the $O(n^2)$ operations.

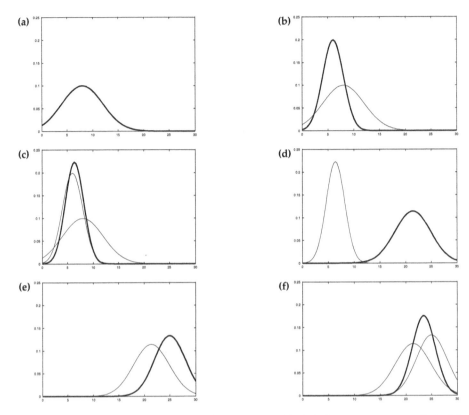

Figure 3.2 Illustration of Kalman filters: (a) initial belief, (b) a measurement (in bold) with the associated uncertainty, (c) belief after integrating the measurement into the belief using the Kalman filter algorithm, (d) belief after motion to the right (which introduces uncertainty), (e) a new measurement with associated uncertainty, and (f) the resulting belief.

3.2.3 Illustration

Figure 3.2 illustrates the Kalman filter algorithm for a simplistic one-dimensional localization scenario. Suppose the robot moves along the horizontal axis in each diagram in Figure 3.2. Let the prior over the robot location be given by the normal distribution shown in Figure 3.2a. The robot queries its sensors on its location (e.g., a GPS system), and those return a measurement that is centered at the peak of the bold Gaussian in Figure 3.2b. This bold Gaussian illustrates this measurement: Its peak is the value predicted by the sensors, and its width (variance) corresponds to the uncertainty in the measurement. Combining the prior with the measurement, via lines 4 through 6 of the Kalman filter algorithm in Table 3.1, yields the bold Gaussian in Figure 3.2c. This belief's mean lies between the two original means, and its uncertainty radius is smaller than both contributing Gaussians. The fact that the residual uncertainty is smaller than the contributing Gaussians may appear counter-intuitive, but it is a general characteristic of information integration in Kalman filters.

Next, assume the robot moves towards the right. Its uncertainty grows due to the fact that the state transition is stochastic. Lines 2 and 3 of the Kalman filter provide us with the Gaussian shown in bold in Figure 3.2d. This Gaussian is shifted by the amount the robot moved, and it is also wider for the reasons just explained. The robot receives a second measurement illustrated by the bold Gaussian in Figure 3.2e, which leads to the posterior shown in bold in Figure 3.2f.

As this example illustrates, the Kalman filter alternates a *measurement update step* (lines 5-7), in which sensor data is integrated into the present belief, with a *prediction step* (or control update step), which modifies the belief in accordance to an action. The update step decreases and the prediction step increases uncertainty in the robot's belief.

3.2.4 Mathematical Derivation of the KF

This section derives the Kalman filter algorithm in Table 3.1. The section can safely be skipped at first reading; it is only included for completeness.

Up front, the derivation of the KF is largely an exercise in manipulating quadratic expressions. When multiplying two Gaussians, for example, the exponents add. Since both original exponents are quadratic, so is the resulting sum. The remaining exercise is then to come up with a factorization of the result into a form that makes it possible to read off the desired parameters.

Part 1: Prediction

Our derivation begins with lines 2 and 3 of the algorithm, in which the belief $\overline{bel}(x_t)$ is calculated from the belief one time step earlier, $bel(x_{t-1})$. Lines 2 and 3 implement the update step described in Equation (2.41), restated here for the reader's convenience:

$$(3.8) \quad \overline{bel}(x_t) = \int \underbrace{p(x_t \mid x_{t-1}, u_t)}_{\sim \mathcal{N}(x_t; A_t x_{t-1} + B_t u_t, R_t)} \underbrace{bel(x_{t-1})}_{\sim \mathcal{N}(x_{t-1}; \mu_{t-1}, \Sigma_{t-1})} dx_{t-1}$$

The belief $bel(x_{t-1})$ is represented by the mean μ_{t-1} and the covariance Σ_{t-1}. The state transition probability $p(x_t \mid x_{t-1}, u_t)$ was given in (3.4) as a normal distribution over x_t with mean $A_t x_{t-1} + B_t u_t$ and covariance R_t. As we shall show now, the outcome of (3.8) is again a Gaussian with mean $\bar{\mu}_t$ and covariance $\bar{\Sigma}_t$ as stated in Table 3.1.

We begin by writing (3.8) in its Gaussian form:

$$(3.9) \quad \overline{bel}(x_t)$$
$$= \eta \int \exp\left\{-\tfrac{1}{2}(x_t - A_t x_{t-1} - B_t u_t)^T R_t^{-1}(x_t - A_t x_{t-1} - B_t u_t)\right\}$$
$$\exp\left\{-\tfrac{1}{2}(x_{t-1} - \mu_{t-1})^T \Sigma_{t-1}^{-1}(x_{t-1} - \mu_{t-1})\right\} dx_{t-1}$$

In short, we have

$$(3.10) \quad \overline{bel}(x_t) = \eta \int \exp\{-L_t\} \, dx_{t-1}$$

with

$$(3.11) \quad L_t = \tfrac{1}{2}(x_t - A_t x_{t-1} - B_t u_t)^T R_t^{-1}(x_t - A_t x_{t-1} - B_t u_t)$$
$$+ \tfrac{1}{2}(x_{t-1} - \mu_{t-1})^T \Sigma_{t-1}^{-1}(x_{t-1} - \mu_{t-1})$$

Notice that L_t is quadratic in x_{t-1}; it is also quadratic in x_t.

Expression (3.10) contains an integral. Solving this integral requires us to reorder the terms in this interval, in a way that might appear counterintuitive at first. In particular, we will decompose L_t into two functions, $L_t(x_{t-1}, x_t)$ and $L_t(x_t)$:

$$(3.12) \quad L_t = L_t(x_{t-1}, x_t) + L_t(x_t)$$

This decomposition will simply be the result of reordering the terms in L_t. A key goal of this decomposition step shall be that the variables in L_t are partitioned into two sets, of which only one will depend on the variable x_{t-1}. The other, $L_t(x_t)$, will not depend on x_{t-1}. As a result, we will be able to move the latter variables out of the integral over the variable x_{t-1}.

This is illustrated by the following transformation:

$$(3.13) \quad \overline{bel}(x_t) = \eta \int \exp\{-L_t\} \, dx_{t-1}$$

$$= \eta \int \exp\{-L_t(x_{t-1}, x_t) - L_t(x_t)\} \, dx_{t-1}$$

$$= \eta \, \exp\{-L_t(x_t)\} \int \exp\{-L_t(x_{t-1}, x_t)\} \, dx_{t-1}$$

Of course, there exist many ways to decompose L_t into two sets that would meet this criterion. The key insight is that we will choose $L_t(x_{t-1}, x_t)$ such that the value of the integral in (3.13) does not depend on x_t. If we succeed in defining such a function $L_t(x_{t-1}, x_t)$, the entire integral over $L_t(x_{t-1}, x_t)$ will simply become a constant relative to the problem of estimating the belief distribution over x_t. Constants are usually captured in the normalization constant η, so under our decomposition we will be able to subsume this constant into η (now for a different actual value of η as above):

(3.14) $\quad \overline{bel}(x_t) \;=\; \eta \, \exp\{-L_t(x_t)\}$

Thus, our decomposition would make it possible to eliminate the integral from the belief (3.10). The result is just a normalized exponential over a quadratic function, which turns out to be a Gaussian.

Let us now perform this decomposition. We are seeking a function $L_t(x_{t-1}, x_t)$ quadratic in x_{t-1}. (This function will also depend on x_t, but that shall not concern us at this point.) To determine the coefficients of this quadratic, we calculate the first two derivatives of L_t:

(3.15) $\quad \dfrac{\partial L_t}{\partial x_{t-1}} \;=\; -A_t^T \, R_t^{-1} \, (x_t - A_t \, x_{t-1} - B_t \, u_t) + \Sigma_{t-1}^{-1} \, (x_{t-1} - \mu_{t-1})$

(3.16) $\quad \dfrac{\partial^2 L_t}{\partial x_{t-1}^2} \;=\; A_t^T \, R_t^{-1} \, A_t + \Sigma_{t-1}^{-1} \;=:\; \Psi_t^{-1}$

Ψ_t defines the curvature of $L_t(x_{t-1}, x_t)$. Setting the first derivative of L_t to 0 gives us the mean:

(3.17) $\quad A_t^T \, R_t^{-1} \, (x_t - A_t \, x_{t-1} - B_t \, u_t) \;=\; \Sigma_{t-1}^{-1} \, (x_{t-1} - \mu_{t-1})$

This expression is now solved for x_{t-1}

(3.18)
$\Longleftrightarrow \quad A_t^T \, R_t^{-1} \, (x_t - B_t \, u_t) - A_t^T \, R_t^{-1} \, A_t \, x_{t-1} \;=\; \Sigma_{t-1}^{-1} \, x_{t-1} - \Sigma_{t-1}^{-1} \, \mu_{t-1}$

$\Longleftrightarrow \quad A_t^T \, R_t^{-1} \, A_t \, x_{t-1} + \Sigma_{t-1}^{-1} \, x_{t-1} \;=\; A_t^T \, R_t^{-1} \, (x_t - B_t \, u_t) + \Sigma_{t-1}^{-1} \, \mu_{t-1}$

$\Longleftrightarrow \quad (A_t^T \, R_t^{-1} \, A_t + \Sigma_{t-1}^{-1}) \, x_{t-1} \;=\; A_t^T \, R_t^{-1} \, (x_t - B_t \, u_t) + \Sigma_{t-1}^{-1} \, \mu_{t-1}$

$\Longleftrightarrow \quad \Psi_t^{-1} \, x_{t-1} \;=\; A_t^T \, R_t^{-1} \, (x_t - B_t \, u_t) + \Sigma_{t-1}^{-1} \, \mu_{t-1}$

$\Longleftrightarrow \quad x_{t-1} \;=\; \Psi_t \, [A_t^T \, R_t^{-1} \, (x_t - B_t \, u_t) + \Sigma_{t-1}^{-1} \, \mu_{t-1}]$

Thus, we now have a quadratic function $L_t(x_{t-1}, x_t)$, defined as follows:

$$(3.19) \quad L_t(x_{t-1}, x_t) = \tfrac{1}{2}(x_{t-1} - \Psi_t \, [A_t^T \, R_t^{-1} \, (x_t - B_t \, u_t) + \Sigma_{t-1}^{-1} \, \mu_{t-1}])^T \, \Psi^{-1}$$
$$(x_{t-1} - \Psi_t \, [A_t^T \, R_t^{-1} \, (x_t - B_t \, u_t) + \Sigma_{t-1}^{-1} \, \mu_{t-1}])$$

Clearly, this is not the only quadratic function satisfying our decomposition in (3.12). However, $L_t(x_{t-1}, x_t)$ is of the common quadratic form of the negative exponent of a normal distribution. In fact the function

$$(3.20) \quad \det(2\pi\Psi)^{-\tfrac{1}{2}} \, \exp\{-L_t(x_{t-1}, x_t)\}$$

is a valid probability density function (PDF) for the variable x_{t-1}. As the reader easily verifies, this function is of the form defined in (3.1). We know from (2.5) that PDFs integrate to 1. Thus, we have

$$(3.21) \quad \int \det(2\pi\Psi)^{-\tfrac{1}{2}} \, \exp\{-L_t(x_{t-1}, x_t)\} \, dx_{t-1} = 1$$

From this it follows that

$$(3.22) \quad \int \exp\{-L_t(x_{t-1}, x_t)\} \, dx_{t-1} = \det(2\pi\Psi)^{\tfrac{1}{2}}$$

The important thing to notice is that the value of this integral is *independent* of x_t, our target variable. Thus, for our problem of calculating a distribution over x_t, this integral is constant. Subsuming this constant into the normalizer η, we get the following expression for Equation (3.13):

$$(3.23) \quad \overline{bel}(x_t) = \eta \, \exp\{-L_t(x_t)\} \int \exp\{-L_t(x_{t-1}, x_t)\} \, dx_{t-1}$$
$$= \eta \, \exp\{-L_t(x_t)\}$$

This decomposition establishes the correctness of (3.14). Notice once again that the normalizers η are *not* the same in both lines.

It remains to determine the function $L_t(x_t)$, which is the difference of L_t, defined in (3.11), and $L_t(x_{t-1}, x_t)$, defined in (3.19):

$$(3.24) \quad L_t(x_t) = L_t - L_t(x_{t-1}, x_t)$$
$$= \tfrac{1}{2} (x_t - A_t \, x_{t-1} - B_t \, u_t)^T \, R_t^{-1} \, (x_t - A_t \, x_{t-1} - B_t \, u_t)$$
$$+ \tfrac{1}{2} (x_{t-1} - \mu_{t-1})^T \, \Sigma_{t-1}^{-1} \, (x_{t-1} - \mu_{t-1})$$
$$- \tfrac{1}{2}(x_{t-1} - \Psi_t \, [A_t^T \, R_t^{-1} \, (x_t - B_t \, u_t) + \Sigma_{t-1}^{-1} \, \mu_{t-1}])^T \, \Psi^{-1}$$
$$(x_{t-1} - \Psi_t \, [A_t^T \, R_t^{-1} \, (x_t - B_t \, u_t) + \Sigma_{t-1}^{-1} \, \mu_{t-1}])$$

Let us quickly verify that $L_t(x_t)$ indeed does not depend on x_{t-1}. To do so, we substitute back $\Psi_t = (A_t^T \, R_t^{-1} \, A_t + \Sigma_{t-1}^{-1})^{-1}$, and multiply out the terms above. For the reader's convenience, terms that contain x_{t-1} are underlined (doubly if they are quadratic in x_{t-1}).

(3.25) $\quad L_t(x_t) \;=\; \tfrac{1}{2}\, \underline{\underline{x_{t-1}^T A_t^T R_t^{-1} A_t\, x_{t-1}}} - \underline{x_{t-1}^T A_t^T R_t^{-1} (x_t - B_t\, u_t)}$
$\qquad + \tfrac{1}{2} (x_t - B_t\, u_t)^T R_t^{-1} (x_t - B_t\, u_t)$
$\qquad + \tfrac{1}{2}\, \underline{\underline{x_{t-1}^T \Sigma_{t-1}^{-1} x_{t-1}}} - \underline{x_{t-1}^T \Sigma_{t-1}^{-1} \mu_{t-1}} + \tfrac{1}{2} \mu_{t-1}^T \Sigma_{t-1}^{-1} \mu_{t-1}$
$\qquad - \tfrac{1}{2}\, \underline{\underline{x_{t-1}^T (A_t^T R_t^{-1} A_t + \Sigma_{t-1}^{-1})\, x_{t-1}}}$
$\qquad + \underline{x_{t-1}^T [A_t^T R_t^{-1} (x_t - B_t\, u_t) + \Sigma_{t-1}^{-1} \mu_{t-1}]}$
$\qquad - \tfrac{1}{2} [A_t^T R_t^{-1} (x_t - B_t\, u_t) + \Sigma_{t-1}^{-1} \mu_{t-1}]^T (A_t^T R_t^{-1} A_t + \Sigma_{t-1}^{-1})^{-1}$
$\qquad\qquad [A_t^T R_t^{-1} (x_t - B_t\, u_t) + \Sigma_{t-1}^{-1} \mu_{t-1}]$

It is now easily seen that all terms that contain x_{t-1} cancel out. This should come at no surprise, since it is a consequence of our construction of $L_t(x_{t-1}, x_t)$.

(3.26) $\quad L_t(x_t) \;=\; +\tfrac{1}{2} (x_t - B_t\, u_t)^T R_t^{-1} (x_t - B_t\, u_t) + \tfrac{1}{2} \mu_{t-1}^T \Sigma_{t-1}^{-1} \mu_{t-1}$
$\qquad - \tfrac{1}{2} [A_t^T R_t^{-1} (x_t - B_t\, u_t) + \Sigma_{t-1}^{-1} \mu_{t-1}]^T (A_t^T R_t^{-1} A_t + \Sigma_{t-1}^{-1})^{-1}$
$\qquad\qquad [A_t^T R_t^{-1} (x_t - B_t\, u_t) + \Sigma_{t-1}^{-1} \mu_{t-1}]$

Furthermore, $L_t(x_t)$ is quadratic in x_t. This observation means that $\overline{bel}(x_t)$ is indeed normal distributed. The mean and covariance of this distribution are of course the minimum and curvature of $L_t(x_t)$, which we now easily obtain by computing the first and second derivatives of $L_t(x_t)$ with respect to x_t:

(3.27) $\quad \dfrac{\partial L_t(x_t)}{\partial x_t} \;=\; R_t^{-1} (x_t - B_t\, u_t) - R_t^{-1} A_t (A_t^T R_t^{-1} A_t + \Sigma_{t-1}^{-1})^{-1}$
$\qquad\qquad [A_t^T R_t^{-1} (x_t - B_t\, u_t) + \Sigma_{t-1}^{-1} \mu_{t-1}]$
$\qquad = [R_t^{-1} - R_t^{-1} A_t (A_t^T R_t^{-1} A_t + \Sigma_{t-1}^{-1})^{-1} A_t^T R_t^{-1}] (x_t - B_t\, u_t)$
$\qquad\quad - R_t^{-1} A_t (A_t^T R_t^{-1} A_t + \Sigma_{t-1}^{-1})^{-1} \Sigma_{t-1}^{-1} \mu_{t-1}$

The *inversion lemma* stated (and shown) in Table 3.2 allows us to express the first factor as follows:

(3.28) $\quad R_t^{-1} - R_t^{-1} A_t (A_t^T R_t^{-1} A_t + \Sigma_{t-1}^{-1})^{-1} A_t^T R_t^{-1} \;=\; (R_t + A_t\, \Sigma_{t-1}\, A_t^T)^{-1}$

Hence the desired derivative is given by the following expression:

(3.29) $\quad \dfrac{\partial L_t(x_t)}{\partial x_t} \;=\; (R_t + A_t\, \Sigma_{t-1}\, A_t^T)^{-1} (x_t - B_t\, u_t)$
$\qquad\qquad - R_t^{-1} A_t (A_t^T R_t^{-1} A_t + \Sigma_{t-1}^{-1})^{-1} \Sigma_{t-1}^{-1} \mu_{t-1}$

Inversion Lemma. For any invertible quadratic matrices R and Q and any matrix P with appropriate dimensions, the following holds true

$$(R + P Q P^T)^{-1} = R^{-1} - R^{-1} P (Q^{-1} + P^T R^{-1} P)^{-1} P^T R^{-1}$$

assuming that all above matrices can be inverted as stated.

Proof. Define $\Psi = (Q^{-1} + P^T R^{-1} P)^{-1}$. It suffices to show that

$$(R^{-1} - R^{-1} P \Psi P^T R^{-1})(R + P Q P^T) = I$$

This is shown through a series of transformations:

$$= \underbrace{R^{-1} R}_{=I} + R^{-1} P Q P^T - R^{-1} P \Psi P^T \underbrace{R^{-1} R}_{=I}$$
$$\quad - R^{-1} P \Psi P^T R^{-1} P Q P^T$$
$$= I + R^{-1} P Q P^T - R^{-1} P \Psi P^T - R^{-1} P \Psi P^T R^{-1} P Q P^T$$
$$= I + R^{-1} P [Q P^T - \Psi P^T - \Psi P^T R^{-1} P Q P^T]$$
$$= I + R^{-1} P [Q P^T - \Psi \underbrace{Q^{-1} Q}_{=I} P^T - \Psi P^T R^{-1} P Q P^T]$$
$$= I + R^{-1} P [Q P^T - \underbrace{\Psi \Psi^{-1}}_{=I} Q P^T]$$
$$= I + R^{-1} P \underbrace{[Q P^T - Q P^T]}_{=0} = I$$

Table 3.2 The (specialized) inversion lemma, sometimes called the *Sherman/Morrison formula*.

The minimum of $L_t(x_t)$ is attained when the first derivative is zero.

$$(3.30) \quad (R_t + A_t \Sigma_{t-1} A_t^T)^{-1} (x_t - B_t u_t)$$
$$= R_t^{-1} A_t (A_t^T R_t^{-1} A_t + \Sigma_{t-1}^{-1})^{-1} \Sigma_{t-1}^{-1} \mu_{t-1}$$

Solving this for the target variable x_t gives us the surprisingly compact result

$$(3.31) \quad x_t = B_t u_t + \underbrace{(R_t + A_t \Sigma_{t-1} A_t^T) R_t^{-1} A_t}_{A_t + A_t \Sigma_{t-1} A_t^T R_t^{-1} A_t} \underbrace{(A_t^T R_t^{-1} A_t + \Sigma_{t-1}^{-1})^{-1}}_{(\Sigma_{t-1} A_t^T R_t^{-1} A_t + I)^{-1}} \Sigma_{t-1}^{-1} \mu_{t-1}$$
$$= B_t u_t + A_t \underbrace{(I + \Sigma_{t-1} A_t^T R_t^{-1} A_t)(\Sigma_{t-1} A_t^T R_t^{-1} A_t + I)^{-1}}_{=I} \mu_{t-1}$$

$$= B_t \, u_t + A_t \, \mu_{t-1}$$

Thus, the mean of the belief $\overline{bel}(x_t)$ after incorporating the motion command u_t is $B_t \, u_t + A_t \, \mu_{t-1}$. This proves the correctness of line 2 of the Kalman filter algorithm in Table 3.1.

Line 3 is now obtained by calculating the second derivative of $L_t(x_t)$:

$$(3.32) \quad \frac{\partial^2 L_t(x_t)}{\partial x_t^2} = (A_t \, \Sigma_{t-1} \, A_t^T + R_t)^{-1}$$

This is the curvature of the quadratic function $L_t(x_t)$, whose inverse is the covariance of the belief $\overline{bel}(x_t)$.

To summarize, we showed that the prediction steps in lines 2 and 3 of the Kalman filter algorithm indeed implement the Bayes filter prediction step. To do so, we first decomposed the exponent of the belief $\overline{bel}(x_t)$ into two functions, $L_t(x_{t-1}, x_t)$ and $L_t(x_t)$. Then we showed that $L_t(x_{t-1}, x_t)$ changes the predicted belief $\overline{bel}(x_t)$ only by a constant factor, which can be subsumed into the normalizing constant η. Finally, we determined the function $L_t(x_t)$ and showed that it results in the mean $\bar{\mu}_t$ and covariance $\bar{\Sigma}_t$ of the Kalman filter prediction $\overline{bel}(x_t)$.

Part 2: Measurement Update

We will now derive the measurement update in lines 4, 5, and 6 (Table 3.1) of our Kalman filter algorithm. We begin with the general Bayes filter mechanism for incorporating measurements, stated in Equation (2.38) and restated here in annotated form:

$$(3.33) \quad bel(x_t) = \eta \underbrace{p(z_t \mid x_t)}_{\sim \mathcal{N}(z_t; C_t x_t, Q_t)} \underbrace{\overline{bel}(x_t)}_{\sim \mathcal{N}(x_t; \bar{\mu}_t, \bar{\Sigma}_t)}$$

The mean and covariance of $\overline{bel}(x_t)$ are obviously given by $\bar{\mu}_t$ and $\bar{\Sigma}_t$. The measurement probability $p(z_t \mid x_t)$ was defined in (3.6) to be normal as well, with mean $C_t \, x_t$ and covariance Q_t. Thus, the product is given by an exponential

$$(3.34) \quad bel(x_t) = \eta \, \exp\{-J_t\}$$

with

$$(3.35) \quad J_t = \tfrac{1}{2} (z_t - C_t x_t)^T \, Q_t^{-1} \, (z_t - C_t x_t) + \tfrac{1}{2} (x_t - \bar{\mu}_t)^T \, \bar{\Sigma}_t^{-1} \, (x_t - \bar{\mu}_t)$$

This function is quadratic in x_t, hence $bel(x_t)$ is a Gaussian. To calculate its parameters, we once again calculate the first two derivatives of J_t with respect to x_t:

$$\text{(3.36)} \quad \frac{\partial J}{\partial x_t} = -C_t^T Q_t^{-1} (z_t - C_t x_t) + \bar{\Sigma}_t^{-1} (x_t - \bar{\mu}_t)$$

$$\text{(3.37)} \quad \frac{\partial^2 J}{\partial x_t^2} = C_t^T Q_t^{-1} C_t + \bar{\Sigma}_t^{-1}$$

The second term is the inverse of the covariance of $bel(x_t)$:

$$\text{(3.38)} \quad \Sigma_t = (C_t^T Q_t^{-1} C_t + \bar{\Sigma}_t^{-1})^{-1}$$

The mean of $bel(x_t)$ is the minimum of this quadratic function, which we now calculate by setting the first derivative of J_t to zero (and substituting μ_t for x_t):

$$\text{(3.39)} \quad C_t^T Q_t^{-1} (z_t - C_t \mu_t) = \bar{\Sigma}_t^{-1} (\mu_t - \bar{\mu}_t)$$

The expression on the left of the equal sign can be transformed as follows:

$$\text{(3.40)} \quad \begin{aligned} & C_t^T Q_t^{-1} (z_t - C_t \mu_t) \\ &= C_t^T Q_t^{-1} (z_t - C_t \mu_t + C_t \bar{\mu}_t - C_t \bar{\mu}_t) \\ &= C_t^T Q_t^{-1} (z_t - C_t \bar{\mu}_t) - C_t^T Q_t^{-1} C_t (\mu_t - \bar{\mu}_t) \end{aligned}$$

Substituting this back into (3.39) gives us

$$\text{(3.41)} \quad C_t^T Q_t^{-1} (z_t - C_t \bar{\mu}_t) = \underbrace{(C_t^T Q_t^{-1} C_t + \bar{\Sigma}_t^{-1})}_{= \Sigma_t^{-1}} (\mu_t - \bar{\mu}_t)$$

and hence we have

$$\text{(3.42)} \quad \Sigma_t C_t^T Q_t^{-1} (z_t - C_t \bar{\mu}_t) = \mu_t - \bar{\mu}_t$$

We now define the *Kalman gain* as

$$\text{(3.43)} \quad K_t = \Sigma_t C_t^T Q_t^{-1}$$

and obtain

$$\text{(3.44)} \quad \mu_t = \bar{\mu}_t + K_t (z_t - C_t \bar{\mu}_t)$$

This proves the correctness of line 5 in the Kalman filter algorithm in Table 3.1.

The Kalman gain, as defined in (3.43), is a function of Σ_t. This is at odds with the fact that we utilize K_t to calculate Σ_t in line 6 of the algorithm. The following transformation shows us how to express K_t in terms of covariances other than Σ_t. It begins with the definition of K_t in (3.43):

$$
\begin{aligned}
(3.45) \quad K_t &= \Sigma_t\, C_t^T\, Q_t^{-1} \\
&= \Sigma_t\, C_t^T\, Q_t^{-1}\, \underbrace{(C_t\, \bar{\Sigma}_t\, C_t^T + Q_t)\, (C_t\, \bar{\Sigma}_t\, C_t^T + Q_t)^{-1}}_{=I} \\
&= \Sigma_t\, (C_t^T\, Q_t^{-1}\, C_t\, \bar{\Sigma}_t\, C_t^T + C_t^T\, \underbrace{Q_t^{-1}\, Q_t}_{=I})\, (C_t\, \bar{\Sigma}_t\, C_t^T + Q_t)^{-1} \\
&= \Sigma_t\, (C_t^T\, Q_t^{-1}\, C_t\, \bar{\Sigma}_t\, C_t^T + C_t^T)\, (C_t\, \bar{\Sigma}_t\, C_t^T + Q_t)^{-1} \\
&= \Sigma_t\, (C_t^T\, Q_t^{-1}\, C_t\, \bar{\Sigma}_t\, C_t^T + \underbrace{\bar{\Sigma}_t^{-1}\, \bar{\Sigma}_t}_{=I}\, C_t^T)\, (C_t\, \bar{\Sigma}_t\, C_t^T + Q_t)^{-1} \\
&= \Sigma_t\, \underbrace{(C_t^T\, Q_t^{-1}\, C_t + \bar{\Sigma}_t^{-1})}_{=\Sigma_t^{-1}}\, \bar{\Sigma}_t\, C_t^T\, (C_t\, \bar{\Sigma}_t\, C_t^T + Q_t)^{-1} \\
&= \underbrace{\Sigma_t\, \Sigma_t^{-1}}_{=I}\, \bar{\Sigma}_t\, C_t^T\, (C_t\, \bar{\Sigma}_t\, C_t^T + Q_t)^{-1} \\
&= \bar{\Sigma}_t\, C_t^T\, (C_t\, \bar{\Sigma}_t\, C_t^T + Q_t)^{-1}
\end{aligned}
$$

This expression proves the correctness of line 4 of our Kalman filter algorithm.

Line 6 is obtained by expressing the covariance using the Kalman gain K_t. The advantage of the calculation in Table 3.1 over the definition in Equation (3.38) lies in the fact that we can avoid inverting the state covariance matrix. This is essential for applications of Kalman filters to high-dimensional state spaces.

Our transformation is once again carried out using the *inversion lemma*, which was already stated in Table 3.2. Here we restate it using the notation of Equation (3.38):

$$(3.46) \quad (\bar{\Sigma}_t^{-1} + C_t^T\, Q_t^{-1}\, C_t)^{-1} = \bar{\Sigma}_t - \bar{\Sigma}_t\, C_t^T\, (Q_t + C_t\, \bar{\Sigma}_t\, C_t^T)^{-1}\, C_t\, \bar{\Sigma}_t$$

This lets us arrive at the following expression for the covariance:

$$
\begin{aligned}
(3.47) \quad \Sigma_t &= (C_t^T\, Q_t^{-1}\, C_t + \bar{\Sigma}_t^{-1})^{-1} \\
&= \bar{\Sigma}_t - \bar{\Sigma}_t\, C_t^T\, (Q_t + C_t\, \bar{\Sigma}_t\, C_t^T)^{-1}\, C_t\, \bar{\Sigma}_t \\
&= [I - \underbrace{\bar{\Sigma}_t\, C_t^T\, (Q_t + C_t\, \bar{\Sigma}_t\, C_t^T)^{-1}\, C_t}_{=K_t,\ \text{see Eq. (3.45)}}]\, \bar{\Sigma}_t \\
&= (I - K_t\, C_t)\, \bar{\Sigma}_t
\end{aligned}
$$

This completes our correctness proof, in that it shows the correctness of line 6 of our Kalman filter algorithm.

3.3 The Extended Kalman Filter

3.3.1 Why Linearize?

The assumptions that observations are linear functions of the state and that the next state is a linear function of the previous state are crucial for the correctness of the Kalman filter. The observation that any linear transformation of a Gaussian random variable results in another Gaussian random variable played an important role in the derivation of the Kalman filter algorithm. The efficiency of the Kalman filter is then due to the fact that the parameters of the resulting Gaussian can be computed in closed form.

Throughout this and the following chapters, we will illustrate properties of different density representations using the transformation of a one-dimensional Gaussian random variable. Figure 3.3a illustrates the *linear* transformation of such a random variable. The graph on the lower right shows the density of the random variable $X \sim \mathcal{N}(x; \mu, \sigma^2)$. Let us assume that X is passed through the linear function $y = ax + b$, shown in the upper right graph. The resulting random variable, Y, is distributed according to a Gaussian with mean $a\mu + b$ and variance $a^2\sigma^2$. This Gaussian is illustrated by the gray area in the upper left graph of Figure 3.3a. The reader may notice that this example is closely related to the next state update of the Kalman filter, with $X = x_{t-1}$ and $Y = x_t$ but without an additive noise variable; see also Equation (3.2).

Unfortunately, state transitions and measurements are rarely linear in practice. For example, a robot that moves with constant translational and rotational velocity typically moves on a circular trajectory, which cannot be described by linear state transitions. This observation, along with the assumption of unimodal beliefs, renders plain Kalman filters, as discussed so far, inapplicable to all but the most trivial robotics problems.

EXTENDED KALMAN FILTER

The *extended Kalman filter*, or *EKF*, relaxes one of these assumptions: the linearity assumption. Here the assumption is that the state transition probability and the measurement probabilities are governed by *nonlinear* functions g and h, respectively:

$$x_t = g(u_t, x_{t-1}) + \varepsilon_t \tag{3.48}$$
$$z_t = h(x_t) + \delta_t \tag{3.49}$$

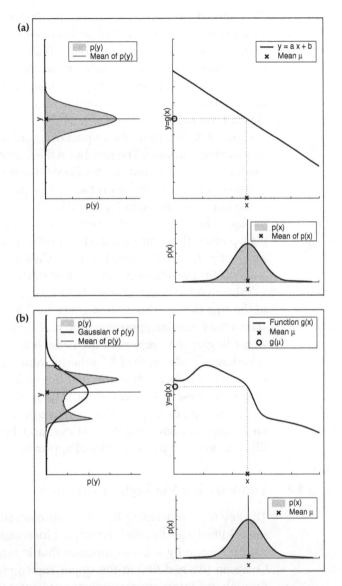

Figure 3.3 (a) Linear and (b) nonlinear transformation of a Gaussian random variable. The lower right plots show the density of the original random variable, X. This random variable is passed through the function displayed in the upper right graphs (the transformation of the mean is indicated by the dotted line). The density of the resulting random variable Y is plotted in the upper left graphs.

This model strictly generalizes the linear Gaussian model underlying Kalman filters, as postulated in Equations (3.2) and (3.5). The function g replaces the matrices A_t and B_t in (3.2), and h replaces the matrix C_t in (3.5). Unfortunately, with arbitrary functions g and h, the belief is no longer a Gaussian. In fact, performing the belief update exactly is usually impossible for nonlinear functions g and h, and the Bayes filter does not possess a closed-form solution.

Figure 3.3b illustrates the impact of a nonlinear transformation on a Gaussian random variable. The graphs on the lower right and upper right plot the random variable X and the nonlinear function g, respectively. The density of the transformed random variable, $Y = g(X)$, is indicated by the gray area in the upper left graph of Figure 3.3b. Since this density cannot be computed in closed form, it was estimated by drawing 500,000 samples according to $p(x)$, passing them through the function g, and then histogramming over the range of g. As can be seen, Y is not a Gaussian because the nonlinearities in g distort the density of X in ways that destroy its Gaussian shape.

The extended Kalman filter (EKF) calculates a Gaussian approximation to the true belief. The dashed curve in the upper left graph of Figure 3.3b shows the Gaussian approximation to the density of the random variable Y. Accordingly, EKFs represent the belief $bel(x_t)$ at time t by a mean μ_t and a covariance Σ_t. Thus, the EKF inherits from the Kalman filter the basic belief representation, but it differs in that this belief is only approximate, not exact as was the case in Kalman filters. The goal of the EKF is thus shifted from computing the exact posterior to efficiently estimating its mean and covariance. However, since these statistics cannot be computed in closed form, the EKF has to resort to an additional approximation.

3.3.2 Linearization Via Taylor Expansion

The key idea underlying the EKF approximation is called *linearization*. Figure 3.4 illustrates the basic concept. Linearization approximates the nonlinear function g by a linear function that is tangent to g at the mean of the Gaussian (dashed line in the upper right graph). Projecting the Gaussian through this linear approximation results in a Gaussian density, as indicated by the dashed line in the upper left graph. The solid line in the upper left graph represents the mean and covariance of the Monte-Carlo approximation. The mismatch between these two Gaussians indicates the error caused by the linear approximation of g.

The key advantage of the linearization, however, lies in its efficiency.

3.3 The Extended Kalman Filter

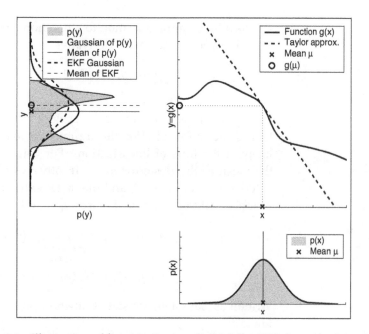

Figure 3.4 Illustration of linearization applied by the EKF. Instead of passing the Gaussian through the nonlinear function g, it is passed through a linear approximation of g. The linear function is tangent to g at the mean of the original Gaussian. The resulting Gaussian is shown as the dashed line in the upper left graph. The linearization incurs an approximation error, as indicated by the mismatch between the linearized Gaussian (dashed) and the Gaussian computed from the highly accurate Monte-Carlo estimate (solid).

The Monte-Carlo estimate of the Gaussian was achieved by passing 500,000 points through g followed by the computation of their mean and covariance. The linearization applied by the EKF, on the other hand, only requires determination of the linear approximation followed by the closed-form computation of the resulting Gaussian. In fact, once g is linearized, the mechanics of the EKF's belief propagation are equivalent to those of the Kalman filter.

This technique also is applied to the multiplication of Gaussians when a measurement function h is involved. Again, the EKF approximates h by a linear function tangent to h, thereby retaining the Gaussian nature of the posterior belief.

There exist many techniques for linearizing nonlinear functions. EKFs utilize a method called (first order) *Taylor expansion*. Taylor expansion con-

TAYLOR EXPANSION

structs a linear approximation to a function g from g's value and slope. The slope is given by the partial derivative

$$(3.50) \quad g'(u_t, x_{t-1}) := \frac{\partial g(u_t, x_{t-1})}{\partial x_{t-1}}$$

Clearly, both the value of g and its slope depend on the argument of g. A logical choice for selecting the argument is to choose the state deemed most likely at the time of linearization. For Gaussians, the most likely state is the mean of the posterior μ_{t-1}. In other words, g is approximated by its value at μ_{t-1} (and at u_t), and the linear extrapolation is achieved by a term proportional to the gradient of g at μ_{t-1} and u_t:

$$(3.51) \quad g(u_t, x_{t-1}) \approx g(u_t, \mu_{t-1}) + \underbrace{g'(u_t, \mu_{t-1})}_{=:\, G_t} (x_{t-1} - \mu_{t-1})$$

$$= g(u_t, \mu_{t-1}) + G_t (x_{t-1} - \mu_{t-1})$$

Written as Gaussian, the state transition probability is approximated as follows:

$$(3.52) \quad p(x_t \mid u_t, x_{t-1})$$
$$\approx \det(2\pi R_t)^{-\frac{1}{2}} \exp\left\{-\tfrac{1}{2} [x_t - g(u_t, \mu_{t-1}) - G_t(x_{t-1} - \mu_{t-1})]^T \right.$$
$$\left. R_t^{-1} [x_t - g(u_t, \mu_{t-1}) - G_t(x_{t-1} - \mu_{t-1})]\right\}$$

JACOBIAN

Notice that G_t is a matrix of size $n \times n$, with n denoting the dimension of the state. This matrix is often called the *Jacobian*. The value of the Jacobian depends on u_t and μ_{t-1}, hence it differs for different points in time.

EKFs implement the exact same linearization for the measurement function h. Here the Taylor expansion is developed around $\bar{\mu}_t$, the state deemed most likely by the robot at the time it linearizes h:

$$(3.53) \quad h(x_t) \approx h(\bar{\mu}_t) + \underbrace{h'(\bar{\mu}_t)}_{=:\, H_t} (x_t - \bar{\mu}_t)$$

$$= h(\bar{\mu}_t) + H_t (x_t - \bar{\mu}_t)$$

with $h'(x_t) = \frac{\partial h(x_t)}{\partial x_t}$. Written as a Gaussian, we have

$$(3.54) \quad p(z_t \mid x_t) = \det(2\pi Q_t)^{-\frac{1}{2}} \exp\left\{-\tfrac{1}{2} [z_t - h(\bar{\mu}_t) - H_t(x_t - \bar{\mu}_t)]^T \right.$$
$$\left. Q_t^{-1} [z_t - h(\bar{\mu}_t) - H_t(x_t - \bar{\mu}_t)]\right\}$$

```
1:      Algorithm Extended_Kalman_filter($\mu_{t-1}, \Sigma_{t-1}, u_t, z_t$):
2:          $\bar{\mu}_t = g(u_t, \mu_{t-1})$
3:          $\bar{\Sigma}_t = G_t \, \Sigma_{t-1} \, G_t^T + R_t$
4:          $K_t = \bar{\Sigma}_t \, H_t^T (H_t \, \bar{\Sigma}_t \, H_t^T + Q_t)^{-1}$
5:          $\mu_t = \bar{\mu}_t + K_t(z_t - h(\bar{\mu}_t))$
6:          $\Sigma_t = (I - K_t \, H_t) \, \bar{\Sigma}_t$
7:          return $\mu_t, \Sigma_t$
```

Table 3.3 The extended Kalman filter algorithm.

3.3.3 The EKF Algorithm

Table 3.3 states the *EKF algorithm*. In many ways, this algorithm is similar to the Kalman filter algorithm stated in Table 3.1. The most important differences are summarized by the following table:

	Kalman filter	EKF
state prediction (line 2)	$A_t \, \mu_{t-1} + B_t \, u_t$	$g(u_t, \mu_{t-1})$
measurement prediction (line 5)	$C_t \, \bar{\mu}_t$	$h(\bar{\mu}_t)$

That is, the linear predictions in Kalman filters are replaced by their nonlinear generalizations in EKFs. Moreover, EKFs use Jacobians G_t and H_t instead of the corresponding linear system matrices A_t, B_t, and C_t in Kalman filters. The Jacobian G_t corresponds to the matrices A_t and B_t, and the Jacobian H_t corresponds to C_t. A detailed example for extended Kalman filters will be given in Chapter 7.

3.3.4 Mathematical Derivation of the EKF

The mathematical derivation of the EKF parallels that of the Kalman filter in Chapter 3.2.4, and hence shall only be sketched here. The prediction is calculated as follows (c.f. (3.8)):

$$(3.55) \quad \overline{bel}(x_t) = \int \underbrace{p(x_t \mid x_{t-1}, u_t)}_{\sim \mathcal{N}(x_t; g(u_t, \mu_{t-1}) + G_t(x_{t-1} - \mu_{t-1}), R_t)} \underbrace{bel(x_{t-1})}_{\sim \mathcal{N}(x_{t-1}; \mu_{t-1}, \Sigma_{t-1})} dx_{t-1}$$

This distribution is the EKF analog of the prediction distribution in the Kalman filter, stated in (3.8). The Gaussian $p(x_t \mid x_{t-1}, u_t)$ can be found in Equation (3.52). The function L_t is given by (c.f. (3.11))

(3.56) $$\begin{aligned}L_t &= \tfrac{1}{2}(x_t - g(u_t, \mu_{t-1}) - G_t(x_{t-1} - \mu_{t-1}))^T \\ &\quad R_t^{-1}(x_t - g(u_t, \mu_{t-1}) - G_t(x_{t-1} - \mu_{t-1})) \\ &\quad + \tfrac{1}{2}(x_{t-1} - \mu_{t-1})^T \Sigma_{t-1}^{-1}(x_{t-1} - \mu_{t-1})\end{aligned}$$

which is quadratic in both x_{t-1} and x_t, as above. As in (3.12), we decompose L_t into $L_t(x_{t-1}, x_t)$ and $L_t(x_t)$:

(3.57) $$\begin{aligned}&L_t(x_{t-1}, x_t) \\ &= \tfrac{1}{2}(x_{t-1} - \Phi_t[G_t^T R_t^{-1}(x_t - g(u_t, \mu_{t-1})+G_t\mu_{t-1}) + \Sigma_{t-1}^{-1}\mu_{t-1}])^T \Phi^{-1} \\ &\quad (x_{t-1} - \Phi_t[G_t^T R_t^{-1}(x_t - g(u_t, \mu_{t-1})+G_t\mu_{t-1}) + \Sigma_{t-1}^{-1}\mu_{t-1}])\end{aligned}$$

with

(3.58) $$\Phi_t = (G_t^T R_t^{-1} G_t + \Sigma_{t-1}^{-1})^{-1}$$

and hence

(3.59) $$\begin{aligned}L_t(x_t) &= \tfrac{1}{2}(x_t - g(u_t, \mu_{t-1}) + G_t\mu_{t-1})^T R_t^{-1}(x_t - g(u_t, \mu_{t-1}) + G_t\mu_{t-1}) \\ &\quad + \tfrac{1}{2}(x_{t-1} - \mu_{t-1})^T \Sigma_{t-1}^{-1}(x_{t-1} - \mu_{t-1}) \\ &\quad - \tfrac{1}{2}[G_t^T R_t^{-1}(x_t - g(u_t, \mu_{t-1}) + G_t\mu_{t-1}) + \Sigma_{t-1}^{-1}\mu_{t-1}]^T \\ &\quad \Phi_t[G_t^T R_t^{-1}(x_t - g(u_t, \mu_{t-1}) + G_t\mu_{t-1}) + \Sigma_{t-1}^{-1}\mu_{t-1}]\end{aligned}$$

As the reader can easily verify, setting the first derivative of $L_t(x_t)$ to zero gives us the update $\mu_t = g(u_t, \mu_{t-1})$, in analogy to the derivation in Equations (3.27) through (3.31). The second derivative is given by $(R_t + G_t \Sigma_{t-1} G_t^T)^{-1}$ (see (3.32)).

The measurement update is also derived analogously to the Kalman filter in Chapter 3.2.4. In analogy to (3.33), we have for the EKF

(3.60) $$bel(x_t) = \eta \underbrace{p(z_t \mid x_t)}_{\sim \mathcal{N}(z_t; h(\bar{\mu}_t) + H_t(x_t - \bar{\mu}_t), Q_t)} \underbrace{\overline{bel}(x_t)}_{\sim \mathcal{N}(x_t; \bar{\mu}_t, \bar{\Sigma}_t)}$$

using the linearized state transition function from (3.53). This leads to the exponent (see (3.35)):

(3.61) $$\begin{aligned}J_t &= \tfrac{1}{2}(z_t - h(\bar{\mu}_t) - H_t(x_t - \bar{\mu}_t))^T Q_t^{-1}(z_t - h(\bar{\mu}_t) - H_t(x_t - \bar{\mu}_t)) \\ &\quad + \tfrac{1}{2}(x_t - \bar{\mu}_t)^T \bar{\Sigma}_t^{-1}(x_t - \bar{\mu}_t)\end{aligned}$$

The resulting mean and covariance is given by

(3.62) $$\mu_t = \bar{\mu}_t + K_t(z_t - h(\bar{\mu}_t))$$
(3.63) $$\Sigma_t = (I - K_t H_t)\bar{\Sigma}_t$$

with the Kalman gain

$$(3.64) \quad K_t = \bar{\Sigma}_t H_t^T (H_t \bar{\Sigma}_{t-1} H_t^T + Q_t)^{-1}$$

The derivation of these equations is analogous to Equations (3.36) through (3.47).

3.3.5 Practical Considerations

The EKF has become just about the most popular tool for state estimation in robotics. Its strength lies in its simplicity and in its computational efficiency. As was the case for the Kalman filter, each update requires time $O(k^{2.4} + n^2)$, where k is the dimension of the measurement vector z_t, and n is the dimension of the state vector x_t. Other algorithms, such as the particle filter discussed further below, may require time exponential in n.

The EKF owes its computational efficiency to the fact that it represents the belief by a multivariate Gaussian distribution. A Gaussian is a unimodal distribution, which can be thought of as a single guess annotated with an uncertainty ellipse. In many practical problems, Gaussians are robust estimators. Applications of the Kalman filter to state spaces with 1,000 dimensions or more will be discussed in later chapters of this book. EKFs have been applied with great success to a number of state estimation problems that violate the underlying assumptions.

An important limitation of the EKF arises from the fact that it approximates state transitions and measurements using linear Taylor expansions. In most robotics problems, state transitions and measurements are nonlinear. The goodness of the linear approximation applied by the EKF depends on two main factors: The degree of uncertainty and the degree of local nonlinearity of the functions that are being approximated. The two graphs in Figure 3.5 illustrate the dependency on the uncertainty. Here, two Gaussian random variables are passed through the same nonlinear function (c.f. also Figure 3.4). While both Gaussians have the same mean, the variable shown in (a) has a higher uncertainty than the one in (b). Since the Taylor expansion only depends on the mean, both Gaussians are passed through the same linear approximation. The gray areas in the upper left plots of the two figures show the densities of the resulting random variable, computed by Monte-Carlo estimation. The density resulting from the wider Gaussian is far more distorted than the density resulting from the narrow, less uncertain Gaussian. The Gaussian approximations of these densities are given by the solid lines in the figures. The dashed graphs show the Gaussians estimated by the

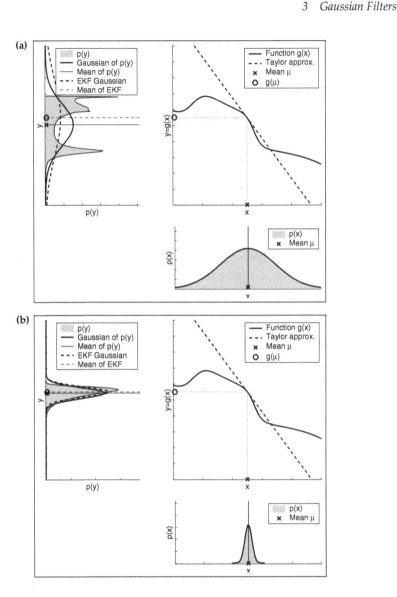

Figure 3.5 Dependency of approximation quality on uncertainty. Both Gaussians (lower right) have the same mean and are passed through the same nonlinear function (upper right). The higher uncertainty of the left Gaussian produces a more distorted density of the resulting random variable (gray area in upper left graph). The solid lines in the upper left graphs show the Gaussians extracted from these densities. The dashed lines represent the Gaussians generated by the linearization applied by the EKF.

linearization. A comparison to the Gaussians resulting from the Monte-Carlo approximations illustrates the fact that higher uncertainty typically results in less accurate estimates of the mean and covariance of the resulting random variable.

The second factor for the quality of the linear Gaussian approximation is the local nonlinearity of the function g, as illustrated in Figure 3.6. Shown there are two Gaussians with the same variance passed through the same nonlinear function. In Panel (a), the mean of the Gaussian falls into a more nonlinear region of the function g than in Panel (b). The mismatch between the accurate Monte-Carlo estimate of the Gaussian (solid line, upper left) and the Gaussian resulting from linear approximation (dashed line) shows that higher nonlinearities result in larger approximation errors. The EKF Gaussian clearly underestimates the spread of the resulting density.

Sometimes, one might want to pursue multiple distinct hypotheses. For example, a robot might have two distinct hypotheses as to where it is, but the arithmetic mean of these hypotheses is not a likely contender. Such situations require multi-modal representations for the posterior belief. EKFs, in the form described here, are incapable of representing such multimodal beliefs. A common extension of EKFs is to represent posteriors using mixtures, or sums, of Gaussians. A mixture of Gaussians may be of the form

MIXTURE OF GAUSSIANS

$$(3.65) \quad bel(x_t) = \frac{1}{\sum_l \psi_{t,l}} \sum_l \psi_{t,l} \det(2\pi \Sigma_{t,l})^{-\frac{1}{2}} \exp\left\{-\frac{1}{2}(x_t - \mu_{t,l})^T \Sigma_{t,l}^{-1} (x_t - \mu_{t,l})\right\}$$

Here $\psi_{t,l}$ are mixture parameters with $\psi_{t,l} \geq 0$. These parameters serve as weights of the mixture components. They are estimated from the likelihoods of the observations conditioned on the corresponding Gaussians. EKFs that utilize such mixture representations are called *multi-hypothesis (extended) Kalman filters*, or *MHEKF*.

MULTI-HYPOTHESIS EKF

To summarize, if the nonlinear functions are approximately linear at the mean of the estimate, then the EKF approximation may generally be a good one, and EKFs may approximate the posterior belief with sufficient accuracy. Furthermore, the less certain the robot, the wider its Gaussian belief, and the more it is affected by nonlinearities in the state transition and measurement functions. In practice, when applying EKFs it is therefore important to keep the uncertainty of the state estimate small.

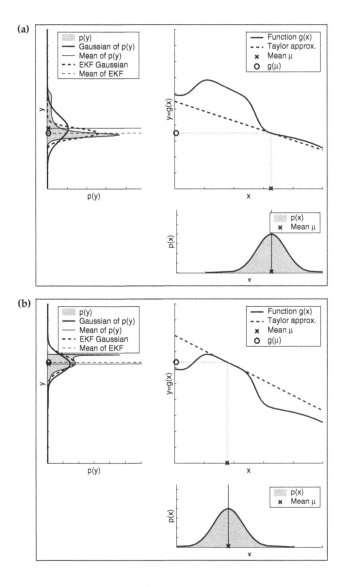

Figure 3.6 Dependence of the approximation quality on local nonlinearity of the function g. Both Gaussians (lower right in each of the two panels) have the same covariance and are passed through the same function (upper right). The linear approximation applied by the EKF is shown as the dashed lines in the upper right graphs. The solid lines in the upper left graphs show the Gaussians extracted from the highly accurate Monte-Carlo estimates. The dashed lines represent the Gaussians generated by the EKF linearization.

3.4 The Unscented Kalman Filter

The Taylor series expansion applied by the EKF is only one way to linearize the transformation of a Gaussian. Two other approaches have often been found to yield superior results. One is known as *moments matching* (and the resulting filter is known as *assumed density filter*, or *ADF*), in which the linearization is calculated in a way that preserves the true mean and the true covariance of the posterior distribution (which is not the case for EKFs). Another linearization method is applied by the *unscented Kalman filter*, or *UKF*, which performs a stochastic linearization through the use of a weighted statistical linear regression process. We now discuss the UKF algorithm without mathematical derivation. The reader is encouraged to read more details in the literature referenced in the bibliographical remarks.

UNSCENTED KALMAN FILTER

3.4.1 Linearization Via the Unscented Transform

Figure 3.7 illustrates the linearization applied by the UKF, called the *unscented transform*. Instead of approximating the function g by a Taylor series expansion, the UKF deterministically extracts so-called *sigma points* from the Gaussian and passes these through g. In the general case, these sigma points are located at the mean and symmetrically along the main axes of the covariance (two per dimension). For an n-dimensional Gaussian with mean μ and covariance Σ, the resulting $2n+1$ sigma points $\mathcal{X}^{[i]}$ are chosen according to the following rule:

SIGMA POINT

$$
\begin{aligned}
\mathcal{X}^{[0]} &= \mu \\
\mathcal{X}^{[i]} &= \mu + \left(\sqrt{(n+\lambda)\Sigma}\right)_i && \text{for } i = 1, \ldots, n \\
\mathcal{X}^{[i]} &= \mu - \left(\sqrt{(n+\lambda)\Sigma}\right)_{i-n} && \text{for } i = n+1, \ldots, 2n
\end{aligned}
\tag{3.66}
$$

Here $\lambda = \alpha^2(n+\kappa) - n$, with α and κ being scaling parameters that determine how far the sigma points are spread from the mean. Each sigma point $\mathcal{X}^{[i]}$ has two weights associated with it. One weight, $w_m^{[i]}$, is used when computing the mean, the other weight, $w_c^{[i]}$, is used when recovering the covariance of the Gaussian.

$$
\begin{aligned}
w_m^{[0]} &= \frac{\lambda}{n+\lambda} \\
w_c^{[0]} &= \frac{\lambda}{n+\lambda} + (1 - \alpha^2 + \beta)
\end{aligned}
\tag{3.67}
$$

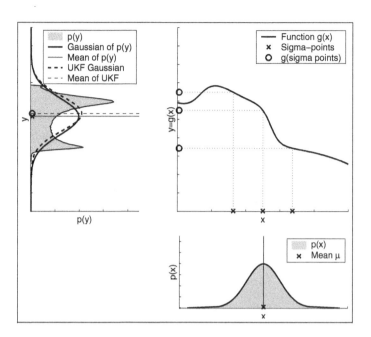

Figure 3.7 Illustration of linearization applied by the UKF. The filter first extracts $2n + 1$ weighted sigma points from the n-dimensional Gaussian ($n = 1$ in this example). These sigma points are passed through the nonlinear function g. The linearized Gaussian is then extracted from the mapped sigma points (small circles in the upper right plot). As for the EKF, the linearization incurs an approximation error, indicated by the mismatch between the linearized Gaussian (dashed) and the Gaussian computed from the highly accurate Monte-Carlo estimate (solid).

$$w_m^{[i]} = w_c^{[i]} = \frac{1}{2(n + \lambda)} \quad \text{for } i = 1, \ldots, 2n.$$

The parameter β can be chosen to encode additional (higher order) knowledge about the distribution underlying the Gaussian representation. If the distribution is an exact Gaussian, then $\beta = 2$ is the optimal choice.

The sigma points are then passed through the function g, thereby probing how g changes the shape of the Gaussian.

$$\mathcal{Y}^{[i]} = g(\mathcal{X}^{[i]}) \tag{3.68}$$

The parameters $(\mu' \ \Sigma')$ of the resulting Gaussian are extracted from the mapped sigma points $\mathcal{Y}^{[i]}$ according to

$$\mu' = \sum_{i=0}^{2n} w_m^{[i]} \, \mathcal{Y}^{[i]} \qquad (3.69)$$

$$\Sigma' = \sum_{i=0}^{2n} w_c^{[i]} \, (\mathcal{Y}^{[i]} - \mu')(\mathcal{Y}^{[i]} - \mu')^T.$$

Figure 3.8 illustrates the dependency of the unscented transform on the uncertainty of the original Gaussian. For comparison, the results using the EKF Taylor series expansion are plotted alongside the UKF results.

Figure 3.9 shows an additional comparison between UKF and EKF approximation, here in dependency of the local nonlinearity of the function g. As can be seen, the unscented transform is more accurate than the first order Taylor series expansion applied by the EKF. In fact, it can be shown that the unscented transform is accurate in the first two terms of the Taylor expansion, while the EKF captures only the first order term. (It should be noted, however, that both the EKF and the UKF can be modified to capture higher order terms.)

3.4.2 The UKF Algorithm

The UKF algorithm utilizing the unscented transform is presented in Table 3.4. The input and output are identical to the EKF algorithm. Line 2 determines the sigma points of the previous belief using Equation (3.66), with γ short for $\sqrt{n + \lambda}$. These points are propagated through the noise-free state prediction in line 3. The predicted mean and variance are then computed from the resulting sigma points (lines 4 and 5). R_t in line 5 is added to the sigma point covariance in order to model the additional prediction noise uncertainty (compare line 3 of the EKF algorithm in Table 3.3). The prediction noise R_t is assumed to be additive. Later, in Chapter 7, we present a version of the UKF algorithm that performs more accurate estimation of the prediction and measurement noise terms.

A new set of sigma points is extracted from the predicted Gaussian in line 6. This sigma point set $\bar{\mathcal{X}}_t$ now captures the overall uncertainty after the prediction step. In line 7, a predicted observation is computed for each sigma point. The resulting observation sigma points $\bar{\mathcal{Z}}_t$ are used to compute the predicted observation $\hat{z}_t$ and its uncertainty, S_t. The matrix Q_t is the covariance matrix of the additive measurement noise. Note that S_t represents the same uncertainty as $H_t \, \bar{\Sigma}_t \, H_t^T + Q_t$ in line 4 of the EKF algorithm in

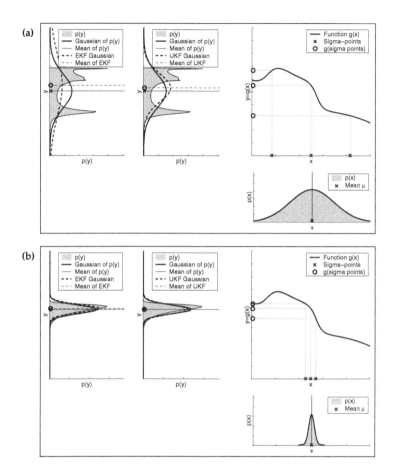

Figure 3.8 Linearization results for the UKF depending on the uncertainty of the original Gaussian. The results of the EKF linearization are also shown for comparison (c.f. Figure3.5). The unscented transform incurs smaller approximation errors, as can be seen by the stronger similarity between the dashed and the solid Gaussians.

Table 3.3. Line 10 determines the cross-covariance between state and observation, which is then used in line 11 to compute the Kalman gain K_t. The cross-covariance $\bar{\Sigma}_t^{x,z}$ corresponds to the term $\bar{\Sigma}_t\, H_t^T$ in line 4 of the EKF algorithm. With this in mind it is straightforward to show that the estimation update performed in lines 12 and 13 is of equivalent form to the update performed by the EKF algorithm.

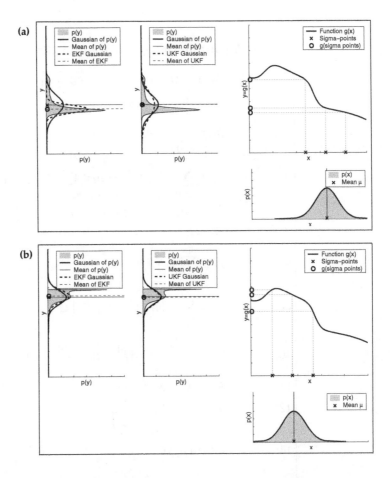

Figure 3.9 Linearization results for the UKF depending on the mean of the original Gaussian. The results of the EKF linearization are also shown for comparison (c.f. Figure3.6). The sigma point linearization incurs smaller approximation errors, as can be seen by the stronger similarity between the dashed and the solid Gaussians.

The asymptotic complexity of the UKF algorithm is the same as for the EKF. In practice, the EKF is often slightly faster than the UKF. The UKF is still highly efficient, even with this slowdown by a constant factor. Furthermore, the UKF inherits the benefits of the unscented transform for linearization. For purely linear systems, it can be shown that the estimates generated by the UKF are identical to those generated by the Kalman filter. For nonlin-

1:      **Algorithm Unscented_Kalman_filter($\mu_{t-1}, \Sigma_{t-1}, u_t, z_t$):**

2:      $\mathcal{X}_{t-1} = (\mu_{t-1} \quad \mu_{t-1} + \gamma\sqrt{\Sigma_{t-1}} \quad \mu_{t-1} - \gamma\sqrt{\Sigma_{t-1}})$

3:      $\bar{\mathcal{X}}_t^* = g(u_t, \mathcal{X}_{t-1})$

4:      $\bar{\mu}_t = \sum_{i=0}^{2n} w_m^{[i]} \bar{\mathcal{X}}_t^{*[i]}$

5:      $\bar{\Sigma}_t = \sum_{i=0}^{2n} w_c^{[i]} (\bar{\mathcal{X}}_t^{*[i]} - \bar{\mu}_t)(\bar{\mathcal{X}}_t^{*[i]} - \bar{\mu}_t)^T + R_t$

6:      $\bar{\mathcal{X}}_t = (\bar{\mu}_t \quad \bar{\mu}_t + \gamma\sqrt{\bar{\Sigma}_t} \quad \bar{\mu}_t - \gamma\sqrt{\bar{\Sigma}_t})$

7:      $\bar{\mathcal{Z}}_t = h(\bar{\mathcal{X}}_t)$

8:      $\hat{z}_t = \sum_{i=0}^{2n} w_m^{[i]} \bar{\mathcal{Z}}_t^{[i]}$

9:      $S_t = \sum_{i=0}^{2n} w_c^{[i]} (\bar{\mathcal{Z}}_t^{[i]} - \hat{z}_t)(\bar{\mathcal{Z}}_t^{[i]} - \hat{z}_t)^T + Q_t$

10:    $\bar{\Sigma}_t^{x,z} = \sum_{i=0}^{2n} w_c^{[i]} (\bar{\mathcal{X}}_t^{[i]} - \bar{\mu}_t)(\bar{\mathcal{Z}}_t^{[i]} - \hat{z}_t)^T$

11:    $K_t = \bar{\Sigma}_t^{x,z} S_t^{-1}$

12:    $\mu_t = \bar{\mu}_t + K_t(z_t - \hat{z}_t)$

13:    $\Sigma_t = \bar{\Sigma}_t - K_t S_t K_t^T$

14:    return μ_t, Σ_t

Table 3.4 The unscented Kalman filter algorithm. The variable n denotes the dimensionality of the state vector.

ear systems the UKF produces equal or better results than the EKF, where the improvement over the EKF depends on the nonlinearities and spread of the prior state uncertainty. In many practical applications, the difference between EKF and UKF is negligible.

Another advantage of the UKF is the fact that it does not require the computation of Jacobians, which are difficult to determine in some domains. The UKF is thus often referred to as a *derivative-free filter*.

DERIVATIVE-FREE FILTER

Finally, the unscented transform has some resemblance to the sample-based representation used by particle filters, which will be discussed in the next chapter. A key difference however, is that the sigma points of the unscented transform are determined deterministically, while particle filters draw samples randomly. This has important implications. If the underlying distribution is approximately Gaussian, then the UKF representation is far more efficient than the particle filter representation. If, on the other hand, the belief is highly non-Gaussian, then the UKF representation is too restrictive and the filter can perform arbitrarily poorly.

3.5 The Information Filter

The dual of the Kalman filter is the *information filter*, or *IF*. Just like the KF and its nonlinear versions, the EKF and the UKF, the information filter represents the belief by a Gaussian. Thus, the standard information filter is subject to the same assumptions underlying the Kalman filter. The key difference between the KF and the IF arises from the way the Gaussian belief is represented. Whereas in the Kalman filter family of algorithms, Gaussians are represented by their moments (mean, covariance), information filters represent Gaussians in their canonical parameterization, which is comprised of an information matrix and an information vector. The difference in parameterization leads to different update equations. In particular, what is computationally complex in one parameterization happens to be simple in the other (and vice versa). The canonical and the moments parameterizations are often considered *dual* to each other, and thus are the IF and the KF.

3.5.1 Canonical Parameterization

CANONICAL PARAMETERIZATION

The *canonical parameterization* of a multivariate Gaussian is given by a matrix Ω and a vector ξ. The matrix Ω is the inverse of the covariance matrix:

$$(3.70) \quad \Omega = \Sigma^{-1}$$

INFORMATION MATRIX

Ω is called the *information matrix*, or sometimes the *precision matrix*. The vector ξ is called the *information vector*. It is defined as

$$(3.71) \quad \xi = \Sigma^{-1} \mu$$

It is easy to see that Ω and ξ are a complete parameterization of a Gaussian. In particular, the mean and covariance of the Gaussian can easily be obtained from the canonical parameterization by the inverse of (3.70) and (3.71):

$$\Sigma = \Omega^{-1} \tag{3.72}$$

$$\mu = \Omega^{-1} \xi \tag{3.73}$$

The canonical parameterization is often derived by multiplying out the exponent of a Gaussian. In (3.1), we defined the multivariate normal distribution as follows:

$$p(x) = \det(2\pi\Sigma)^{-\frac{1}{2}} \exp\left\{-\tfrac{1}{2}(x-\mu)^T \Sigma^{-1}(x-\mu)\right\} \tag{3.74}$$

A straightforward sequence of transformations leads to the following parameterization:

$$\begin{aligned} p(x) &= \det(2\pi\Sigma)^{-\frac{1}{2}} \exp\left\{-\tfrac{1}{2}x^T \Sigma^{-1} x + x^T \Sigma^{-1}\mu - \tfrac{1}{2}\mu^T \Sigma^{-1}\mu\right\} \\ &= \underbrace{\det(2\pi\Sigma)^{-\frac{1}{2}} \exp\left\{-\tfrac{1}{2}\mu^T \Sigma^{-1}\mu\right\}}_{\text{const.}} \exp\left\{-\tfrac{1}{2}x^T \Sigma^{-1} x + x^T \Sigma^{-1}\mu\right\} \end{aligned} \tag{3.75}$$

The term labeled "const." does not depend on the target variable x. Hence, it can be subsumed into the normalizer η.

$$p(x) = \eta \exp\left\{-\tfrac{1}{2}x^T \Sigma^{-1} x + x^T \Sigma^{-1}\mu\right\} \tag{3.76}$$

This form motivates the parameterization of a Gaussian by its canonical parameters Ω and ξ.

$$p(x) = \eta \exp\left\{-\tfrac{1}{2}x^T \Omega\, x + x^T \xi\right\} \tag{3.77}$$

In many ways, the canonical parameterization is more elegant than the moments parameterization. In particular, the negative logarithm of the Gaussian is a quadratic function in x, with the canonical parameters Ω and ξ:

$$-\log p(x) = \text{const.} + \tfrac{1}{2}x^T \Omega\, x - x^T \xi \tag{3.78}$$

Here "const." is a constant. The reader may notice that we cannot use the symbol η to denote this constant, since negative logarithms of probabilities do not normalize to 1. The negative logarithm of our distribution $p(x)$ is quadratic in x, with the quadratic term parameterized by Ω and the linear term by ξ. In fact, for Gaussians, Ω must be positive semidefinite, hence $-\log p(x)$ is a quadratic distance function with mean $\mu = \Omega^{-1} \xi$. This is easily verified by setting the first derivative of (3.78) to zero:

$$\frac{\partial[-\log p(x)]}{\partial x} = 0 \iff \Omega x - \xi = 0 \iff x = \Omega^{-1}\xi \tag{3.79}$$

The matrix Ω determines the rate at which the distance function increases in the different dimensions of the variable x. A quadratic distance that is weighted by a matrix Ω is called a *Mahalanobis distance*.

MAHALANOBIS DISTANCE

1:	**Algorithm Information_filter**($\xi_{t-1}, \Omega_{t-1}, u_t, z_t$):
2:	$\bar{\Omega}_t = (A_t \, \Omega_{t-1}^{-1} \, A_t^T + R_t)^{-1}$
3:	$\bar{\xi}_t = \bar{\Omega}_t (A_t \, \Omega_{t-1}^{-1} \, \xi_{t-1} + B_t \, u_t)$
4:	$\Omega_t = C_t^T \, Q_t^{-1} \, C_t + \bar{\Omega}_t$
5:	$\xi_t = C_t^T \, Q_t^{-1} \, z_t + \bar{\xi}_t$
6:	return ξ_t, Ω_t

Table 3.5 The information filter algorithm.

3.5.2 The Information Filter Algorithm

Table 3.5 states the update algorithm known as *information filter*. Its input is a Gaussian in its canonical parameterization ξ_{t-1} and Ω_{t-1}, representing the belief at time $t-1$. Just like all Bayes filters, its input includes the control u_t and the measurement z_t. The output are the parameters ξ_t and Ω_t of the updated Gaussian.

The update involves matrices A_t, B_t, C_t, R_t, and Q_t. Those were defined in Chapter 3.2. The information filter assumes that the state transition and measurement probabilities are governed by the following linear Gaussian equations, originally defined in (3.2) and (3.5):

$$\text{(3.80)} \quad x_t = A_t x_{t-1} + B_t u_t + \varepsilon_t$$
$$\text{(3.81)} \quad z_t = C_t x_t + \delta_t$$

Here R_t and Q_t are the covariances of the zero-mean noise variables ε_t and δ_t, respectively.

Just like the Kalman filter, the information filter is updated in two steps, a prediction step and a measurement update step. The prediction step is implemented in lines 2 and 3 in Table 3.5. The parameters $\bar{\xi}_t$ and $\bar{\Omega}_t$ describe the Gaussian belief over x_t after incorporating the control u_t, but before incorporating the measurement z_t. The latter is done through lines 4 and 5. Here the belief is updated based on the measurement z_t.

These two update steps can be vastly different in complexity, especially if the state space possesses many dimensions. The prediction step, as stated in Table 3.5, involves the inversion of two matrices of the size $n \times n$, where n is the dimension of the state space. This inversion requires approximately

$O(n^{2.4})$ time. In Kalman filters, the update step is additive and requires at most $O(n^2)$ time; it requires less time if only a subset of variables is affected by a control, or if variables transition independently of each other. These roles are reversed for the measurement update step. Measurement updates are additive in the information filter. They require at most $O(n^2)$ time, and they are even more efficient if measurements carry only information about a subset of all state variables at a time. The measurement update is the difficult step in Kalman filters. It requires matrix inversion whose worst case complexity is $O(n^{2.4})$. This illustrates the dual character of Kalman and information filters.

3.5.3 Mathematical Derivation of the Information Filter

The derivation of the information filter is analogous to that of the Kalman filter.

PREDICTION STEP   To derive the *prediction step* (lines 2 and 3 in Table 3.5), we begin with the corresponding update equations of the Kalman filters, which can be found in lines 2 and 3 of the algorithm in Table 3.1 and are restated here for the reader's convenience:

$$\bar{\mu}_t = A_t \mu_{t-1} + B_t u_t \tag{3.82}$$
$$\bar{\Sigma}_t = A_t \Sigma_{t-1} A_t^T + R_t \tag{3.83}$$

The information filter prediction step follows now directly by substituting the moments μ and Σ by the canonical parameters ξ and Ω according to their definitions in (3.72) and (3.73):

$$\mu_{t-1} = \Omega_{t-1}^{-1} \xi_{t-1} \tag{3.84}$$
$$\Sigma_{t-1} = \Omega_{t-1}^{-1} \tag{3.85}$$

Substituting these expressions in (3.82) and (3.83) gives us the set of prediction equations

$$\bar{\Omega}_t = (A_t \Omega_{t-1}^{-1} A_t^T + R_t)^{-1} \tag{3.86}$$
$$\bar{\xi}_t = \bar{\Omega}_t (A_t \Omega_{t-1}^{-1} \xi_{t-1} + B_t u_t) \tag{3.87}$$

These equations are identical to those in Table 3.5. As is easily seen, the prediction step involves two nested inversions of a potentially large matrix. These nested inversions can be avoided when only a small number of state variables is affected by the motion update, a topic that will be discussed later in this book.

MEASUREMENT UPDATE

The derivation of the *measurement update* is even simpler. We begin with the Gaussian of the belief at time t, which was provided in Equation (3.35) and is restated here once again:

$$(3.88) \quad bel(x_t) = \eta \exp\left\{-\tfrac{1}{2}(z_t - C_t x_t)^T Q_t^{-1}(z_t - C_t x_t) - \tfrac{1}{2}(x_t - \bar{\mu}_t)^T \bar{\Sigma}_t^{-1}(x_t - \bar{\mu}_t)\right\}$$

For Gaussians represented in their canonical form this distribution is given by

$$(3.89) \quad bel(x_t) = \eta \exp\left\{-\tfrac{1}{2} x_t^T C_t^T Q_t^{-1} C_t\, x_t + x_t^T C_t^T Q_t^{-1} z_t - \tfrac{1}{2} x_t^T \bar{\Omega}_t x_t + x_t^T \bar{\xi}_t\right\}$$

which, by reordering the terms in the exponent, resolves to

$$(3.90) \quad bel(x_t) = \eta \exp\left\{-\tfrac{1}{2} x_t^T [C_t^T Q_t^{-1} C_t + \bar{\Omega}_t]\, x_t + x_t^T [C_t^T Q_t^{-1} z_t + \bar{\xi}_t]\right\}$$

We can now read off the measurement update equations, by collecting the terms in the squared brackets:

$$(3.91) \quad \xi_t = C_t^T Q_t^{-1} z_t + \bar{\xi}_t$$
$$(3.92) \quad \Omega_t = C_t^T Q_t^{-1} C_t + \bar{\Omega}_t$$

These equations are identical to the measurement update equations in lines 4 and 5 of Table 3.5.

3.5.4 The Extended Information Filter Algorithm

The *extended information filter*, or *EIF*, extends the information filter to the nonlinear case, very much in the same way the EKF is the non-linear extension of the Kalman filter. Table 3.6 depicts the EIF algorithm. The prediction is realized in lines 2 through 4, and the measurement update in lines 5 through 7. These update equations are largely analog to the linear information filter, with the functions g and h (and their Jacobian G_t and H_t) replacing the parameters of the linear model A_t, B_t, and C_t. As before, g and h specify the nonlinear state transition function and measurement function, respectively. Those were defined in (3.48) and (3.49) and are restated here:

$$(3.93) \quad x_t = g(u_t, x_{t-1}) + \varepsilon_t$$
$$(3.94) \quad z_t = h(x_t) + \delta_t$$

Unfortunately, both g and h require a state as an input. This mandates the recovery of a state estimate μ from the canonical parameters. The recovery

---

1:      **Algorithm Extended_information_filter**($\xi_{t-1}, \Omega_{t-1}, u_t, z_t$):
2:      $\mu_{t-1} = \Omega_{t-1}^{-1} \xi_{t-1}$
3:      $\bar{\Omega}_t = (G_t \, \Omega_{t-1}^{-1} \, G_t^T + R_t)^{-1}$
4:      $\bar{\xi}_t = \bar{\Omega}_t \, g(u_t, \mu_{t-1})$
5:      $\bar{\mu}_t = g(u_t, \mu_{t-1})$
6:      $\Omega_t = \bar{\Omega}_t + H_t^T \, Q_t^{-1} \, H_t$
7:      $\xi_t = \bar{\xi}_t + H_t^T \, Q_t^{-1} \, [z_t - h(\bar{\mu}_t) + H_t \, \bar{\mu}_t]$
8:      *return* ξ_t, Ω_t

Table 3.6 The extended information filter (EIF) algorithm.

takes place in line 2, in which the state μ_{t-1} is calculated from Ω_{t-1} and ξ_{t-1} in the obvious way. Line 5 computes the state $\bar{\mu}_t$ using the equation familiar from the EKF (line 2 in Table 3.3). The necessity to recover the state estimate seems at odds with the desire to represent the filter using its canonical parameters. We will revisit this topic when discussing the use of extended information filters in the context of robotic mapping.

3.5.5 Mathematical Derivation of the Extended Information Filter

The extended information filter is easily derived by essentially performing the same linearization that led to the extended Kalman filter above. As in (3.51) and (3.53), the extended information filter approximates g and h by a Taylor expansion:

(3.95) $\quad g(u_t, x_{t-1}) \approx g(u_t, \mu_{t-1}) + G_t \, (x_{t-1} - \mu_{t-1})$
(3.96) $\quad h(x_t) \approx h(\bar{\mu}_t) + H_t \, (x_t - \bar{\mu}_t)$

Here G_t and H_t are the Jacobians of g and h at μ_{t-1} and $\bar{\mu}_t$, respectively:

(3.97) $\quad G_t = g'(u_t, \mu_{t-1})$
(3.98) $\quad H_t = h'(\bar{\mu}_t)$

These definitions are equivalent to those in the EKF. The prediction step is now derived from lines 2 and 3 of the EKF algorithm (Table 3.3), which are restated here:

$$\bar{\Sigma}_t = G_t \, \Sigma_{t-1} \, G_t^T + R_t \tag{3.99}$$

$$\bar{\mu}_t = g(u_t, \mu_{t-1}) \tag{3.100}$$

Substituting Σ_{t-1} by Ω_{t-1}^{-1} and $\bar{\mu}_t$ by $\bar{\Omega}_t^{-1} \bar{\xi}_t$ gives us the prediction equations of the extended information filter:

$$\bar{\Omega}_t = (G_t \, \Omega_{t-1}^{-1} \, G_t^T + R_t)^{-1} \tag{3.101}$$

$$\bar{\xi}_t = \bar{\Omega}_t \, g(u_t, \Omega_{t-1}^{-1} \, \xi_{t-1}) \tag{3.102}$$

The measurement update is derived from Equations (3.60) and (3.61). In particular, (3.61) defines the following Gaussian posterior:

$$bel(x_t) = \eta \exp\left\{-\tfrac{1}{2}(z_t - h(\bar{\mu}_t) - H_t(x_t - \bar{\mu}_t))^T Q_t^{-1} \right. \tag{3.103}$$
$$\left. (z_t - h(\bar{\mu}_t) - H_t(x_t - \bar{\mu}_t)) - \tfrac{1}{2}(x_t - \bar{\mu}_t)^T \bar{\Sigma}_t^{-1}(x_t - \bar{\mu}_t)\right\}$$

Multiplying out the exponent and reordering the terms gives us the following expression for the posterior:

$$\begin{aligned} bel(x_t) &= \eta \, \exp\left\{-\tfrac{1}{2} x_t^T H_t^T Q_t^{-1} H_t \, x_t + x_t^T H_t^T Q_t^{-1} [z_t - h(\bar{\mu}_t) + H_t \, \bar{\mu}_t] \right. \\ &\quad \left. -\tfrac{1}{2} x_t^T \bar{\Sigma}_t^{-1} x_t + x_t^T \bar{\Sigma}_t^{-1} \bar{\mu}_t\right\} \\ &= \eta \, \exp\left\{-\tfrac{1}{2} x_t^T \left[H_t^T Q_t^{-1} H_t + \bar{\Sigma}_t^{-1}\right] x_t \right. \\ &\quad \left. + x_t^T \left[H_t^T Q_t^{-1} [z_t - h(\bar{\mu}_t) + H_t \, \bar{\mu}_t] + \bar{\Sigma}_t^{-1} \bar{\mu}_t\right]\right\} \end{aligned} \tag{3.104}$$

With $\bar{\Sigma}_t^{-1} = \bar{\Omega}_t$ this expression resolves to the following information form:

$$\begin{aligned} bel(x_t) &= \eta \, \exp\left\{-\tfrac{1}{2} x_t^T \left[H_t^T Q_t^{-1} H_t + \bar{\Omega}_t\right] x_t \right. \\ &\quad \left. + x_t^T \left[H_t^T Q_t^{-1} [z_t - h(\bar{\mu}_t) + H_t \, \bar{\mu}_t] + \bar{\xi}_t\right]\right\} \end{aligned} \tag{3.105}$$

We can now read off the measurement update equations by collecting the terms in the squared brackets:

$$\Omega_t = \bar{\Omega}_t + H_t^T \, Q_t^{-1} \, H_t \tag{3.106}$$

$$\xi_t = \bar{\xi}_t + H_t^T \, Q_t^{-1} \, [z_t - h(\bar{\mu}_t) + H_t \, \bar{\mu}_t] \tag{3.107}$$

3.5.6 Practical Considerations

When applied to robotics problems, the information filter possesses several advantages over the Kalman filter. For example, representing global uncertainty is simple in the information filter: simply set $\Omega = 0$. When using moments, such global uncertainty amounts to a covariance of infinite magnitude. This is especially problematic when sensor measurements carry information about a strict subset of all state variables, a situation often encountered in robotics. Special provisions have to be made to handle such situations in EKFs. The information filter tends to be numerically more stable than the Kalman filter in many of the applications discussed later in this book.

As we shall see in later chapters of this book, the information filter and several extensions enable a robot to integrate information without immediately resolving it into probabilities. This can be of great advantage in complex estimation problems, involving hundreds or even millions of variables. For such large problems, the integration á la Kalman filter induces severe computational problems, since any new piece of information requires propagation through a large system of variables. The information filter, with appropriate modification, can side-step this issue by simply adding the new information locally into the system. However, this is *not* a property yet of the simple information filter discussed here; we will extend this filter in Chapter 12.

Another advantage of the information filter over the Kalman filter arises from its natural fit for multi-robot problems. Multi-robot problems often involve the integration of sensor data collected decentrally. Such integration is commonly performed through Bayes rule. When represented in logarithmic form, Bayes rule becomes an addition. As noted above, the canonical parameters of information filters represent a probability in logarithmic form. Thus, information integration is achieved by summing up information from multiple robots. Addition is commutative. Because of this, information filters can often integrate information in arbitrary order, with arbitrary delays, and in a completely decentralized manner. While the same is possible using the moments parameterization—after all, they represent the same information—the necessary overhead for doing so is much higher. Despite this advantage, the use of information filters in multi-robot systems remains largely under-explored. We will revisit the multi-robot topic in Chapter 12.

These advantages of the information filter are offset by important limitations. A primary disadvantage of the extended information filter is the need to recover a state estimate in the update step, when applied to nonlinear systems. This step, if implemented as stated here, requires the inversion of the information matrix. Further matrix inversions are required for the prediction step of the information filters. In many robotics problems, the EKF does not involve the inversion of matrices of comparable size. For high dimensional state spaces, the information filter is generally believed to be computationally inferior to the Kalman filter. In fact, this is one of the reasons why the

EKF has been vastly more popular than the extended information filter.

As we will see later in this book, these limitations do not necessarily apply to problems in which the information matrix possesses structure. In many robotics problems, the interaction of state variables is local; as a result, the information matrix may be sparse. Such sparseness does *not* translate to sparseness of the covariance.

MARKOV RANDOM FIELD

Information filters can be thought of as graphs, where states are connected whenever the corresponding off-diagonal element in the information matrix is non-zero. Sparse information matrices correspond to sparse graphs; in fact, such graphs are commonly known as Gaussian *Markov random fields*. A flurry of algorithms exist to perform the basic update and estimation equations efficiently for such fields, under names like *loopy belief propagation*. In this book, we will encounter a mapping problem in which the information matrix is (approximately) sparse, and develop an extended information filter that is significantly more efficient than both Kalman filters and non-sparse information filters.

3.6 Summary

In this section, we introduced efficient Bayes filter algorithms that represent the posterior by multivariate Gaussians. We noted that

- Gaussians can be represented in two different ways: The moments parameterization and the canonical parameterization. The moments parameterization consists of the mean (first moment) and the covariance (second moment) of the Gaussian. The canonical, or natural, parameterization consists of an information matrix and an information vector. Both parameterizations are duals of each other, and each can be recovered from the other via matrix inversion.

- Bayes filters can be implemented for both parameterizations. When using the moments parameterization, the resulting filter is called Kalman filter. The dual of the Kalman filter is the information filter, which represents the posterior in the canonical parameterization. Updating a Kalman filter based on a control is computationally simple, whereas incorporating a measurement is more difficult. The opposite is the case for the information filter, where incorporating a measurement is simple, but updating the filter based on a control is difficult.

- For both filters to calculate the correct posterior, three assumptions have to be fulfilled. First, the initial belief must be Gaussian. Second, the state transition probability must be composed of a function that is linear in its argument with added independent Gaussian noise. Third, the same applies to the measurement probability. It must also be linear in its argument, with added Gaussian noise. Systems that meet these assumptions are called linear Gaussian systems.

- Both filters can be extended to nonlinear problems. One technique described in this chapter calculates a tangent to the nonlinear function. Tangents are linear, making the filters applicable. The technique for finding a tangent is called Taylor expansion. Performing a Taylor expansion involves calculating the first derivative of the target function, and evaluating it at a specific point. The result of this operation is a matrix known as the Jacobian. The resulting filters are called 'extended.'

- The unscented Kalman filter uses a different linearization technique, called unscented transform. It probes the function to be linearized at selected points and calculates a linearized approximation based on the outcomes of these probes. This filter can be implemented without the need for any Jacobians, it is thus often referred to as derivative-free. The unscented Kalman filter is equivalent to the Kalman filter for linear systems but often provides improved estimates for nonlinear systems. The computational complexity of this filter is the same as for the extended Kalman filter.

- The accuracy of Taylor series expansions and unscented transforms depends on two factors: The degree of nonlinearity in the system, and the width of the posterior. Extended filters tend to yield good results if the state of the system is known with relatively high accuracy, so that the remaining covariance is small. The larger the uncertainty, the higher the error introduced by the linearization.

- One of the primary advantages of Gaussian filters is computational: The update requires time polynomial in the dimensionality of the state space. This is not the case of some of the techniques described in the next chapter. The primary disadvantage is their confinement to unimodal Gaussian distributions.

- An extension of Gaussians to multimodal posteriors is known as multi-hypothesis Kalman filter. This filter represents the posterior by a mixture of Gaussians, which is nothing else but a weighted sum of Gaussians.

The mechanics of updating this filter require mechanisms for splitting and fusing or pruning individual Gaussians. Multi-hypothesis Kalman filters are particularly well suited for problems with discrete data association, which commonly occur in robotics.

- Within the multivariate Gaussian regime, both filters, the Kalman filter and the information filter, have orthogonal strengths and weaknesses. However, the Kalman filter and its nonlinear extension, the extended Kalman filter, are vastly more popular than the information filter.

The selection of the material in this chapter is based on today's most popular techniques in robotics. There exists a huge number of variations and extensions of Gaussian filters, which address the various limitations and shortcomings of the individual filters.

A good number of algorithms in this book are based on Gaussian filters. Many practical robotics problems require extensions that exploit sparse structures or factorizations of the posterior.

3.7 Bibliographical Remarks

The Kalman filter was invented by Swerling (1958) and Kalman (1960). It is usually introduced as an optimal estimator under the least-squares assumption, and less frequently as a method for calculating posterior distributions—although under the appropriate assumptions both views are identical. There exists a number of excellent textbooks on Kalman filters and information filters, including the ones by Maybeck (1990) and Jazwinsky (1970). Contemporary treatments of Kalman filters with data association are provided by Bar-Shalom and Fortmann (1988); Bar-Shalom and Li (1998).

The inversion lemma can be found in Golub and Loan (1986). Matrix inversion can be carried out in $O(n^{2.376})$ time, according to Coppersmith and Winograd (1990). This result is the most recent one in a series of papers that provided improvements over the $O(n^3)$ complexity of the variable elimination algorithm. The series started with Strassen's (1969) seminal paper, in which he gave an algorithm requiring $O(n^{2.807})$. Cover and Thomas (1991) provides a survey of information theory, but with a focus on discrete systems. The unscented Kalman filter is due to Julier and Uhlmann (1997). A comparison of UKF to the EKF in the context of various state estimation problems can be found in van der Merwe (2004). Minka (2001) provided a recent treatment of moments matching and assumed density filtering for Gaussian mixtures.

3.8 Exercises

1. In this and the following exercise, you are asked to design a Kalman filter for a simple dynamical system: a car with linear dynamics moving in a linear environment. Assume $\Delta t = 1$ for simplicity. The position of the

car at time t is given by x_t. Its velocity is $\dot{x}_t$, and its acceleration is $\ddot{x}_t$. Suppose the acceleration is set randomly at each point in time, according to a Gaussian with zero mean and covariance $\sigma^2 = 1$.

(a) What is a minimal state vector for the Kalman filter (so that the resulting system is Markovian)?

(b) For your state vector, design the state transition probability $p(x_t \mid u_t, x_{t-1})$. Hint: this transition function will possess linear matrices A and B and a noise covariance R (c.f., Equation (3.4) and Table 3.1).

(c) Implement the state prediction step of the Kalman filter. Assuming we know at time $t = 0$, $x_0 = \dot{x}_0 = \ddot{x}_0 = 0$. Compute the state distributions for times $t = 1, 2, \ldots, 5$.

(d) For each value of t, plot the joint posterior over x and $\dot{x}$ in a diagram, where x is the horizontal and $\dot{x}$ is the vertical axis. For each posterior, you are asked to plot an *uncertainty ellipse*, which is the ellipse of points that are one standard deviation away from the mean. Hint: If you do not have access to a mathematics library, you can create those ellipses by analyzing the eigenvalues of the covariance matrix.

UNCERTAINTY ELLIPSE

(e) What will happen to the correlation between x_t and $\dot{x}_t$ as $t \uparrow \infty$?

2. We will now add measurements to our Kalman filter. Suppose at time t, we can receive a noisy observation of x. In expectation, our sensor measures the true location. However, this measurement is corrupted by Gaussian noise with covariance $\sigma^2 = 10$.

   (a) Define the measurement model. Hint: You need to define a matrix C and another matrix Q (c.f., Equation (3.6) and Table 3.1).

   (b) Implement the measurement update. Suppose at time $t = 5$, we observe a measurement $z = 5$. State the parameters of the Gaussian estimate before and after updating the KF. Plot the uncertainty ellipse before and after incorporating the measurement (see above for instructions as to how to plot an uncertainty ellipse).

3. In Chapter 3.2.4, we derived the prediction step of the KF. This step is often derived with Z transforms or Fourier transforms, using the Convolution Theorem. Re-derive the prediction step using transforms. *Notice: This exercise requires knowledge of transforms and convolution, which goes beyond the material in this book.*

3.8 Exercises

4. We noted in the text that the EKF linearization is an approximation. To see how bad this approximation is, we ask you to work out an example. Suppose we have a mobile robot operating in a planar environment. Its state is its x-y-location and its global heading direction θ. Suppose we know x and y with high certainty, but the orientation θ is unknown. This is reflected by our initial estimate

$$\mu = \begin{pmatrix} 0 & 0 & 0 \end{pmatrix} \quad \text{and} \quad \Sigma = \begin{pmatrix} 0.01 & 0 & 0 \\ 0 & 0.01 & 0 \\ 0 & 0 & 10000 \end{pmatrix}$$

   (a) Draw, graphically, your best model of the posterior over the robot pose after the robot moves $d = 1$ units forward. For this exercise, we assume the robot moves flawlessly without any noise. Thus, the expected location of the robot after motion will be

$$\begin{pmatrix} x' \\ y' \\ \theta' \end{pmatrix} = \begin{pmatrix} x + \cos\theta \\ y + \sin\theta \\ \theta \end{pmatrix}$$

   For your drawing, you can ignore θ and only draw the posterior in x-y-coordinates.

   (b) Now develop this motion into a prediction step for the EKF. For that, you have to define a state transition function and linearize it. You then have to generate a new Gaussian estimate of the robot pose using the linearized model. You should give the exact mathematical equations for each of these steps, and state the Gaussian that results.

   (c) Draw the uncertainty ellipse of the Gaussian and compare it with your intuitive solution.

   (d) Now incorporate a measurement. Our measurement shall be a noisy projection of the x-coordinate of the robot, with covariance $Q = 0.01$. Specify the measurement model. Now apply the measurement both to your intuitive posterior, and formally to the EKF estimate using the standard EKF machinery. Give the exact result of the EKF, and compare it with the result of your intuitive analysis.

   (e) Discuss the difference between your estimate of the posterior, and the Gaussian produced by the EKF. How significant are those differences? What can be changed to make the approximation more accurate? What would have happened if the initial orientation had been known, but not the robot's y-coordinate?

5. The Kalman filter Table 3.1 lacked a constant additive term in the motion and the measurement models. Extend this algorithm to contain such terms.

6. Prove (via example) the existence of a sparse information matrix in multivariate Gaussians (of dimension d) that correlate all d variables with correlation coefficient that are ε-close to 1. We say an information matrix is sparse if all but a constant number of elements in each row and each column are zero.

4 *Nonparametric Filters*

A popular alternative to Gaussian techniques are *nonparametric filters*. Nonparametric filters do not rely on a fixed functional form of the posterior, such as Gaussians. Instead, they approximate posteriors by a finite number of values, each roughly corresponding to a region in state space. Some nonparametric Bayes filters rely on a decomposition of the state space, in which each such value corresponds to the cumulative probability of the posterior density in a compact subregion of the state space. Others approximate the state space by random samples drawn from the posterior distribution. In all cases, the number of parameters used to approximate the posterior can be varied. The quality of the approximation depends on the number of parameters used to represent the posterior. As the number of parameters goes to infinity, nonparametric techniques tend to converge uniformly to the correct posterior—under specific smoothness assumptions.

This chapter discusses two nonparametric approaches for approximating posteriors over continuous spaces with finitely many values. The first decomposes the state space into finitely many regions, and represents the posterior by a histogram. A histogram assigns to each region a single cumulative probability; they are best thought of as piecewise constant approximations to a continuous density. The second technique represents posteriors by finitely many samples. The resulting filter is known as *particle filter* and has become immensely popular in robotics.

Both types of techniques, histograms and particle filters, do not make strong parametric assumptions on the posterior density. In particular, they are well-suited to represent complex multimodal beliefs. For this reason, they are often the method of choice when a robot has to cope with phases of global uncertainty, and when it faces hard data association problems that yield separate, distinct hypotheses. However, the representational power of these techniques comes at the price of added computational complexity.

Fortunately, both nonparametric techniques described in this chapter make it possible to adapt the number of parameters to the (suspected) complexity of the posterior. When the posterior is of low complexity (e.g., focused on a single state with a small margin of uncertainty), they use only small numbers of parameters. For complex posteriors, e.g., posteriors with many modes scattered across the state space, the number of parameters grows larger.

Techniques that can adapt the number of parameters to represent the posterior online are called *adaptive*. They are called *resource-adaptive* if they can adapt based on the computational resources available for belief computation. Resource-adaptive techniques play an important role in robotics. They enable robots to make decisions in real time, regardless of the computational resources available. Particle filters are often implemented as a resource-adaptive algorithm, by adapting the number of particles online based on the available computational resources.

RESOURCE-ADAPTIVE ALGORITHMS

4.1 The Histogram Filter

Histogram filters decompose the state space into finitely many regions and represent the cumulative posterior for each region by a single probability value. When applied to finite spaces, such filters are known as *discrete Bayes filters*; when applied to continuous spaces, they are commonly called *histogram filters*. We will first describe the discrete Bayes filter and then discuss its use in continuous state spaces.

4.1.1 The Discrete Bayes Filter Algorithm

Discrete Bayes filters apply to problems with *finite* state spaces, where the random variable X_t can take on finitely many values. We already encountered a discrete Bayes filter in Chapter 2.4.2, when discussing the example of a robot estimating the probability that a door is open. Some of the robotic mapping problems discussed in later chapters also involve discrete random variables. For example, occupancy grid mapping algorithms assume that each location in the environment is either occupied or free. The corresponding random variable is binary. It can take on two different values. Thus, finite state spaces play an important role in robotics.

Table 4.1 provides pseudo-code for the discrete Bayes filter. This code is derived from the general Bayes filter in Table 2.1 by replacing the integration

4.1 The Histogram Filter

1:         **Algorithm Discrete_Bayes_filter($\{p_{k,t-1}\}, u_t, z_t$):**
2:            for all k do
3:                 $\bar{p}_{k,t} = \sum_i p(X_t = x_k \mid u_t, X_{t-1} = x_i)\, p_{i,t-1}$
4:                 $p_{k,t} = \eta\, p(z_t \mid X_t = x_k)\, \bar{p}_{k,t}$
5:            endfor
6:            return $\{p_{k,t}\}$

Table 4.1 The discrete Bayes filter. Here x_i, x_k denote individual states.

with a finite sum. The variables x_i and x_k denote individual states, of which there may only be finitely many. The belief at time t is an assignment of a probability to each state x_k, denoted $p_{k,t}$. Thus, the input to the algorithm is a discrete probability distribution $\{p_{k,t}\}$, along with the most recent control u_t and measurement z_t. Line 3 calculates the prediction, the belief for the new state based on the control alone. This prediction is then updated in line 4, so as to incorporate the measurement. The discrete Bayes filter algorithm is popular in many areas of signal processing, where it is often referred to as the forward pass of a *hidden Markov model*, or *HMM*.

HIDDEN MARKOV MODEL

4.1.2 Continuous State

Of particular interest will be the use of discrete Bayes filters as an approximate inference tool for *continuous* state spaces. As noted above, such filters are called *histogram filters*. Figure 4.1 illustrates how a histogram filter represents a random variable and its nonlinear transform. Shown there is the projection of a histogrammed Gaussian through a nonlinear function. The original Gaussian distribution possesses 10 bins. So does the projected probability distribution, but in two of the resulting bins the probability is so close to zero that they cannot be seen in this figure. Figure 4.1 also shows the correct continuous distributions for comparison.

Histogram filters decompose a continuous state space into finitely many *bins*, or *regions*:

$$\text{(4.1)} \quad \text{dom}(X_t) \;=\; \mathbf{x}_{1,t} \cup \mathbf{x}_{2,t} \cup \ldots \mathbf{x}_{K,t}$$

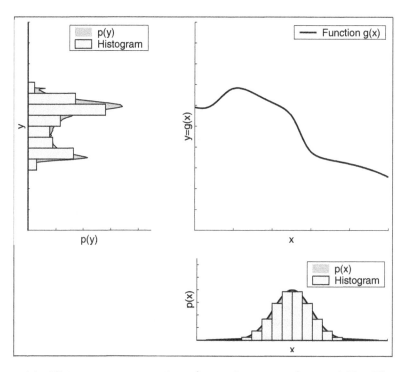

Figure 4.1 Histogram representation of a continuous random variable. The gray shaded area in the lower right plot shows the density of the continuous random variable, X. The histogram approximation of this density is overlaid in light-gray. The random variable is passed through the function displayed in the upper right graph. The density and the histogram approximation of the resulting random variable, Y, are plotted in the upper left graph. The histogram of the transformed random variable was computed by passing multiple points from each histogram bin of X through the nonlinear function.

Here X_t is the familiar random variable describing the state of the robot at time t. The function $\text{dom}(X_t)$ denotes the state space, which is the universe of possible values that X_t might assume. Each $\mathbf{x}_{k,t}$ describes a convex region. These regions together form a partitioning of the state space. For each $i \neq k$ we have $\mathbf{x}_{i,t} \cap \mathbf{x}_{k,t} = \emptyset$ and $\bigcup_k \mathbf{x}_{k,t} = \text{dom}(X_t)$.

A straightforward decomposition of a continuous state space is a multi-dimensional grid, where each $\mathbf{x}_{k,t}$ is a grid cell. Through the granularity of the decomposition, we can trade off accuracy and computational efficiency. Fine-grained decompositions infer smaller approximation errors than coarse ones, but at the expense of increased computational complexity.

As we already discussed, the discrete Bayes filter assigns to each region $\mathbf{x}_{k,t}$ a probability, $p_{k,t}$. Within each region, the discrete Bayes filter carries no further information on the belief distribution. Thus, the posterior becomes a piecewise constant PDF, which assigns a uniform probability to each state x_t within each region $\mathbf{x}_{k,t}$:

$$(4.2) \quad p(x_t) = \frac{p_{k,t}}{|\mathbf{x}_{k,t}|}$$

Here $|\mathbf{x}_{k,t}|$ is the volume of the region $\mathbf{x}_{k,t}$.

If the state space is truly discrete, the conditional probabilities $p(\mathbf{x}_{k,t} \mid u_t, \mathbf{x}_{i,t-1})$ and $p(z_t \mid \mathbf{x}_{k,t})$ are well-defined, and the algorithm can be implemented as stated. In continuous state spaces, one is usually given the densities $p(x_t \mid u_t, x_{t-1})$ and $p(z_t \mid x_t)$, which are defined for individual states (and not for regions in state space). For cases where each region $\mathbf{x}_{k,t}$ is small and of the same size, these densities are usually approximated by substituting $\mathbf{x}_{k,t}$ by a representative of this region. For example, we might simply "probe" using the mean state in $\mathbf{x}_{k,t}$

$$(4.3) \quad \hat{x}_{k,t} = |\mathbf{x}_{k,t}|^{-1} \int_{\mathbf{x}_{k,t}} x_t \, dx_t$$

One then simply replaces

$$(4.4) \quad p(z_t \mid \mathbf{x}_{k,t}) \approx p(z_t \mid \hat{x}_{k,t})$$
$$(4.5) \quad p(\mathbf{x}_{k,t} \mid u_t, \mathbf{x}_{i,t-1}) \approx \eta \, |\mathbf{x}_{k,t}| \, p(\hat{x}_{k,t} \mid u_t, \hat{x}_{i,t-1})$$

These approximations are the result of the piecewise uniform interpretation of the discrete Bayes filter stated in (4.2), and a Taylor-approximation analogous to the one used by EKFs.

4.1.3 Mathematical Derivation of the Histogram Approximation

To see that (4.4) is a reasonable approximation, we note that $p(z_t \mid \mathbf{x}_{k,t})$ can be expressed as the following integral:

$$(4.6) \quad p(z_t \mid \mathbf{x}_{k,t}) = \frac{p(z_t, \mathbf{x}_{k,t})}{p(\mathbf{x}_{k,t})}$$
$$= \frac{\int_{\mathbf{x}_{k,t}} p(z_t, x_t) \, dx_t}{\int_{\mathbf{x}_{k,t}} p(x_t) \, dx_t}$$

$$
= \frac{\int_{\mathbf{x}_{k,t}} p(z_t \mid x_t)\, p(x_t)\, dx_t}{\int_{\mathbf{x}_{k,t}} p(x_t)\, dx_t}
$$

$$
\stackrel{(4.2)}{=} \frac{\int_{\mathbf{x}_{k,t}} p(z_t \mid x_t)\, \frac{p_{k,t}}{|\mathbf{x}_{k,t}|}\, dx_t}{\int_{\mathbf{x}_{k,t}} \frac{p_{k,t}}{|\mathbf{x}_{k,t}|}\, dx_t}
$$

$$
= \frac{\frac{p_{k,t}}{|\mathbf{x}_{k,t}|} \int_{\mathbf{x}_{k,t}} p(z_t \mid x_t)\, dx_t}{\frac{p_{k,t}}{|\mathbf{x}_{k,t}|} \int_{\mathbf{x}_{k,t}} 1\, dx_t}
$$

$$
= \frac{\int_{\mathbf{x}_{k,t}} p(z_t \mid x_t)\, dx_t}{\int_{\mathbf{x}_{k,t}} 1\, dx_t}
$$

$$
= |\mathbf{x}_{k,t}|^{-1} \int_{\mathbf{x}_{k,t}} p(z_t \mid x_t)\, dx_t
$$

This expression is an exact description of the desired probability under the piecewise uniform distribution model in (4.2). If we now approximate $p(z_t \mid x_t)$ by $p(z_t \mid \hat{x}_{k,t})$ for $x_t \in \mathbf{x}_{k,t}$, we obtain

$$
\begin{align}
(4.7)\quad p(z_t \mid \mathbf{x}_{k,t}) &\approx |\mathbf{x}_{k,t}|^{-1} \int_{\mathbf{x}_{k,t}} p(z_t \mid \hat{x}_{k,t})\, dx_t \\
&= |\mathbf{x}_{k,t}|^{-1} p(z_t \mid \hat{x}_{k,t}) \int_{\mathbf{x}_{k,t}} 1\, dx_t \\
&= |\mathbf{x}_{k,t}|^{-1} p(z_t \mid \hat{x}_{k,t})\, |\mathbf{x}_{k,t}| \\
&= p(z_t \mid \hat{x}_{k,t})
\end{align}
$$

which is the approximation stated above in (4.4).

The derivation of the approximation to $p(\mathbf{x}_{k,t} \mid u_t, \mathbf{x}_{i,t-1})$ in (4.5) is slightly more involved, since regions occur on both sides of the conditioning bar. In analogy to our transformation above, we obtain:

$$
\begin{align}
(4.8)\quad & p(\mathbf{x}_{k,t} \mid u_t, \mathbf{x}_{i,t-1}) \\
&= \frac{p(\mathbf{x}_{k,t}, \mathbf{x}_{i,t-1} \mid u_t)}{p(\mathbf{x}_{i,t-1} \mid u_t)}
\end{align}
$$

4.1 The Histogram Filter

$$= \frac{\int_{\mathbf{x}_{k,t}} \int_{\mathbf{x}_{i,t-1}} p(x_t, x_{t-1} \mid u_t) \, dx_t \, dx_{t-1}}{\int_{\mathbf{x}_{i,t-1}} p(x_{t-1} \mid u_t) \, dx_{t-1}}$$

$$= \frac{\int_{\mathbf{x}_{k,t}} \int_{\mathbf{x}_{i,t-1}} p(x_t \mid u_t, x_{t-1}) \, p(x_{t-1} \mid u_t) \, dx_t \, dx_{t-1}}{\int_{\mathbf{x}_{i,t-1}} p(x_{t-1} \mid u_t) \, dx_{t-1}}$$

We now exploit the Markov assumption, which implies independence between x_{t-1} and u_t, and thus $p(x_{t-1} \mid u_t) = p(x_{t-1})$:

(4.9) $p(\mathbf{x}_{k,t} \mid u_t, \mathbf{x}_{i,t-1})$

$$= \frac{\int_{\mathbf{x}_{k,t}} \int_{\mathbf{x}_{i,t-1}} p(x_t \mid u_t, x_{t-1}) \, p(x_{t-1}) \, dx_t \, dx_{t-1}}{\int_{\mathbf{x}_{i,t-1}} p(x_{t-1}) \, dx_{t-1}}$$

$$= \frac{\int_{\mathbf{x}_{k,t}} \int_{\mathbf{x}_{i,t-1}} p(x_t \mid u_t, x_{t-1}) \, \frac{p_{i,t-1}}{|\mathbf{x}_{i,t-1}|} \, dx_t \, dx_{t-1}}{\int_{\mathbf{x}_{i,t-1}} \frac{p_{i,t-1}}{|\mathbf{x}_{i,t-1}|} \, dx_{t-1}}$$

$$= \frac{\int_{\mathbf{x}_{k,t}} \int_{\mathbf{x}_{i,t-1}} p(x_t \mid u_t, x_{t-1}) \, dx_t \, dx_{t-1}}{\int_{\mathbf{x}_{i,t-1}} 1 \, dx_{t-1}}$$

$$= |\mathbf{x}_{i,t-1}|^{-1} \int_{\mathbf{x}_{k,t}} \int_{\mathbf{x}_{i,t-1}} p(x_t \mid u_t, x_{t-1}) \, dx_t \, dx_{t-1}$$

If we now approximate $p(x_t \mid u_t, x_{t-1})$ by $p(\hat{x}_{k,t} \mid u_t, \hat{x}_{i,t-1})$ as before, we obtain the following approximation. Note that the normalizer η becomes necessary to ensure that the approximation is a valid probability distribution:

(4.10) $p(\mathbf{x}_{k,t} \mid u_t, \mathbf{x}_{i,t-1})$

$$\approx \eta \, |\mathbf{x}_{i,t-1}|^{-1} \int_{\mathbf{x}_{k,t}} \int_{\mathbf{x}_{i,t-1}} p(\hat{x}_{k,t} \mid u_t, \hat{x}_{i,t-1}) \, dx_t \, dx_{t-1}$$

$$= \eta \, |\mathbf{x}_{i,t-1}|^{-1} \, p(\hat{x}_{k,t} \mid u_t, \hat{x}_{i,t-1}) \int_{\mathbf{x}_{k,t}} \int_{\mathbf{x}_{i,t-1}} 1 \, dx_t \, dx_{t-1}$$

$$= \eta \, |\mathbf{x}_{i,t-1}|^{-1} \, p(\hat{x}_{k,t} \mid u_t, \hat{x}_{i,t-1}) |\mathbf{x}_{k,t}| \, |\mathbf{x}_{i,t-1}|$$

$$= \eta \, |\mathbf{x}_{k,t}| \, p(\hat{x}_{k,t} \mid u_t, \hat{x}_{i,t-1})$$

If all regions are of equal size (meaning that $|\mathbf{x}_{k,t}|$ is the same for all k), we can simply omit the factor $|\mathbf{x}_{k,t}|$, since it is subsumed by the normalizer. The

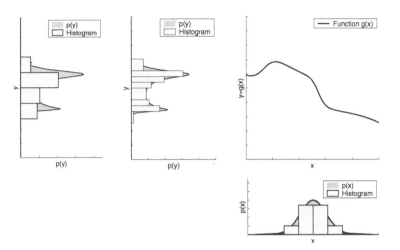

Figure 4.2 Dynamic vs. static decomposition. The upper left graph shows the static histogram approximation of the random variable Y, using 10 bins for covering the domain of Y (of which 6 are of nearly zero probability). The upper middle graph presents a tree representation of the same random variable, using the same number of bins.

resulting discrete Bayes filter is then equivalent to the algorithm outlined in Table 4.1. If implemented as stated there, the auxiliary parameters $\bar{p}_k$ do not constitute a probability distribution, since they are not normalized (compare line 3 to (4.10)). However, normalization takes place in line 4, so that the output parameters are indeed a valid probability distribution.

4.1.4 Decomposition Techniques

In robotics, decomposition techniques of continuous state spaces come in two basic flavors: *static* and *dynamic*. Static techniques rely on a fixed decomposition that is chosen in advance, irrespective of the shape of the posterior that is being approximated. Dynamic techniques adapt the decomposition to the specific shape of the posterior distribution. Static techniques are usually easier to implement, but they can be wasteful with regards to computational resources.

DENSITY TREES

A primary example of a dynamic decomposition technique is the family of *density trees*. Density trees decompose the state space recursively, in ways that adapt the resolution to the posterior probability mass. The intuition behind this decomposition is that the level of detail in the decomposition is a function of the posterior probability: The less likely a region, the coarser the decomposition. Figure 4.2 illustrates the difference between a static grid representation and a density tree representation. Due to its more compact representation, the density tree achieves a higher approximation quality using the same number of bins. Dynamic techniques like density trees can often cut the computation complexity by orders of magnitude over static ones, yet they require additional implementation effort.

SELECTIVE UPDATING   An effect similar to that of dynamic decompositions can be achieved by *selective updating*. When updating a posterior represented by a grid, selective techniques update a fraction of all grid cells only. A common implementation of this idea updates only those grid cells whose posterior probability exceeds a user-specified threshold.

Selective updating can be viewed as a hybrid decomposition, which decomposes the state space into a fine-grained grid and one large set that contains all regions not chosen by the selective update procedure. In this light, it can be thought of as a dynamic decomposition technique, since the decision as to which grid cells to consider during the update is made online, based on the shape of the posterior distribution. Selective updating techniques can reduce the computational effort involved in updating beliefs by orders of magnitude. They make it possible to use grid decompositions in spaces of three or more dimensions.

The mobile robotics literature often distinguishes *topological* from *metric* representations of space. While no clear definition of these terms exist, topological representations are often thought of as coarse graph-like representations, where nodes in the graph correspond to significant places (or features) in the environment. For indoor environments, such places may correspond to intersections, T-junctions, dead ends, and so on. The resolution of such decompositions, thus, depends on the structure of the environment. Alternatively, one might decompose the state space using regularly-spaced grids. Such a decomposition does not depend on the shape and location of the environmental features. Grid representations are often thought of as metric although, strictly speaking, it is the embedding space that is metric, not the decomposition. In mobile robotics, the spatial resolution of grid representations tends to be higher than that of topological representations. For instance, some of the examples in Chapter 7 use grid decompositions with cell sizes of 10 centimeters or less. This increased accuracy comes at the expense of increased computational costs.

1:   **Algorithm binary_Bayes_filter**(l_{t-1}, z_t):
2: $l_t = l_{t-1} + \log \frac{p(x|z_t)}{1-p(x|z_t)} - \log \frac{p(x)}{1-p(x)}$
3:   return l_t

Table 4.2 The binary Bayes filter in log odds form with an inverse measurement model. Here l_t is the log odds of the posterior belief over a binary state variable that does not change over time.

4.2 Binary Bayes Filters with Static State

Certain problems in robotics are best formulated as estimation problems with binary state that does not change over time. Those problems are addressed by the *binary Bayes filter*. Problems of this type arise if a robot estimates a fixed binary quantity in the environment from a sequence of sensor measurements. For example, a robot might want to know if a door is open or closed, in a context where the door state does not change during sensing. Another example of binary Bayes filters with static state are *occupancy grid maps*, which we will encounter in Chapter 9.

OCCUPANCY GRID MAPS

When the state is static, the belief is a function only of the measurements:

(4.11) $\quad bel_t(x) \;=\; p(x \mid z_{1:t}, u_{1:t}) \;=\; p(x \mid z_{1:t})$

where the state is chosen from two possible values, denoted by x and $\neg x$. In particular, we have $bel_t(\neg x) = 1 - bel_t(x)$. The lack of a time index for the state x reflects the fact that the state does not change.

Naturally, binary estimation problems of this type can be tackled using the discrete Bayes filter in Table 4.1. However, the belief is commonly implemented as a *log odds ratio*. The *odds* of a state x is defined as the ratio of the probability of this event divided by the probability of its negate

LOG ODDS RATIO

(4.12) $\quad \frac{p(x)}{p(\neg x)} \;=\; \frac{p(x)}{1-p(x)}$

The log odds is the logarithm of this expression

(4.13) $\quad l(x) \;:=\; \log \frac{p(x)}{1-p(x)}$

4.2 Binary Bayes Filters with Static State

Log odds assume values from $-\infty$ to ∞. The Bayes filter for updating beliefs in log odds representation is computationally elegant. It avoids truncation problems that arise for probabilities close to 0 or 1.

Table 4.2 states the basic update algorithm. This algorithm is additive; in fact, any algorithm that increments and decrements a variable in response to measurements can be interpreted as a Bayes filter in log odds form. This binary Bayes filter uses an *inverse measurement model* $p(x \mid z_t)$, instead of the familiar forward model $p(z_t \mid x)$. The inverse measurement model specifies a distribution over the (binary) state variable as a function of the measurement z_t.

INVERSE MEASUREMENT MODEL

Inverse models are often used in situations where measurements are more complex than the binary state. An example of such a situation is the problem of estimating whether or not a door is closed from camera images. Here the state is extremely simple, but the space of all measurements is huge. It is easier to devise a function that calculates a probability of a door being closed from a camera image, than describing the distribution over all camera images that show a closed door. In other words, it is easier to implement an inverse than a forward sensor model.

As the reader easily verifies from our definition of the log odds (4.13), the belief $bel_t(x)$ can be recovered from the log odds ratio l_t by the following equation:

$$(4.14) \quad bel_t(x) = 1 - \frac{1}{1 + \exp\{l_t\}}$$

To verify the correctness of our binary Bayes filter algorithm, we briefly restate the basic filter equation with the Bayes normalizer made explicit:

$$p(x \mid z_{1:t}) = \frac{p(z_t \mid x, z_{1:t-1}) \, p(x \mid z_{1:t-1})}{p(z_t \mid z_{1:t-1})}$$

$$(4.15) \quad = \frac{p(z_t \mid x) \, p(x \mid z_{1:t-1})}{p(z_t \mid z_{1:t-1})}$$

We now apply Bayes rule to the measurement model $p(z_t \mid x)$:

$$(4.16) \quad p(z_t \mid x) = \frac{p(x \mid z_t) \, p(z_t)}{p(x)}$$

and obtain

$$(4.17) \quad p(x \mid z_{1:t}) = \frac{p(x \mid z_t) \, p(z_t) \, p(x \mid z_{1:t-1})}{p(x) \, p(z_t \mid z_{1:t-1})}.$$

By analogy, we have for the opposite event $\neg x$:

$$p(\neg x \mid z_{1:t}) = \frac{p(\neg x \mid z_t)\, p(z_t)\, p(\neg x \mid z_{1:t-1})}{p(\neg x)\, p(z_t \mid z_{1:t-1})} \tag{4.18}$$

Dividing (4.17) by (4.18) leads to cancellation of various difficult-to-calculate probabilities:

$$\begin{aligned}
\frac{p(x \mid z_{1:t})}{p(\neg x \mid z_{1:t})} &= \frac{p(x \mid z_t)}{p(\neg x \mid z_t)} \frac{p(x \mid z_{1:t-1})}{p(\neg x \mid z_{1:t-1})} \frac{p(\neg x)}{p(x)} \\
&= \frac{p(x \mid z_t)}{1 - p(x \mid z_t)} \frac{p(x \mid z_{1:t-1})}{1 - p(x \mid z_{1:t-1})} \frac{1 - p(x)}{p(x)}
\end{aligned} \tag{4.19}$$

We denote the log odds ratio of the belief $bel_t(x)$ by $l_t(x)$. The log odds belief at time t is given by the logarithm of (4.19).

$$\begin{aligned}
l_t(x) &= \log \frac{p(x \mid z_t)}{1 - p(x \mid z_t)} + \log \frac{p(x \mid z_{1:t-1})}{1 - p(x \mid z_{1:t-1})} + \log \frac{1 - p(x)}{p(x)} \\
&= \log \frac{p(x \mid z_t)}{1 - p(x \mid z_t)} - \log \frac{p(x)}{1 - p(x)} + l_{t-1}(x)
\end{aligned} \tag{4.20}$$

Here $p(x)$ is the *prior* probability of the state x. As in (4.20), each measurement update involves the addition of the prior (in log odds form). The prior also defines the log odds of the initial belief before processing any sensor measurement:

$$l_0(x) = \log \frac{p(x)}{1 - p(x)} \tag{4.21}$$

4.3 The Particle Filter

4.3.1 Basic Algorithm

The *particle filter* is an alternative nonparametric implementation of the Bayes filter. Just like histogram filters, particle filters approximate the posterior by a finite number of parameters. However, they differ in the way these parameters are generated, and in which they populate the state space. The key idea of the particle filter is to represent the posterior $bel(x_t)$ by a set of random state samples drawn from this posterior. Figure 4.3 illustrates this idea for a Gaussian. Instead of representing the distribution by a parametric form—which would have been the exponential function that defines the density of a normal distribution—particle filters represent a distribution by a set of samples drawn from this distribution. Such a representation is approximate,

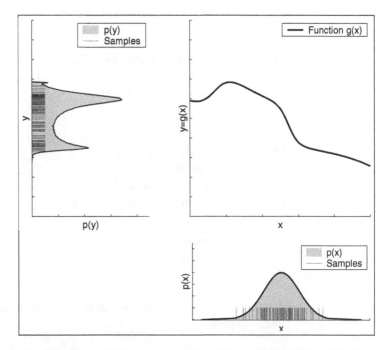

Figure 4.3 The "particle" representation used by particle filters. The lower right graph shows samples drawn from a Gaussian random variable, X. These samples are passed through the nonlinear function shown in the upper right graph. The resulting samples are distributed according to the random variable Y.

but it is nonparametric, and therefore can represent a much broader space of distributions than, for example, Gaussians. Another advantage of the sample based representation is its ability to model nonlinear transformations of random variables, as shown in Figure 4.3.

In particle filters, the samples of a posterior distribution are called *particles* and are denoted

(4.22) $$\mathcal{X}_t := x_t^{[1]}, x_t^{[2]}, \ldots, x_t^{[M]}$$

Each particle $x_t^{[m]}$ (with $1 \leq m \leq M$) is a concrete instantiation of the state at time t. Put differently, a particle is a hypothesis as to what the true world state may be at time t. Here M denotes the number of particles in the particle set $\mathcal{X}_t$. In practice, the number of particles M is often a large number, e.g., $M = 1,000$. In some implementations M is a function of t or other quantities related to the belief $bel(x_t)$.

```
1:      Algorithm Particle_filter($\mathcal{X}_{t-1}, u_t, z_t$):
2:          $\bar{\mathcal{X}}_t = \mathcal{X}_t = \emptyset$
3:          for $m = 1$ to $M$ do
4:              sample $x_t^{[m]} \sim p(x_t \mid u_t, x_{t-1}^{[m]})$
5:              $w_t^{[m]} = p(z_t \mid x_t^{[m]})$
6:              $\bar{\mathcal{X}}_t = \bar{\mathcal{X}}_t + \langle x_t^{[m]}, w_t^{[m]} \rangle$
7:          endfor
8:          for $m = 1$ to $M$ do
9:              draw $i$ with probability $\propto w_t^{[i]}$
10:             add $x_t^{[i]}$ to $\mathcal{X}_t$
11:         endfor
12:         return $\mathcal{X}_t$
```

Table 4.3 The particle filter algorithm, a variant of the Bayes filter based on importance sampling.

The intuition behind particle filters is to approximate the belief $bel(x_t)$ by the set of particles $\mathcal{X}_t$. Ideally, the likelihood for a state hypothesis x_t to be included in the particle set $\mathcal{X}_t$ shall be proportional to its Bayes filter posterior $bel(x_t)$:

$$x_t^{[m]} \sim p(x_t \mid z_{1:t}, u_{1:t}) \tag{4.23}$$

As a consequence of (4.23), the denser a subregion of the state space is populated by samples, the more likely it is that the true state falls into this region. As we will discuss below, the property (4.23) holds only asymptotically for $M \uparrow \infty$ for the standard particle filter algorithm. For finite M, particles are drawn from a slightly different distribution. In practice, this difference is negligible as long as the number of particles is not too small (e.g., $M \geq 100$).

Just like all other Bayes filter algorithms discussed thus far, the particle filter algorithm constructs the belief $bel(x_t)$ recursively from the belief $bel(x_{t-1})$ one time step earlier. Since beliefs are represented by sets of particles, this means that particle filters construct the particle set $\mathcal{X}_t$ recursively from the set $\mathcal{X}_{t-1}$.

The most basic variant of the particle filter algorithm is stated in Table 4.3. The input of this algorithm is the particle set $\mathcal{X}_{t-1}$, along with the most re-

4.3 The Particle Filter

cent control u_t and the most recent measurement z_t. The algorithm then first constructs a temporary particle set $\bar{\mathcal{X}}$ that represented the belief $\overline{bel}(x_t)$. It does this by systematically processing each particle $x_{t-1}^{[m]}$ in the input particle set $\mathcal{X}_{t-1}$. Subsequently, it transforms these particles into the set $\mathcal{X}_t$, which approximates the posterior distribution $bel(x_t)$. In detail:

1. Line 4 generates a hypothetical state $x_t^{[m]}$ for time t based on the particle $x_{t-1}^{[m]}$ and the control u_t. The resulting sample is indexed by m, indicating that it is generated from the m-th particle in $\mathcal{X}_{t-1}$. This step involves sampling from the state transition distribution $p(x_t \mid u_t, x_{t-1})$. To implement this step, one needs to be able to sample from this distribution. The set of particles obtained after M iterations is the filter's representation of $\overline{bel}(x_t)$.

IMPORTANCE FACTOR

2. Line 5 calculates for each particle $x_t^{[m]}$ the so-called *importance factor*, denoted $w_t^{[m]}$. Importance factors are used to incorporate the measurement z_t into the particle set. The importance, thus, is the probability of the measurement z_t under the particle $x_t^{[m]}$, given by $w_t^{[m]} = p(z_t \mid x_t^{[m]})$. If we interpret $w_t^{[m]}$ as the *weight* of a particle, the set of weighted particles represents (in approximation) the Bayes filter posterior $bel(x_t)$.

RESAMPLING

3. The real "trick" of the particle filter algorithm occurs in lines 8 through 11 in Table 4.3. These lines implemented what is known as *resampling* or *importance sampling*. The algorithm draws with replacement M particles from the temporary set $\bar{\mathcal{X}}_t$. The probability of drawing each particle is given by its importance weight. Resampling transforms a particle set of M particles into another particle set of the same size. By incorporating the importance weights into the resampling process, the distribution of the particles change: Whereas before the resampling step, they were distributed according to $\overline{bel}(x_t)$, after the resampling they are distributed (approximately) according to the posterior $bel(x_t) = \eta \, p(z_t \mid x_t^{[m]}) \overline{bel}(x_t)$. In fact, the resulting sample set usually possesses many duplicates, since particles are drawn with replacement. More important are the particles *not* contained in $\mathcal{X}_t$: Those tend to be the particles with lower importance weights.

The resampling step has the important function to force particles back to the posterior $bel(x_t)$. In fact, an alternative (and usually inferior) version of the particle filter would never resample, but instead would maintain for each particle an importance weight that is initialized by 1 and updated multiplicatively:

$$(4.24) \quad w_t^{[m]} = p(z_t \mid x_t^{[m]}) \, w_{t-1}^{[m]}$$

Such a particle filter algorithm would still approximate the posterior, but many of its particles would end up in regions of low posterior probability. As a result, it would require many more particles; how many depends on the shape of the posterior. The resampling step is a probabilistic implementation of the Darwinian idea of *survival of the fittest*: It refocuses the particle set to regions in state space with high posterior probability. By doing so, it focuses the computational resources of the filter algorithm to regions in the state space where they matter the most.

4.3.2 Importance Sampling

For the derivation of the particle filter, it shall prove useful to discuss the resampling step in more detail.

Intuitively, we are faced with the problem of computing an expectation over a probability density function f, but we are only given samples generated from a different probability density function, g. For example, we might be interested in the expectation that $x \in A$. We can express this probability as an expectation over g. Here I is the indicator function, which is 1 if its argument is true, and 0 otherwise.

$$(4.25) \quad \begin{aligned} E_f[I(x \in A)] &= \int f(x) \, I(x \in A) \, dx \\ &= \int \underbrace{\frac{f(x)}{g(x)}}_{=:w(x)} g(x) \, I(x \in A) \, dx \\ &= E_g[w(x) \, I(x \in A)] \end{aligned}$$

Here $w(x) = \frac{f(x)}{g(x)}$ is a weighting factor that accounts for the "mismatch" between f and g. For this equation to be correct, we need $f(x) > 0 \longrightarrow g(x) > 0$.

TARGET DISTRIBUTION

The importance sampling algorithm utilizes this transformation. Figure 4.4a shows a density function f of a probability distribution, which henceforth will be called the *target distribution*. As before, what we would like to achieve is to obtain a sample from f. However, sampling from f directly shall be impossible. We instead generate particles from a density g in

4.3 The Particle Filter

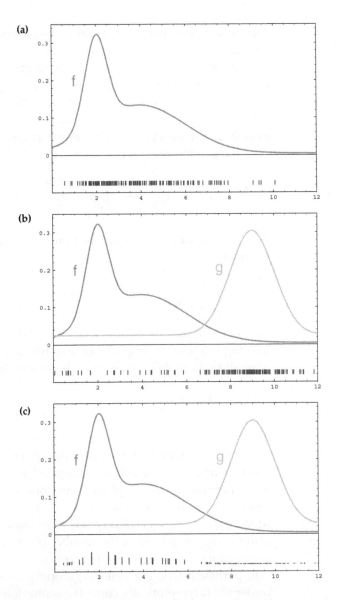

Figure 4.4 Illustration of importance factors in particle filters: (a) We seek to approximate the target density f. (b) Instead of sampling from f directly, we can only generate samples from a different density, g. Samples drawn from g are shown at the bottom of this diagram. (c) A sample of f is obtained by attaching the weight $f(x)/g(x)$ to each sample x. In particle filters, f corresponds to the belief $bel(x_t)$ and g to the belief $\overline{bel}(x_t)$.

PROPOSAL DISTRIBUTION

Figure 4.4b. The distribution that corresponds to the density g is called *proposal distribution*. The density g must be such that $f(x) > 0$ implies $g(x) > 0$, so that there is a non-zero probability to generate a particle when sampling from g for any state that might be generated by sampling from f. However, the resulting particle set, shown at the bottom of Figure 4.4b, is distributed according to g, not to f. In particular, for any interval $A \subseteq \text{dom}(X)$ (or more generally, any Borel set A) the empirical count of particles that fall into A converges to the integral of g under A:

$$(4.26) \quad \frac{1}{M} \sum_{m=1}^{M} I(x^{[m]} \in A) \longrightarrow \int_A g(x)\, dx$$

To offset this difference between f and g, particles $x^{[m]}$ are weighted by the quotient

$$(4.27) \quad w^{[m]} = \frac{f(x^{[m]})}{g(x^{[m]})}$$

This is illustrated by Figure 4.4c: The vertical bars in this figure indicate the magnitude of the importance weights. Importance weights are the non-normalized probability mass of each particle. In particular, we have

$$(4.28) \quad \left[\sum_{m=1}^{M} w^{[m]}\right]^{-1} \sum_{m=1}^{M} I(x^{[m]} \in A)\, w^{[m]} \longrightarrow \int_A f(x)\, dx$$

where the first term serves as the normalizer for all importance weights. In other words, even though we generated the particles from the density g, the appropriately weighted particles converge to the density f. It can be shown that under mild conditions, this approximation converges to the desired $E_f[I(x \in A)]$ for arbitrary sets A. In most cases, the rate of convergence is in $O(\frac{1}{\sqrt{M}})$, where M is the number of samples. The constant factor depends on the similarity of $f(x)$ and $g(x)$.

In particle filters, the density f corresponds to the target belief $bel(x_t)$. Under the (asymptotically correct) assumption that the particles in $\mathcal{X}_{t-1}$ are distributed according to $bel(x_{t-1})$, the density g corresponds to the product distribution:

$$(4.29) \quad p(x_t \mid u_t, x_{t-1})\, bel(x_{t-1})$$

Once again, this distribution is the proposal distribution.

4.3.3 Mathematical Derivation of the PF

To derive particle filters mathematically, it shall prove useful to think of particles as samples of state sequences

$$(4.30) \quad x_{0:t}^{[m]} = x_0^{[m]}, x_1^{[m]}, \ldots, x_t^{[m]}$$

It is easy to modify the algorithm accordingly: Simply append to the particle $x_t^{[m]}$ the sequence of state samples from which it was generated $x_{0:t-1}^{[m]}$. This particle filter calculates the posterior over all state sequences:

$$(4.31) \quad bel(x_{0:t}) = p(x_{0:t} \mid u_{1:t}, z_{1:t})$$

instead of the belief $bel(x_t) = p(x_t \mid u_{1:t}, z_{1:t})$. Admittedly, the space over all state sequences is huge, and covering it with particles is usually not such a good idea. However, this shall not deter us here, as this definition serves only as the means to derive the particle filter algorithm in Table 4.3.

The posterior $bel(x_{0:t})$ is obtained analogously to the derivation of $bel(x_t)$ in Chapter 2.4.3. In particular, we have

$$(4.32) \quad \begin{aligned} p(x_{0:t} \mid z_{1:t}, u_{1:t}) &\stackrel{\text{Bayes}}{=} \eta\, p(z_t \mid x_{0:t}, z_{1:t-1}, u_{1:t})\, p(x_{0:t} \mid z_{1:t-1}, u_{1:t}) \\ &\stackrel{\text{Markov}}{=} \eta\, p(z_t \mid x_t)\, p(x_{0:t} \mid z_{1:t-1}, u_{1:t}) \\ &= \eta\, p(z_t \mid x_t)\, p(x_t \mid x_{0:t-1}, z_{1:t-1}, u_{1:t})\, p(x_{0:t-1} \mid z_{1:t-1}, u_{1:t}) \\ &\stackrel{\text{Markov}}{=} \eta\, p(z_t \mid x_t)\, p(x_t \mid x_{t-1}, u_t)\, p(x_{0:t-1} \mid z_{1:t-1}, u_{1:t-1}) \end{aligned}$$

Notice the absence of integral signs in this derivation, which is the result of maintaining all states in the posterior, not just the most recent one as in Chapter 2.4.3.

The derivation is now carried out by induction. The initial condition is trivial to verify, assuming that our first particle set is obtained by sampling the prior $p(x_0)$. Let us assume that the particle set at time $t-1$ is distributed according to $bel(x_{0:t-1})$. For the m-th particle $x_{0:t-1}^{[m]}$ in this set, the sample $x_t^{[m]}$ generated in Step 4 of our algorithm is generated from the proposal distribution:

$$(4.33) \quad p(x_t \mid x_{t-1}, u_t)\, bel(x_{0:t-1}) = p(x_t \mid x_{t-1}, u_t)\, p(x_{0:t-1} \mid z_{1:t-1}, u_{1:t-1})$$

with

$$(4.34) \quad w_t^{[m]} = \frac{\text{target distribution}}{\text{proposal distribution}}$$

$$= \frac{\eta \, p(z_t \mid x_t) \, p(x_t \mid x_{t-1}, u_t) \, p(x_{0:t-1} \mid z_{1:t-1}, u_{1:t-1})}{p(x_t \mid x_{t-1}, u_t) \, p(x_{0:t-1} \mid z_{0:t-1}, u_{0:t-1})}$$
$$= \eta \, p(z_t \mid x_t)$$

The constant η plays no role since the resampling takes place with probabilities *proportional* to the importance weights. By resampling particles with probability proportional to $w_t^{[m]}$, the resulting particles are indeed distributed according to the product of the proposal and the importance weights $w_t^{[m]}$:

$$(4.35) \quad \eta \, w_t^{[m]} \, p(x_t \mid x_{t-1}, u_t) \, p(x_{0:t-1} \mid z_{0:t-1}, u_{0:t-1}) = bel(x_{0:t})$$

(Notice that the constant factor η here differs from the one in (4.34).) The algorithm in Table 4.4 follows now from the simple observation that if $x_{0:t}^{[m]}$ is distributed according to $bel(x_{0:t})$, then the state sample $x_t^{[m]}$ is (trivially) distributed according to $bel(x_t)$.

As we will argue below, this derivation is only correct for $M \uparrow \infty$, due to a laxness in our consideration of the normalization constants. However, even for finite M it explains the intuition behind the particle filter.

4.3.4 Practical Considerations and Properties of Particle Filters

Density Extraction

The sample sets maintained by particle filters represent discrete approximations of continuous beliefs. Many applications, however, require the availability of continuous estimates, that is, estimates not only at the states represented by particles, but at any point in the state space. The problem of extracting a continuous density from such samples is called density estimation. We will only informally illustrate some approaches to density estimation.

Figure 4.5 illustrates different ways of extracting a density from particles. The leftmost graph shows the particles and density of the transformed Gaussian from our standard example (c.f. Figure 4.3). A simple and highly efficient approach to extracting a density from such particles is to compute a *Gaussian approximation*, as illustrated by the dashed Gaussian in Figure 4.5(b). In this case, the Gaussian extracted from the particles is virtually identical to the Gaussian approximation of the true density (solid line).

Obviously, a Gaussian approximation captures only basic properties of a density, and it is only appropriate if the density is unimodal. Multimodal sample distributions require more complex techniques such as *k-means clus-*

K-MEANS ALGORITHM

tering, which approximates a density using mixtures of Gaussians. An alternative approach is illustrated in Figure 4.5(c). Here, a discrete *histogram* is superimposed over the state space and the probability of each bin is computed by summing the weights of the particles that fall into its range. As with histogram filters, an important shortcoming of this technique is the fact that the space complexity is exponential in the number of dimensions. On the other hand, histograms can represent multi-modal distributions, they can be computed extremely efficiently, and the density at any state can be extracted in time independent of the number of particles.

DENSITY TREE

The space complexity of histogram representations can be reduced significantly by generating a *density tree* from the particles, as discussed in Chapter 4.1.4. However, density trees come at the cost of more expensive lookups when extracting the density at any point in the state space (logarithmic in the depth of the tree).

KERNEL DENSITY ESTIMATION

Kernel density estimation is another way of converting a particle set into a continuous density. Here, each particle is used as the center of a so-called kernel, and the overall density is given by a mixture of the kernel densities. Figure 4.5(d) shows such a mixture density resulting from placing a Gaussian kernel at each particle. The advantage of kernel density estimates is their smoothness and algorithmic simplicity. However, the complexity of computing the density at any point is linear in the number of particles, or kernels.

Which of these density extraction techniques should be used in practice? This depends on the problem at hand. For example, in many robotics applications, processing power is very limited and the mean of the particles provides enough information to control the robot. Other applications, such as active localization, depend on more complex information about the uncertainty in the state space. In such situations, histograms or mixtures of Gaussians are a better choice. The combination of data collected by multiple robots sometimes requires the multiplication of densities underlying different sample sets. Density trees or kernel density estimates are well suited for this purpose.

Sampling Variance

An important source of error in the particle filter relates to the variation inherent in random sampling. Whenever a finite number of samples is drawn from a probability density, statistics extracted from these samples differ slightly from the statistics of the original density. For instance, if we draw

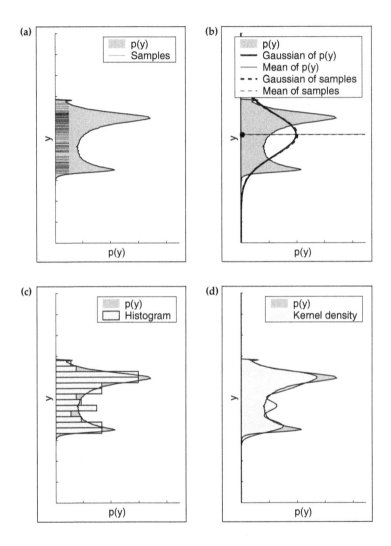

Figure 4.5 Different ways of extracting densities from particles. (a) Density and sample set approximation, (b) Gaussian approximation (mean and variance), (c) histogram approximation, (d) kernel density estimate. The choice of approximation strongly depends on the specific application and the computational resources.

4.3 The Particle Filter

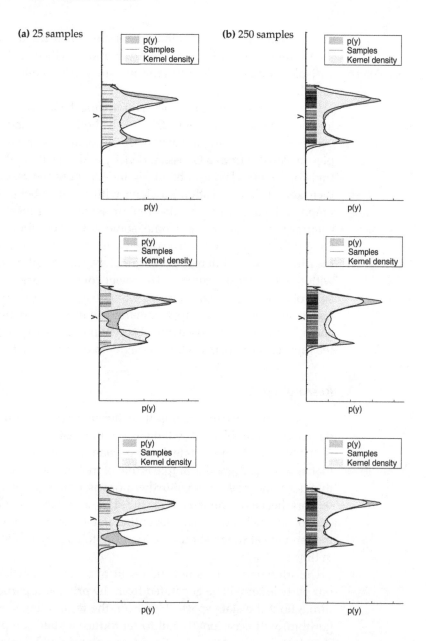

Figure 4.6 Variance due to random sampling. Samples are drawn from a Gaussian and passed through a nonlinear function. Samples and kernel estimates resulting from repeated sampling of 25 (left column) and 250 (right column) samples are shown. Each row shows one random experiment.

SAMPLE VARIANCE

samples from a Gaussian random variable, then the mean and variance of the samples will differ from the mean and variance of the original random variable. Variability due to random sampling is called the *variance* of the sampler.

Imagine two identical robots with identical, Gaussian beliefs performing identical, noise-free actions. Obviously, both robots should have the same belief after performing the action. To simulate this situation, we draw samples repeatedly from a Gaussian density and pass them through a nonlinear transformation. The graphs in Figure 4.6 show the resulting samples and their kernel density estimates along with the true belief (gray area). Each graph in the upper row results from drawing 25 samples from the Gaussian. Contrary to the desired outcome, some of the kernel density estimates differ substantially from the true density, and there is a large variability among the different kernel densities. Fortunately, the sampling variance decreases with the number of samples. The lower row in Figure 4.6 shows typical results obtained with 250 samples. Obviously, the higher number of samples results in more accurate approximations with less variability. In practice, if enough samples are chosen, the observations made by a robot typically keep the sample based belief "close enough" to the true belief.

Resampling

The sampling variance is amplified through repetitive resampling. To understand this problem, it will be useful to consider the extreme case, which is that of a robot whose state does not change. Sometimes, we know for a fact that $x_t = x_{t-1}$. A good example is that of mobile robot localization for a robot that does not move. Let us furthermore assume that the robot possesses no sensors, hence it cannot estimate the state, and that it is unaware of the state. Obviously, such a robot can never find out anything about its location, hence the estimate at time t should be identical to its initial estimate, for any point in time t.

Unfortunately, this is not the result of a vanilla particle filters. Initially, our particle set will be generated from the prior, and particles will be spread throughout the state space. However, the resampling step (line 8 in the algorithm) will occasionally fail to reproduce a state sample $x^{[m]}$. Since our state transition is deterministic, no new states will be introduced in the forward sampling step (line 4). As time goes on, more and more particles are erased simply due to the random nature of the resampling step, without the creation of any new particles. The result is quite daunting: With probability one, M identical copies of a single particle will survive; the diversity will disappear due to the repetitive resampling. To an outside observer, it may appear that the robot has uniquely determined the world state—an apparent contradiction to the fact that the robot possesses no sensors.

This example hints at another limitation of particle filters with important practical ramifications. In particular, the resampling process induces a loss of diversity in the particle population, which in fact manifests itself as approximation error: Even though the variance of the particle set itself decreases, the variance of the particle set as an estimator of the true belief increases. Controlling this variance, or error, of the particle filter is essential for any practical implementation.

VARIANCE REDUCTION

There exist two major strategies for *variance reduction*. First, one may reduce the frequency at which resampling takes place. When the state is known to be static ($x_t = x_{t-1}$) one should never resample. This is the case, for example, in mobile robot localization: When the robot stops, resampling should be suspended (and in fact it is usually a good idea to suspend the integration of measurements as well). Even if the state changes, it is often a good idea to reduce the frequency of resampling. Multiple measurements can always be integrated via multiplicatively updating the importance factor as noted above. More specifically, it maintains the importance weight in memory and updates them as follows:

$$(4.36) \quad w_t^{[m]} = \begin{cases} 1 & \text{if resampling took place} \\ p(z_t \mid x_t^{[m]})\, w_{t-1}^{[m]} & \text{if no resampling took place} \end{cases}$$

The choice of when to resample is intricate and requires practical experience: Resampling too often increases the risk of losing diversity. If one samples too infrequently, many samples might be wasted in regions of low probability. A standard approach to determining whether or not resampling should be performed is to measure the variance of the importance weights. The variance of the weights relates to the efficiency of the sample based representation. If all weights are identical, then the variance is zero and no resampling should be performed. If, on the other hand, the weights are concentrated on a small number of samples, then the weight variance is high and resampling should be performed.

LOW VARIANCE SAMPLING

The second strategy for reducing the sampling error is known as *low variance sampling*. Table 4.4 depicts an implementation of a low variance sampler. The basic idea is that instead of selecting samples independently of each other in the resampling process (as is the case for the basic particle filter in Table 4.3), the selection involves a sequential stochastic process.

```
1:      Algorithm Low_variance_sampler($\mathcal{X}_t, \mathcal{W}_t$):
2:          $\bar{\mathcal{X}}_t = \emptyset$
3:          $r = \text{rand}(0; M^{-1})$
4:          $c = w_t^{[1]}$
5:          $i = 1$
6:          for $m = 1$ to $M$ do
7:              $U = r + (m - 1) \cdot M^{-1}$
8:              while $U > c$
9:                  $i = i + 1$
10:                 $c = c + w_t^{[i]}$
11:             endwhile
12:             add $x_t^{[i]}$ to $\bar{\mathcal{X}}_t$
13:         endfor
14:         return $\bar{\mathcal{X}}_t$
```

Table 4.4 Low variance resampling for the particle filter. This routine uses a single random number to sample from the particle set $\mathcal{X}$ with associated weights $\mathcal{W}$, yet the probability of a particle to be resampled is still proportional to its weight. Furthermore, the sampler is efficient: Sampling M particles requires $O(M)$ time.

Instead of choosing M random numbers and selecting those particles that correspond to these random numbers, this algorithm computes a single random number and selects samples according to this number but still with a probability proportional to the sample weight. This is achieved by drawing a random number r in the interval $[0; M^{-1}]$, where M is the number of samples to be drawn at time t. The algorithm in Table 4.4 then selects particles by repeatedly adding the fixed amount M^{-1} to r and by choosing the particle that corresponds to the resulting number. Any number U in $[0; 1]$ points to exactly one particle, namely the particle i for which

$$(4.37) \quad i = \operatorname*{argmin}_{j} \sum_{m=1}^{j} w_t^{[m]} \geq U$$

The while loop in Table 4.4 serves two tasks, it computes the sum in the right-hand side of this equation and additionally checks whether i is the index of the first particle such that the corresponding sum of weights exceeds U.

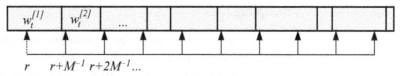

Figure 4.7 Principle of the low variance resampling procedure. We choose a random number r and then select those particles that correspond to $u = r + (m-1) \cdot M^{-1}$ where $m = 1, \ldots, M$.

The selection is then carried out in line 12. This process is also illustrated in Figure 4.7.

The advantage of the low-variance sampler is threefold. First, it covers the space of samples in a more systematic fashion than the independent random sampler. This should be obvious from the fact that the dependent sampler cycles through all particles systematically, rather than choosing them independently at random. Second, if all the samples have the same importance factors, the resulting sample set $\bar{\mathcal{X}}_t$ is equivalent to $\mathcal{X}_t$ so that no samples are lost if we resample without having integrated an observation into $\mathcal{X}_t$. Third, the low-variance sampler has a complexity of $O(M)$. Achieving the same complexity for independent sampling is difficult; obvious implementations require a $O(\log M)$ search for each particle once a random number has been drawn, which results in a complexity of $O(M \log M)$ for the entire resampling process. Computation time is of essence when using particle filters, and often an efficient implementation of the resampling process can make a huge difference in the practical performance. For these reasons, implementations of particle filters in robotics tend to rely on mechanisms like the one just discussed.

STRATIFIED SAMPLING

In general, the literature on efficient sampling is huge. Another popular option is *stratified sampling*, in which particles are grouped into subsets. Sampling from these sets is performed in a two stage procedure. First, the number of samples drawn from each subset is determined based on the total weight of the particles contained in the subset. In the second stage, individual samples are drawn randomly from each subset using, for example, low variance resampling. Such a technique has lower sampling variance and tends to perform well when a robot tracks multiple, distinct hypotheses with a single particle filter.

Sampling Bias

The fact that only finitely many particles are used also introduces a systematic *bias* in the posterior estimate. Consider the extreme case of $M = 1$ particle. In this case, the loop in lines 3 through 7 in Table 4.3 will only be executed once, and $\bar{\mathcal{X}}_t$ will contain only a single particle, sampled from the motion model. The key insight is that the resampling step (lines 8 through 11 in Table 4.3) will now *deterministically* accept this sample, regardless of its importance factor $w_t^{[m]}$. Hence the measurement probability $p(z_t \mid x_t^{[m]})$ plays no role in the result of the update, and neither does z_t. Thus, if $M = 1$, the particle filter generates particles from the probability $p(x_t \mid u_{1:t})$ instead of the desired posterior $p(x_t \mid u_{1:t}, z_{1:t})$. It flatly ignores all measurements. How can this happen?

The culprit is the normalization, implicit in the resampling step. When sampling in proportion to the importance weights (line 9 of the algorithm), $w_t^{[m]}$ becomes its own normalizer if $M = 1$:

$$(4.38) \quad p(\text{draw } x_t^{[m]} \text{ in line 9}) = \frac{w_t^{[m]}}{w_t^{[m]}} = 1$$

In general, the problem is that the non-normalized values $w_t[m]$ are drawn from an M-dimensional space, but after normalization they reside in a space of dimension $M - 1$. This is because after normalization, the m-th weight can be recovered from the $M - 1$ other weights by subtracting those from 1. Fortunately, for larger values of M, the effect of loss of dimensionality, or degrees of freedom, becomes less and less pronounced.

Particle Deprivation

Even with a large number of particles, it may happen that there are no particles in the vicinity of the correct state. This problem is known as the *particle deprivation problem*. It occurs mostly when the number of particles is too small to cover all relevant regions with high likelihood. However, one might argue that this ultimately must happen in any particle filter, regardless of the particle set size M.

Particle deprivation occurs as the result of the variance in random sampling; an unlucky series of random numbers can wipe out all particles near the true state. At each sampling step, the probability for this to happen is larger than zero (although it is usually exponentially small in M). Thus, we only have to run the particle filter long enough. Eventually we will generate an estimate that is arbitrarily incorrect.

In practice, problems of this nature only tend to arise when M is small relative to the space of all states with high likelihood. A popular solution to the particle deprivation problem is to add a small number of randomly generated particles into the set after each resampling process, regardless of the actual sequence of motion and measurement commands. Such a methodology can reduce (but not fix) the deprivation problem, but at the expense of an incorrect posterior estimate. The advantage of adding random samples lies in its simplicity: The software modification necessary to add random samples in a particle filter is minimal. As a rule of thumb, adding random samples should be considered a measure of last resort, which should only be applied if all other techniques for fixing a deprivation problem have failed. Alternative approaches to dealing with particle deprivation will be discussed in Chapter 8, in the context of robot localization.

This discussion showed that the quality of the sample based representation increases with the number of samples. An important question is therefore how many samples should be used for a specific estimation problem. Unfortunately, there is no perfect answer to this question and it is often left to the user to determine the required number of samples. As a rule of thumb, the number of samples strongly depends on the dimensionality of the state space and the uncertainty of the distributions approximated by the particle filter. For example, uniform distributions require many more samples than distributions focused on a small region of the state space. A more detailed discussion on sample sizes will be given in the context of robot localization and mapping in future chapters of this book.

4.4 Summary

This section introduced two nonparametric Bayes filters, histogram filters and particle filters. Nonparametric filters approximate the posterior by a finite number of values. Under mild assumptions on the system model and the shape of the posterior, both have the property that the approximation error converges uniformly to zero as the the number of values used to represent the posterior goes to infinity.

- The histogram filter decomposes the state space into finitely many convex regions. It represents the cumulative posterior probability of each region by a single numerical value.

- There exist many decomposition techniques in robotics. In particular, the granularity of a decomposition may or may not depend on the structure of the environment. When it does, the resulting algorithms are often called 'topological.'

- Decomposition techniques can be divided into static and dynamic. Static decompositions are made in advance, irrespective of the shape of the belief. Dynamic decompositions rely on specifics of the robot's belief when decomposing the state space, often attempting to increase spatial resolution in proportion to the posterior probability. Dynamic decompositions tend to give better results, but they are also more difficult to implement.

- An alternative nonparametric technique is known as particle filter. Particle filters represent posteriors by a random sample of states, drawn from the posterior. Such samples are called particles. Particle filters are extremely easy to implement, and they are the most versatile of all Bayes filter algorithms represented in this book.

- Specific strategies exist to reduce the error in particle filters. Among the most popular ones are techniques for reducing the variance of the estimate that arises from the randomness of the algorithm, and techniques for adapting the number of particles in accordance with the complexity of the posterior.

The filter algorithms discussed in this and the previous chapter lay the groundwork for most probabilistic robotics algorithms discussed throughout the remainder of this book. The material presented here represents many of today's most popular algorithms and representations in probabilistic robotics.

4.5 Bibliographical Remarks

West and Harrison (1997) provides an in-depth treatment of several techniques discussed in this and the previous chapter. Histograms have been used in statistics for many decades. Sturges (1926) provides one of the early rules for selecting the resolution of a histogram approximation, and more recent treatment is by Freedman and Diaconis (1981). A contemporary analysis can be found in Scott (1992). Once a state space is mapped into a discrete histogram, the resulting temporal inference problem becomes an instance of a discrete Hidden Markov model, of the type made popular by Rabiner and Juang (1986). Two contemporary texts can be found in MacDonald and Zucchini (1997) and Elliott et al. (1995).

Particle filters can be traced back to Metropolis and Ulam (1949), the inventors of Monte Carlo methods; see Rubinstein (1981) for a more contemporary introduction. The sampling importance resampling technique, which is part of the particle filter, goes back to two seminal papers by Rubin (1988) and Smith and Gelfand (1992). Stratified sampling was first invented by Neyman (1934). In the past few years, particle filters have been studied extensively in the field of Bayesian statistics (Doucet 1998; Kitagawa 1996; Liu and Chen 1998; Pitt and Shephard 1999). In AI, particle filters were reinvented under the name *survival of the fittest* (Kanazawa et al. 1995); in computer vision, an algorithm called *condensation* by Isard and Blake (1998) applies them to tracking problems. A good contemporary text on particle filters is due to Doucet et al. (2001).

4.6 Exercises

1. In this exercise, you will be asked to implement a histogram filter for a linear dynamical system studied in the previous chapter.

    (a) Implement a histogram filter for the dynamical system described in Exercise 1 of the previous chapter (see page 81). Use the filter to predict a sequence of posterior distributions for $t = 1, 2, \ldots, 5$. For each value of t, plot the joint posterior over x and $\dot{x}$ into a diagram, where x is the horizontal and $\dot{x}$ is the vertical axis.

    (b) Now implement the measurement update step into your histogram filter, as described in Exercise 2 of the previous chapter (page 82). Suppose at time $t = 5$, we observe a measurement $z = 5$. State and plot the posterior before and after updating the histogram filter.

2. You are now asked to implement the histogram filter for the nonlinear studied in Exercise 4 in the previous chapter (page 83). There, we studied a nonlinear system defined over three state variables, and with the deterministic state transition

$$\begin{pmatrix} x' \\ y' \\ \theta' \end{pmatrix} = \begin{pmatrix} x + \cos\theta \\ y + \sin\theta \\ \theta \end{pmatrix}$$

The initial state estimate was as follows:

$$\mu = \begin{pmatrix} 0 & 0 & 0 \end{pmatrix} \text{ and } \Sigma = \begin{pmatrix} 0.01 & 0 & 0 \\ 0 & 0.01 & 0 \\ 0 & 0 & 10000 \end{pmatrix}$$

    (a) Propose a suitable initial estimate for a histogram filter, which reflects the state of knowledge in the Gaussian prior.

(b) Implement a histogram filter and run its prediction step. Compare the resulting posterior with the one from the EKF and from your intuitive analysis. What can you learn about the resolution of the x-y coordinates and the orientation θ in your histogram filter?

(c) Now incorporate a measurement into your estimate. As before, the measurement shall be a noisy projection of the x-coordinate of the robot, with covariance $Q = 0.01$. Implement the step, compute the result, plot it, and compare it with the result of the EKF and your intuitive drawing.

Notice: When plotting the result of a histogram filter, you can show multiple density plots, one for each discrete slice in the space of all θ-values.

3. We talked about the effect of using a single particle. What is the effect of using $M = 2$ particles in particle filtering? Can you give an example where the posterior will be biased? If so, by what amount?

4. Implement Exercise 1 using particle filters instead of histograms, and plot and discuss the results.

5. Implement Exercise 2 using particle filters instead of histograms, and plot and discuss the results. Investigate the effect of varying numbers of particles on the result.

5 Robot Motion

5.1 Introduction

This and the next chapter describe the two remaining components for implementing the filter algorithms described thus far: the motion and the measurement models. This chapter focuses on the motion model. *Motion models* comprise the state transition probability $p(x_t \mid u_t, x_{t-1})$, which plays an essential role in the prediction step of the Bayes filter. This chapter provides in-depth examples of probabilistic motion models as they are being used in actual robotics implementations. The subsequent chapter will describe probabilistic models of sensor measurements $p(z_t \mid x_t)$, which are essential for the measurement update step. The material presented here will be essential for *implementing* any of the algorithms described in subsequent chapters.

Robot kinematics, which is the central topic of this chapter, has been studied thoroughly in past decades. However, it has almost exclusively been addressed in deterministic form. Probabilistic robotics generalizes kinematic equations to the fact that the outcome of a control is uncertain, due to control noise or unmodeled exogenous effects. Following the theme of this book, our description will be probabilistic: The outcome of a control will be described by a posterior probability. In doing so, the resulting models will be amenable to the probabilistic state estimation techniques described in the previous chapters.

Our exposition focuses entirely on mobile robot kinematics for robots operating in planar environments. In this way, it is much more specific than most contemporary treatments of kinematics. No model of manipulator kinematics will be provided, neither will we discuss models of robot dynamics. However, this restricted choice of material is by no means to be interpreted that probabilistic ideas are limited to simple kinematic models of mobile robots. Rather, it is descriptive of the present state of the art, as probabilistic techniques have enjoyed their biggest successes in mobile robotics using relatively basic models of the types described in this chapter. The use of more sophisticated probabilistic models (e.g., probabilistic models of robot dynamics) remains largely unexplored in the literature. Such extensions, however, are not infeasible. As this chapter illustrates, deterministic robot actuator models are "probilified" by adding noise variables that characterize the types of uncertainty that exist in robotic actuation.

In theory, the goal of a proper probabilistic model may appear to accurately model the specific types of uncertainty that exist in robot actuation and perception. In practice, the exact shape of the model often seems to be less important than the fact that some provisions for uncertain outcomes are provided in the first place. In fact, many of the models that have proven most successful in practical applications vastly overestimate the amount of uncertainty. By doing so, the resulting algorithms are more robust to violations of the Markov assumptions (Chapter 2.4.4), such as unmodeled state and the effect of algorithmic approximations. We will point out such findings in later chapters, when discussing actual implementations of probabilistic robotic algorithms.

5.2 Preliminaries

5.2.1 Kinematic Configuration

CONFIGURATION

Kinematics is the calculus describing the effect of control actions on the configuration of a robot. The *configuration* of a rigid mobile robot is commonly described by six variables, its three-dimensional Cartesian coordinates and its three Euler angles (roll, pitch, yaw) relative to an external coordinate frame. The material presented in this book is largely restricted to mobile robots operating in planar environments, whose kinematic state is summarized by three variables, referred to as pose in this text.

POSE

The *pose* of a mobile robot operating in a plane is illustrated in Figure 5.1. It comprises its two-dimensional planar coordinates relative to an external coordinate frame, along with its angular orientation. Denoting the former as x and y (not to be confused with the state variable x_t), and the latter by θ, the

5.2 Preliminaries

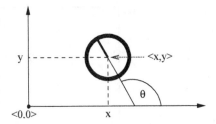

Figure 5.1 Robot pose, shown in a global coordinate system.

pose of the robot is described by the following vector:

(5.1) $$\begin{pmatrix} x \\ y \\ \theta \end{pmatrix}$$

BEARING   The orientation of a robot is often called *bearing*, or *heading direction*. As shown in Figure 5.1, we postulate that a robot with orientation $\theta = 0$ points into the direction of its x-axis. A robot with orientation $\theta = .5\pi$ points into the direction of its y-axis.

LOCATION   Pose without orientation will be called *location*. The concept of location will be important in the next chapter, when we discuss measures to describe robot environments. For simplicity, locations in this book are usually described by two-dimensional vectors, which refer to the x-y coordinates of an object:

(5.2) $$\begin{pmatrix} x \\ y \end{pmatrix}$$

The pose and the locations of objects in the environment may constitute the kinematic state x_t of the robot-environment system.

5.2.2 Probabilistic Kinematics

The probabilistic kinematic model, or *motion model* plays the role of the state transition model in mobile robotics. This model is the familiar conditional density

(5.3) $p(x_t \mid u_t, x_{t-1})$

Here x_t and x_{t-1} are both robot poses (and not just its x-coordinates), and u_t is a motion command. This model describes the posterior distribution

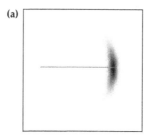 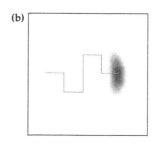

Figure 5.2 The motion model: Posterior distributions of the robot's pose upon executing the motion command illustrated by the solid line. The darker a location, the more likely it is. This plot has been projected into 2-D. The original density is three-dimensional, taking the robot's heading direction θ into account.

over kinematic states that a robot assumes when executing the motion command u_t at x_{t-1}. In implementations, u_t is sometimes provided by a robot's odometry. However, for conceptual reasons we will refer to u_t as control.

Figure 5.2 shows two examples that illustrate the kinematic model for a rigid mobile robot operating in a planar environment. In both cases, the robot's initial pose is x_{t-1}. The distribution $p(x_t \mid u_t, x_{t-1})$ is visualized by the shaded area: The darker a pose, the more likely it is. In this figure, the posterior pose probability is projected into x-y-space; the figure lacks a dimension corresponding to the robot's orientation. In Figure 5.2a, a robot moves forward some distance, during which it may accrue translational and rotational error as indicated. Figure 5.2b shows the resulting distribution of a more complicated motion command, which leads to a larger spread of uncertainty.

This chapter provides in detail two specific probabilistic motion models $p(x_t \mid u_t, x_{t-1})$, both for mobile robots operating in the plane. Both models are somewhat complementary in the type of motion information that is being processed. The first assumes that the motion data u_t specifies the velocity commands given to the robot's motors. Many commercial mobile robots (e.g., differential drive, synchro drive) are actuated by independent translational and rotational velocities, or are best thought of being actuated in this way. The second model assumes that one has access to odometry information. Most commercial bases provide odometry using kinematic information (distance traveled, angle turned). The resulting probabilistic model for integrating such information is somewhat different from the velocity model.

In practice, odometry models tend to be more accurate than velocity models, for the simple reason that most commercial robots do not execute velocity commands with the level of accuracy that can be obtained by measuring the revolution of the robot's wheels. However, odometry is only available after executing a motion command. Hence it cannot be used for motion planning. Planning algorithms such as collision avoidance have to predict the effects of motion. Thus, odometry models are usually applied for estimation, whereas velocity models are used for probabilistic motion planning.

5.3 Velocity Motion Model

The *velocity motion model* assumes that we can control a robot through two velocities, a rotational and a translational velocity. Many commercial robots offer control interfaces where the programmer specifies velocities. Drive trains commonly controlled in this way include differential drives, Ackerman drives, and synchro-drives. Drive systems not covered by our model are those without non-holonomic constraints, such as robots equipped with Mecanum wheels or legged robots.

We will denote the *translational velocity* at time t by v_t, and the *rotational velocity* by ω_t. Hence, we have

$$(5.4) \quad u_t = \begin{pmatrix} v_t \\ \omega_t \end{pmatrix}$$

We arbitrarily postulate that positive rotational velocities ω_t induce a counterclockwise rotation (left turns). Positive translational velocities v_t correspond to forward motion.

5.3.1 Closed Form Calculation

A possible algorithm for computing the probability $p(x_t \mid u_t, x_{t-1})$ is shown in Table 5.1. It accepts as input an initial pose $x_{t-1} = (x \ y \ \theta)^T$, a control $u_t = (v \ \omega)^T$, and a hypothesized successor pose $x_t = (x' \ y' \ \theta')^T$. It outputs the probability $p(x_t \mid u_t, x_{t-1})$ of being at x_t after executing control u_t beginning in state x_{t-1}, assuming that the control is carried out for the fixed duration Δt. The parameters α_1 to α_6 are robot-specific motion error parameters. The algorithm in Table 5.1 first calculates the controls of an error-free robot; the meaning of the individual variables in this calculation will become more apparent below, when we derive it. These parameters are given by $\hat{v}$ and $\hat{\omega}$.

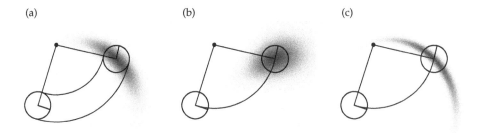

Figure 5.3 The velocity motion model, for different noise parameter settings.

The function **prob**(x, b^2) models the motion error. It computes the probability of its parameter x under a zero-centered random variable with variance b^2. Two possible implementations are shown in Table 5.2, for error variables with normal distribution and triangular distribution, respectively.

Figure 5.3 shows graphical examples of the velocity motion model, projected into x-y-space. In all three cases, the robot sets the same translational and angular velocity. Figure 5.3a shows the resulting distribution with moderate error parameters α_1 to α_6. The distribution shown in Figure 5.3b is obtained with smaller angular error (parameters α_3 and α_4) but larger translational error (parameters α_1 and α_2). Figure 5.3c shows the distribution under large angular and small translational error.

5.3.2 Sampling Algorithm

For particle filters (c.f. Chapter 4.3), it suffices to sample from the motion model $p(x_t \mid u_t, x_{t-1})$, instead of computing the posterior for arbitrary x_t, u_t and x_{t-1}. *Sampling* from a conditional density is different than calculating the density: In sampling, one is given u_t and x_{t-1} and seeks to generate a random x_t drawn according to the motion model $p(x_t \mid u_t, x_{t-1})$. When calculating the density, one is also given x_t generated through other means, and one seeks to compute the probability of x_t under $p(x_t \mid u_t, x_{t-1})$.

The algorithm **sample_motion_model_velocity** in Table 5.3 generates random samples from $p(x_t \mid u_t, x_{t-1})$ for a fixed control u_t and pose x_{t-1}. It accepts x_{t-1} and u_t as input and generates a random pose x_t according to the distribution $p(x_t \mid u_t, x_{t-1})$. Line 2 through 4 "perturb" the commanded control parameters by noise, drawn from the error parameters of the kinematic motion model. The noise values are then used to generate the sample's

5.3 Velocity Motion Model

1:	**Algorithm motion_model_velocity**(x_t, u_t, x_{t-1}):
2:	$\mu = \dfrac{1}{2} \dfrac{(x - x') \cos\theta + (y - y') \sin\theta}{(y - y') \cos\theta - (x - x') \sin\theta}$
3:	$x^* = \dfrac{x + x'}{2} + \mu(y - y')$
4:	$y^* = \dfrac{y + y'}{2} + \mu(x' - x)$
5:	$r^* = \sqrt{(x - x^*)^2 + (y - y^*)^2}$
6:	$\Delta\theta = \operatorname{atan2}(y' - y^*, x' - x^*) - \operatorname{atan2}(y - y^*, x - x^*)$
7:	$\hat{v} = \dfrac{\Delta\theta}{\Delta t} r^*$
8:	$\hat{\omega} = \dfrac{\Delta\theta}{\Delta t}$
9:	$\hat{\gamma} = \dfrac{\theta' - \theta}{\Delta t} - \hat{\omega}$
10:	**return** $\mathbf{prob}(v - \hat{v}, \alpha_1 v^2 + \alpha_2 \omega^2) \cdot \mathbf{prob}(\omega - \hat{\omega}, \alpha_3 v^2 + \alpha_4 \omega^2)$ $\cdot \mathbf{prob}(\hat{\gamma}, \alpha_5 v^2 + \alpha_6 \omega^2)$

Table 5.1 Algorithm for computing $p(x_t \mid u_t, x_{t-1})$ based on velocity information. Here we assume x_{t-1} is represented by the vector $(x \; y \; \theta)^T$; x_t is represented by $(x' \; y' \; \theta')^T$; and u_t is represented by the velocity vector $(v \; \omega)^T$. The function $\mathbf{prob}(a, b^2)$ computes the probability of its argument a under a zero-centered distribution with variance b^2. It may be implemented using any of the algorithms in Table 5.2.

1:	**Algorithm prob_normal_distribution**(a, b^2):
2:	**return** $\dfrac{1}{\sqrt{2\pi b^2}} \exp\left\{-\dfrac{1}{2}\dfrac{a^2}{b^2}\right\}$
3:	**Algorithm prob_triangular_distribution**(a, b^2):
4:	**return** $\max\left\{0, \dfrac{1}{\sqrt{6}\,b} - \dfrac{|a|}{6\,b^2}\right\}$

Table 5.2 Algorithms for computing densities of a zero-centered normal distribution and a triangular distribution with variance b^2.

1:	**Algorithm sample_motion_model_velocity**(u_t, x_{t-1}):
2:	$\hat{v} = v + \mathbf{sample}(\alpha_1 v^2 + \alpha_2 \omega^2)$
3:	$\hat{\omega} = \omega + \mathbf{sample}(\alpha_3 v^2 + \alpha_4 \omega^2)$
4:	$\hat{\gamma} = \mathbf{sample}(\alpha_5 v^2 + \alpha_6 \omega^2)$
5:	$x' = x - \frac{\hat{v}}{\hat{\omega}} \sin\theta + \frac{\hat{v}}{\hat{\omega}} \sin(\theta + \hat{\omega}\Delta t)$
6:	$y' = y + \frac{\hat{v}}{\hat{\omega}} \cos\theta - \frac{\hat{v}}{\hat{\omega}} \cos(\theta + \hat{\omega}\Delta t)$
7:	$\theta' = \theta + \hat{\omega}\Delta t + \hat{\gamma}\Delta t$
8:	return $x_t = (x', y', \theta')^T$

Table 5.3 Algorithm for sampling poses $x_t = (x' \ y' \ \theta')^T$ from a pose $x_{t-1} = (x \ y \ \theta)^T$ and a control $u_t = (v \ \omega)^T$. Note that we are perturbing the final orientation by an additional random term, $\hat{\gamma}$. The variables α_1 through α_6 are the parameters of the motion noise. The function **sample**(b^2) generates a random sample from a zero-centered distribution with variance b^2. It may, for example, be implemented using the algorithms in Table 5.4.

1:	**Algorithm sample_normal_distribution**(b^2):
2:	return $\frac{1}{2} \sum_{i=1}^{12} \mathbf{rand}(-b, b)$
3:	**Algorithm sample_triangular_distribution**(b^2):
4:	return $\frac{\sqrt{6}}{2} [\mathbf{rand}(-b, b) + \mathbf{rand}(-b, b)]$

Table 5.4 Algorithm for sampling from (approximate) normal and triangular distributions with zero mean and variance b^2; see Winkler (1995: p293). The function **rand**(x, y) is assumed to be a pseudo random number generator with uniform distribution in $[x, y]$.

5.3 Velocity Motion Model

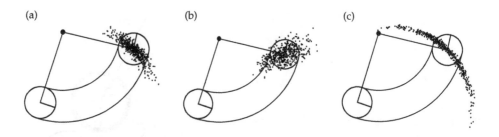

Figure 5.4 Sampling from the velocity motion model, using the same parameters as in Figure 5.3. Each diagram shows 500 samples.

new pose, in lines 5 through 7. Thus, the sampling procedure implements a simple physical robot motion model that incorporates control noise in its prediction, in just about the most straightforward way. Figure 5.4 illustrates the outcome of this sampling routine. It depicts 500 samples generated by **sample_motion_model_velocity**. The reader might want to compare this figure with the density depicted in Figure 5.3.

We note that in many cases, it is easier to sample x_t than calculate the density of a given x_t. This is because samples require only a forward simulation of the physical motion model. To compute the probability of a hypothetical pose amounts to retro-guessing of the error parameters, which requires us to calculate the inverse of the physical motion model. The fact that particle filters rely on sampling makes them specifically attractive from an implementation point of view.

5.3.3 Mathematical Derivation of the Velocity Motion Model

We will now derive the algorithms **motion_model_velocity** and **sample_motion_model_velocity**. As usual, the reader not interested in the mathematical details is invited to skip this section at first reading, and continue in Chapter 5.4 (page 132). The derivation begins with a generative model of robot motion, and then derives formulae for sampling and computing $p(x_t \mid u_t, x_{t-1})$ for arbitrary x_t, u_t, and x_{t-1}.

Exact Motion

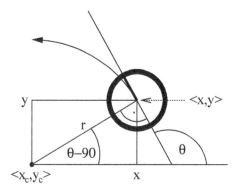

Figure 5.5 Motion carried out by a noise-free robot moving with constant velocities v and ω and starting at $(x\ y\ \theta)^T$.

Before turning to the probabilistic case, let us begin by stating the kinematics for an ideal, noise-free robot. Let $u_t = (v\ \omega)^T$ denote the control at time t. If both velocities are kept at a fixed value for the entire time interval $(t-1, t]$, the robot moves on a circle with radius

$$(5.5) \quad r = \left|\frac{v}{\omega}\right|$$

This follows from the general relationship between the translational and rotational velocities v and ω for an arbitrary object moving on a circular trajectory with radius r:

$$(5.6) \quad v = \omega \cdot r$$

Equation (5.5) encompasses the case where the robot does not turn at all (i.e., $\omega = 0$), in which case the robot moves on a straight line. A straight line corresponds to a circle with infinite radius, hence we note that r may be infinite.

Let $x_{t-1} = (x, y, \theta)^T$ be the initial pose of the robot, and suppose we keep the velocity constant at $(v\ \omega)^T$ for some time Δt. As one easily shows, the center of the circle is at

$$(5.7) \quad x_c = x - \frac{v}{\omega}\sin\theta$$
$$(5.8) \quad y_c = y + \frac{v}{\omega}\cos\theta$$

The variables $(x_c \ y_c)^T$ denote this coordinate. After Δt time of motion, our ideal robot will be at $x_t = (x', y', \theta')^T$ with

$$
(5.9) \quad \begin{pmatrix} x' \\ y' \\ \theta' \end{pmatrix} = \begin{pmatrix} x_c + \frac{v}{\omega} \sin(\theta + \omega \Delta t) \\ y_c - \frac{v}{\omega} \cos(\theta + \omega \Delta t) \\ \theta + \omega \Delta t \end{pmatrix}
$$

$$
= \begin{pmatrix} x \\ y \\ \theta \end{pmatrix} + \begin{pmatrix} -\frac{v}{\omega} \sin\theta + \frac{v}{\omega} \sin(\theta + \omega \Delta t) \\ \frac{v}{\omega} \cos\theta - \frac{v}{\omega} \cos(\theta + \omega \Delta t) \\ \omega \Delta t \end{pmatrix}
$$

The derivation of this expression follows from simple trigonometry: After Δt units of time, the noise-free robot has progressed $v \cdot \Delta t$ along the circle, which caused its heading direction to turn by $\omega \cdot \Delta t$. At the same time, its x and y coordinate is given by the intersection of the circle about $(x_c \ y_c)^T$, and the ray starting at $(x_c \ y_c)^T$ at the angle perpendicular to $\omega \cdot \Delta t$. The second transformation simply substitutes (5.8) into the resulting motion equations.

Of course, real robots cannot jump from one velocity to another, and keep velocity constant in each time interval. To compute the kinematics with non-constant velocities, it is therefore common practice to use small values for Δt, and to approximate the actual velocity by a constant within each time interval. The (approximate) final pose is then obtained by concatenating the corresponding cyclic trajectories using the mathematical equations just stated.

Real Motion

In reality, robot motion is subject to noise. The actual velocities differ from the commanded ones (or measured ones, if the robot possesses a sensor for measuring velocity). We will model this difference by a zero-centered random variable with finite variance. More precisely, let us assume the actual velocities are given by

$$
(5.10) \quad \begin{pmatrix} \hat{v} \\ \hat{\omega} \end{pmatrix} = \begin{pmatrix} v \\ \omega \end{pmatrix} + \begin{pmatrix} \varepsilon_{\alpha_1 v^2 + \alpha_2 \omega^2} \\ \varepsilon_{\alpha_3 v^2 + \alpha_4 \omega^2} \end{pmatrix}
$$

Here ε_{b^2} is a zero-mean error variable with variance b^2. Thus, the true velocity equals the commanded velocity plus some small, additive error (noise). In our model, the standard deviation of the error is proportional to the commanded velocity. The parameters α_1 to α_4 (with $\alpha_i \geq 0$ for $i = 1, \ldots, 4$) are robot-specific error parameters. They model the accuracy of the robot. The less accurate a robot, the larger these parameters.

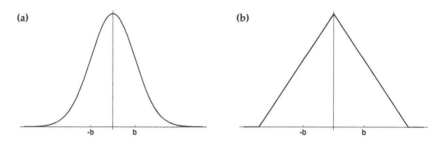

Figure 5.6 Probability density functions with variance b^2: (a) Normal distribution, (b) triangular distribution.

Two common choices for the error ε_{b^2} are the normal and the triangular distribution.

NORMAL DISTRIBUTION

The *normal distribution* with zero mean and variance b^2 is given by the density function

$$(5.11) \quad \varepsilon_{b^2}(a) = \frac{1}{\sqrt{2\pi b^2}} e^{-\frac{1}{2}\frac{a^2}{b^2}}$$

Figure 5.6a shows the density function of a normal distribution with variance b^2. Normal distributions are commonly used to model noise in continuous stochastic processes. Its support, which is the set of points a with $p(a) > 0$, is $\Re$.

TRIANGULAR DISTRIBUTION

The density of a *triangular distribution* with zero mean and variance b^2 is given by

$$(5.12) \quad \varepsilon_{b^2}(a) = \max\left\{0, \frac{1}{\sqrt{6}\,b} - \frac{|a|}{6\,b^2}\right\}$$

which is non-zero only in $(-\sqrt{6b}; \sqrt{6b})$. As Figure 5.6b suggests, the density resembles the shape of a symmetric triangle—hence the name.

A better model of the actual pose $x_t = (x'\ y'\ \theta')^T$ after executing the motion command $u_t = (v\ \omega)^T$ at $x_{t-1} = (x\ y\ \theta)^T$ is thus

$$(5.13) \quad \begin{pmatrix} x' \\ y' \\ \theta' \end{pmatrix} = \begin{pmatrix} x \\ y \\ \theta \end{pmatrix} + \begin{pmatrix} -\frac{\hat{v}}{\hat{\omega}}\sin\theta + \frac{\hat{v}}{\hat{\omega}}\sin(\theta + \hat{\omega}\Delta t) \\ \frac{\hat{v}}{\hat{\omega}}\cos\theta - \frac{\hat{v}}{\hat{\omega}}\cos(\theta + \hat{\omega}\Delta t) \\ \hat{\omega}\Delta t \end{pmatrix}$$

This equation is obtained by substituting the commanded velocity $u_t = (v\ \omega)^T$ with the noisy motion $(\hat{v}\ \hat{\omega})^T$ in (5.9). However, this model is still not very realistic, for reasons discussed in turn.

Final Orientation

The two equations given above exactly describe the final location of the robot given that the robot actually moves on an exact circular trajectory with radius $r = \frac{\hat{v}}{\hat{\omega}}$. While the radius of this circular segment and the distance traveled is influenced by the control noise, the very fact that the trajectory is circular is not. The assumption of circular motion leads to an important degeneracy. In particular, the support of the density $p(x_t \mid u_t, x_{t-1})$ is two-dimensional, within a three-dimensional embedding pose space. The fact that all posterior poses are located on a two-dimensional manifold within the three-dimensional pose space is a direct consequence of the fact that we used only two noise variables, one for v and one for ω. Unfortunately, this degeneracy has important ramifications when applying Bayes filters for state estimation.

In reality, any meaningful posterior distribution is of course not degenerate, and poses can be found within a three-dimensional space of variations in x, y, and θ. To generalize our motion model accordingly, we will assume that the robot performs a rotation $\hat{\gamma}$ when it arrives at its final pose. Thus, instead of computing θ' according to (5.13), we model the final orientation by

$$\theta' = \theta + \hat{\omega}\Delta t + \hat{\gamma}\Delta t \tag{5.14}$$

with

$$\hat{\gamma} = \varepsilon_{\alpha_5 v^2 + \alpha_6 \omega^2} \tag{5.15}$$

Here α_5 and α_6 are additional robot-specific parameters that determine the variance of the additional rotational noise. Thus, the resulting motion model is as follows:

$$\begin{pmatrix} x' \\ y' \\ \theta' \end{pmatrix} = \begin{pmatrix} x \\ y \\ \theta \end{pmatrix} + \begin{pmatrix} -\frac{\hat{v}}{\hat{\omega}}\sin\theta + \frac{\hat{v}}{\hat{\omega}}\sin(\theta + \hat{\omega}\Delta t) \\ \frac{\hat{v}}{\hat{\omega}}\cos\theta - \frac{\hat{v}}{\hat{\omega}}\cos(\theta + \hat{\omega}\Delta t) \\ \hat{\omega}\Delta t + \hat{\gamma}\Delta t \end{pmatrix} \tag{5.16}$$

Computation of $p(x_t \mid u_t, x_{t-1})$

The algorithm **motion_model_velocity** in Table 5.1 implements the computation of $p(x_t \mid u_t, x_{t-1})$ for given values of $x_{t-1} = (x \ y \ \theta)^T$, $u_t = (v \ \omega)^T$, and $x_t = (x' \ y' \ \theta')^T$. The derivation of this algorithm is somewhat involved, as it effectively implements an inverse motion model. In particular, **motion_model_velocity** determines motion parameters $\hat{u}_t = (\hat{v} \ \hat{\omega})^T$ from the poses x_{t-1} and x_t, along with an appropriate final rotation $\hat{\gamma}$. Our derivation makes it obvious as to why a final rotation is needed: For almost all values of x_{t-1}, u_t, and x_t, the motion probability would simply be zero without allowing for a final rotation.

Let us calculate the probability $p(x_t \mid u_t, x_{t-1})$ of control action $u_t = (v \; w)^T$ carrying the robot from the pose $x_{t-1} = (x \; y \; \theta)^T$ to the pose $x_t = (x' \; y' \; \theta')^T$ within Δt time units. To do so, we will first determine the control $\hat{u} = (\hat{v} \; \hat{\omega})^T$ required to carry the robot from x_{t-1} to position $(x' \; y')$, regardless of the robot's final orientation. Subsequently, we will determine the final rotation $\hat{\gamma}$ necessary for the robot to attain the orientation θ'. Based on these calculations, we can then easily calculate the desired probability $p(x_t \mid u_t, x_{t-1})$.

The reader may recall that our model assumes that the robot travels with a fixed velocity during Δt, resulting in a circular trajectory. For a robot that moved from $x_{t-1} = (x \; y \; \theta)^T$ to $x_t = (x' \; y')^T$, the center of the circle is defined as $(x^* \; y^*)^T$ and given by

$$(5.17) \quad \begin{pmatrix} x^* \\ y^* \end{pmatrix} = \begin{pmatrix} x \\ y \end{pmatrix} + \begin{pmatrix} -\lambda \sin \theta \\ \lambda \cos \theta \end{pmatrix} = \begin{pmatrix} \frac{x+x'}{2} + \mu(y - y') \\ \frac{y+y'}{2} + \mu(x' - x) \end{pmatrix}$$

for some unknown $\lambda, \mu \in \Re$. The first equality is the result of the fact that the circle's center is orthogonal to the initial heading direction of the robot; the second is a straightforward constraint that the center of the circle lies on a ray that lies on the half-way point between $(x \; y)^T$ and $(x' \; y')^T$ and is orthogonal to the line between these coordinates.

Usually, Equation (5.17) has a unique solution—except in the degenerate case of $\omega = 0$, in which the center of the circle lies at infinity. As the reader might want to verify, the solution is given by

$$(5.18) \quad \mu = \frac{1}{2} \frac{(x - x') \cos \theta + (y - y') \sin \theta}{(y - y') \cos \theta - (x - x') \sin \theta}$$

and hence

$$(5.19) \quad \begin{pmatrix} x^* \\ y^* \end{pmatrix} = \begin{pmatrix} \frac{x+x'}{2} + \frac{1}{2} \frac{(x-x')\cos\theta + (y-y')\sin\theta}{(y-y')\cos\theta - (x-x')\sin\theta} (y - y') \\ \frac{y+y'}{2} + \frac{1}{2} \frac{(x-x')\cos\theta + (y-y')\sin\theta}{(y-y')\cos\theta - (x-x')\sin\theta} (x' - x) \end{pmatrix}$$

The radius of the circle is now given by the Euclidean distance

$$(5.20) \quad r^* = \sqrt{(x - x^*)^2 + (y - y^*)^2} = \sqrt{(x' - x^*)^2 + (y' - y^*)^2}$$

Furthermore, we can now calculate the change of heading direction

$$(5.21) \quad \Delta \theta = \operatorname{atan2}(y' - y^*, x' - x^*) - \operatorname{atan2}(y - y^*, x - x^*)$$

5.3 Velocity Motion Model

Here atan2 is the common extension of the arcus tangens of y/x extended to the $\Re^2$ (most programming languages provide an implementation of this function):

$$(5.22) \quad \text{atan2}(y, x) = \begin{cases} \text{atan}(y/x) & \text{if } x > 0 \\ \text{sign}(y)\,(\pi - \text{atan}(|y/x|)) & \text{if } x < 0 \\ 0 & \text{if } x = y = 0 \\ \text{sign}(y)\,\pi/2 & \text{if } x = 0, y \neq 0 \end{cases}$$

Since we assume that the robot follows a circular trajectory, the translational distance between x_t and x_{t-1} along this circle is

$$(5.23) \quad \Delta\text{dist} = r^* \cdot \Delta\theta$$

From Δdist and $\Delta\theta$, it is now easy to compute the velocities $\hat{v}$ and $\hat{\omega}$:

$$(5.24) \quad \hat{u}_t = \begin{pmatrix} \hat{v} \\ \hat{\omega} \end{pmatrix} = \Delta t^{-1} \begin{pmatrix} \Delta\text{dist} \\ \Delta\theta \end{pmatrix}$$

The rotational velocity $\hat{\gamma}$ needed to achieve the final heading θ' of the robot in $(x' y')$ within Δt can be determined according to (5.14) as:

$$(5.25) \quad \hat{\gamma} = \Delta t^{-1}(\theta' - \theta) - \hat{\omega}$$

The *motion error* is the deviation of $\hat{u}_t$ and $\hat{\gamma}$ from the commanded velocity $u_t = (v\ \ \omega)^T$ and $\gamma = 0$, as defined in Equations (5.24) and (5.25).

$$(5.26) \quad v_{\text{err}} = v - \hat{v}$$
$$(5.27) \quad \omega_{\text{err}} = \omega - \hat{\omega}$$
$$(5.28) \quad \gamma_{\text{err}} = \hat{\gamma}$$

Under our error model, specified in Equations (5.10), and (5.15), these errors have the following probabilities:

$$(5.29) \quad \varepsilon_{\alpha_1 v^2 + \alpha_2 \omega^2}(v_{\text{err}})$$
$$(5.30) \quad \varepsilon_{\alpha_3 v^2 + \alpha_4 \omega^2}(\omega_{\text{err}})$$
$$(5.31) \quad \varepsilon_{\alpha_5 v^2 + \alpha_6 \omega^2}(\gamma_{\text{err}})$$

where ε_{b^2} denotes a zero-mean error variable with variance b^2, as before. Since we assume independence between the different sources of error, the desired probability $p(x_t \mid u_t, x_{t-1})$ is the product of these individual errors:

$$(5.32) \quad p(x_t \mid u_t, x_{t-1}) = \varepsilon_{\alpha_1 v^2 + \alpha_2 \omega^2}(v_{\text{err}}) \cdot \varepsilon_{\alpha_3 v^2 + \alpha_4 \omega^2}(\omega_{\text{err}}) \cdot \varepsilon_{\alpha_5 v^2 + \alpha_6 \omega^2}(\gamma_{\text{err}})$$

To see the correctness of the algorithm **motion_model_velocity** in Table 5.1, the reader may notice that this algorithm implements this expression. More specifically, lines 2 to 9 are equivalent to Equations (5.18), (5.19), (5.20), (5.21), (5.24), and (5.25). Line 10 implements (5.32), substituting the error terms as specified in Equations (5.29) to (5.31).

Sampling from $p(x' \mid u, x)$

The sampling algorithm **sample_motion_model_velocity** in Table 5.3 implements a forward model, as discussed earlier in this section. Lines 5 through 7 correspond to Equation (5.16). The noisy values calculated in lines 2 through 4 correspond to Equations (5.10) and (5.15).

The algorithm **sample_normal_distribution** in Table 5.4 implements a common approximation to sampling from a normal distribution. This approximation exploits the central limit theorem, which states that any average of non-degenerate random variables converges to a normal distribution. By averaging 12 uniform distributions, **sample_normal_distribution** generates values that are approximately normal distributed; though technically the resulting values lie always in $[-2b, 2b]$. Finally, **sample_triangular_distribution** in Table 5.4 implements a sampler for triangular distributions.

5.4 Odometry Motion Model

The velocity motion model discussed thus far uses the robot's velocity to compute posteriors over poses. Alternatively, one might want to use the odometry measurements as the basis for calculating the robot's motion over time. Odometry is commonly obtained by integrating wheel encoder information; most commercial robots make such integrated pose estimation available in periodic time intervals (e.g., every tenth of a second). This leads to a second motion model discussed in this chapter, the *odometry motion model*. The odometry motion model uses odometry measurements in lieu of controls.

Practical experience suggests that odometry, while still erroneous, is usually more accurate than velocity. Both suffer from drift and slippage, but velocity additionally suffers from the mismatch between the actual motion controllers and its (crude) mathematical model. However, odometry is only available in retrospect, after the robot moved. This poses no problem for fil-

5.4 Odometry Motion Model

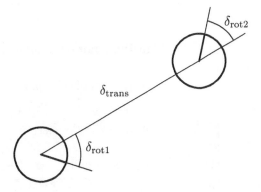

Figure 5.7 Odometry model: The robot motion in the time interval $(t-1, t]$ is approximated by a rotation δ_{rot1}, followed by a translation δ_{trans} and a second rotation δ_{rot2}. The turns and translations are noisy.

ter algorithms, such as the localization and mapping algorithms discussed in later chapters. But it makes this information unusable for accurate motion planning and control.

5.4.1 Closed Form Calculation

Technically, odometric information are sensor measurements, not controls. To model odometry as measurements, the resulting Bayes filter would have to include the actual velocity as state variables—which increases the dimension of the state space. To keep the state space small, it is therefore common to consider odometry data as if it were control signals. In this section, we will treat odometry measurements just like controls. The resulting model is at the core of many of today's best probabilistic robot systems.

Let us define the format of our control information. At time t, the correct pose of the robot is modeled by the random variable x_t. The robot odometry estimates this pose; however, due to drift and slippage there is no fixed coordinate transformation between the coordinates used by the robot's internal odometry and the physical world coordinates. In fact, knowing this transformation would solve the robot localization problem!

The odometry model uses the *relative motion information*, as measured by the robot's internal odometry. More specifically, in the time interval $(t-1, t]$, the robot advances from a pose x_{t-1} to pose x_t. The odometry reports back to us a related advance from $\bar{x}_{t-1} = (\bar{x}\ \bar{y}\ \bar{\theta})^T$ to $\bar{x}_t = (\bar{x}'\ \bar{y}'\ \bar{\theta}')^T$. Here the

---

1:         **Algorithm motion_model_odometry**(x_t, u_t, x_{t-1}):

2:         $\delta_{\text{rot1}} = \text{atan2}(\bar{y}' - \bar{y}, \bar{x}' - \bar{x}) - \bar{\theta}$
3:         $\delta_{\text{trans}} = \sqrt{(\bar{x} - \bar{x}')^2 + (\bar{y} - \bar{y}')^2}$
4:         $\delta_{\text{rot2}} = \bar{\theta}' - \bar{\theta} - \delta_{\text{rot1}}$

5:         $\hat{\delta}_{\text{rot1}} = \text{atan2}(y' - y, x' - x) - \theta$
6:         $\hat{\delta}_{\text{trans}} = \sqrt{(x - x')^2 + (y - y')^2}$
7:         $\hat{\delta}_{\text{rot2}} = \theta' - \theta - \hat{\delta}_{\text{rot1}}$

8:         $p_1 = \textbf{prob}(\delta_{\text{rot1}} - \hat{\delta}_{\text{rot1}}, \alpha_1 \hat{\delta}_{\text{rot1}}^2 + \alpha_2 \hat{\delta}_{\text{trans}}^2)$
9:         $p_2 = \textbf{prob}(\delta_{\text{trans}} - \hat{\delta}_{\text{trans}}, \alpha_3 \hat{\delta}_{\text{trans}}^2 + \alpha_4 \hat{\delta}_{\text{rot1}}^2 + \alpha_4 \hat{\delta}_{\text{rot2}}^2)$
10:        $p_3 = \textbf{prob}(\delta_{\text{rot2}} - \hat{\delta}_{\text{rot2}}, \alpha_1 \hat{\delta}_{\text{rot2}}^2 + \alpha_2 \hat{\delta}_{\text{trans}}^2)$

11:        **return** $p_1 \cdot p_2 \cdot p_3$

Table 5.5 Algorithm for computing $p(x_t \mid u_t, x_{t-1})$ based on odometry information. Here the control u_t is given by $(\bar{x}_{t-1} \ \bar{x}_t)^T$, with $\bar{x}_{t-1} = (\bar{x} \ \bar{y} \ \bar{\theta})$ and $\bar{x}_t = (\bar{x}' \ \bar{y}' \ \bar{\theta}')$.

bar indicates that these are odometry measurements embedded in a robot-internal coordinate whose relation to the global world coordinates is unknown. The key insight for utilizing this information in state estimation is that the relative difference between $\bar{x}_{t-1}$ and $\bar{x}_t$, under an appropriate definition of the term "difference," is a good estimator for the difference of the true poses x_{t-1} and x_t. The motion information u_t is, thus, given by the pair

$$(5.33) \quad u_t = \begin{pmatrix} \bar{x}_{t-1} \\ \bar{x}_t \end{pmatrix}$$

To extract relative odometry, u_t is transformed into a sequence of three steps: a rotation, followed by a straight line motion (translation), and another rotation. Figure 5.7 illustrates this decomposition: the initial turn is called δ_{rot1}, the translation δ_{trans}, and the second rotation δ_{rot2}. As the reader easily verifies, each pair of positions $(\bar{s} \ \bar{s}')$ has a unique parameter vector $(\delta_{\text{rot1}} \ \delta_{\text{trans}} \ \delta_{\text{rot2}})^T$, and these parameters are sufficient to reconstruct the

5.4 Odometry Motion Model

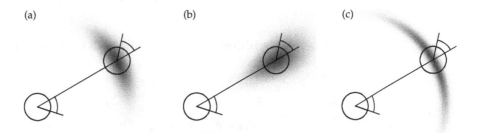

Figure 5.8 The odometry motion model, for different noise parameter settings.

relative motion between $\bar{s}$ and $\bar{s}'$. Thus, $\delta_{\text{rot1}}, \delta_{\text{trans}}, \delta_{\text{rot2}}$ form together a sufficient statistics of the relative motion encoded by the odometry.

The probabilistic motion model assumes that these three parameters are corrupted by independent noise. The reader may note that odometry motion uses one more parameter than the velocity vector defined in the previous section, for which reason we will not face the same degeneracy that led to the definition of a "final rotation."

Before delving into mathematical detail, let us state the basic algorithm for calculating this density in closed form. Table 5.5 depicts the algorithm for computing $p(x_t \mid u_t, x_{t-1})$ from odometry. This algorithm accepts as an input an initial pose x_{t-1}, a pair of poses $u_t = (\bar{x}_{t-1} \;\; \bar{x}_t)^T$ obtained from the robot's odometry, and a hypothesized final pose x_t. It outputs the numerical probability $p(x_t \mid u_t, x_{t-1})$.

Lines 2 to 4 in Table 5.5 recover relative motion parameters $(\delta_{\text{rot1}} \;\; \delta_{\text{trans}} \;\; \delta_{\text{rot2}})^T$ from the odometry readings. As before, they implement an *inverse motion model*. The corresponding relative motion parameters $(\hat{\delta}_{\text{rot1}} \;\; \hat{\delta}_{\text{trans}} \;\; \hat{\delta}_{\text{rot2}})^T$ for the given poses x_{t-1} and x_t are calculated in lines 5 through 7 of this algorithm. Lines 8 to 10 compute the error probabilities for the individual motion parameters. As above, the function **prob**(a, b^2) implements an error distribution over a with zero mean and variance b^2. Here the implementer must observe that all angular differences must lie in $[-\pi, \pi]$. Hence the outcome of $\delta_{\text{rot2}} - \bar{\delta}_{\text{rot2}}$ has to be truncated correspondingly—a common error that tends to be difficult to debug. Finally, line 11 returns the combined error probability, obtained by multiplying the individual error probabilities p_1, p_2, and p_3. This last step assumes independence between the different error sources. The variables α_1 through α_4 are robot-specific parameters that specify the noise in robot motion.

1:	**Algorithm sample_motion_model_odometry**(u_t, x_{t-1}):
2:	$\delta_{\text{rot1}} = \text{atan2}(\bar{y}' - \bar{y}, \bar{x}' - \bar{x}) - \bar{\theta}$
3:	$\delta_{\text{trans}} = \sqrt{(\bar{x} - \bar{x}')^2 + (\bar{y} - \bar{y}')^2}$
4:	$\delta_{\text{rot2}} = \bar{\theta}' - \bar{\theta} - \delta_{\text{rot1}}$
5:	$\hat{\delta}_{\text{rot1}} = \delta_{\text{rot1}} - \textbf{sample}(\alpha_1 \delta_{\text{rot1}}^2 + \alpha_2 \delta_{\text{trans}}^2)$
6:	$\hat{\delta}_{\text{trans}} = \delta_{\text{trans}} - \textbf{sample}(\alpha_3 \delta_{\text{trans}}^2 + \alpha_4 \delta_{\text{rot1}}^2 + \alpha_4 \delta_{\text{rot2}}^2)$
7:	$\hat{\delta}_{\text{rot2}} = \delta_{\text{rot2}} - \textbf{sample}(\alpha_1 \delta_{\text{rot2}}^2 + \alpha_2 \delta_{\text{trans}}^2)$
8:	$x' = x + \hat{\delta}_{\text{trans}} \cos(\theta + \hat{\delta}_{\text{rot1}})$
9:	$y' = y + \hat{\delta}_{\text{trans}} \sin(\theta + \hat{\delta}_{\text{rot1}})$
10:	$\theta' = \theta + \hat{\delta}_{\text{rot1}} + \hat{\delta}_{\text{rot2}}$
11:	**return** $x_t = (x', y', \theta')^T$

Table 5.6 Algorithm for sampling from $p(x_t \mid u_t, x_{t-1})$ based on odometry information. Here the pose at time t is represented by $x_{t-1} = (x\ y\ \theta)^T$. The control is a differentiable set of two pose estimates obtained by the robot's odometer, $u_t = (\bar{x}_{t-1}\ \bar{x}_t)^T$, with $\bar{x}_{t-1} = (\bar{x}\ \bar{y}\ \bar{\theta})$ and $\bar{x}_t = (\bar{x}'\ \bar{y}'\ \bar{\theta}')$.

Figure 5.8 shows examples of our odometry motion model for different values of the error parameters α_1 to α_4. The distribution in Figure 5.8a is a typical one, whereas the ones shown in Figures 5.8b and 5.8c indicate unusually large translational and rotational errors, respectively. The reader may want to carefully compare these diagrams with those in Figure 5.3 on page 122. The smaller the time between two consecutive measurements, the more similar those different motion models. Thus, if the belief is updated frequently e.g., every tenth of a second for a conventional indoor robot, the difference between these motion models is not very significant.

5.4 Odometry Motion Model

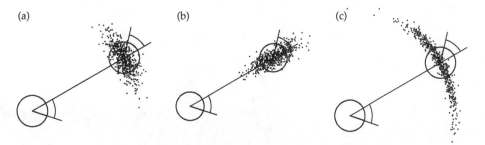

Figure 5.9 Sampling from the odometry motion model, using the same parameters as in Figure 5.8. Each diagram shows 500 samples.

5.4.2 Sampling Algorithm

If particle filters are used for localization, we would also like to have an algorithm for *sampling* from $p(x_t \mid u_t, x_{t-1})$. Recall that particle filters (Chapter 4.3) require samples of $p(x_t \mid u_t, x_{t-1})$, rather than a closed-form expression for computing $p(x_t \mid u_t, x_{t-1})$ for any x_{t-1}, u_t, and x_t. The algorithm **sample_motion_model_odometry**, shown in Table 5.6, implements the sampling approach. It accepts an initial pose x_{t-1} and an odometry reading u_t as input, and outputs a random x_t distributed according to $p(x_t \mid u_t, x_{t-1})$. It differs from the previous algorithm in that it randomly guesses a pose x_t (lines 5-10), instead of computing the probability of a given x_t. As before, the sampling algorithm **sample_motion_model_odometry** is somewhat easier to implement than the closed-form algorithm **motion_model_odometry**, since it side-steps the need for an inverse model.

Figure 5.9 shows examples of sample sets generated by **sample_motion_model_odometry**, using the same parameters as in the model shown in Figure 5.8. Figure 5.10 illustrates the motion model "in action" by superimposing sample sets from multiple time steps. This data has been generated using the motion update equations of the algorithm **particle_filter** (Table 4.3), assuming the robot's odometry follows the path indicated by the solid line. The figure illustrates how the uncertainty grows as the robot moves. The samples are spread across an increasingly large space.

5.4.3 Mathematical Derivation of the Odometry Motion Model

The derivation of the algorithms is relatively straightforward, and once again may be skipped at first reading. To derive a probabilistic motion model using

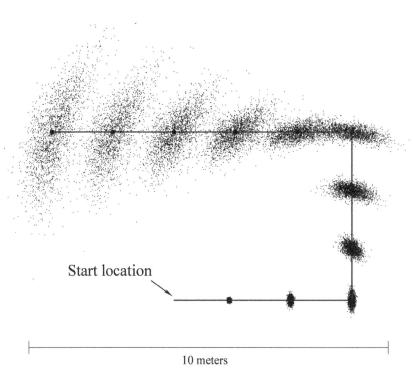

Figure 5.10 Sampling approximation of the position belief for a non-sensing robot. The solid line displays the actions, and the samples represent the robot's belief at different points in time.

odometry, we recall that the relative difference between any two poses is represented by a concatenation of three basic motions: a rotation, a straight-line motion (translation), and another rotation. The following equations show how to calculate the values of the two rotations and the translation from the odometry reading $u_t = (\bar{x}_{t-1}\ \bar{x}_t)^T$, with $\bar{x}_{t-1} = (\bar{x}\ \bar{y}\ \bar{\theta})$ and $\bar{x}_t = (\bar{x}'\ \bar{y}'\ \bar{\theta}')$:

$$\begin{align}
(5.34)\quad \delta_{\text{rot1}} &= \text{atan2}(\bar{y}' - \bar{y}, \bar{x}' - \bar{x}) - \bar{\theta} \\
(5.35)\quad \delta_{\text{trans}} &= \sqrt{(\bar{x} - \bar{x}')^2 + (\bar{y} - \bar{y}')^2} \\
(5.36)\quad \delta_{\text{rot2}} &= \bar{\theta}' - \bar{\theta} - \delta_{\text{rot1}}
\end{align}$$

To model the motion error, we assume that the "true" values of the rotation and translation are obtained from the measured ones by subtracting independent noise ε_{b^2} with zero mean and variance b^2:

$$(5.37) \quad \hat{\delta}_{\text{rot1}} = \delta_{\text{rot1}} - \varepsilon_{\alpha_1 \delta_{\text{rot1}}^2 + \alpha_2 \delta_{\text{trans}}^2}$$

$$(5.38) \quad \hat{\delta}_{\text{trans}} = \delta_{\text{trans}} - \varepsilon_{\alpha_3 \delta_{\text{trans}}^2 + \alpha_4 \delta_{\text{rot1}}^2 + \alpha_4 \delta_{\text{rot2}}^2}$$

$$(5.39) \quad \hat{\delta}_{\text{rot2}} = \delta_{\text{rot2}} - \varepsilon_{\alpha_1 \delta_{\text{rot2}}^2 + \alpha_2 \delta_{\text{trans}}^2}$$

As in the previous section, ε_{b^2} is a zero-mean noise variable with variance b^2. The parameters α_1 to α_4 are robot-specific error parameters, which specify the error accrued with motion.

Consequently, the true position, x_t, is obtained from x_{t-1} by an initial rotation with angle $\hat{\delta}_{\text{rot1}}$, followed by a translation with distance $\hat{\delta}_{\text{trans}}$, followed by another rotation with angle $\hat{\delta}_{\text{rot2}}$. Thus,

$$(5.40) \quad \begin{pmatrix} x' \\ y' \\ \theta' \end{pmatrix} = \begin{pmatrix} x \\ y \\ \theta \end{pmatrix} + \begin{pmatrix} \hat{\delta}_{\text{trans}} \cos(\theta + \hat{\delta}_{\text{rot1}}) \\ \hat{\delta}_{\text{trans}} \sin(\theta + \hat{\delta}_{\text{rot1}}) \\ \hat{\delta}_{\text{rot1}} + \hat{\delta}_{\text{rot2}} \end{pmatrix}$$

Notice that algorithm **sample_motion_model_odometry** implements Equations (5.34) through (5.40).

The algorithm **motion_model_odometry** is obtained by noticing that lines 5-7 compute the motion parameters $\hat{\delta}_{\text{rot1}}$, $\hat{\delta}_{\text{trans}}$, and $\hat{\delta}_{\text{rot2}}$ for the hypothesized pose x_t, relative to the initial pose x_{t-1}. The difference of both,

$$(5.41) \quad \delta_{\text{rot1}} - \hat{\delta}_{\text{rot1}}$$

$$(5.42) \quad \delta_{\text{trans}} - \hat{\delta}_{\text{trans}}$$

$$(5.43) \quad \delta_{\text{rot2}} - \hat{\delta}_{\text{rot2}}$$

is the *error* in odometry, assuming of course that x_t is the true final pose. The error model (5.37) to (5.39) implies that the probability of these errors is given by

$$(5.44) \quad p_1 = \varepsilon_{\alpha_1 \delta_{\text{rot1}}^2 + \alpha_2 \delta_{\text{trans}}^2}(\delta_{\text{rot1}} - \hat{\delta}_{\text{rot1}})$$

$$(5.45) \quad p_2 = \varepsilon_{\alpha_3 \delta_{\text{trans}}^2 + \alpha_4 \delta_{\text{rot1}}^2 + \alpha_4 \delta_{\text{rot2}}^2}(\delta_{\text{trans}} - \hat{\delta}_{\text{trans}})$$

$$(5.46) \quad p_3 = \varepsilon_{\alpha_1 \delta_{\text{rot2}}^2 + \alpha_2 \delta_{\text{trans}}^2}(\delta_{\text{rot2}} - \hat{\delta}_{\text{rot2}})$$

with the distributions ε defined as above. These probabilities are computed in lines 8-10 of our algorithm **motion_model_odometry**, and since the errors are assumed to be independent, the joint error probability is the product $p_1 \cdot p_2 \cdot p_3$ (c.f., line 11).

5.5 Motion and Maps

By considering $p(x_t \mid u_t, x_{t-1})$, we defined robot motion in a vacuum. In particular, this model describes robot motion in the absence of any knowledge about the nature of the environment. In many cases, we are also given a map m, which may contain information pertaining to the places that a robot may or may not be able to navigate. For example, *occupancy maps*, which will be explained in Chapter 9, distinguish *free* (traversable) from *occupied* terrain. The robot's pose must always be in the free space. Therefore, knowing m gives us further information about the robot pose x_t before, during, and after executing a control u_t.

This consideration calls for a motion model that takes the map m into account. We will denote this model by $p(x_t \mid u_t, x_{t-1}, m)$, indicating that it considers the map m in addition to the standard variables. If m carries information relevant to pose estimation, we have

$$(5.47) \quad p(x_t \mid u_t, x_{t-1}) \neq p(x_t \mid u_t, x_{t-1}, m)$$

MAP-BASED MOTION MODEL

The motion model $p(x_t \mid u_t, x_{t-1}, m)$ should give better results than the map-free motion model $p(x_t \mid u_t, x_{t-1})$. We will refer to $p(x_t \mid u_t, x_{t-1}, m)$ as *map-based motion model*. The map-based motion model computes the likelihood that a robot placed in a world with map m arrives at pose x_t upon executing action u_t at pose x_{t-1}. Unfortunately, computing this motion model in closed form is difficult. This is because to compute the likelihood of being at x_t after executing action u_t, one has to incorporate the probability that an unoccupied path exists between x_{t-1} and x_t and that the robot might have followed this unoccupied path when executing the control u_t—a complex operation.

Luckily, there exists an efficient approximation for the map-based motion model, which works well if the distance between x_{t-1} and x_t is small (e.g., smaller than half a robot diameter). The approximation factorizes the map-based motion model into two components:

$$(5.48) \quad p(x_t \mid u_t, x_{t-1}, m) = \eta \, \frac{p(x_t \mid u_t, x_{t-1}) \, p(x_t \mid m)}{p(x_t)}$$

where η is the usual normalizer. Usually, $p(x_t)$ is also uniform and can be subsumed into the constant normalizer. One then simply multiplies the map-free estimate $p(x_t \mid u_t, x_{t-1})$ with a second term, $p(x_t \mid m)$, which expresses the "consistency" of pose x_t with the map m. In the case of occupancy maps, $p(x_t \mid m) = 0$ if and only if the robot would be placed in an occupied grid cell

5.5 Motion and Maps

1:         **Algorithm motion_model_with_map**(x_t, u_t, x_{t-1}, m):
2:            return $p(x_t \mid u_t, x_{t-1}) \cdot p(x_t \mid m)$

1:         **Algorithm sample_motion_model_with_map**(u_t, x_{t-1}, m):
2:            do
3:                 $x_t =$ **sample_motion_model**(u_t, x_{t-1})
3:                 $\pi = p(x_t \mid m)$
4:            until $\pi > 0$
5:            return $\langle x_t, \pi \rangle$

Table 5.7 Algorithm for computing $p(x_t \mid u_t, x_{t-1}, m)$, which utilizes a map m of the environment. This algorithms bootstraps previous motion models (Tables 5.1, 5.3, 5.5, and 5.6) to models that take into account that robots cannot be placed in occupied space in the map m.

in the map; otherwise it assumes a constant value. By multiplying $p(x_t \mid m)$ and $p(x_t \mid u_t, x_{t-1})$, we obtain a distribution that assigns all probability mass to poses x_t consistent with the map, which otherwise has the same shape as $p(x_t \mid u_t, x_{t-1})$. As η can be computed by normalization, this approximation of a map-based motion model can be computed efficiently without any significant overhead compared to a map-free motion model.

Table 5.7 states the basic algorithms for computing and for sampling from the map-based motion model. Notice that the sampling algorithm returns a weighted sample, which includes an importance factor proportional to $p(x_t \mid m)$. Care has to be taken in the implementation of the sample version, to ensure termination of the inner loop. An example of the motion model is illustrated in Figure 5.11. The density in Figure 5.11a is $p(x_t \mid u_t, x_{t-1})$, computed according to the velocity motion model. Now suppose the map m possesses a long rectangular obstacle, as indicated in Figure 5.11b. The probability $p(x_t \mid m)$ is zero at all poses x_t where the robot would intersect the obstacle. Since our example robot is circular, this region is equivalent to the obstacle grown by a robot radius—this is equivalent to mapping the obstacle from *workspace* to the robot's *configuration space* or *pose space*. The resulting

CONFIGURATION SPACE

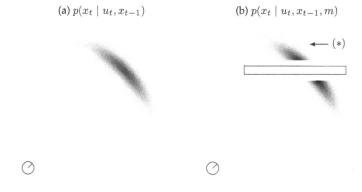

Figure 5.11 Velocity motion model (a) without a map and (b) conditioned on a map m.

probability $p(x_t \mid u_t, x_{t-1}, m)$, shown in Figure 5.11b, is the normalized product of $p(x_t \mid m)$ and $p(x_t \mid u_t, x_{t-1})$. It is zero in the extended obstacle area, and proportional to $p(x_t \mid u_t, x_{t-1})$ everywhere else.

Figure 5.11 also illustrates a problem with our approximation. The region marked (∗) possesses non-zero likelihood, since both $p(x_t \mid u_t, x_{t-1})$ and $p(x_t \mid m)$ are non-zero in this region. However, for the robot to be in this particular area it must have gone through the wall, which is impossible in the real world. This error is the result of checking model consistency at the final pose x_t only, instead of verifying the consistency of the robot's path to the goal. In practice, however, such errors only occur for relatively large motions u_t, and it can be neglected for higher update frequencies.

To shed light onto the nature of the approximation, let us briefly derive it. Equation (5.48) can be obtained by applying Bayes rule:

$$(5.49) \quad p(x_t \mid u_t, x_{t-1}, m) = \eta\, p(m \mid x_t, u_t, x_{t-1})\, p(x_t \mid u_t, x_{t-1})$$

If we approximate $p(m \mid x_t, u_t, x_{t-1})$ by $p(m \mid x_t)$ and observe that $p(m)$ is a constant relative to the desired posterior, we obtain the desired equation as follows:

$$(5.50) \quad \begin{aligned} p(x_t \mid u_t, x_{t-1}, m) &= \eta\, p(m \mid x_t)\, p(x_t \mid u_t, x_{t-1}) \\ &= \eta\, \frac{p(x_t \mid m)\, p(m)}{p(x_t)}\, p(x_t \mid u_t, x_{t-1}) \\ &= \eta\, \frac{p(x_t \mid m)\, p(x_t \mid u_t, x_{t-1})}{p(x_t)} \end{aligned}$$

Here η is the normalizer (notice that the value of η is different for the different steps in our transformation). This brief analysis shows that our map-based model is justified under the rough assumption that

$$(5.51) \quad p(m \mid x_t, u_t, x_{t-1}) = p(m \mid x_t)$$

Obviously, these expressions are not equal. When computing the conditional over m, our approximation omits two terms: u_t and x_{t-1}. By omitting these terms, we discard any information relating to the robot's path leading up to x_t. All we know is that its final pose is x_t. We already noticed the consequences of this omission in our example above, when we observed that poses behind a wall may possess non-zero likelihood. Our approximate map-based motion model may falsely assume that the robot just went through a wall, as long as the initial and final poses are in the unoccupied space. How damaging can this be? As noted above, this depends on the update interval. In fact, for sufficiently high update rates, and assuming that the noise variables in the motion model are bounded, we can guarantee that the approximation is tight and this effect will not occur.

This analysis illustrates a subtle insight pertaining to the implementation of the algorithm. In particular, one has to pay attention to the update frequency. A Bayes filter that is updated frequently might yield fundamentally different results than one that is updated only occasionally.

5.6 Summary

This section derived the two principal probabilistic motion models for mobile robots operating on the plane.

- We derived an algorithm for the probabilistic motion model $p(x_t \mid u_t, x_{t-1})$ that represents control u_t by a translational and angular velocity, executed over a fixed time interval Δt. In implementing this model, we realized that two control noise parameters, one for the translational and one for the rotational velocity, are insufficient to generate a space-filling (non-generate) posterior. We therefore added a third noise parameter, expressed as a noisy "final rotation."

- We presented an alternative motion model that uses the robot's odometry as input. Odometric measurements were expressed by three parameters, an initial rotation, followed by a translation, and a final rotation. The probabilistic motion model was implemented by assuming that all three of these parameters are subject to noise. We noted that odometry readings are technically not controls; however, by using them just like controls we arrived at a simpler formulation of the estimation problem.

- For both motion models, we presented two types of implementations, one in which the probability $p(x_t \mid u_t, x_{t-1})$ is calculated in closed form, and one that enables us to generate samples from $p(x_t \mid u_t, x_{t-1})$. The closed-form expression accepts as an input x_t, u_t, and x_{t-1}, and outputs a numerical probability value. To calculate this probability, the algorithms effectively invert the motion model, to compare the *actual* with the *commanded* control parameters. The sampling model does not require such an inversion. Instead, it implements a forward model of the motion model $p(x_t \mid u_t, x_{t-1})$. It accepts as an input the values u_t and x_{t-1} and outputs a random x_t drawn according to $p(x_t \mid u_t, x_{t-1})$. Closed-form models are required for some probabilistic algorithms. Others, most notably particle filters, utilize sampling models.

- Finally we extended all motion models to incorporate a map of the environment. The resulting probability $p(x_t \mid u_t, x_{t-1}, m)$ incorporates a map m in its conditional. This extension followed the intuition that the map specifies where a robot may be, which has an effect of the ability to move from pose x_{t-1} to x_t. The resulting algorithm was approximate, in that we only checked for the validity of the final pose.

The motion models discussed here are only examples: Clearly, the field of robotic actuators is much richer than just mobile robots operating in flat terrain. Even within the field of mobile robotics, there exist a number of devices that are not covered by the models discussed here. Examples include holonomic robots which can move sideways, or cars with suspension. Our description also does not consider robot dynamics, which are important for fast-moving vehicles such as cars on highways. Most of these robots can be modeled analogously; simply specify the physical laws of robot motion, and specify appropriate noise parameters. For dynamic models, this will require extending the robot state by a velocity vector that captures the dynamic state of the vehicle. In many ways, these extensions are straightforward.

As far as measuring ego-motion is concerned, many robots rely on inertial sensors to measure motion, as a supplement to or in place of odometry. Entire books have been dedicated to filter design using inertial sensors. Readers are encouraged to include richer models and sensors when odometry is insufficient.

5.7 Bibliographical Remarks

The present material extends the basic kinematic equations of specific types of mobile robots (Cox and Wilfong 1990) by a probabilistic component. Drives covered by our model are the differential drive, the Ackerman drive, and synchro-drive (Borenstein et al. 1996). Drive systems not covered by our model are those without non-holonomic constraints (Latombe 1991) like robots equipped with Mecanum wheels (Ilon 1975) or even legged robots, as described in pioneering papers by Raibert et al. (1986); Raibert (1991); Saranli and Koditschek (2002).

The field of robotics has studied robot motion and interaction with a robotic environment in much more depth. Contemporary *texts on mobile robots* covering aspects of kinematics and dynamics are due to Murphy (2000c); Dudek and Jenkin (2000); Siegwart and Nourbakhsh (2004). Cox and Wilfong (1990) provides a collection of articles by leading researchers at the time of publication; see also Kortenkamp et al. (1998). Classical treatments of robotic kinematics and dynamics can be found in Craig (1989); Vukobratović (1989); Paul (1981); and Yoshikawa (1990). A more modern text addressing robotic dynamics is the one by Featherstone (1987). Compliant motion as one form of environment interaction has been studied by Mason (2001). Terramechanics, which refers to the interaction of wheeled robots with the ground, has been studied in seminal texts by Bekker (1956, 1969) and Wong (1989). A contemporary text on wheel-ground interaction can be found in Iagnemma and Dubowsky (2004). Generalizing such models into a probabilistic framework is a promising direction for future research.

5.8 Exercises

DYNAMICS

1. All robot models in this chapter were kinematic. In this exercise, you will consider a robot with *dynamics*. Consider a robot that lives in a 1-D coordinate system. Its location will be denoted by x, its velocity by $\dot{x}$, and its acceleration by $\ddot{x}$. Suppose we can only control the acceleration $\ddot{x}$. Develop a mathematical motion model that computes the posterior over the pose x' and the velocity $\dot{x}'$ from an initial pose x and velocity $\dot{x}$, assuming that the acceleration $\ddot{x}$ is the sum of a commanded acceleration and a zero-mean Gaussian noise term with variance σ^2 (and assume that the actual acceleration remains constant in the simulation interval Δt). Are x' and $\dot{x}'$ correlated in the posterior? Explain why/why not.

2. Consider again the dynamic robot from Exercise 1. Provide a mathematical formula for computing the posterior distribution over the final velocity $\dot{x}'$, from the initial robot location x, the initial velocity $\dot{x}$, and the final pose x'. What is remarkable about this posterior?

3. Suppose we control this robot with random accelerations for T time intervals, for some large value of T. Will the final location x and the velocity $\dot{x}$ be correlated? If yes, will they be *fully* correlated as $T \uparrow \infty$, so that one variable becomes a deterministic function of the other?

4. Now consider a simple kinematic model of an idealized *bicycle*. Both tires are of diameter d, and are mounted to a frame of length l. The front tire can swivel around a vertical axis, and its steering angle will be denoted α. The rear tire is always parallel to the bicycle frame and cannot swivel.

   For the sake of this exercise, the pose of the bicycle shall be defined through three variables: the x-y location of the center of the front tire, and the angular orientation θ (yaw) of the bicycle frame relative to an external coordinate frame. The controls are the forward velocity v of the bicycle, and the steering angle α, which we will assume to be constant during each prediction cycle.

   Provide the mathematical prediction model for a time interval Δt, assuming that it is subject to Gaussian noise in the steering angle α and the forward velocity v. The model will have to predict the posterior of the bicycle state after Δt time, starting from a known state. If you cannot find an exact model, approximate it, and explain your approximations.

5. Consider the kinematic bicycle model from Exercise 4. Implement a sampling function for posterior poses of the bicycles under the same noise assumptions.

   For your simulation, you might assume $l = 100cm$, $d = 80cm$, $\Delta t = 1sec$, $|\alpha| \leq 80°$, $v \in [0; 100]cm/sec$. Assume further that the variance of the steering angle is $\sigma_\alpha^2 = 25°^2$ and the variance of the velocity is $\sigma_v^2 = 50cm^2/sec^2 \cdot v^2$. Notice that the variance of the velocity depends on the commanded velocity.

   For a bicycle starting at the origin, plot the resulting sample sets for the following values of the control parameters:

problem number	α	v
1	25°	20cm/sec
2	−25°	20cm/sec
3	25°	90cm/sec
4	80°	10cm/sec
1	85°	90cm/sec

   All your plots should show coordinate axes with units.

6. Consider once again the kinematic bicycle model from Exercise 4. Given an initial state x, y, θ and a final x' and y' (but no final θ'), provide a mathematical formula for determining the most likely values of α, v, and θ'. If

5.8 Exercises

you cannot find a closed form solution, you could instead give a technique for approximating the desired values.

HOLONOMIC

7. A common drive for indoor robots is *holonomic*. A holonomic robot has as many controllable degrees of freedom as the dimension of its configuration (or pose) space. In this exercise, you are asked to generalize the velocity model to a holonomic robot operating in the plane. Assume the robot can control its forward velocity, an orthogonal sidewards velocity, and a rotational velocity. Let us arbitrarily give sidewards motion to the left positive values, and motion to the right negative values.

   - State a mathematical model for such a robot, assuming that its controls are subject to independent Gaussian noise.
   - Provide a procedure for calculating $p(x_t \mid u_t, x_{t-1})$.
   - Provide a sampling procedure for sampling $x_t \sim p(x_t \mid u_t, x_{t-1})$.

8. Prove that the triangular distribution in Equation (5.12) has mean 0 and variance b^2. Prove the same for the sampling algorithm in Table 5.4.

6 Robot Perception

6.1 Introduction

Environment measurement models comprise the second domain-specific model in probabilistic robotics, next to motion models. Measurement models describe the formation process by which sensor measurements are generated in the physical world. Today's robots use a variety of different sensor modalities, such as tactile sensors, range sensors, or cameras. The specifics of the model depends on the sensor: Imaging sensors are best modeled by projective geometry, whereas sonar sensors are best modeled by describing the sound wave and its reflection on surfaces in the environment.

Probabilistic robotics explicitly models the noise in sensor measurements. Such models account for the inherent uncertainty in the robot's sensors. Formally, the measurement model is defined as a conditional probability distribution $p(z_t \mid x_t, m)$, where x_t is the robot pose, z_t is the measurement at time t, and m is the map of the environment. Although we mainly address range-sensors throughout this chapter, the underlying principles and equations are not limited to this type of sensors. Instead the basic principle can be applied to any kind of sensor, such as a camera or a bar-code operated landmark detector.

SONAR RANGE SCAN

To illustrate the basic problem of mobile robots that use their sensors to perceive their environment, Figure 6.1a shows a typical *sonar range scan* obtained in a corridor with a mobile robot equipped with a cyclic array of 24 ultrasound sensors. The distances measured by the individual sensors are depicted in light gray and the map of the environment is shown in black. Most of these measurements correspond to the distance of the nearest object in the measurement cone; some measurements, however, have failed to detect any object.

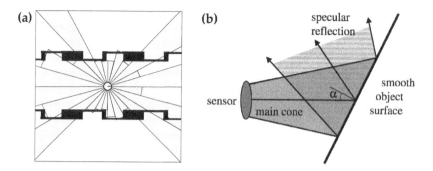

Figure 6.1 (a) Typical ultrasound scan of a robot in its environment. (b) A misreading in ultrasonic sensing. This effect occurs when firing a sonar signal towards a reflective surface at an angle α that exceeds half the opening angle of the sensor.

SPECULAR REFLECTION

The inability for sonar to reliably measure range to nearby objects is often paraphrased as sensor noise. Technically, this noise is quite predictable: When measuring smooth surfaces (such as walls), the reflection is usually *specular*, and the wall effectively becomes a mirror for the sound wave. This can be problematic when hitting a smooth surface at an angle. Here the echo may travel into a direction other than the sonar sensor, as illustrated in Figure 6.1b. This effect often leads to overly large range measurements when compared to the true distance to the nearest object in the main cone. The likelihood of this to happen depends on a number of properties, such as the surface material, the angle between the surface normal and the direction of the sensor cone, the range of the surface, the width of the main sensor cone, and the sensitivity of the sonar sensor. Other errors, such as short readings, may be caused by cross-talk between different sensors (sound is slow!) or by unmodeled objects in the proximity of the robot, such as people.

LASER RANGE SCAN

Figure 6.2 shows a typical *laser range scan*, acquired with a 2-D laser range finder. Laser is similar to sonar in that it also actively emits a signal and records its echo, but in the case of laser the signal is a light beam. A key difference to sonars is that lasers provide much more focused beams. The specific laser in Figure 6.2 is based on a time-of-flight measurement, and measurements are spaced in one degree increments.

As a rule of thumb, the more accurate a sensor model, the better the results—though there are some important caveats that were already discussed in Chapter 2.4.4. In practice, however, it is often impossible to model

6.1 Introduction

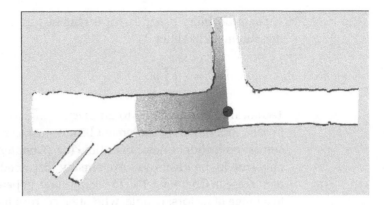

Figure 6.2 A typical laser range scan, acquired with a SICK LMS laser. The environment shown here is a coal mine. Image courtesy of Dirk Hähnel, University of Freiburg.

a sensor accurately, primarily due to the complexity of physical phenomena.

Often, the response characteristics of a sensor depends on variables we prefer not to make explicit in a probabilistic robotics algorithm (such as the surface material of walls, which for no particular reason is commonly not considered in robotic mapping). Probabilistic robotics accommodates inaccuracies of sensor models in the stochastic aspects: By modeling the measurement process as a conditional probability density, $p(z_t \mid x_t)$, instead of a deterministic function $z_t = f(x_t)$, the uncertainty in the sensor model can be accommodated in the non-deterministic aspects of the model. Herein lies a key advantage of probabilistic techniques over classical robotics: in practice, we can get away with extremely crude models. However, when devising a probabilistic model, care has to be taken to capture the different types of uncertainties that may affect a sensor measurement.

Many sensors generate more than one numerical measurement value when queried. For example, cameras generate entire arrays of values (brightness, saturation, color); similarly, range finders usually generate entire scans of ranges. We will denote the number of such measurement values within a measurement z_t by K, hence we can write:

(6.1) $\quad z_t \;=\; \{z_t^1, \ldots, z_t^K\}$

We will use z_t^k to refer to an individual measurement (e.g., one range value).

The probability $p(z_t \mid x_t, m)$ is obtained as the product of the individual measurement likelihoods

$$(6.2) \quad p(z_t \mid x_t, m) = \prod_{k=1}^{K} p(z_t^k \mid x_t, m)$$

Technically, this amounts to an *independence assumption* between the noise in each individual measurement beam—just as our Markov assumption assumes independent noise over time (c.f., Chapter 2.4.4). This assumption is only true in the ideal case. We already discussed possible causes of dependent noise in Chapter 2.4.4. To recapitulate, dependencies typically exist due to a range of factors: people, who often corrupt measurements of several adjacent sensors; errors in the model m; approximations in the posterior; and so on. For now, however, we will simply not worry about violations of the independence assumption, as we will return to this issue in later chapters.

6.2 Maps

To express the process of generating measurements, we need to specify the environment in which a measurement is generated. A *map* of the environment is a list of objects in the environment and their locations. We have already informally discussed maps in the previous chapter, where we developed robot motion models that took into consideration the occupancy of different locations in the world. Formally, a map m is a list of objects in the environment along with their properties:

$$(6.3) \quad m = \{m_1, m_2, \ldots, m_N\}$$

Here N is the total number of objects in the environment, and each m_n with $1 \leq n \leq N$ specifies a property. Maps are usually indexed in one of two ways, known as *feature-based* and *location-based*. In feature-based maps, n is a feature index. The value of m_n contains, next to the properties of a feature, the Cartesian location of the feature. In location-based maps, the index n corresponds to a specific location. In planar maps, it is common to denote a map element by $m_{x,y}$ instead of m_n, to make explicit that $m_{x,y}$ is the property of a specific world coordinate, $(x\ y)$.

VOLUMETRIC MAPS

Both types of maps have advantages and disadvantages. Location-based maps are *volumetric*, in that they offer a label for any location in the world. Volumetric maps contain information not only about objects in the environment, but also about the absence of objects (e.g., free-space). This is quite different in feature-based maps. *Feature-based maps* only specify the shape of the environment at the specific locations, namely the locations of the objects contained in the map. Feature representation makes it easier to adjust the position of an object; e.g., as a result of additional sensing. For this reason, feature-based maps are popular in the robotic mapping field, where maps are constructed from sensor data. In this book, we will encounter both types of maps—in fact, we will occasionally move from one representation to the other.

A classical map representation is known as *occupancy grid map*, which will be discussed in detail in Chapter 9. Occupancy maps are location-based: They assign to each x-y coordinate a binary occupancy value that specifies whether or not a location is occupied with an object. Occupancy grid maps are great for mobile robot navigation: They make it easy to find paths through the unoccupied space.

Throughout this book, we will drop the distinction between the physical world and the map. Technically, sensor measurements are caused by physical objects, not the map of those objects. However, it is tradition to condition sensor models on the map m; hence we will adopt a notation that suggests measurements depend on the map.

6.3 Beam Models of Range Finders

Range finders are among the most popular sensors in robotics. Our first *measurement model* in this chapter is therefore an approximative physical model of range finders. Range finders measure the range to nearby objects. Range may be measured along a beam—which is a good model of the workings of laser range finders—or within a cone—which is the preferable model of ultrasonic sensors.

6.3.1 The Basic Measurement Algorithm

Our model incorporates four types of measurement errors, all of which are essential to making this model work: small measurement noise, errors due to unexpected objects, errors due to failures to detect objects, and random unexplained noise. The desired model $p(z_t \mid x_t, m)$ is therefore a mixture of four densities, each of which corresponds to a particular type of error:

1. **Correct range with local measurement noise.** In an ideal world, a range finder would always measure the correct range to the nearest object in its

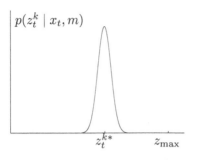
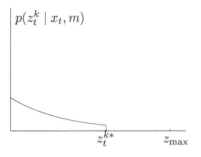
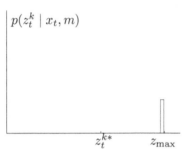
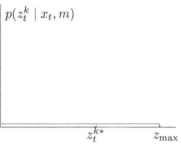

Figure 6.3 Components of the range finder sensor model. In each diagram the horizontal axis corresponds to the measurement z_t^k, the vertical to the likelihood.

measurement field. Let us use z_t^{k*} to denote the "true" range of the object measured by z_t^k. In location-based maps, the range z_t^{k*} can be determined using *ray casting*; in feature-based maps, it is usually obtained by searching for the closest feature within a measurement cone. However, even if the sensor correctly measures the range to the nearest object, the value it returns is subject to error. This error arises from the limited resolution of range sensors, atmospheric effect on the measurement signal, and so on. This *measurement noise* is usually modeled by a narrow Gaussian with mean z_t^{k*} and standard deviation σ_{hit}. We will denote the Gaussian by p_{hit}. Figure 6.3a illustrates this density p_{hit}, for a specific value of z_t^{k*}.

MEASUREMENT NOISE

In practice, the values measured by the range sensor are limited to the interval $[0; z_{\text{max}}]$, where z_{max} denotes the maximum sensor range. Thus, the measurement probability is given by

(6.4) $$p_{\text{hit}}(z_t^k \mid x_t, m) = \begin{cases} \eta \, \mathcal{N}(z_t^k; z_t^{k*}, \sigma_{\text{hit}}^2) & \text{if } 0 \leq z_t^k \leq z_{\max} \\ 0 & \text{otherwise} \end{cases}$$

where z_t^{k*} is calculated from x_t and m via ray casting, and $\mathcal{N}(z_t^k; z_t^{k*}, \sigma_{\text{hit}}^2)$ denotes the univariate normal distribution with mean z_t^{k*} and standard deviation σ_{hit}:

(6.5) $$\mathcal{N}(z_t^k; z_t^{k*}, \sigma_{\text{hit}}^2) = \frac{1}{\sqrt{2\pi\sigma_{\text{hit}}^2}} e^{-\frac{1}{2}\frac{(z_t^k - z_t^{k*})^2}{\sigma_{\text{hit}}^2}}$$

The normalizer η evaluates to

(6.6) $$\eta = \left(\int_0^{z_{\max}} \mathcal{N}(z_t^k; z_t^{k*}, \sigma_{\text{hit}}^2) \, dz_t^k \right)^{-1}$$

The standard deviation σ_{hit} is an intrinsic noise parameter of the measurement model. Below we will discuss strategies for setting this parameter.

2. **Unexpected objects.** Environments of mobile robots are dynamic, whereas maps m are static. As a result, objects not contained in the map can cause range finders to produce surprisingly short ranges—at least when compared to the map. A typical example of moving objects are people that share the operational space of the robot. One way to deal with such objects is to treat them as part of the state vector and estimate their location; another, much simpler approach, is to treat them as sensor noise. Treated as sensor noise, unmodeled objects have the property that they cause ranges to be shorter than z_t^{k*}, not longer.

The likelihood of sensing unexpected objects decreases with range. To see, imagine there are two people that independently and with the same fixed likelihood show up in the perceptual field of a proximity sensor. One person's range is r_1, and the second person's range is r_2. Let us further assume that $r_1 < r_2$, without loss of generality. Then we are more likely to measure r_1 than r_2. Whenever the first person is present, our sensor measures r_1. However, for it to measure r_2, the second person must be present *and* the first must be absent.

Mathematically, the probability of range measurements in such situations is described by an *exponential distribution*. The parameter of this distribution, λ_{short}, is an intrinsic parameter of the measurement model. According to the definition of an exponential distribution we obtain the following equation for $p_{\text{short}}(z_t^k \mid x_t, m)$:

(6.7) $$p_{\text{short}}(z_t^k \mid x_t, m) = \begin{cases} \eta \, \lambda_{\text{short}} \, e^{-\lambda_{\text{short}} z_t^k} & \text{if } 0 \leq z_t^k \leq z_t^{k*} \\ 0 & \text{otherwise} \end{cases}$$

As in the previous case, we need a normalizer η since our exponential is limited to the interval $[0; z_t^{k*}]$. Because the cumulative probability in this interval is given as

(6.8) $$\int_0^{z_t^{k*}} \lambda_{\text{short}} \, e^{-\lambda_{\text{short}} z_t^k} \, dz_t^k = -e^{-\lambda_{\text{short}} z_t^{k*}} + e^{-\lambda_{\text{short}} 0}$$
$$= 1 - e^{-\lambda_{\text{short}} z_t^{k*}}$$

the value of η can be derived as:

(6.9) $$\eta = \frac{1}{1 - e^{-\lambda_{\text{short}} z_t^{k*}}}$$

Figure 6.3b depicts this density graphically. This density falls off exponentially with the range z_t^k.

3. **Failures.** Sometimes, obstacles are missed altogether. For example, this happens frequently for sonar sensors as a result of specular reflections. Failures also occur with laser range finders when sensing black, light-absorbing objects, or for some laser systems when measuring objects in bright sunlight. A typical result of a *sensor failure* is a *max-range measurement*: the sensor returns its maximum allowable value z_{max}. Since such events are quite frequent, it is necessary to explicitly model max-range measurements in the measurement model.

SENSOR FAILURE

We will model this case with a point-mass distribution centered at z_{max}:

(6.10) $$p_{\text{max}}(z_t^k \mid x_t, m) = I(z = z_{\text{max}}) = \begin{cases} 1 & \text{if } z = z_{\text{max}} \\ 0 & \text{otherwise} \end{cases}$$

Here I denotes the indicator function that takes on the value 1 if its argument is true, and is 0 otherwise. Technically, p_{max} does not possess a probability density function. This is because p_{max} is a discrete distribution. However, this shall not worry us here, as our mathematical model of evaluating the probability of a sensor measurement is not affected by the non-existence of a density function. (In our diagrams, we simply draw p_{max} as a very narrow uniform distribution centered at z_{max}, so that we can pretend a density exists).

6.3 Beam Models of Range Finders

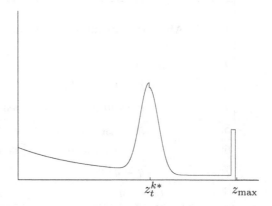

Figure 6.4 "Pseudo-density" of a typical mixture distribution $p(z_t^k \mid x_t, m)$.

UNEXPLAINABLE MEASUREMENTS

4. **Random measurements.** Finally, range finders occasionally produce entirely *unexplainable measurements*. For example, sonars often generate phantom readings when they bounce off walls, or when they are subject to cross-talk between different sensors. To keep things simple, such measurements will be modeled using a uniform distribution spread over the entire sensor measurement range $[0; z_{\max}]$:

$$(6.11) \quad p_{\text{rand}}(z_t^k \mid x_t, m) = \begin{cases} \frac{1}{z_{\max}} & \text{if } 0 \leq z_t^k < z_{\max} \\ 0 & \text{otherwise} \end{cases}$$

Figure 6.3d shows the density of the distribution p_{rand}.

These four different distributions are now mixed by a weighted average, defined by the parameters z_{hit}, z_{short}, $z_{\max}$, and z_{rand} with $z_{\text{hit}} + z_{\text{short}} + z_{\max} + z_{\text{rand}} = 1$.

$$(6.12) \quad p(z_t^k \mid x_t, m) = \begin{pmatrix} z_{\text{hit}} \\ z_{\text{short}} \\ z_{\max} \\ z_{\text{rand}} \end{pmatrix}^T \cdot \begin{pmatrix} p_{\text{hit}}(z_t^k \mid x_t, m) \\ p_{\text{short}}(z_t^k \mid x_t, m) \\ p_{\max}(z_t^k \mid x_t, m) \\ p_{\text{rand}}(z_t^k \mid x_t, m) \end{pmatrix}$$

A typical density resulting from this linear combination of the individual densities is shown in Figure 6.4 (with our visualization of the point-mass distribution $p_{\max}$ as a small uniform density). As the reader may notice, the basic characteristics of all four basic models are still present in this combined density.

```
1:      Algorithm beam_range_finder_model($z_t, x_t, m$):

2:          $q = 1$
3:          for $k = 1$ to $K$ do
4:              compute $z_t^{k*}$ for the measurement $z_t^k$ using ray casting
5:              $p = z_{\text{hit}} \cdot p_{\text{hit}}(z_t^k \mid x_t, m) + z_{\text{short}} \cdot p_{\text{short}}(z_t^k \mid x_t, m)$
6:                  $+ z_{\text{max}} \cdot p_{\text{max}}(z_t^k \mid x_t, m) + z_{\text{rand}} \cdot p_{\text{rand}}(z_t^k \mid x_t, m)$
7:              $q = q \cdot p$
8:          return $q$
```

Table 6.1 Algorithm for computing the likelihood of a range scan z_t, assuming conditional independence between the individual range measurements in the scan.

The range finder model is implemented by the algorithm **beam_range_finder_model** in Table 6.1. The input of this algorithm is a complete range scan z_t, a robot pose x_t, and a map m. Its outer loop (lines 2 and 7) multiplies the likelihood of individual sensor beams z_t^k, following Equation (6.2). Line 4 applies ray casting to compute the noise-free range for a particular sensor measurement. The likelihood of each individual range measurement z_t^k is computed in line 5, which implements the mixing rule for densities stated in (6.12). After iterating through all sensor measurements z_t^k in z_t, the algorithm returns the desired probability $p(z_t \mid x_t, m)$.

6.3.2 Adjusting the Intrinsic Model Parameters

In our discussion so far we have not addressed the question of how to choose the various parameters of the sensor model. These parameters include the mixing parameters z_{hit}, z_{short}, z_{max}, and z_{rand}. They also include the parameters σ_{hit} and λ_{short}. We will refer to the set of all intrinsic parameters as Θ. Clearly, the likelihood of any sensor measurement is a function of Θ. Thus, we will now discuss an algorithm for *adjusting model parameters*.

One way to determine the intrinsic parameters is to rely on data. Figure 6.5 depicts two series of 10,000 measurements obtained with a mobile robot traveling through a typical office environment. Both plots show only range measurements for which the expected range was approximately 3 me-

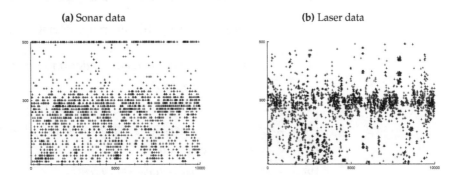

Figure 6.5 Typical data obtained with (a) a sonar sensor and (b) a laser-range sensor in an office environment for a "true" range of 300 cm and a maximum range of 500 cm.

ters (between 2.9m and 3.1m). The left plot depicts the data for sonar sensors, and the right plot the corresponding data for laser sensors. In both plots, the x-axis shows the number of the reading (from 1 to 10,000), and the y-axis is the range measured by the sensor.

Whereas most of the measurements are close to the correct range for both sensors, the behaviors of the sensors differ substantially. The ultrasound sensor appears to suffer from many more measurement noise and detection errors. Quite frequently it fails to detect an obstacle, and instead reports maximum range. In contrast, the laser range finder is more accurate. However, it also occasionally reports false ranges.

A perfectly acceptable way to set the intrinsic parameters Θ is by hand: simply eyeball the resulting density until it agrees with your experience. Another, more principled way is to learn these parameters from actual data. This is achieved by maximizing the likelihood of a reference data set $Z = \{z_i\}$ with associated positions $X = \{x_i\}$ and map m, where each z_i is an actual measurement, x_i is the pose at which the measurement was taken, and m is the map. The likelihood of the data Z is given by

$$(6.13) \quad p(Z \mid X, m, \Theta)$$

Our goal is to identify intrinsic parameters Θ that maximize this likelihood. Any estimator, or algorithm, that maximizes the likelihood of data is known as a *maximum likelihood estimator*, or *ML estimator*.

MAXIMUM LIKELIHOOD ESTIMATOR

Table 6.2 depicts the algorithm **learn_intrinsic_parameters**, which is an algorithm for calculating the maximum likelihood estimate for the intrinsic

1:	**Algorithm learn_intrinsic_parameters**(Z, X, m):
2:	repeat until convergence criterion satisfied
3:	for all z_i in Z do
4:	$\eta = [\, p_{\text{hit}}(z_i \mid x_i, m) + p_{\text{short}}(z_i \mid x_i, m)$ $\qquad + p_{\text{max}}(z_i \mid x_i, m) + p_{\text{rand}}(z_i \mid x_i, m)\,]^{-1}$
5:	calculate z_i^*
6:	$e_{i,\text{hit}} = \eta\, p_{\text{hit}}(z_i \mid x_i, m)$
7:	$e_{i,\text{short}} = \eta\, p_{\text{short}}(z_i \mid x_i, m)$
8:	$e_{i,\text{max}} = \eta\, p_{\text{max}}(z_i \mid x_i, m)$
9:	$e_{i,\text{rand}} = \eta\, p_{\text{rand}}(z_i \mid x_i, m)$
10:	$z_{\text{hit}} = \lvert Z \rvert^{-1} \sum_i e_{i,\text{hit}}$
11:	$z_{\text{short}} = \lvert Z \rvert^{-1} \sum_i e_{i,\text{short}}$
12:	$z_{\text{max}} = \lvert Z \rvert^{-1} \sum_i e_{i,\text{max}}$
13:	$z_{\text{rand}} = \lvert Z \rvert^{-1} \sum_i e_{i,\text{rand}}$
14:	$\sigma_{\text{hit}} = \sqrt{\dfrac{1}{\sum_i e_{i,\text{hit}}} \sum_i e_{i,\text{hit}} (z_i - z_i^*)^2}$
15:	$\lambda_{\text{short}} = \dfrac{\sum_i e_{i,\text{short}}}{\sum_i e_{i,\text{short}}\, z_i}$
16:	return $\Theta = \{z_{\text{hit}}, z_{\text{short}}, z_{\text{max}}, z_{\text{rand}}, \sigma_{\text{hit}}, \lambda_{\text{short}}\}$

Table 6.2 Algorithm for learning the intrinsic parameters of the beam-based sensor model from data.

parameters. As we shall see below, the algorithm is an instance of the *expectation maximization* (EM) algorithm, an iterative procedure for estimating ML parameters.

Initially, the algorithm **learn_intrinsic_parameters** in Table 6.2 requires a good initialization of the intrinsic parameters σ_{hit} and λ_{short}. In lines 3 through 9, it estimates auxiliary variables: Each $e_{i,\text{xxx}}$ is the probability that the measurement z_i is caused by "xxx," where "xxx" is chosen from the four aspects of the sensor model, hit, short, max, and random. Subsequently, it estimates the intrinsic parameters in lines 10 through 15. The intrinsic parameters, however, are a function of the expectations calculated before. Adjusting the intrinsic parameters causes the expectations to change, for which reason the algorithm has to be iterated. However, in practice the iteration converges quickly, and a dozen iterations are usually sufficient to give good results.

Figure 6.6 graphically depicts four examples of data and the ML measurement model calculated by **learn_intrinsic_parameters**. The first row shows approximations to data recorded with the ultrasound sensor. The second row contains plots of two functions generated for laser range data. The columns correspond to different "true" ranges. The data is organized in histograms. One can clearly see the differences between the different graphs. The smaller the range z_t^{k*} the more accurate the measurement. For both sensors the Gaussians are narrower for the shorter range than they are for the longer measurement. Furthermore, the laser range finder is more accurate than the ultrasound sensor, as indicated by the narrower Gaussians and the smaller number of maximum range measurements. The other important thing to notice is the relatively high likelihood of short and random measurements. This large error likelihood has a disadvantage and an advantage: On the negative side, it reduces the information in each sensor reading, since the difference in likelihood between a hit and a random measurement is small. On the positive side this model is less susceptible to unmodeled *systematic* perturbations, such as people who block the robot's path for long periods of time.

Figure 6.7 illustrates the learned sensor model in action. Shown in Figure 6.7a is a 180 degree range scan. The robot is placed in a previously acquired occupancy grid map at its true pose. Figure 6.7b plots a map of the environment along with the likelihood $p(z_t \mid x_t, m)$ of this range scan projected into x-y-space (by maximizing over the orientation θ). The darker a location, the more likely it is. As is easily seen, all regions with high likelihood are located in the corridor. This comes at little surprise, as the specific scan is geometrically more consistent with corridor locations than with locations inside any of the rooms. The fact that the probability mass is spread out throughout the corridor suggests that a single sensor scan is insufficient to determine the robot's exact pose. This is largely due to the symmetry of the corridor. The fact that the posterior is organized in two narrow horizontal bands is due to the fact that the orientation of the robot is unknown: each of these bands corresponds to one of the two surviving heading directions of the robot.

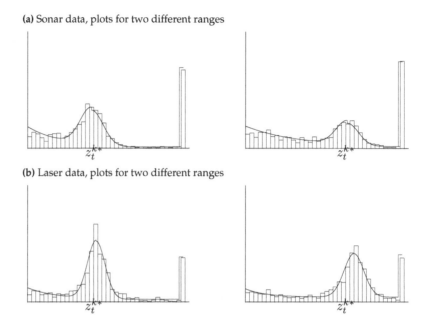

Figure 6.6 Approximation of the beam model based on (a) sonar data and (b) laser range data. The sensor models depicted on the left were obtained by a maximum likelihood approximation to the data sets depicted in Figure 6.5.

6.3.3 Mathematical Derivation of the Beam Model

To derive the ML estimator, it shall prove useful to introduce auxiliary variables c_i, the so-called correspondence variable. Each c_i can take on one of four values, hit, short, max, and random, corresponding to the four possible mechanisms that might have produced a measurement z_i.

Let us first consider the case in which the c_i's are known. We know which of the four mechanisms described above caused each measurement z_i. Based on the values of the c_i's, we can decompose Z into four disjoint sets, Z_{hit}, Z_{short}, Z_{max}, and Z_{rand}, which together comprise the set Z. The ML estimators for the intrinsic parameters z_{hit}, z_{short}, z_{max}, and z_{rand} are simply the

6.3 Beam Models of Range Finders

(a) Laser scan and part of the map

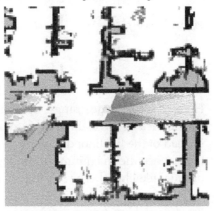

(b) Likelihood for different positions

Figure 6.7 Probabilistic model of perception: (a) Laser range scan, projected into a previously acquired map m. (b) The likelihood $p(z_t \mid x_t, m)$, evaluated for all positions x_t and projected into the map (shown in gray). The darker a position, the larger $p(z_t \mid x_t, m)$.

normalized ratios:

$$(6.14) \quad \begin{pmatrix} z_{\text{hit}} \\ z_{\text{short}} \\ z_{\text{max}} \\ z_{\text{rand}} \end{pmatrix} = |Z|^{-1} \begin{pmatrix} |Z_{\text{hit}}| \\ |Z_{\text{short}}| \\ |Z_{\text{max}}| \\ |Z_{\text{rand}}| \end{pmatrix}$$

The remaining intrinsic parameters, σ_{hit} and λ_{short}, are obtained as follows.

For the data set Z_{hit}, we get from (6.5)

(6.15) $$p(Z_{\text{hit}} \mid X, m, \Theta) = \prod_{z_i \in Z_{\text{hit}}} p_{\text{hit}}(z_i \mid x_i, m, \Theta)$$

$$= \prod_{z_i \in Z_{\text{hit}}} \frac{1}{\sqrt{2\pi\sigma_{\text{hit}}^2}} e^{-\frac{1}{2}\frac{(z_i - z_i^*)^2}{\sigma_{\text{hit}}^2}}$$

Here z_i^* is the "true" range, computed from the pose x_i and the map m. A classic trick of ML estimation is to maximize the logarithm of the likelihood, instead of the likelihood directly. The logarithm is a strictly monotonic function, hence the maximum of the log-likelihood is also the maximum of the original likelihood. The log-likelihood is given by

(6.16) $$\log p(Z_{\text{hit}} \mid X, m, \Theta) = \sum_{z_i \in Z_{\text{hit}}} \left[-\frac{1}{2} \log 2\pi\sigma_{\text{hit}}^2 - \frac{1}{2} \frac{(z_i - z_i^*)^2}{\sigma_{\text{hit}}^2} \right]$$

which is now easily transformed as follows

(6.17) $$\log p(Z_{\text{hit}} \mid X, m, \Theta)$$

$$= -\frac{1}{2} \sum_{z_i \in Z_{\text{hit}}} \left[\log 2\pi\sigma_{\text{hit}}^2 + \frac{(z_i - z_i^*)^2}{\sigma_{\text{hit}}^2} \right]$$

$$= -\frac{1}{2} \left[|Z_{\text{hit}}| \log 2\pi + 2|Z_{\text{hit}}| \log \sigma_{\text{hit}} + \sum_{z_i \in Z_{\text{hit}}} \frac{(z_i - z_i^*)^2}{\sigma_{\text{hit}}^2} \right]$$

$$= \text{const.} - |Z_{\text{hit}}| \log \sigma_{\text{hit}} - \frac{1}{2\sigma_{\text{hit}}^2} \sum_{z_i \in Z_{\text{hit}}} (z_i - z_i^*)^2$$

The derivative of this expression in the intrinsic parameter σ_{hit} is as follows:

(6.18) $$\frac{\partial \log p(Z_{\text{hit}} \mid X, m, \Theta)}{\partial \sigma_{\text{hit}}} = -\frac{|Z_{\text{hit}}|}{\sigma_{\text{hit}}} + \frac{1}{\sigma_{\text{hit}}^3} \sum_{z_i \in Z_{\text{hit}}} (z_i - z_i^*)^2$$

The maximum of the log-likelihood is now obtained by setting this derivative to zero. From that we get the solution to our ML estimation problem.

(6.19) $$\sigma_{\text{hit}} = \sqrt{\frac{1}{|Z_{\text{hit}}|} \sum_{z_i \in Z_{\text{hit}}} (z_i - z_i^*)^2}$$

The estimation of the remaining intrinsic parameter λ_{short} proceeds just about in the same way. The posterior over the data Z_{short} is given by

(6.20) $$p(Z_{\text{short}} \mid X, m, \Theta) = \prod_{z_i \in Z_{\text{short}}} p_{\text{short}}(z_i \mid x_i, m)$$

6.3 Beam Models of Range Finders

$$= \prod_{z_i \in Z_{\text{short}}} \lambda_{\text{short}} \, e^{-\lambda_{\text{short}} z_i}$$

The logarithm is

(6.21) $\log p(Z_{\text{short}} \mid X, m, \Theta) = \sum_{z_i \in Z_{\text{short}}} \log \lambda_{\text{short}} - \lambda_{\text{short}} z_i$

$$= |Z_{\text{short}}| \log \lambda_{\text{short}} - \lambda_{\text{short}} \sum_{z_i \in Z_{\text{short}}} z_i$$

The first derivative of this expression with respect to the intrinsic parameter λ_{short} is as follows:

(6.22) $\dfrac{\partial \log p(Z_{\text{short}} \mid X, m, \Theta)}{\partial \lambda_{\text{short}}} = \dfrac{|Z_{\text{short}}|}{\lambda_{\text{short}}} - \sum_{z_i \in Z_{\text{short}}} z_i$

Setting this to zero gives us the ML estimate for the intrinsic parameter λ_{short}

(6.23) $\lambda_{\text{short}} = \dfrac{|Z_{\text{short}}|}{\sum_{z_i \in Z_{\text{short}}} z_i}$

This derivation assumed knowledge of the parameters c_i. We now extend it to the case where the c_i's are unknown. As we shall see, the resulting ML estimation problem lacks a closed-form solution. However, we can devise a technique that iterates two steps, one that calculates an expectation for the c_i's and one that computes the intrinsic model parameters under these expectations. As noted, the resulting algorithm is an instance of the *expectation maximization* algorithm, usually abbreviated as EM.

EM ALGORITHM

To derive EM, it will be beneficial to define the likelihood of the data Z first:

(6.24) $\log p(Z \mid X, m, \Theta)$

$$= \sum_{z_i \in Z} \log p(z_i \mid x_i, m, \Theta)$$

$$= \sum_{z_i \in Z_{\text{hit}}} \log p_{\text{hit}}(z_i \mid x_i, m) + \sum_{z_i \in Z_{\text{short}}} \log p_{\text{short}}(z_i \mid x_i, m)$$

$$+ \sum_{z_i \in Z_{\text{max}}} \log p_{\text{max}}(z_i \mid x_i, m) + \sum_{z_i \in Z_{\text{rand}}} \log p_{\text{rand}}(z_i \mid x_i, m)$$

This expression can be rewritten using the variables c_i:

(6.25) $\log p(Z \mid X, m, \Theta) = \sum_{z_i \in Z} I(c_i = \text{hit}) \log p_{\text{hit}}(z_i \mid x_i, m)$

$$+I(c_i = \text{short}) \log p_{\text{short}}(z_i \mid x_i, m)$$
$$+I(c_i = \text{max}) \log p_{\text{max}}(z_i \mid x_i, m)$$
$$+I(c_i = \text{rand}) \log p_{\text{rand}}(z_i \mid x_i, m)$$

where I is the indicator function. Since the values for c_i are unknown, it is common to integrate them out. Put differently, EM maximizes the expectation $E[\log p(Z \mid X, m, \Theta)]$, where the expectation is taken over the unknown variables c_i:

(6.26)
$$E[\log p(Z \mid X, m, \Theta)]$$
$$= \sum_i p(c_i = \text{hit}) \log p_{\text{hit}}(z_i \mid x_i, m) + p(c_i = \text{short}) \log p_{\text{short}}(z_i \mid x_i, m)$$
$$+ p(c_i = \text{max}) \log p_{\text{max}}(z_i \mid x_i, m) + p(c_i = \text{rand}) \log p_{\text{rand}}(z_i \mid x_i, m)$$
$$=: \sum_i e_{i,\text{hit}} \log p_{\text{hit}}(z_i \mid x_i, m) + e_{i,\text{short}} \log p_{\text{short}}(z_i \mid x_i, m)$$
$$+ e_{i,\text{max}} \log p_{\text{max}}(z_i \mid x_i, m) + e_{i,\text{rand}} \log p_{\text{rand}}(z_i \mid x_i, m)$$

With the definition of the variable e as indicated. This expression is maximized in two steps. In a first step, we consider the intrinsic parameters σ_{hit} and λ_{short} given and calculate the expectation over the variables c_i.

(6.27)
$$\begin{pmatrix} e_{i,\text{hit}} \\ e_{i,\text{short}} \\ e_{i,\text{max}} \\ e_{i,\text{rand}} \end{pmatrix} := \begin{pmatrix} p(c_i = \text{hit}) \\ p(c_i = \text{short}) \\ p(c_i = \text{max}) \\ p(c_i = \text{rand}) \end{pmatrix} = \eta \begin{pmatrix} p_{\text{hit}}(z_i \mid x_i, m) \\ p_{\text{short}}(z_i \mid x_i, m) \\ p_{\text{max}}(z_i \mid x_i, m) \\ p_{\text{rand}}(z_i \mid x_i, m) \end{pmatrix}$$

The normalizer is given by

(6.28)
$$\eta = [\, p_{\text{hit}}(z_i \mid x_i, m) + p_{\text{short}}(z_i \mid x_i, m)$$
$$+ p_{\text{max}}(z_i \mid x_i, m) + p_{\text{rand}}(z_i \mid x_i, m) \,]^{-1}$$

This step is called the "E-step," indicating that we calculate expectations over the latent variables c_i. The remaining step is now straightforward, since the expectations decouple the dependencies between the different components of the sensor model. First, we note that the ML mixture parameters are simply the normalized expectations

(6.29)
$$\begin{pmatrix} z_{\text{hit}} \\ z_{\text{short}} \\ z_{\text{max}} \\ z_{\text{rand}} \end{pmatrix} = |Z|^{-1} \sum_i \begin{pmatrix} e_{i,\text{hit}} \\ e_{i,\text{short}} \\ e_{i,\text{max}} \\ e_{i,\text{rand}} \end{pmatrix}$$

The ML parameters σ_{hit} and λ_{short} are then obtained analogously, by replacing the hard assignments in (6.19) and (6.23) by soft assignments weighted by the expectations.

$$(6.30) \quad \sigma_{\text{hit}} = \sqrt{\frac{1}{\sum_{z_i \in Z} e_{i,\text{hit}}} \sum_{z_i \in Z} e_{i,\text{hit}} (z_i - z_i^*)^2}$$

and

$$(6.31) \quad \lambda_{\text{short}} = \frac{\sum_{z_i \in Z} e_{i,\text{short}}}{\sum_{z_i \in Z} e_{i,\text{short}} \, z_i}$$

6.3.4 Practical Considerations

In practice, computing the densities of all sensor readings can be quite involved from a computational perspective. For example, laser range scanners often return hundreds of values per scan, at a rate of several scans per second. Since one has to perform a ray casting operation for each beam of the scan and every possible pose considered, the integration of the whole scan into the current belief cannot always be carried out in real-time. One typical approach to solve this problem is to incorporate only a small subset of all measurements (e.g., 8 equally spaced measurements per laser range scan instead of 360). This approach has an important additional benefit. Since adjacent beams of a range scan are often not independent, the state estimation process becomes less susceptible to correlated noise in adjacent measurements.

When dependencies between adjacent measurements are strong, the ML model may make the robot overconfident and yield suboptimal results. One simple remedy is to replace $p(z_t^k \mid x_t, m)$ by a "weaker" version $p(z_t^k \mid x_t, m)^\alpha$ for $\alpha < 1$. The intuition here is to reduce, by a factor of α, the information extracted from a sensor measurement (the log of this probability is given by $\alpha \log p(z_t^k \mid x_t, m)$). Another possibility—which we will only mention here—is to learn the intrinsic parameters in the context of the application: For example, in mobile localization it is possible to train the intrinsic parameters via gradient descent to yield good localization results over multiple time steps. Such a multi-time step methodology is significantly different from the single time step ML estimator described above. In practical implementations it can yield superior results; see Thrun (1998a).

The main drain of computing time for beam-based models is the ray casting operation. The runtime costs of computing $p(z_t \mid x_t, m)$ can be substantially reduced by *pre-cashing* the ray casting algorithm, and storing the result in memory—so that the ray casting operation can be replaced by a (much faster) table lookup. An obvious implementation of this idea is to decompose the state space into a fine-grained three-dimensional grid, and to pre-compute the ranges z_t^{k*} for each grid cell. This idea was already investigated in Chapter 4.1. Depending on the resolution of the grid, the memory requirements can be significant. In mobile robot localization, we find that pre-computing the range with a grid resolution of 15 centimeters and 2 degrees works well for indoor localization problems. It fits well into the RAM for moderate-sized computers, yielding speed-ups by an order of magnitude over the plain implementation that casts rays online.

6.3.5 Limitations of the Beam Model

The beam-based sensor model, while closely linked to the geometry and physics of range finders, suffers two major drawbacks.

In particular, the beam-based model exhibits a *lack of smoothness*. In cluttered environments with many small obstacles, the distribution $p(z_t^k \mid x_t, m)$ can be very unsmooth in x_t. Consider, for example, an environment with many chairs and tables (like a typical conference room). A robot like the ones shown in Chapter 1 will sense the legs of those obstacles. Obviously, small changes of a robot's pose x_t can have a tremendous impact on the correct range of a sensor beam. As a result, the measurement model $p(z_t^k \mid x_t, m)$ is highly discontinuous in x_t. The heading direction θ_t is particularly affected, since small changes in heading can cause large displacements in x-y-space at a range.

Lack of smoothness has two problematic consequences. First, any approximate belief representation runs the danger of missing the correct state, as nearby states might have drastically different posterior likelihoods. This poses constraints on the accuracy of the approximation which, if not met, increase the resulting error in the posterior. Second, hill climbing methods for finding the most likely state are prone to local minima, due to the large number of local maxima in such unsmooth models.

The beam-based model is also computational involved. Evaluating $p(z_t^k \mid x_t, m)$ for each single sensor measurement z_t^k involves ray casting, which is computationally expensive. As noted above, the problem can be partially remedied by pre-computing the ranges over a discrete grid in pose space. Such an approach shifts the computation into an initial off-line phase, with the benefit that the algorithm is faster at run time. However, the resulting tables are very large, since they cover a large three-dimensional space. Thus, pre-computing ranges is computationally expensive and requires substantial memory.

6.4 Likelihood Fields for Range Finders

6.4.1 Basic Algorithm

LIKELIHOOD FIELD

We will now describe an alternative model, called *likelihood field*, which overcomes these limitations. This model lacks a plausible physical explanation. In fact, it is an "ad hoc" algorithm that does not necessarily compute a conditional probability relative to any meaningful generative model of the physics of sensors. However, the approach works well in practice. The resulting posteriors are much smoother even in cluttered space, and the computation is more efficient.

The key idea is to first project the end points of a sensor scan z_t into the global coordinate space of the map. To do so, we need to know where relative to the global coordinate frame the robot's local coordinate system is located, where on the robot the sensor beam z_k originates, and where the sensor points. As usual let $x_t = (x \ y \ \theta)^T$ denote a robot pose at time t. Keeping with our two-dimensional view of the world, we denote the relative location of the sensor in the robot's fixed, local coordinate system by $(x_{k,\text{sens}} \ y_{k,\text{sens}})^T$, and the angular orientation of the sensor beam relative to the robot's heading direction by $\theta_{k,\text{sens}}$. These values are sensor-specific. The end point of the measurement z_t^k is now mapped into the global coordinate system via the obvious trigonometric transformation.

$$(6.32) \quad \begin{pmatrix} x_{z_t^k} \\ y_{z_t^k} \end{pmatrix} = \begin{pmatrix} x \\ y \end{pmatrix} + \begin{pmatrix} \cos\theta & -\sin\theta \\ \sin\theta & \cos\theta \end{pmatrix} \begin{pmatrix} x_{k,\text{sens}} \\ y_{k,\text{sens}} \end{pmatrix} + z_t^k \begin{pmatrix} \cos(\theta + \theta_{k,\text{sens}}) \\ \sin(\theta + \theta_{k,\text{sens}}) \end{pmatrix}$$

These coordinates are only meaningful when the sensor detects an obstacle. If the range sensor takes on its maximum value $z_t^k = z_{\max}$, these coordinates have no meaning in the physical world (even though the measurement does carry information). The likelihood field measurement model simply discards max-range readings.

Similar to the beam model discussed before, we assume three types of sources of noise and uncertainty:

1. **Measurement noise.** Noise arising from the measurement process is modeled using Gaussians. In x-y-space, this involves finding the nearest obstacle in the map. Let $dist$ denote the Euclidean distance between the

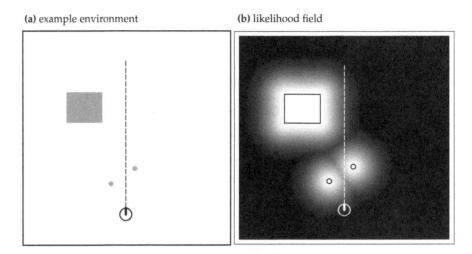

Figure 6.8 (a) Example environment with three obstacles (gray). The robot is located towards the bottom of the figure, and takes a measurement z_t^k as indicated by the dashed line. (b) Likelihood field for this obstacle configuration: the darker a location, the less likely it is to perceive an obstacle there. The probability $p(z_t^k \mid x_t, m)$ for the specific sensor beam is shown in Figure 6.9.

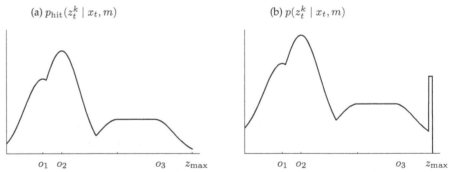

Figure 6.9 (a) Probability $p_{\text{hit}}(z_t^k)$ as a function of the measurement z_t^k, for the situation depicted in Figure 6.8. Here the sensor beam passes by three obstacles, with respective nearest points o_1, o_2, and o_3. (b) Sensor probability $p(z_t^k \mid x_t, m)$, obtained for the situation depicted in Figure 6.8, obtained by adding two uniform distributions.

6.4 Likelihood Fields for Range Finders

measurement coordinates $(x_{z_t^k} \ y_{z_t^k})^T$ and the nearest object in the map m. Then the probability of a sensor measurement is given by a zero-centered Gaussian, which captures the sensor noise:

$$(6.33) \qquad p_{\text{hit}}(z_t^k \mid x_t, m) = \varepsilon_{\sigma_{\text{hit}}}(dist)$$

Figure 6.8a depicts a map, and Figure 6.8b shows the corresponding Gaussian likelihood for measurement points $(x_{z_t^k} \ y_{z_t^k})^T$ in 2-D space. The brighter a location, the more likely it is to measure an object with a range finder. The density p_{hit} is now obtained by intersecting (and normalizing) the likelihood field by the sensor axis, indicated by the dashed line in Figure 6.8. The resulting function is the one shown in Figure 6.9a.

2. **Failures.** As before, we assume that max-range readings have a distinct large likelihood. As before, this is modeled by a point-mass distribution p_{max}.

3. **Unexplained random measurements.** Finally, a uniform distribution p_{rand} is used to model random noise in perception.

Just as for the beam-based sensor model, the desired probability $p(z_t^k \mid x_t, m)$ integrates all three distributions:

$$(6.34) \qquad z_{\text{hit}} \cdot p_{\text{hit}} + z_{\text{rand}} \cdot p_{\text{rand}} + z_{\text{max}} \cdot p_{\text{max}}$$

using the familiar mixing weights z_{hit}, z_{rand}, and z_{max}. Figure 6.9b shows an example of the resulting distribution $p(z_t^k \mid x_t, m)$ along a measurement beam. It should be easy to see that this distribution combines p_{hit}, as shown in Figure 6.9a, and the distributions p_{max} and p_{rand}. Much of what we said about adjusting the mixing parameters transfers over to our new sensor model. They can be adjusted by hand or learned using the ML estimator. A representation like the one in Figure 6.8b, which depicts the likelihood of an obstacle detection as a function of global x-y-coordinates, is called the *likelihood field*.

Table 6.3 provides an algorithm for calculating the measurement probability using the likelihood field. The reader should already be familiar with the outer loop, which multiplies the individual values of $p(z_t^k \mid x_t, m)$, assuming independence between the noise in different sensor beams. Line 4 checks if the sensor reading is a max range reading, in which case it is simply ignored. Lines 5 to 8 handle the interesting case: Here the distance to the nearest obstacle in x-y-space is computed (line 7), and the resulting likelihood is obtained in line 8 by mixing a normal and a uniform distribution. As

---

1: **Algorithm likelihood_field_range_finder_model**(z_t, x_t, m):

2: $\quad q = 1$
3: $\quad$ for all k do
4: $\quad\quad$ if $z_t^k \neq z_{\max}$
5: $\quad\quad\quad x_{z_t^k} = x + x_{k,\text{sens}} \cos\theta - y_{k,\text{sens}} \sin\theta + z_t^k \cos(\theta + \theta_{k,\text{sens}})$
6: $\quad\quad\quad y_{z_t^k} = y + y_{k,\text{sens}} \cos\theta + x_{k,\text{sens}} \sin\theta + z_t^k \sin(\theta + \theta_{k,\text{sens}})$
7: $\quad\quad\quad dist = \min_{x',y'} \left\{ \sqrt{(x_{z_t^k} - x')^2 + (y_{z_t^k} - y')^2} \,\middle|\, \langle x', y' \rangle \text{ occupied in } m \right\}$
8: $\quad\quad\quad q = q \cdot \left(z_{\text{hit}} \cdot \mathbf{prob}(dist, \sigma_{\text{hit}}) + \frac{z_{\text{random}}}{z_{\max}} \right)$
9: $\quad$ return q

Table 6.3 Algorithm for computing the likelihood of a range finder scan using Euclidean distance to the nearest neighbor. The function **prob**($dist, \sigma_{\text{hit}}$) computes the probability of the distance under a zero-centered Gaussian distribution with standard deviation σ_{hit}.

before, the function **prob**($dist, \sigma_{\text{hit}}$) computes the probability of $dist$ under a zero-centered Gaussian distribution with standard deviation σ_{hit}.

The search for the nearest neighbor in the map (line 7) is the most costly operation in algorithm **likelihood_field_range_finder_model**. To speed up this search, it is advantageous to pre-compute the likelihood field, so that calculating the probability of a measurement amounts to a coordinate transformation followed by a table lookup. Of course, if a discrete grid is used, the result of the lookup is only approximate, in that it might return the wrong obstacle coordinates. However, the effect on the probability $p(z_t^k \mid x_t, m)$ is typically small even for moderately course grids.

6.4.2 Extensions

A key advantage of the likelihood field model over the beam-based model discussed before is smoothness. Due to the smoothness of the Euclidean distance, small changes in the robot's pose x_t only have small effects on the resulting distribution $p(z_t^k \mid x_t, m)$. Another key advantage is that the pre-computation takes place in 2-D, instead of 3-D, increasing the compactness of the pre-computed information.

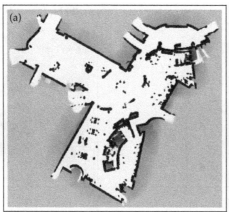 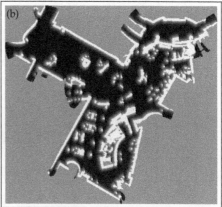

Figure 6.10 (a) Occupancy grid map of the San Jose Tech Museum, (b) pre-processed likelihood field.

However, the current model has three key disadvantages: First, it does not explicitly model people and other dynamics that might cause short readings. Second, it treats sensors as if they can "see through walls." This is because the ray casting operation was replaced by a nearest neighbor function, which is incapable of determining whether a path to a point is intercepted by an obstacle in the map. And third, our approach does not take map uncertainty into account. In particular, it cannot handle *unexplored* areas, for which the map is highly uncertain or unspecified.

The basic algorithm **likelihood_field_range_finder_model** can be extended to diminish the effect of these limitations. For example, one might sort map occupancy values into three categories: *occupied*, *free*, and *unknown*, instead of just the first two. When a sensor measurement z_t^k falls into the category *unknown*, its probability $p(z_t^k \mid x_t, m)$ is assumed to be the constant value $\frac{1}{z_{\max}}$. The resulting probabilistic model is crude. It assumes that in the unexplored space every sensor measurement is equally likely.

Figure 6.10 shows a map and the corresponding likelihood field. Here again the gray-level of an x-y-location indicates the likelihood of receiving a sensor reading there. The reader may notice that the distance to the nearest obstacle is only employed *inside* the map, which corresponds to the explored terrain. Outside, the likelihood $p(z_t^k \mid x_t, m)$ is a constant. For computational efficiency, it is worthwhile to pre-compute the nearest neighbor for a fine-grained 2-D grid.

Likelihood fields over the visible space can also be defined for the most

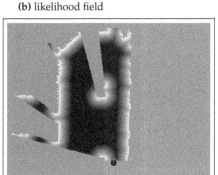

Figure 6.11 (a) Sensor scan, from a bird's eye perspective. The robot is placed at the bottom of this figure, generating a proximity scan that consists of the 180 dots in front of the robot. (b) Likelihood function generated from this sensor scan. The darker a region, the smaller the likelihood for sensing an object there. Notice that occluded regions are white, hence infer no penalty.

recent scan, which in fact defines a local map. Figure 6.11 shows such a likelihood field. It plays an important role in techniques that align individual scans.

6.5 Correlation-Based Measurement Models

MAP MATCHING

There exist a number of range sensor models in the literature that measure correlations between a measurement and the map. A common technique is known as *map matching*. Map matching requires techniques discussed in later chapters of this book, namely the ability to transform scans into occupancy maps. Typically, map matching compiles small numbers of consecutive scans into *local maps*, denoted m_{local}. Figure 6.12 shows such a local map, here in the form of an occupancy grid map. The sensor measurement model compares the local map m_{local} to the global map m, such that the more similar m and m_{local}, the larger $p(m_{\text{local}} \mid x_t, m)$. Since the local map is represented relative to the robot location, this comparison requires that the cells of the local map are transformed into the coordinate framework of the global map. Such a transformation can be done similar to the coordinate transform (6.32) of sensor measurements used in the likelihood field model. If the robot is at location x_t, we denote by $m_{x,y,\text{local}}(x_t)$ the grid cell in the local map that

6.5 Correlation-Based Measurement Models

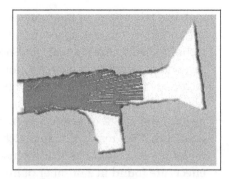

Figure 6.12 Example of a local map generated from 10 range scans, one of which is shown.

corresponds to $(x \ y)^T$ in global coordinates. Once both maps are in the same reference frame, they can be compared using the map correlation function, which is defined as follows:

$$\rho_{m,m_{\text{local}},x_t} = \frac{\sum_{x,y}(m_{x,y} - \bar{m}) \cdot (m_{x,y,\text{local}}(x_t) - \bar{m})}{\sqrt{\sum_{x,y}(m_{x,y} - \bar{m})^2 \sum_{x,y}(m_{x,y,\text{local}}(x_t) - \bar{m})^2}} \quad (6.35)$$

Here the sum is evaluated over cells defined in both maps, and $\bar{m}$ is the average map value:

$$\bar{m} = \frac{1}{2N} \sum_{x,y}(m_{x,y} + m_{x,y,\text{local}}) \quad (6.36)$$

where N denotes the number of elements in the overlap between the local and global map. The correlation $\rho_{m,m_{\text{local}},x_t}$ scales between ± 1. Map matching interprets the value

$$p(m_{\text{local}} \mid x_t, m) = \max\{\rho_{m,m_{\text{local}},x_t}, 0\} \quad (6.37)$$

as the probability of the local map conditioned on the global map m and the robot pose x_t. If the local map is generated from a single range scan z_t, this probability substitutes the measurement probability $p(z_t \mid x_t, m)$.

Map matching has a number of nice properties: just like the likelihood field model, it is easy to compute, though it does not yield smooth probabilities in the pose parameter x_t. One way to approximate the likelihood field (and to obtain smoothness) is to convolve the map m with a Gaussian smoothness kernel, and to run map matching on this smoothed map.

A key advantage of map matching over likelihood fields is that it explicitly considers the free-space in the scoring of two maps; the likelihood field technique only considers the end point of the scans, which by definition correspond to occupied space (or noise). On the other hand, many mapping techniques build local maps beyond the reach of the sensors. For example, many techniques build circular maps around the robot, setting to 0.5 areas beyond the range of actual sensor measurements. In such cases, there is a danger that the result of map matching incorporates areas beyond the actual measurement range, as if the sensor can "see through walls." Such side-effects are found in a number of implemented map matching techniques.

A further disadvantage is that map matching does not possess a plausible physical explanation. Correlations are the normalized quadratic distance between maps, which is *not* the noise characteristic of range sensors.

6.6 Feature-Based Measurement Models

6.6.1 Feature Extraction

FEATURES

The sensor models discussed thus far are all based on raw sensor measurements. An alternative approach is to extract *features* from the measurements. If we denote the feature extractor as a function f, the features extracted from a range measurement are given by $f(z_t)$. Most feature extractors extract a small number of features from high-dimensional sensor measurements. A key advantage of this approach is the enormous reduction of computational complexity: While inference in the high-dimensional measurement space can be costly, inference in the low-dimensional feature space can be orders of magnitude more efficient.

The discussion of specific algorithms for feature extraction is beyond the scope of this book. The literature offers a wide range of features for a number of different sensors. For range sensors, it is common to identify lines, corners, or local minima in range scans, which correspond to walls, corners, or objects such as tree trunks. When cameras are used for navigation, the processing of camera images falls into the realm of computer vision. Computer vision has devised a myriad of feature extraction techniques from camera images. Popular features include edges, corners, distinct patterns, and objects of distinct appearance. In robotics, it is also common to define places as features, such as hallways and intersections.

6.6.2 Landmark Measurements

In many robotics applications, features correspond to distinct objects in the physical world. For example, in indoor environments features may be door posts or windowsills; outdoors they may correspond to tree trunks or corners of buildings. In robotics, it is common to call those physical objects *landmarks*, to indicate that they are being used for robot navigation.

LANDMARKS

The most common model for processing landmarks assumes that the sensor can measure the range and the bearing of the landmark relative to the robot's local coordinate frame. Such sensors are called *range and bearing sensors*. The existence of a range-bearing sensor is not an implausible assumption: Any local feature extracted from range scans come with range and bearing information, as do visual features detected by stereo vision. In addition, the feature extractor may generate a *signature*. In this book, we assume a signature is a numerical value (e.g., an average color); it may equally be an integer that characterizes the type of the observed landmark, or a multi-dimensional vector characterizing a landmark (e.g., height and color).

RANGE AND BEARING SENSOR

SIGNATURE OF A LANDMARK

If we denote the range by r, the bearing by ϕ, and the signature by s, the feature vector is given by a collection of triplets

$$(6.38) \quad f(z_t) = \{f_t^1, f_t^2, \ldots\} = \left\{ \begin{pmatrix} r_t^1 \\ \phi_t^1 \\ s_t^1 \end{pmatrix}, \begin{pmatrix} r_t^2 \\ \phi_t^1 \\ s_t^2 \end{pmatrix}, \ldots \right\}$$

The number of features identified at each time step is variable. However, many probabilistic robotic algorithms assume conditional independence between features

$$(6.39) \quad p(f(z_t) \mid x_t, m) = \prod_i p(r_t^i, \phi_t^i, s_t^i \mid x_t, m)$$

Conditional independence applies if the noise in each individual measurement $(r_t^i \; \phi_t^i \; s_t^i)^T$ is independent of the noise in other measurements $(r_t^j \; \phi_t^j \; s_t^j)^T$ (for $i \neq j$). Under the conditional independence assumption, we can process one feature at a time, just as we did in several of our range measurement models. This makes it much easier to develop algorithms that implement probabilistic measurement models.

Let us now devise a sensor model for features. In the beginning of this chapter, we distinguished between two types of maps: *feature-based* and *location-based*. Landmark measurement models are usually defined only for feature-based maps. The reader may recall that those maps consist of lists of features, $m = \{m_1, m_2, \ldots\}$. Each feature may possess a signature and a *location coordinate*. The location of a feature, denoted $m_{i,x}$ and $m_{i,y}$, is simply its coordinate in the global coordinate frame of the map.

The measurement vector for a noise-free landmark sensor is easily specified by the standard geometric laws. We will model noise in landmark perception by independent Gaussian noise on the range, bearing, and the signature. The resulting measurement model is formulated for the case where the i-th feature at time t corresponds to the j-th landmark in the map. As usual, the robot pose is given by $x_t = (x \; y \; \theta)^T$.

$$(6.40) \quad \begin{pmatrix} r_t^i \\ \phi_t^i \\ s_t^i \end{pmatrix} = \begin{pmatrix} \sqrt{(m_{j,x} - x)^2 + (m_{j,y} - y)^2} \\ \operatorname{atan2}(m_{j,y} - y, m_{j,x} - x) - \theta \\ s_j \end{pmatrix} + \begin{pmatrix} \varepsilon_{\sigma_r^2} \\ \varepsilon_{\sigma_\phi^2} \\ \varepsilon_{\sigma_s^2} \end{pmatrix}$$

Here ε_{σ_r}, $\varepsilon_{\sigma_\phi}$, and ε_{σ_s} are zero-mean Gaussian error variables with standard deviations σ_r, σ_ϕ, and σ_s, respectively.

6.6.3 Sensor Model with Known Correspondence

DATA ASSOCIATION PROBLEM

CORRESPONDENCE VARIABLE

A key problem for range/bearing sensors is known as the *data association problem*. This problem arises when landmarks cannot be uniquely identified, so that some residual uncertainty exists with regards to the identity of a landmark. For developing a range/bearing sensor model, it shall prove useful to introduce a *correspondence variable* between the feature f_t^i and the landmark m_j in the map. This variable will be denoted by c_t^i with $c_t^i \in \{1, \ldots, N+1\}$; N is the number of landmarks in the map m. If $c_t^i = j \leq N$, then the i-th feature observed at time t corresponds to the j-th landmark in the map. In other words, c_t^i is the true identity of an observed feature. The only exception occurs with $c_t^i = N + 1$: Here a feature observation does not correspond to any feature in the map m. This case is important for handling spurious landmarks; it is also of great relevance for the topic of robotic mapping, in which the robot may encounter previously unobserved landmarks.

Table 6.4 depicts the algorithm for calculating the probability of a feature f_t^i with known correspondence $c_t^i \leq N$. Lines 3 and 4 calculate the true range and bearing to the landmark. The probability of the measured ranges and bearing is then calculated in line 5, assuming independence in the noise. As the reader easily verifies, this algorithm implements Equation (6.40).

6.6 Feature-Based Measurement Models

1:     **Algorithm landmark_model_known_correspondence**(f_t^i, c_t^i, x_t, m):

2:     $j = c_t^i$

3:     $\hat{r} = \sqrt{(m_{j,x} - x)^2 + (m_{j,y} - y)^2}$

4:     $\hat{\phi} = \operatorname{atan2}(m_{j,y} - y, m_{j,x} - x)$

5:     $q = \mathbf{prob}(r_t^i - \hat{r}, \sigma_r) \cdot \mathbf{prob}(\phi_t^i - \hat{\phi}, \sigma_\phi) \cdot \mathbf{prob}(s_t^i - s_j, \sigma_s)$

6:     return q

Table 6.4 Algorithm for computing the likelihood of a landmark measurement. The algorithm requires as input an observed feature $f_t^i = (r_t^i \; \phi_t^i \; s_t^i)^T$, and the true identity of the feature c_t^i, the robot pose $x_t = (x \; y \; \theta)^T$, and the map m. Its output is the numerical probability $p(f_t^i \mid c_t^i, m, x_t)$.

6.6.4 Sampling Poses

Sometimes it is desirable to sample robot poses x_t that correspond to a measurement f_t^i with feature identity c_t^i. We already encountered such sampling algorithms in the previous chapter, where we discussed robot motion models. Such sampling models are also desirable for sensor models. For example, when localizing a robot globally, it shall become useful to generate sample poses that incorporate a sensor measurement to generate initial guesses for the robot pose.

While in the general case, sampling poses x_t that correspond to a sensor measurement z_t is difficult, for our landmark model we can actually provide an efficient sampling algorithm. However, such sampling is only possible under further assumptions. In particular, we have to know the prior $p(x_t \mid c_t^i, m)$. For simplicity, let us assume this prior is uniform (it generally is not!). Bayes rule then suggests that

$$\begin{aligned}(6.41) \quad p(x_t \mid f_t^i, c_t^i, m) &= \eta \, p(f_t^i \mid c_t^i, x_t, m) \, p(x_t \mid c_t^i, m) \\ &= \eta \, p(f_t^i \mid c_t^i, x_t, m)\end{aligned}$$

Sampling from $p(x_t \mid f_t^i, c_t^i, m)$ can now be achieved from the "inverse" of the sensor model $p(f_t^i \mid c_t^i, x_t, m)$. Table 6.5 depicts an algorithm that samples poses x_t. The algorithm is tricky: Even in the noise-free case, a landmark observation does not uniquely determine the location of the robot. Instead, the robot may be on a circle around the landmark, whose diameter is the range to the landmark. The indeterminacy of the robot pose also follows

---

1:   Algorithm sample_landmark_model_known_correspondence(f_t^i, c_t^i, m):

2:   $j = c_t^i$
3:   $\hat{\gamma} = \text{rand}(0, 2\pi)$
4:   $\hat{r} = r_t^i + \textbf{sample}(\sigma_r)$
5:   $\hat{\phi} = \phi_t^i + \textbf{sample}(\sigma_\phi)$
6:   $x = m_{j,x} + \hat{r} \cos \hat{\gamma}$
7:   $y = m_{j,y} + \hat{r} \sin \hat{\gamma}$
8:   $\theta = \hat{\gamma} - \pi - \hat{\phi}$
9:   return $(x \ y \ \theta)^T$

Table 6.5 Algorithm for sampling poses from a landmark measurement $f_t^i = (r_t^i \ \phi_t^i \ s_t^i)^T$ with known identity c_t^i.

from the fact that the range and bearing provide two constraints in a three-dimensional space of robot poses.

To implement a pose sampler, we have to sample the remaining free parameter, which determines where on the circle around the landmark the robot is located. This parameter is called $\hat{\gamma}$ in Table 6.5, and is chosen at random in line 3. Lines 4 and 5 perturb the measured range and bearing, exploiting the fact that the mean and the measurement are treated symmetrically in Gaussians. Finally, lines 6 through 8 recover the pose that corresponds to $\hat{\gamma}$, $\hat{r}$, and $\hat{\phi}$.

Figure 6.13 illustrates the pose distribution $p(x_t \mid f_t^i, c_t^i, m)$ (left diagram) and also shows a sample drawn with our algorithm **sample_landmark_model_known_correspondence** (right diagram). The posterior is projected into x-y-space, where it becomes a ring around the measured range r_t^i. In 3-D pose space, it is a spiral that unfolds the ring with the angle θ.

6.6.5 Further Considerations

Both algorithms for landmark-based measurements assume known correspondence. The case of unknown correspondence will be discussed in detail in later chapters, when we address algorithms for localization and mapping under unknown correspondence.

6.6 Feature-Based Measurement Models

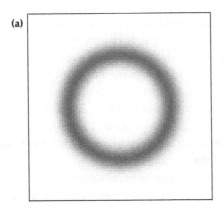 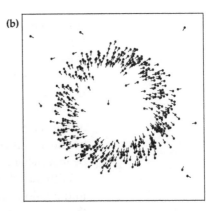

Figure 6.13 Landmark detection model: (a) Posterior distribution of the robot's pose given that it detected a landmark in 5m distance and 30deg relative bearing (projected onto 2-D). (b) Sample robot poses generated from such a detection. The lines indicate the orientation of the poses.

A comment is in order on the topic of landmark signature. Most published algorithms do not make the use of appearance features explicit. When the signature is not provided, all landmarks look equal, and the data association problem of estimating the correspondence variables is harder. We have included the signature in our model because it is a valuable source of information that can often be easily extracted from the sensor measurements.

As noted above, the main motivation for using features instead of the full measurement vector is computational in nature: It is much easier to manage a few hundred features than a few billion range measurements. Our model presented here is extremely crude, and it clearly does not capture the physical laws that underlie the sensor formation process. Nevertheless, the model tends to work well in a great number of applications.

It is important to notice that the reduction of measurements into features comes at a price. In the robotics literature, features are often (mis)taken for SUFFICIENT STATISTICS *sufficient statistics* of the measurement vector z_t. Let X be a variable of interest (e.g., a map, a pose), and Y some other information that we might bring to bear (e.g., past sensor measurements). Then f is a sufficient statistics of z_t if

(6.42) $\quad p(X \mid z_t, Y) \;=\; p(X \mid f(z_t), Y)$

In practice, however, a lot of information is sacrificed by using features instead of the full measurement vector. This lost information makes certain problems more difficult, such as the data association problem of determining whether or not the robot just revisited a previously explored location. It is easy to understand the effects of feature extraction by introspection: When you open your eyes, the visual image of your environment is probably sufficient to tell you unambiguously where you are—even if you were globally uncertain before. If, on the other hand, you only sense certain features, such as the relative location of door posts and windowsills, you would probably be much less certain as to where you are. Quite likely the information may be insufficient for global localization.

With the advent of fast computers, features have gradually lost importance in the field of robotics. Especially when using range sensors, most state-of-the-art algorithms rely on dense measurement vectors, and they use dense location-based maps to represent the environment. Nevertheless, features are still great for educational purposes. They enable us to introduce the basic concepts in probabilistic robotics, and with proper treatment of problems such as the correspondence problem they can be brought to bear even in cases where maps are composed of dense sets of scan points. For this reason, a number of algorithms in this book are first described for feature representations, and then extended into algorithms using raw sensor measurements.

6.7 Practical Considerations

This section surveyed a range of measurement models. We placed a strong emphasis on models for range finders, due to their great importance in robotics. However, the models discussed here are only representatives of a much broader class of probabilistic models. In choosing the right model, it is important to trade off physical realism with properties that might be desirable for an algorithm using these models. For example, we noted that a physically realistic model of range sensors may yield probabilities that are not smooth in the alleged robot pose—which in turn causes problems for algorithms such as particle filters. Physical realism is therefore not the only criterion in choosing the right sensor model; an equally important criterion is the goodness of a model for the algorithm that utilizes it.

As a general rule of thumb, the more accurate a model, the better. In particular, the more information we can extract from a sensor measurement, the better. Feature-based models extract relatively little information, by virtue of the fact that feature extractors project high-dimensional sensor measurements into lower dimensional space. As a result, feature-based methods tend to produce inferior results. This disadvantage is offset by superior computational properties of feature-based representations.

When adjusting the intrinsic parameters of a measurement model, it is often useful to artificially inflate the uncertainty. This is because of a key limitation of the probabilistic approach: to make probabilistic techniques computationally tractable, we have to ignore dependencies that exist in the physical world, along with a myriad of latent variables that cause these dependencies. When such dependencies are not modeled, algorithms that integrate evidence from multiple measurements quickly become overconfident. Such *overconfidence* can ultimately lead to wrong conclusions, which negatively affects the results. In practice, it is therefore a good rule of thumb to reduce the information conveyed by a sensor. Doing so by projecting the measurement into a low-dimensional feature space is one way of achieving this. However, it suffers the limitations mentioned above. Uniformly decaying the information by exponentiating a measurement model with a parameter α, as discussed in Chapter 6.3.4, is a much better way, in that it does not introduce additional variance in the outcome of a probabilistic algorithm.

OVERCONFIDENCE

6.8 Summary

This section described probabilistic measurement models.

- Starting with models for range finders—and lasers in particular—we discussed measurement models $p(z_t^k \mid x_t, m)$. The first such model used ray casting to determine the shape of $p(z_t^k \mid x_t, m)$ for particular maps m and poses x_t. We devised a mixture model that addressed the various types of noise that can affect range measurements.

- We devised a maximum likelihood technique for identifying the intrinsic noise parameters of the measurement model. Since the measurement model is a mixture model, we provided an iterative procedure for maximum likelihood estimation. Our approach was an instance of the expectation maximization algorithm, which alternates a phase that calculates expectations over the type of error underlying a measurement, with a maximization phase that finds in closed form the best set of intrinsic parameters relative to these expectations.

- An alternative measurement model for range finders is based on likelihood fields. This technique used the nearest distance in 2-D coordinates to model the probability $p(z_t^k \mid x_t, m)$. We noted that this approach tends to yield smoother distributions $p(z_t^k \mid x_t, m)$. This comes at the expense of undesired side effects: The likelihood field technique ignores information pertaining to free-space, and it fails to consider occlusions in the interpretation of range measurements.

- A third measurement model is based on map matching. Map matching maps sensor scans into local maps, and correlates those maps with global maps. This approach lacks a physical motivation, but can be implemented very efficiently.

- We discussed how pre-computation can reduce the computational burden at runtime. In the beam-based measurement model, the pre-computation takes place in 3-D; the likelihood field requires only a 2-D pre-computation.

- We presented a feature-based sensor model, in which the robot extracts the range, bearing, and signature of nearby landmarks. Feature-based techniques extract from the raw sensor measurements distinct features. In doing so, they reduce the dimensionality of the sensor measurements by several orders of magnitude.

- At the end of the chapter, a discussion on practical issues pointed out some of the pitfalls that may arise in concrete implementations.

6.9 Bibliographical Remarks

This chapter only skims the rich literature on physical modeling of sensors. More accurate models of sonar range sensor can be found in Blahut et al. (1991); Grunbaum et al. (1992) and in Etter (1996). Models of laser range finders are described by Rees (2001). An empirical discussion of proper noise models can be found in Sahin et al. (1998). Relative to these models, the models in this chapter are extremely crude.

An early work on beam models for range sensors can be found in the seminal work by Moravec (1988). A similar model was later applied to mobile robot localization by Burgard et al. (1996). A beam-based model like the one described in this chapter together with the pre-caching of range measurements has first been described by Fox et al. (1999b). The likelihood fields have first been published by Thrun (2001), although they are closely related to the rich literature on scan matching techniques (Besl and McKay 1992). They in fact can be regarded as a soft variant of the correlation model described by Konolige and Chou (1999). Methods for computing the correlation between occupancy grid maps have also been quite popular. Thrun (1993) computes the sum of squared errors between the individual cells of two grid maps. Schiele and Crowley (1994) present a comparison of different models including correlation-based approaches. Yamauchi and Langley (1997) analyzed the robustness of map matching to dynamic environments. Duckett and Nehmzow (2001) transform local occupancy grids into histograms that can be matched more efficiently.

Range and bearings measurements for point landmarks are commonplace in the SLAM literature. Possibly the first mention is by Leonard and Durrant-Whyte (1991). In earlier work, Crowley (1989) devised measurement models for straight line objects.

6.10 Exercises

1. Many early robots navigating using features used artificial landmarks in the environment that were easy to recognize. A good place to mount such markers is a ceiling (why?). A classical example is a visual marker: Suppose we attach the following marker to the ceiling:

Let the world coordinates of the marker be x_m and y_m, and its orientation relative to the global coordinate system θ_m. We will denote the robot's pose by x_r, y_r, and θ_r.

Now assume that we are given a routine that can detect the marker in the image plane of a perspective camera. Let x_i and y_i denote the coordinates of the marker in the image plane, and θ_i its angular orientation. The camera has a focal length of f. From projective geometry, we know that each displacement d in x-y-space gets projected to a proportional displacement of $d \cdot \frac{f}{h}$ in the image plane. (You have to make some choices on your coordinate systems; make these choices explicit).

Your questions:

(a) Describe mathematically where to expect the marker (in global coordinates x_m, y_m, θ_m) when its image coordinates are x_i, y_i, θ_i, and the robot is at x_r, y_r, θ_r.

(b) Provide a mathematical equation for computing the image coordinates x_i, y_i, θ_i from the robot pose x_r, y_r, θ_r and the marker coordinates x_m, y_m, θ_m.

(c) Now give a mathematical equation for determining the robot coordinates x_r, y_r, θ_r assuming we know the true marker coordinates x_m, y_m, θ_m and the image coordinates x_i, y_i, θ_i.

(d) So far we assumed there is only a single marker. Now suppose there are multiple (indistinguishable) markers of the type shown above. How many such markers must a robot be able to see to uniquely identify its pose? Draw such a configuration, and argue why it is sufficient.

Hint: You don't need to consider the uncertainty in the measurement for answering this question. Also, note that the marker is symmetrical. This has an impact on the answer of these questions!

2. In this exercise, you will be asked to extend our calculation in the previous exercise to include error covariances. To simplify the calculation, we now assume a non-symmetric marker whose absolute orientation can be estimated:

Also for simplicity, we assume there shall be no noise in the orientation. However, the x-y estimates in the image plane will be noisy. Specifically, let the measurements be subject to zero-mean Gaussian noise with covariance

$$\Sigma = \begin{pmatrix} \sigma^2 & 0 & 0 \\ 0 & \sigma^2 & 0 \\ 0 & 0 & 0 \end{pmatrix}$$

for some positive value of σ^2.

Calculate for the three questions above the corresponding covariances. In particular,

(a) Given the image coordinates x_i, y_i, θ_i and the robot coordinates x_r, y_r, θ_r, what is the error covariance for the values of x_m, y_m, θ_m.

(b) Given the robot coordinates x_r, y_r, θ_r and the marker covariance x_m, y_m, θ_m, what is your error covariance for the values of x_i, y_i, θ_i?

(c) Given the marker covariance x_m, y_m, θ_m and the image coordinates x_i, y_i, θ_i, what is your error covariance for the values of x_r, y_r, θ_r?

Notice that not all those distributions may be Gaussian. For this exercise, it is fine to apply a Taylor series expansion to attain a Gaussian posterior, but you have to explain how you did this.

3. Now you are being asked to implement a routine **sample_marker_model**, which accepts as an input the location of a marker x_m, y_m, θ_m and the location of the perceived marker in the image plane x_i, y_i, θ_i, and generates as an output samples of the robot pose x_r, y_r, θ_r. The marker is the same ambiguous marker as in Exercise 1:

Generate a plot of samples for the robot coordinates x_r and y_r, for the following parameters (you can ignore the orientation θ_r in your plot).

problem #	x_m	y_m	θ_m	x_i	y_i	θ_i	h/f	σ^2
#1	0cm	0cm	0°	0cm	0cm	0°	200	$0.1cm^2$
#2	0cm	0cm	0°	1cm	0cm	0°	200	$0.1cm^2$
#3	0cm	0cm	0°	2cm	0cm	45°	200	$0.1cm^2$
#4	0cm	0cm	0°	2cm	0cm	45°	200	$1.0cm^2$
#5	50cm	150cm	10°	1cm	6cm	200°	250	$0.5cm^2$

All your plots should show coordinate axes with units. Notice: If you cannot devise an exact sampler, provide an approximate one and explain your approximations.

4. For this exercise you need access to a robot with a sonar sensor, of the type often used in indoor robotics. Place the sensor in front of a flat wall, at a range d and an angle ϕ. Measure the frequency at which the sensor detects the wall. Plot this value for different values of d (in 0.5 meter increments) and different values of ϕ (in 5 degree increments). What do you find?

PART II

Localization

7 Mobile Robot Localization: Markov and Gaussian

This chapter introduces a number of concrete algorithms for mobile robot localization. Mobile robot localization is the problem of determining the pose of a robot relative to a given map of the environment. It is often called *position estimation*. Mobile robot localization is an instance of the general localization problem, which is the most basic perceptual problem in robotics. Nearly all robotics tasks require knowledge of the location of objects that are being manipulated. The techniques described in this and subsequent chapter are equally applicable for object localization tasks.

Figure 7.1 depicts a graphical model for the mobile robot localization problem. The robot is given a map of its environment and its goal is to determine its position relative to this map given the perceptions of the environment and its movements.

Mobile robot localization can be seen as a problem of coordinate transformation. Maps are described in a global coordinate system, which is independent of a robot's pose. Localization is the process of establishing correspondence between the map coordinate system and the robot's local coordinate system. Knowing this coordinate transformation enables the robot to express the location of objects of interest within its own coordinate frame—a necessary prerequisite for robot navigation. As the reader easily verifies, knowing the pose $x_t = (x \ y \ \theta)^T$ of the robot is sufficient to determine this coordinate transformation, assuming that the pose is expressed in the same coordinate frame as the map.

Unfortunately—and herein lies the problem of mobile robot localization—the pose can usually *not* be sensed directly. Put differently, most robots do not possess a noise-free sensor for measuring pose. The pose therefore has to be inferred from data. A key difficulty arises from the fact that a single sensor measurement is usually insufficient to determine the pose. Instead,

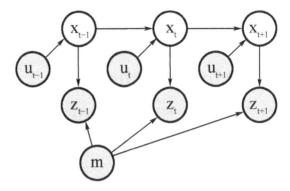

Figure 7.1 Graphical model of mobile robot localization. The value of shaded nodes are known: the map m, the measurements z, and the controls u. The goal of localization is to infer the robot pose variables x.

the robot has to integrate data over time to determine its pose. To see why this is necessary, just picture a robot located inside a building where many corridors look alike. Here a single sensor measurement (e.g., a range scan) is usually insufficient to identify the specific corridor.

Localization techniques have been developed for a broad set of map representations. We already discussed two types of maps: *feature-based* and *location-based*. An example of the latter was occupancy grid maps, which are subject to a later chapter in this book. Some other types of maps are shown in Figure 7.2. This figure shows a hand-drawn metric 2-D map, a graph-like topological map, an occupancy grid map, and an image mosaic of a ceiling (which can also be used as a map). Later chapters will investigate specific map types and discuss algorithms for acquiring maps from data. Localization assumes that an accurate map is available.

In this and the subsequent chapter, we present some basic probabilistic algorithms for mobile localization. All of these algorithms are variants of the basic Bayes filter described in Chapter 2. We discuss the advantages and shortcomings of each representation and associated algorithms. The chapter also goes through a series of extensions that address different localization problems, as defined through the following taxonomy.

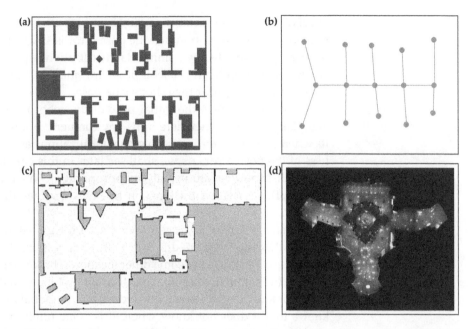

Figure 7.2 Example maps used for robot localization: (a) a manually constructed 2-D metric layout, (b) a graph-like topological map, (c) an occupancy grid map, and (d) an image mosaic of a ceiling. (d) courtesy of Frank Dellaert, Georgia Institute of Technology.

7.1 A Taxonomy of Localization Problems

Not every localization problem is equally hard. To understand the difficulty of a localization problem, let us first briefly discuss a taxonomy of localization problems. This taxonomy divides localization problems along a number of important dimensions pertaining to the nature of the environment and the initial knowledge that a robot may possess relative to the localization problem.

Local Versus Global Localization Localization problems are characterized by the type of knowledge that is available initially and at run-time. We distinguish three types of localization problems with an increasing degree of difficulty.

POSITION TRACKING

Position tracking assumes that the *initial* robot pose is known. Localizing the robot can be achieved by accommodating the noise in robot motion. The effect of such noise is usually small. Hence, methods for position tracking often rely on the assumption that the pose error is small. The pose uncertainty is often approximated by a unimodal distribution (e.g., a Gaussian). The position tracking problem is a *local* problem, since the uncertainty is local and confined to region near the robot's true pose.

GLOBAL LOCALIZATION

In *global localization*, the initial pose of the robot is unknown. The robot is initially placed somewhere in its environment, but it lacks knowledge of its whereabouts. Approaches to global localization cannot assume boundedness of the pose error. As we shall see later in this chapter, unimodal probability distributions are usually inappropriate. Global localization is more difficult than position tracking; in fact, it subsumes the position tracking problem.

KIDNAPPED ROBOT PROBLEM

The *kidnapped robot problem* is a variant of the global localization problem, but one that is even more difficult. During operation, the robot can get kidnapped and teleported to some other location. The kidnapped robot problem is more difficult than the global localization problems, in that the robot might believe it knows where it is while it does not. In global localization, the robot knows that it does not know where it is. One might argue that robots are rarely kidnapped in practice. The practical importance of this problem, however, arises from the observation that most state-of-the-art localization algorithms cannot be guaranteed never to fail. The ability to recover from failures is essential for truly autonomous robots. Testing a localization algorithm by kidnapping it measures its ability to recover from global localization failures.

Static Versus Dynamic Environments A second dimension that has a substantial impact on the difficulty of localization is the environment. Environments can be static or dynamic.

STATIC ENVIRONMENT

Static environments are environments where the only variable quantity (state) is the robot's pose. Put differently, only the robot moves in static environment. All other objects in the environments remain at the same location forever. Static environments have some nice mathematical properties that make them amenable to efficient probabilistic estimation.

DYNAMIC ENVIRONMENT

Dynamic environments possess objects other than the robot whose location or configuration changes over time. Of particular interest are changes that persist over time, and that impact more than a single sensor reading. Changes that are not measurable are of course of no relevance to localization, and those that affect only a single measurement are best treated as noise (*cf.* Chapter 2.4.4). Examples of more persistent changes are: people, daylight (for robots equipped with cameras), movable furniture, or doors. Clearly, most real environments are dynamic, with state changes occurring at a range of different speeds.

Obviously, localization in dynamic environments is more difficult than localization in static ones. There are two principal approaches for accommodating dynamics: First, dynamic entities might be included in the state vector. As a result, the Markov assumption might now be justified, but such an approach carries the burden of additional computational and modeling complexity. Second, in certain situations sensor data can be filtered so as to eliminate the damaging effect of unmodeled dynamics. Such an approach is described further below in Chapter 8.4.

Passive Versus Active Approaches A third dimension that characterizes different localization problems pertains to the fact whether or not the localization algorithm controls the motion of the robot. We distinguish two cases:

PASSIVE LOCALIZATION

In *passive localization*, the localization module only *observes* the robot operating. The robot is controlled through some other means, and the robot's motion is not aimed at facilitating localization. For example, the robot might move randomly or perform its everyday tasks.

ACTIVE LOCALIZATION

Active localization algorithms control the robot so as to minimize the localization error and/or the costs arising from moving a poorly localized robot into a hazardous place.

Active approaches to localization usually yield better localization results than passive ones. We already discussed an example in the introduction to this book: *coastal navigation*. A second example situation is shown in Figure 7.3. Here the robot is located in a symmetric corridor, and its belief after navigating the corridor for a while is centered at two (symmetric) poses. The local symmetry of the environment makes it impossible to localize the robot while in the corridor. Only if it moves into a room will it be able to eliminate the ambiguity and to determine its pose. It is situations like these where active localization gives much better results: Instead of merely waiting until the robot accidentally moves into a room, active localization can recognize the impasse and escape from it.

However, a key limitation of active approaches is that they require control over the robot. Thus, in practice, an active localization technique alone tends to be insufficient: The robot has to be able to localize itself even when carrying out some other task than localization. Some active localization techniques are built on top of a passive technique. Others combine task perfor-

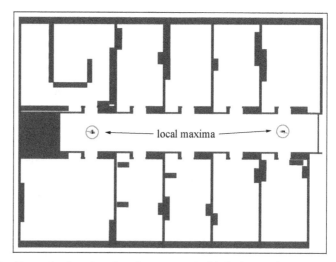

Figure 7.3 Example situation that shows a typical belief state during global localization in a locally symmetric environment. The robot has to move into one of the rooms to determine its location.

mance goals with localization goals when controlling a robot.

This chapter exclusively considers passive localization algorithms. Active localization will be discussed in Chapter 17.

Single-Robot Versus Multi-Robot A fourth dimension of the localization problem is related to the number of robots involved.

SINGLE-ROBOT LOCALIZATION

Single-robot localization is the most commonly studied approach to localization. It deals with a single robot only. Single robot localization offers the convenience that all data is collected at a single robot platform, and there is no communication issue.

MULTI-ROBOT LOCALIZATION

The *multi-robot localization* problem arises in teams of robots. At first glance, each robot could localize itself individually, hence the multi-robot localization problem can be solved through single-robot localization. If robots are able to detect each other, however, there is the opportunity to do better. This is because one robot's belief can be used to bias another robot's belief if knowledge of the relative location of both robots is available. The issue of multi-robot localization raises interesting, non-trivial issues on the representation of beliefs and the nature of the communication between them.

These four dimensions capture the four most important characteristics of the mobile robot localization problem. There exist a number of other charac-

---

1:　　　**Algorithm Markov_localization**($bel(x_{t-1}), u_t, z_t, m$):
2:　　　　　for all x_t do
3:　　　　　　　$\overline{bel}(x_t) = \int p(x_t \mid u_t, x_{t-1}, m)\, bel(x_{t-1})\, dx_{t-1}$
4:　　　　　　　$bel(x_t) = \eta\, p(z_t \mid x_t, m)\, \overline{bel}(x_t)$
5:　　　　　endfor
6:　　　　　return $bel(x_t)$

Table 7.1 Markov localization.

terizations that impact the hardness of the problem, such as the information provided by robot measurements and the information lost through motion. Also, symmetric environments are more difficult than asymmetric ones, due to the higher degree of ambiguity.

7.2 Markov Localization

Probabilistic localization algorithms are variants of the Bayes filter. The straightforward application of Bayes filters to the localization problem is called *Markov localization*. Table 7.1 depicts the basic algorithm. This algorithm is derived from the algorithm **Bayes_filter** (Table 2.1 on page 27). Notice that **Markov_localization** also requires a map m as input. The map plays a role in the measurement model $p(z_t \mid x_t, m)$ (line 4). It often, but not always, is incorporated in the motion model $p(x_t \mid u_t, x_{t-1}, m)$ as well (line 3). Just like the Bayes filter, Markov localization transforms a probabilistic belief at time $t-1$ into a belief at time t. Markov localization addresses the global localization problem, the position tracking problem, and the kidnapped robot problem in static environments.

The initial belief, $bel(x_0)$, reflects the initial knowledge of the robot's pose. It is set differently depending on the type of localization problem.

- **Position tracking.** If the initial pose is known, $bel(x_0)$ is initialized by a point-mass distribution. Let $\bar{x}_0$ denote the (known) initial pose. Then

$$(7.1) \quad bel(x_0) = \begin{cases} 1 & \text{if } x_0 = \bar{x}_0 \\ 0 & \text{otherwise} \end{cases}$$

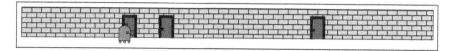

Figure 7.4 Example environment used to illustrate mobile robot localization: One-dimensional hallway environment with three indistinguishable doors. Initially the robot does not know its location except for its heading direction. Its goal is to find out where it is.

Point-mass distributions are discrete and therefore do not possess a density.

In practice the initial pose is often just known in approximation. The belief $bel(x_0)$ is then usually initialized by a narrow Gaussian distribution centered around $\bar{x}_0$. Gaussians were defined in Equation (2.4) on page 15.

$$(7.2) \quad bel(x_0) = \underbrace{\det(2\pi\Sigma)^{-\frac{1}{2}} \exp\left\{-\tfrac{1}{2}(x_0 - \bar{x}_0)^T \Sigma^{-1} (x_0 - \bar{x}_0)\right\}}_{\sim \mathcal{N}(x_0; \bar{x}_0, \Sigma)}$$

Σ is the covariance of the initial pose uncertainty.

- **Global localization.** If the initial pose is unknown, $bel(x_0)$ is initialized by a uniform distribution over the space of all legal poses in the map:

$$(7.3) \quad bel(x_0) = \frac{1}{|X|}$$

where $|X|$ stands for the volume (Lebesgue measure) of the space of all poses within the map.

- **Other.** Partial knowledge of the robot's position can usually easily be transformed into an appropriate initial distribution. For example, if the robot is known to start next to a door, one might initialize $bel(x_0)$ using a density that is zero except for places near doors, where it may be uniform. If it is known to be located in a specific corridor, one might initialize $bel(x_0)$ by a uniform distribution in the area of the corridor and zero anywhere else.

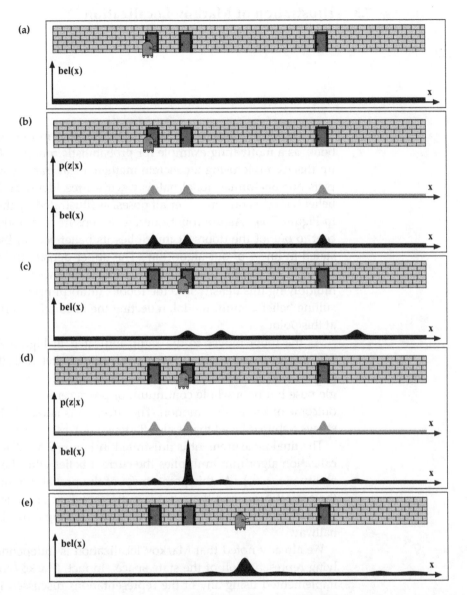

Figure 7.5 Illustration of the Markov localization algorithm. Each picture depicts the position of the robot in the hallway and its current belief $bel(x)$. (b) and (d) additionally depict the observation model $p(z_t \mid x_t)$, which describes the probability of observing a door at the different locations in the hallway.

7.3 Illustration of Markov Localization

We have already discussed Markov localization in the introduction to this book, as a motivating example for probabilistic robotics. Now we can back up this example using a concrete mathematical framework. Figure 7.4 depicts our one-dimensional hallway with three identical doors. The initial belief $bel(x_0)$ is uniform over all poses, as illustrated by the uniform density in Figure 7.5a. As the robot queries its sensors and notices that it is adjacent to one of the doors, it multiplies its belief $bel(x_0)$ by $p(z_t \mid x_t, m)$, as stated in line 4 of our algorithm. The upper density in Figure 7.5b visualizes $p(z_t \mid x_t, m)$ for the hallway example. The lower density is the result of multiplying this density into the robot's uniform prior belief. Again, the resulting belief is multi-modal, reflecting the residual uncertainty of the robot at this point.

As the robot moves to the right, indicated in Figure 7.5c, line 3 of the Markov localization algorithm convolves its belief with the motion model $p(x_t \mid u_t, x_{t-1})$. The motion model $p(x_t \mid u_t, x_{t-1})$ is not focused on a single pose but on a whole continuum of poses centered around the expected outcome of a noise-free motion. The effect is visualized in Figure 7.5c, which shows a shifted belief that is also flattened out, as a result of the convolution.

The final measurement is illustrated in Figure 7.5d. Here the Markov localization algorithm multiplies the current belief with the perceptual probability $p(z_t \mid x_t)$. At this point, most of the probability mass is focused on the correct pose, and the robot is quite confident of having localized itself. Figure 7.5e illustrates the robot's belief after having moved further down the hallway.

We already noted that Markov localization is independent of the underlying representation of the state space. In fact, Markov localization can be implemented using any of the representations discussed in Chapter 2. We now consider three different representations and devise practical algorithms that can localize mobile robots in real time. We begin with Kalman filters, which represent beliefs by their first and second moment. We then continue with discrete, grid representations and finally introduce algorithms using particle filters.

7.4 EKF Localization

The *extended Kalman filter localization* algorithm, or *EKF localization*, is a special case of Markov localization. EKF localization represents beliefs $bel(x_t)$ by their first and second moment, the mean μ_t and the covariance Σ_t. The basic EKF algorithm was stated in Table 3.3 in Chapter 3.3 (page 59). EKF localization shall be our first concrete implementation of an EKF in the context of an actual robotics problem.

Our EKF localization algorithm assumes that the map is represented by a collection of features. At any point in time t, the robot gets to observe a vector of ranges and bearings to nearby features: $z_t = \{z_t^1, z_t^2, \ldots\}$. We begin with a localization algorithm in which all features are uniquely identifiable. The existence of uniquely identifiable features may not be a bad assumption: For example, the Eiffel Tower in Paris is a landmark that is rarely confused with other landmarks, and it is widely visible throughout Paris. The identity of a feature is expressed by a set of *correspondence variables*, denoted c_t^i, one for each feature vector z_t^i. Correspondence variables were already discussed in Chapter 6.6. Let us first assume that the correspondences are known. We then progress to a more general version that allows for ambiguity among features. The second, more general version applies a maximum likelihood estimator to estimate the value of the latent correspondence variable, and uses the result of this estimation as if it were ground truth.

CORRESPONDENCE
VARIABLE

7.4.1 Illustration

Figure 7.6 illustrates the EKF localization algorithm using our example of mobile robot localization in the one-dimensional corridor environment (*cf.* Figure 7.4). To accommodate the unimodal shape of the belief in EKFs, we make two convenient assumptions: First, we assume that the correspondences are known. We attach unique labels to each door (1, 2, and 3), and we denote the measurement model by $p(z_t \mid x_t, m, c_t)$, where m is the map and $c_t \in \{1, 2, 3\}$ is the identity of the door observed at time t. Second, we assume that the initial pose is relatively well known. A typical initial belief is represented by the Gaussian distribution shown in Figure 7.6a, centered on the area near Door 1 and with a Gaussian uncertainty as indicated in that figure. As the robot moves to the right, its belief is convolved with the motion model. The resulting belief is a shifted Gaussian of increased width, as shown in Figure 7.6b.

Now suppose the robot detects that it is in front of door $c_t = 2$. The upper

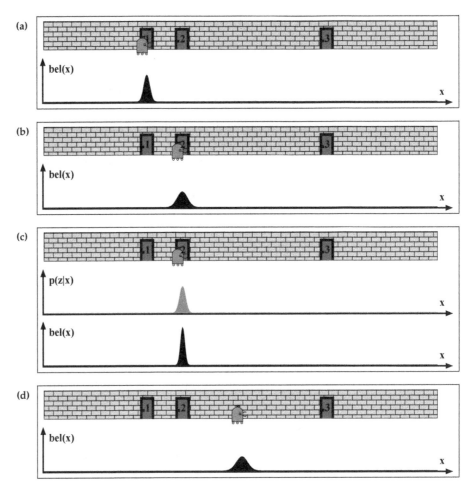

Figure 7.6 Application of the Kalman filter algorithm to mobile robot localization. All densities are represented by unimodal Gaussians.

density in Figure 7.6c visualizes $p(z_t \mid x_t, m, c_t)$ for this observation—again a Gaussian. Folding this measurement probability into the robot's belief yields the posterior shown in Figure 7.6c. Note that the variance of the resulting belief is smaller than the variances of both the robot's previous belief and the observation density. This is natural, since integrating two independent estimates should make the robot more certain than each estimate in isolation. After moving down the hallway, the robot's uncertainty in its position increases again, since the EKF continues to incorporate motion uncertainty into the robot's belief. Figure 7.6d shows one of these beliefs. This example illustrates the Kalman filter in our limited setting.

7.4.2 The EKF Localization Algorithm

The discussion thus far has been fairly abstract: We have silently assumed the availability of an appropriate motion and measurement model, and have left unspecified a number of key variables in the EKF update. We now discuss a concrete implementation of the EKF for feature-based maps. Our feature-based maps consist of point landmarks, as already discussed in Chapter 6.2. For such point landmarks, we use the common measurement model discussed in Chapter 6.6. We also adopt the velocity motion model defined in Chapter 5.3. The reader may take a moment to briefly reacquire the basic measurement and motion equations discussed in these chapters before reading on.

Table 7.2 describes **EKF_localization_known_correspondences**, the EKF algorithm for localization with known correspondences. This algorithm is derived from the EKF in Table 3.3 in Chapter 3. It requires as its input a Gaussian estimate of the robot pose at time $t-1$, with mean μ_{t-1} and covariance Σ_{t-1}. Further, it requires a control u_t, a map m, and a set of features $z_t = \{z_t^1, z_t^2, \ldots\}$ measured at time t, along with the correspondence variables $c_t = \{c_t^1, c_t^2, \ldots\}$. Its output is a new, revised estimate μ_t, Σ_t, along with the likelihood of the feature observation, p_{z_t}. The algorithm does not handle the case of straight motion for which $\omega_t = 0$. The treatment of this special case is left as an exercise.

The individual calculations in this algorithm are explained further below. Lines 3 and 4 compute the Jacobians needed for the linearized motion model. Line 5 determines the motion noise covariance matrix from the control. Lines 6 and 7 implement the familiar motion update. The predicted pose after the motion is calculated as $\bar{\mu}_t$ in line 6, and line 7 computes the corresponding uncertainty ellipse. The measurement update (correction step) is realized through Lines 8 to 21. The core of this update is a loop through all features i observed at time t. In line 10, the algorithm assigns to j the correspondence of the i-th feature in the measurement vector. It then calculates a predicted measurement $\hat{z}_t^i$ and the Jacobian H_t^i of the measurement model. Using this Jacobian, the algorithm determines S_t^i, the uncertainty corresponding to the predicted measurement $\hat{z}_t^i$. The Kalman gain K_t^i is then calculated in line 15. The estimate is updated in lines 16 and 17, once for each feature. Lines

1: **Algorithm EKF_localization_known_correspondences**$(\mu_{t-1}, \Sigma_{t-1}, u_t, z_t, c_t, m)$:

2: $\quad \theta = \mu_{t-1,\theta}$

3: $\quad G_t = \begin{pmatrix} 1 & 0 & -\frac{v_t}{\omega_t}\cos\theta + \frac{v_t}{\omega_t}\cos(\theta + \omega_t \Delta t) \\ 0 & 1 & -\frac{v_t}{\omega_t}\sin\theta + \frac{v_t}{\omega_t}\sin(\theta + \omega_t \Delta t) \\ 0 & 0 & 1 \end{pmatrix}$

4: $\quad V_t = \begin{pmatrix} \frac{-\sin\theta + \sin(\theta + \omega_t \Delta t)}{\omega_t} & \frac{v_t(\sin\theta - \sin(\theta + \omega_t \Delta t))}{\omega_t^2} + \frac{v_t \cos(\theta + \omega_t \Delta t)\Delta t}{\omega_t} \\ \frac{\cos\theta - \cos(\theta + \omega_t \Delta t)}{\omega_t} & -\frac{v_t(\cos\theta - \cos(\theta + \omega_t \Delta t))}{\omega_t^2} + \frac{v_t \sin(\theta + \omega_t \Delta t)\Delta t}{\omega_t} \\ 0 & \Delta t \end{pmatrix}$

5: $\quad M_t = \begin{pmatrix} \alpha_1 v_t^2 + \alpha_2 \omega_t^2 & 0 \\ 0 & \alpha_3 v_t^2 + \alpha_4 \omega_t^2 \end{pmatrix}$

6: $\quad \bar{\mu}_t = \mu_{t-1} + \begin{pmatrix} -\frac{v_t}{\omega_t}\sin\theta + \frac{v_t}{\omega_t}\sin(\theta + \omega_t \Delta t) \\ \frac{v_t}{\omega_t}\cos\theta - \frac{v_t}{\omega_t}\cos(\theta + \omega_t \Delta t) \\ \omega_t \Delta t \end{pmatrix}$

7: $\quad \bar{\Sigma}_t = G_t \Sigma_{t-1} G_t^T + V_t M_t V_t^T$

8: $\quad Q_t = \begin{pmatrix} \sigma_r^2 & 0 & 0 \\ 0 & \sigma_\phi^2 & 0 \\ 0 & 0 & \sigma_s^2 \end{pmatrix}$

9: $\quad$ for all observed features $z_t^i = (r_t^i \; \phi_t^i \; s_t^i)^T$ do

10: $\quad\quad j = c_t^i$

11: $\quad\quad q = (m_{j,x} - \bar{\mu}_{t,x})^2 + (m_{j,y} - \bar{\mu}_{t,y})^2$

12: $\quad\quad \hat{z}_t^i = \begin{pmatrix} \sqrt{q} \\ \text{atan2}(m_{j,y} - \bar{\mu}_{t,y}, m_{j,x} - \bar{\mu}_{t,x}) - \bar{\mu}_{t,\theta} \\ m_{j,s} \end{pmatrix}$

13: $\quad\quad H_t^i = \begin{pmatrix} -\frac{m_{j,x} - \bar{\mu}_{t,x}}{\sqrt{q}} & -\frac{m_{j,y} - \bar{\mu}_{t,y}}{\sqrt{q}} & 0 \\ \frac{m_{j,y} - \bar{\mu}_{t,y}}{q} & -\frac{m_{j,x} - \bar{\mu}_{t,x}}{q} & -1 \\ 0 & 0 & 0 \end{pmatrix}$

14: $\quad\quad S_t^i = H_t^i \bar{\Sigma}_t [H_t^i]^T + Q_t$

15: $\quad\quad K_t^i = \bar{\Sigma}_t [H_t^i]^T [S_t^i]^{-1}$

16: $\quad\quad \bar{\mu}_t = \bar{\mu}_t + K_t^i (z_t^i - \hat{z}_t^i)$

17: $\quad\quad \bar{\Sigma}_t = (I - K_t^i H_t^i) \bar{\Sigma}_t$

18: $\quad$ endfor

19: $\quad \mu_t = \bar{\mu}_t$

20: $\quad \Sigma_t = \bar{\Sigma}_t$

21: $\quad p_{z_t} = \prod_i \det(2\pi S_t^i)^{-\frac{1}{2}} \exp\left\{-\frac{1}{2}(z_t^i - \hat{z}_t^i)^T [S_t^i]^{-1}(z_t^i - \hat{z}_t^i)\right\}$

22: $\quad$ return μ_t, Σ_t, p_{z_t}

Table 7.2 The extended Kalman filter (EKF) localization algorithm, formulated here for a feature-based map and a robot equipped with sensors for measuring range and bearing. This version assumes knowledge of the exact correspondences.

19 and 20 set the new pose estimate, followed by the computation of the measurement likelihood in line 21. In this algorithm, care has to be taken when computing the difference of two angles, since the result may be off by 2π.

7.4.3 Mathematical Derivation of EKF Localization

Prediction Step (Lines 3–7) The EKF localization algorithm uses the motion model defined in Equation (5.13). Let us briefly restate the definition:

$$(7.4) \quad \begin{pmatrix} x' \\ y' \\ \theta' \end{pmatrix} = \begin{pmatrix} x \\ y \\ \theta \end{pmatrix} + \begin{pmatrix} -\frac{\hat{v}_t}{\hat{\omega}_t} \sin\theta + \frac{\hat{v}_t}{\hat{\omega}_t} \sin(\theta + \hat{\omega}_t \Delta t) \\ \frac{\hat{v}_t}{\hat{\omega}_t} \cos\theta - \frac{\hat{v}_t}{\hat{\omega}_t} \cos(\theta + \hat{\omega}_t \Delta t) \\ \hat{\omega}_t \Delta t \end{pmatrix}$$

Here $x_{t-1} = (x\ y\ \theta)^T$ and $x_t = (x'\ y'\ \theta')^T$ are the state vectors at time $t-1$ and t, respectively. The true motion is described by a translational velocity, $\hat{v}_t$, and a rotational velocity, $\hat{\omega}_t$. As already stated in Equation (5.10), these velocities are generated by the motion control, $u_t = (v_t\ \omega_t)^T$, with additive Gaussian noise:

$$(7.5) \quad \begin{pmatrix} \hat{v}_t \\ \hat{\omega}_t \end{pmatrix} = \begin{pmatrix} v_t \\ \omega_t \end{pmatrix} + \begin{pmatrix} \varepsilon_{\alpha_1 v_t^2 + \alpha_2 \omega_t^2} \\ \varepsilon_{\alpha_3 v_t^2 + \alpha_4 \omega_t^2} \end{pmatrix} = \begin{pmatrix} v_t \\ \omega_t \end{pmatrix} + \mathcal{N}(0, M_t)$$

We already know from Chapter 3 that EKF localization maintains a local posterior estimate of the state, represented by the mean μ_{t-1} and covariance Σ_{t-1}. We also recall that the "trick" of the EKF lies in linearizing the motion and measurement model. For that, we decompose the motion model into a noise-free part and a random noise component:

$$(7.6) \quad \underbrace{\begin{pmatrix} x' \\ y' \\ \theta' \end{pmatrix}}_{x_t} = \underbrace{\begin{pmatrix} x \\ y \\ \theta \end{pmatrix} + \begin{pmatrix} -\frac{v_t}{\omega_t} \sin\theta + \frac{v_t}{\omega_t} \sin(\theta + \omega_t \Delta t) \\ \frac{v_t}{\omega_t} \cos\theta - \frac{v_t}{\omega_t} \cos(\theta + \omega_t \Delta t) \\ \omega_t \Delta t \end{pmatrix}}_{g(u_t, x_{t-1})} + \mathcal{N}(0, R_t)$$

Equation (7.6) approximates Equation (7.4) by replacing the true motion $(\hat{v}_t\ \hat{\omega}_t)^T$ by the executed control $(v_t\ \omega_t)^T$, and capturing the motion noise in an additive Gaussian with zero mean. Thus the left term in Equation (7.6) treats the control as if it were the true motion of the robot. We recall from Chapter 3.3 that EKF linearization approximates the function g through a Taylor expansion:

$$(7.7) \quad g(u_t, x_{t-1}) \approx g(u_t, \mu_{t-1}) + G_t\,(x_{t-1} - \mu_{t-1})$$

The function $g(u_t, \mu_{t-1})$ is simply obtained by replacing the exact state x_{t-1}—which we do not know—by our expectation μ_{t-1}—which we know. The Jacobian G_t is the derivative of the function g with respect to x_{t-1} evaluated at u_t and μ_{t-1}:

$$(7.8) \quad G_t = \frac{\partial g(u_t, \mu_{t-1})}{\partial x_{t-1}} = \begin{pmatrix} \frac{\partial x'}{\partial \mu_{t-1,x}} & \frac{\partial x'}{\partial \mu_{t-1,y}} & \frac{\partial x'}{\partial \mu_{t-1,\theta}} \\ \frac{\partial y'}{\partial \mu_{t-1,x}} & \frac{\partial y'}{\partial \mu_{t-1,y}} & \frac{\partial y'}{\partial \mu_{t-1,\theta}} \\ \frac{\partial \theta'}{\partial \mu_{t-1,x}} & \frac{\partial \theta'}{\partial \mu_{t-1,y}} & \frac{\partial \theta'}{\partial \mu_{t-1,\theta}} \end{pmatrix}$$

Here $\mu_{t-1} = (\mu_{t-1,x} \; \mu_{t-1,y} \; \mu_{t-1,\theta})^T$ denotes the mean estimate factored into its individual three values, and $\frac{\partial x'}{\partial \mu_{t-1,x}}$ is short for the derivative of g along the x' dimension, taken with respect to x at μ_{t-1}. Calculating these derivatives from Equation (7.6) gives us the following matrix:

$$(7.9) \quad G_t = \begin{pmatrix} 1 & 0 & \frac{v_t}{\omega_t}(-\cos \mu_{t-1,\theta} + \cos(\mu_{t-1,\theta} + \omega_t \Delta t)) \\ 0 & 1 & \frac{v_t}{\omega_t}(-\sin \mu_{t-1,\theta} + \sin(\mu_{t-1,\theta} + \omega_t \Delta t)) \\ 0 & 0 & 1 \end{pmatrix}$$

To derive the covariance of the additional motion noise, $\mathcal{N}(0, R_t)$, we first determine the covariance matrix M_t of the noise in *control space*. This follows directly from the motion model in Equation (7.5):

$$(7.10) \quad M_t = \begin{pmatrix} \alpha_1 v_t^2 + \alpha_2 \omega_t^2 & 0 \\ 0 & \alpha_3 v_t^2 + \alpha_3 \omega_t^2 \end{pmatrix}$$

The motion model in (7.6) requires this motion noise to be mapped into *state space*. The transformation from control space to state space is performed by another linear approximation. The Jacobian needed for this approximation, denoted V_t, is the derivative of the motion function g with respect to the motion parameters, evaluated at u_t and μ_{t-1}:

$$(7.11) \quad V_t = \frac{\partial g(u_t, \mu_{t-1})}{\partial u_t}$$

$$= \begin{pmatrix} \frac{\partial x'}{\partial v_t} & \frac{\partial x'}{\partial \omega_t} \\ \frac{\partial y'}{\partial v_t} & \frac{\partial y'}{\partial \omega_t} \\ \frac{\partial \theta'}{\partial v_t} & \frac{\partial \theta'}{\partial \omega_t} \end{pmatrix}$$

$$= \begin{pmatrix} \frac{-\sin\theta + \sin(\theta + \omega_t \Delta t)}{\omega_t} & \frac{v_t(\sin\theta - \sin(\theta + \omega_t \Delta t))}{\omega_t^2} + \frac{v_t \cos(\theta + \omega_t \Delta t)\Delta t}{\omega_t} \\ \frac{\cos\theta - \cos(\theta + \omega_t \Delta t)}{\omega_t} & -\frac{v_t(\cos\theta - \cos(\theta + \omega_t \Delta t))}{\omega_t^2} + \frac{v_t \sin(\theta + \omega_t \Delta t)\Delta t}{\omega_t} \\ 0 & \Delta t \end{pmatrix}$$

The multiplication $V_t \, M_t \, V_t^T$ then provides an approximate mapping between the motion noise in control space to the motion noise in state space.

7.4 EKF Localization

With this derivation, lines 6 and 7 of the EKF localization algorithm correspond exactly to the prediction updates of the general EKF algorithm, described in Table 3.3.

Correction Step (Lines 8–20) To perform the correction step, EKF localization also requires a linearized measurement model with additive Gaussian noise. The measurement model for our feature-based maps shall be a variant of Equation (6.40) in Chapter 6.6, which presupposes knowledge of the landmark identity via the correspondence variable c_t. Let $j = c_t^i$ be the identity of the landmark that corresponds to the i-th component in the measurement vector. Then we have

$$(7.12) \quad \underbrace{\begin{pmatrix} r_t^i \\ \phi_t^i \\ s_t^i \end{pmatrix}}_{z_t^i} = \underbrace{\begin{pmatrix} \sqrt{(m_{j,x} - x)^2 + (m_{j,y} - y)^2} \\ \text{atan2}(m_{j,y} - y, m_{j,x} - x) - \theta \\ m_{j,s} \end{pmatrix}}_{h(x_t, j, m)} + \mathcal{N}(0, Q_t),$$

where $(m_{j,x} \ m_{j,y})^T$ are the coordinates of the i-th landmark detection at time t, and $m_{j,s}$ is its (correct) signature. The Taylor approximation of this measurement model is

$$(7.13) \quad h(x_t, j, m) \approx h(\bar{\mu}_t, j, m) + H_t^i (x_t - \bar{\mu}_t).$$

H_t^i is the Jacobian of h with respect to the robot location, computed at the predicted mean $\bar{\mu}_t$:

$$(7.14) \quad H_t^i = \frac{\partial h(\bar{\mu}_t, j, m)}{\partial x_t} = \begin{pmatrix} \frac{\partial r_t^i}{\partial \bar{\mu}_{t,x}} & \frac{\partial r_t^i}{\partial \bar{\mu}_{t,y}} & \frac{\partial r_t^i}{\partial \bar{\mu}_{t,\theta}} \\ \frac{\partial \phi_t^i}{\partial \bar{\mu}_{t,x}} & \frac{\partial \phi_t^i}{\partial \bar{\mu}_{t,y}} & \frac{\partial \phi_t^i}{\partial \bar{\mu}_{t,\theta}} \\ \frac{\partial s_t^i}{\partial \bar{\mu}_{t,x}} & \frac{\partial s_t^i}{\partial \bar{\mu}_{t,y}} & \frac{\partial s_t^i}{\partial \bar{\mu}_{t,\theta}} \end{pmatrix}$$

$$= \begin{pmatrix} -\frac{m_{j,x} - \bar{\mu}_{t,x}}{\sqrt{q}} & -\frac{m_{j,y} - \bar{\mu}_{t,y}}{\sqrt{q}} & 0 \\ \frac{m_{j,y} - \bar{\mu}_{t,y}}{q} & -\frac{m_{j,x} - \bar{\mu}_{t,x}}{q} & -1 \\ 0 & 0 & 0 \end{pmatrix}$$

with q short for $(m_{j,x} - \bar{\mu}_{t,x})^2 + (m_{j,y} - \bar{\mu}_{t,y})^2$. Notice that the last row of H_t^i is all zero. This is because the signature does not depend on the robot pose. The effect of this degeneracy is that the observed signature s_t^i has no effect on the result of the EKF update. This should come at no surprise: knowledge of the correct correspondence c_t^i renders the observed signature entirely uninformative.

The covariance Q_t of the additional measurement noise in Equation (7.12) follows directly from (6.40):

$$(7.15) \quad Q_t = \begin{pmatrix} \sigma_r^2 & 0 & 0 \\ 0 & \sigma_\phi^2 & 0 \\ 0 & 0 & \sigma_s^2 \end{pmatrix}$$

Finally, we note that our feature-based localizer processes multiple measurements at a time, whereas the EKF discussed in Chapter 3.2 only processed a single sensor item. Our algorithm relies on an implicit conditional independence assumption, which we briefly discussed in Chapter 6.6, Equation (6.39). Essentially, we assume that all feature measurement probabilities are independent given the pose x_t, the landmark identities c_t, and the map m:

$$(7.16) \quad p(z_t \mid x_t, c_t, m) = \prod_i p(z_t^i \mid x_t, c_t^i, m)$$

This is usually a good assumption, especially if the world is static. It enables us to incrementally add the information from multiple features into our filter, as specified in lines 9 through 18 in Table 7.2. Care has to be taken that the pose estimate is updated in each iteration of the loop, since otherwise the algorithm computes incorrect observation predictions (intuitively, this loop corresponds to multiple observation updates with zero motion in between). With this in mind it is straightforward to see that lines 8–20 are indeed an implementation of the general EKF correction step.

MEASUREMENT LIKELIHOOD

Measurement Likelihood (Line 21) Line 21 computes the *likelihood* $p(z_t \mid c_{1:t}, m, z_{1:t-1}, u_{1:t})$ of a measurement z_t. This likelihood is not essential for the EKF update but is useful for the purpose of outlier rejection or in the case of unknown correspondences. Assuming independence between the individual feature vectors, we can restrict the derivation to individual feature vectors z_i^t and compute the overall likelihood analogous to (7.16). For known data associations $c_{1:t}$, the likelihood can be computed from the predicted belief $\overline{bel}(x_t) = \mathcal{N}(x_t; \bar{\mu}_t, \bar{\Sigma}_t)$ by integrating over the pose x_t, and omitting irrelevant conditioning variables:

$$(7.17) \quad p(z_t^i \mid c_{1:t}, m, z_{1:t-1}, u_{1:t})$$
$$= \int p(z_t^i \mid x_t, c_{1:t}, m, z_{1:t-1}, u_{1:t}) \, p(x_t \mid c_{1:t}, m, z_{1:t-1}, u_{1:t}) \, dx_t$$
$$= \int p(z_t^i \mid x_t, c_t^i, m) \, p(x_t \mid c_{1:t-1}, m, z_{1:t-1}, u_{1:t}) \, dx_t$$

7.4 EKF Localization

$$= \int p(z_t^i \mid x_t, c_t^i, m) \, \overline{bel}(x_t) \, dx_t$$

The left term in the final integral is the measurement likelihood assuming knowledge of the robot location x_t. This likelihood is given by a Gaussian with mean at the measurement that is expected at location x_t. This measurement, denoted $\hat{z}_t^i$, is provided by the measurement function h. The covariance of the Gaussian is given by the measurement noise Q_t.

(7.18) $\quad p(z_t^i \mid x_t, c_t^i, m) \;\sim\; \mathcal{N}(z_t^i; \, h(x_t, c_t^i, m), \, Q_t)$
$\qquad\qquad\qquad\quad \approx \; \mathcal{N}(z_t^i; \, h(\bar{\mu}_t, c_t^i, m) + H_t \,(x_t - \bar{\mu}_t), \, Q_t)$

(7.18) follows by applying our Taylor expansion (7.13) to h. Plugging this equation back into (7.17), and replacing $\overline{bel}(x_t)$ by its Gaussian form, we get the following measurement likelihood:

(7.19) $\quad p(z_t^i \mid c_{1:t}, m, z_{1:t-1}, u_{1:t})$
$\qquad \approx \; \mathcal{N}(z_t^i; \, h(\bar{\mu}_t, c_t^i, m) + H_t\,(x_t - \bar{\mu}_t), \, Q_t) \;\otimes\; \mathcal{N}(x_t; \, \bar{\mu}_t, \bar{\Sigma}_t)$

where $\otimes$ denotes the familiar convolution over the variable x_t. This equation reveals that the likelihood function is a convolution of two Gaussians; one representing the measurement noise, the other representing the state uncertainty. We already encountered integrals of this form in Chapter 3.2, where we derived the motion update of the Kalman filter and the EKF. The closed-form solution to this integral is derived completely analogously to those derivations. In particular, the Gaussian defined by (7.19) has mean $h(\bar{\mu}_t, c_t^i, m)$ and covariance $H_t \, \bar{\Sigma}_t \, H_t^T + Q_t$. Thus, we have under our linear approximation the following expression for the measurement likelihood:

(7.20) $\quad p(z_t^i \mid c_{1:t}, m, z_{1:t-1}, u_{1:t}) \;\sim\; \mathcal{N}(z_t^i; \, h(\bar{\mu}_t, c_t^i, m), \, H_t \bar{\Sigma}_t H_t^T + Q_t)$

That is,

(7.21) $\quad p(z_t^i \mid c_{1:t}, m, z_{1:t-1}, u_{1:t})$
$\qquad = \; \eta \, \exp\left\{-\frac{1}{2}(z_t^i - h(\bar{\mu}_t, c_t^i, m))^T \, [H_t \bar{\Sigma}_t H_t^T + Q_t]^{-1} \, (z_t^i - h(\bar{\mu}_t, c_t^i, m))\right\}$

By replacing the mean and covariance of this expression by $\hat{z}_t^i$ and S_t, respectively, we get line 21 of the EKF algorithm in Table 7.2.

The EKF localization algorithm can now easily be modified to accommodate outliers. The standard approach is to only accept landmarks for which the likelihood passes a threshold test. This is generally a good idea: Gaussians fall off exponentially, and a single outlier can have a huge effect on the pose estimate. In practice, thresholding adds an important layer of robustness to the algorithm without which EKF localization tends to be brittle.

Figure 7.7 AIBO robots on the RoboCup soccer field. Six landmarks are placed at the corners and the midlines of the field.

7.4.4 Physical Implementation

We now illustrate the EKF algorithm using simulations of a four-legged AIBO robot localizing on a RoboCup soccer field. Here, the robot localizes using six uniquely colored markers placed around the field (see Figure 7.7). Just like in the EKF algorithm given in Table 7.2, motion control $u_t = (v_t \; \omega_t)^T$ is modeled by translational and rotational velocity, and observations $z_t = (r_t \; \phi_t \; s_t)^T$ measure relative distance and bearing to a marker. For simplicity we assume that the robot detects only one landmark at a time.

Prediction Step (Lines 3–7) Figure 7.8 illustrates the prediction step of the EKF localization algorithm. Shown there are the prediction uncertainties resulting from different motion noise parameters, $\alpha_1 - \alpha_4$, used in line 5 of the algorithm. The parameters α_2 and α_3 are set to 5% in all visualizations. The main translational and rotational noise parameters α_1 and α_4 vary between $\langle 10\%, 10\% \rangle$, $\langle 30\%, 10\% \rangle$, $\langle 10\%, 30\% \rangle$, and $\langle 30\%, 30\% \rangle$ (from upper left to lower right in Figure 7.8). In each of the plots the robot executes the control $u_t = \langle 10\text{cm/sec}, 5°/\text{sec}\rangle$ for 9 seconds, resulting in a circular arc of length 90cm and rotation 45°. The robot's previous location estimate is represented by the ellipse centered at the mean $\mu_{t-1} = \langle 80, 100, 0 \rangle$.

The EKF algorithm computes the predicted mean $\bar{\mu}_t$ by shifting the previous estimate under the assumption of noise free motion (line 6). The corresponding uncertainty ellipse, $\bar{\Sigma}_t$, consists of two components; one estimating

7.4 EKF Localization

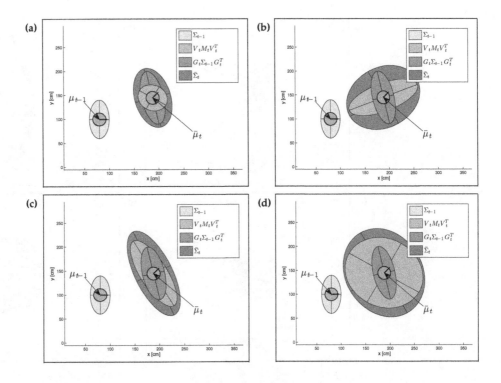

Figure 7.8 Prediction step of the EKF algorithm. The panels were generated with different motion noise parameters. The robot's initial estimate is represented by the ellipse centered at μ_{t-1}. After moving on a circular arc of 90cm length while turning 45 degrees to the left, the predicted position is centered at $\bar{\mu}_t$. In panel (a), the motion noise is relatively small in both translation and rotation. The other panels represent (b) high translational noise, (c) high rotational noise, and (d) high noise in both translation and rotation.

uncertainty due to the initial location uncertainty, the other estimating uncertainty due to motion noise (line 7). The first component, $G_t \Sigma_{t-1} G_t^T$, ignores motion noise and projects the previous uncertainty Σ_{t-1} through a linear approximation of the motion function. Recall from Equations (7.8) and (7.9) that this linear approximation is represented by the matrix G_t, which is the Jacobian of the motion function w.r.t. the previous robot location.

The resulting noise ellipses are identical in the four panels since they do not consider motion noise. Uncertainty due to motion noise is modeled by the second component of $\bar{\Sigma}_t$, which is given by $V_t \, M_t \, V_t^T$ (line 7). The matrix

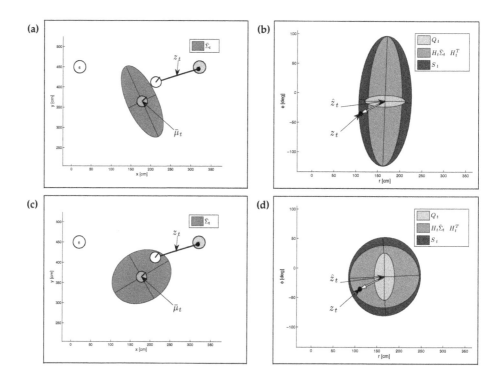

Figure 7.9 Measurement prediction. The left plots show two predicted robot locations along with their uncertainty ellipses. The true robot and the observation are indicated by the white circle and the bold line, respectively. The panels on the right show the resulting measurement predictions. The white arrows indicate the innovations, the differences between observed and predicted measurements.

M_t represents the motion noise in control space (line 5). This motion noise matrix is mapped into state space by multiplication with V_t, which is the Jacobian of the motion function w.r.t. motion control (line 4). As can be seen, the resulting ellipse represents large translational velocity error ($\alpha_1 = 30\%$) by large uncertainty along the motion direction (right plots in Figure 7.8). Large rotational error ($\alpha_4 = 30\%$) results in large uncertainty orthogonal to the motion direction (lower plots in Figure 7.8). The overall uncertainty of the prediction, $\bar{\Sigma}_t$, is then given by adding the two uncertainty components.

Correction Step: Measurement Prediction (Lines 8–14) In the first part of the correction step, the EKF algorithm predicts the measurement, $\bar{z}_t$, using

7.4   EKF Localization   213

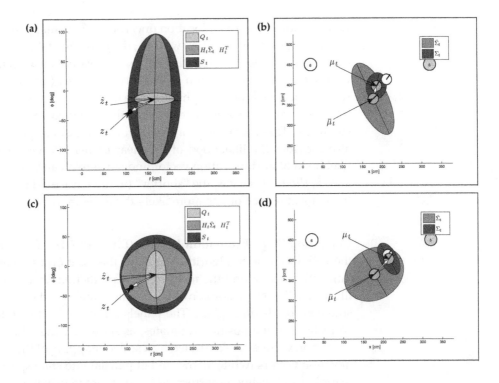

Figure 7.10 Correction step of the EKF algorithm. The panels on the left show the measurement prediction, and the panels on the right the resulting corrections, which update the mean estimate and reduce the position uncertainty ellipses.

the predicted robot location and its uncertainty. Figure 7.9 illustrates the measurement prediction. The left plots show the predicted robot locations along with their uncertainty ellipses. The true robot location is indicated by the white circle. Now assume that the robot observes the landmark ahead to its right, as indicated by the bold line. The panels on the right show the corresponding predicted and actual measurements in measurement space. The predicted measurement, $\bar{z}_t$, is computed from the relative distance and bearing between the predicted mean, $\bar{\mu}_t$, and the observed landmark (line 12). The uncertainty in this prediction is represented by the ellipse S_t. Similar to state prediction, this uncertainty results from a convolution of two Gaussians. The ellipse Q_t represents uncertainty due to measurement noise (line 8), and the ellipse $H_t \bar{\Sigma}_t H_t^T$ represents uncertainty due to uncertainty in the

MEASUREMENT INNOVATION

robot location. The robot location uncertainty $\bar{\Sigma}_t$ is mapped into observation uncertainty by multiplication with H_t, the Jacobian of the measurement function w.r.t. the robot location (line 13). S_t, the overall measurement prediction uncertainty, is then the sum of these two ellipses (line 14). The white arrows in the panels on the right illustrate the so-called *innovation vector* $z_t - \bar{z}_t$, which is simply the difference between the observed and the predicted measurement. This vector plays a crucial role in the subsequent update step. It also provides the likelihood of the measurement z_t, which is given by the likelihood of the innovation vector under a zero mean Gaussian with covariance S_t (line 21). That is, the "shorter" (in the sense of Mahalanobis distance) the innovation vector, the more likely the measurement.

Correction Step: Estimation Update (Lines 15–21) The correction step of the EKF localization algorithm updates the location estimate based on the innovation vector and the measurement prediction uncertainty. Figure 7.10 illustrates this step. For convenience, the panels on the left show the measurement prediction again. The panels on the right illustrate the resulting corrections in the position estimates, as shown by the white arrows. These correction vectors are computed by a scaled mapping of the measurement innovation vectors (white arrows in left panels) into state space (line 16). This mapping and scaling is performed by the Kalman gain matrix, K_t, computed in line 15. Intuitively, the measurement innovation gives the offset between predicted and observed measurement. This offset is then mapped into state space and used to move the location estimate in the direction that would reduce the measurement innovation. The Kalman gain additionally scales the innovation vector, thereby considering the uncertainty in the measurement prediction. The more certain the observation, the higher the Kalman gain, and hence the stronger the resulting location correction. The uncertainty ellipse of the location estimate is updated by similar reasoning (line 17).

Example Sequence Figure 7.11 shows two sequences of EKF updates, using different observation uncertainties. The left panels show the robot's trajectories according to the motion control (dashed lines) and the resulting true trajectories (solid lines). Landmark detections are indicated by thin lines, with the measurements in the upper panel being less noisy. The dashed lines in the right panels plot the paths as estimated by the EKF localization algorithm. As expected, the smaller measurement uncertainty in the upper row results in smaller uncertainty ellipses and in smaller estimation errors.

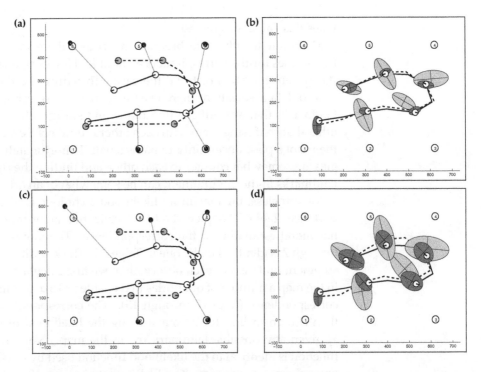

Figure 7.11 EKF-based localization with an accurate (upper row) and a less accurate (lower row) landmark detection sensor. The dashed lines in the left panel indicate the robot trajectories as estimated from the motion controls. The solid lines represent the true robot motion resulting from these controls. Landmark detections at five locations are indicated by the thin lines. The dashed lines in the right panels show the corrected robot trajectories, along with uncertainty before (light gray, $\bar{\Sigma}_t$) and after (dark gray, Σ_t) incorporating a landmark detection.

7.5 Estimating Correspondences

7.5.1 EKF Localization with Unknown Correspondences

The EKF localization discussed thus far is only applicable when landmark correspondences can be determined with absolute certainty. In practice, this is rarely the case. Most implementations therefore determine the identity of the landmark during localization. Throughout this book, we will encounter a number of strategies to cope with the correspondence problem. The most simple of all is known as *maximum likelihood correspondence*, in which one first determines the most likely value of the correspondence variable, and then

MAXIMUM LIKELIHOOD
CORRESPONDENCE

takes this value for granted.

Maximum likelihood techniques are brittle if there are many equally likely hypotheses for the correspondence variable. However, one can often design the system for this not to be the case. To reduce the danger of asserting a false data association, there exist essentially two techniques: First, select landmarks that are sufficiently unique and sufficiently far apart from each other that confusing them with each other is unlikely. Second, make sure that the robot's pose uncertainty remains small. Unfortunately, these two strategies are somewhat counter to each other, and finding the right granularity of landmarks in the environment can be somewhat of an art.

Nevertheless, the maximum likelihood technique is of great practical importance. Table 7.3 depicts the EKF localization algorithm with a maximum likelihood estimator for the correspondence. The motion update in lines 2 through 7 is identical to the one in Table 7.2. The key difference is in the measurement update: For each observation, we first calculate for all landmarks k in the map a number of quantities that enable us to determine the most likely correspondence (lines 10 through 15). The correspondence variable $j(i)$ is then chosen in line 16, by maximizing the likelihood of the measurement z_t^i given any possible landmark m_k in the map. Note that this likelihood function is identical to the likelihood function used by the EKF algorithm for known correspondences. The EKF update in lines 18 and 19 only incorporates the most likely correspondences.

We note that the algorithm in Table 7.3 may not quite be as efficient as it could be. It can be improved through a more thoughtful selection of landmarks in line 10. In most settings, the robot only sees a small number of landmarks at a time in its immediate vicinity; and simple tests can reject a large number of unlikely landmarks in the map.

7.5.2 Mathematical Derivation of the ML Data Association

The maximum likelihood estimator determines the correspondence that maximizes the data likelihood.

$$(7.22) \quad \hat{c}_t = \underset{c_t}{\operatorname{argmax}}\, p(z_t \mid c_{1:t}, m, z_{1:t-1}, u_{1:t})$$

Here c_t is the correspondence vector at time t. As before, the vector $z_t = \{z_t^1, z_t^2, \ldots\}$ is the measurement vector that contains the list of features, or landmarks, z_t^i, observed at time t.

The argmax operator in (7.22) selects the correspondence vector $\hat{c}_t$ that maximizes the likelihood of the measurement. Note that this expression is

1: **Algorithm EKF_localization**($\mu_{t-1}, \Sigma_{t-1}, u_t, z_t, m$):

2: $\theta = \mu_{t-1,\theta}$

3: $G_t = \begin{pmatrix} 1 & 0 & -\frac{v_t}{\omega_t}\cos\theta + \frac{v_t}{\omega_t}\cos(\theta + \omega_t \Delta t) \\ 0 & 1 & -\frac{v_t}{\omega_t}\sin\theta + \frac{v_t}{\omega_t}\sin(\theta + \omega_t \Delta t) \\ 0 & 0 & 1 \end{pmatrix}$

4: $V_t = \begin{pmatrix} \frac{-\sin\theta + \sin(\theta + \omega_t \Delta t)}{\omega_t} & \frac{v_t(\sin\theta - \sin(\theta + \omega_t \Delta t))}{\omega_t^2} + \frac{v_t \cos(\theta + \omega_t \Delta t)\Delta t}{\omega_t} \\ \frac{\cos\theta - \cos(\theta + \omega_t \Delta t)}{\omega_t} & -\frac{v_t(\cos\theta - \cos(\theta + \omega_t \Delta t))}{\omega_t^2} + \frac{v_t \sin(\theta + \omega_t \Delta t)\Delta t}{\omega_t} \\ 0 & \Delta t \end{pmatrix}$

5: $M_t = \begin{pmatrix} \alpha_1 v_t^2 + \alpha_2 \omega_t^2 & 0 \\ 0 & \alpha_3 v_t^2 + \alpha_4 \omega_t^2 \end{pmatrix}$

6: $\bar{\mu}_t = \mu_{t-1} + \begin{pmatrix} -\frac{v_t}{\omega_t}\sin\theta + \frac{v_t}{\omega_t}\sin(\theta + \omega_t \Delta t) \\ \frac{v_t}{\omega_t}\cos\theta - \frac{v_t}{\omega_t}\cos(\theta + \omega_t \Delta t) \\ \omega_t \Delta t \end{pmatrix}$

7: $\bar{\Sigma}_t = G_t \, \Sigma_{t-1} \, G_t^T + V_t \, M_t \, V_t^T$

8: $Q_t = \begin{pmatrix} \sigma_r^2 & 0 & 0 \\ 0 & \sigma_\phi^2 & 0 \\ 0 & 0 & \sigma_s^2 \end{pmatrix}$

9: for all observed features $z_t^i = (r_t^i \; \phi_t^i \; s_t^i)^T$ do

10:     for all landmarks k in the map m do

11:         $q = (m_{k,x} - \bar{\mu}_{t,x})^2 + (m_{k,y} - \bar{\mu}_{t,y})^2$

12:         $\hat{z}_t^k = \begin{pmatrix} \sqrt{q} \\ \text{atan2}(m_{k,y} - \bar{\mu}_{t,y}, m_{k,x} - \bar{\mu}_{t,x}) - \bar{\mu}_{t,\theta} \\ m_{k,s} \end{pmatrix}$

13:         $H_t^k = \begin{pmatrix} -\frac{m_{k,x} - \bar{\mu}_{t,x}}{\sqrt{q}} & -\frac{m_{k,y} - \bar{\mu}_{t,y}}{\sqrt{q}} & 0 \\ \frac{m_{k,y} - \bar{\mu}_{t,y}}{q} & -\frac{m_{k,x} - \bar{\mu}_{t,x}}{q} & -1 \\ 0 & 0 & 0 \end{pmatrix}$

14:         $S_t^k = H_t^k \, \bar{\Sigma}_t \, [H_t^k]^T + Q_t$

15:     endfor

16:     $j(i) = \underset{k}{\text{argmax}} \; \det\left(2\pi S_t^k\right)^{-\frac{1}{2}} \exp\left\{-\frac{1}{2}(z_t^i - \hat{z}_t^k)^T [S_t^k]^{-1}(z_t^i - \hat{z}_t^k)\right\}$

17:     $K_t^i = \bar{\Sigma}_t \, [H_t^{j(i)}]^T [S_t^{j(i)}]^{-1}$

18:     $\bar{\mu}_t = \bar{\mu}_t + K_t^i (z_t^i - \hat{z}_t^{j(i)})$

19:     $\bar{\Sigma}_t = (I - K_t^i H_t^{j(i)}) \, \bar{\Sigma}_t$

20: endfor

21: $\mu_t = \bar{\mu}_t$

22: $\Sigma_t = \bar{\Sigma}_t$

23: return μ_t, Σ_t

Table 7.3 The extended Kalman filter (EKF) localization algorithm with unknown correspondences. The correspondences $j(i)$ are estimated via a maximum likelihood estimator.

conditioned on prior correspondences $c_{1:t-1}$. While those have been estimated in previous update steps, the maximum likelihood approach treats them as if they are always correct. This has two important ramifications: It makes it possible to update the filter incrementally. But it also introduces brittleness in the filter, which tends to diverge when correspondence estimates are erroneous.

Even under the assumption of known prior correspondences, there are exponentially many terms in the maximization (7.22). When the number of detected landmarks per measurement is large, the number of possible correspondences may grow too large for practical implementations. The most common technique to avoid such an exponential complexity performs the maximization separately for each individual feature z_t^i in the measurement vector z_t. We already derived the likelihood function for individual features in the derivation of the EKF localization algorithm for known correspondences. Following Equations (7.17) through (7.20), the correspondence of each feature follows as:

$$\begin{aligned} (7.23) \quad \hat{c}_t^i &= \operatorname*{argmax}_{c_t^i} p(z_t^i \mid c_{1:t}, m, z_{1:t-1}, u_{1:t}) \\ &\approx \operatorname*{argmax}_{c_t^i} \mathcal{N}(z_t^i;\, h(\bar{\mu}_t, c_t^i, m),\, H_t \bar{\Sigma}_t H_t^T + Q_t) \end{aligned}$$

This calculation is implemented in line 16 in Table 7.3. This component-wise optimization is "justified" only when we happen to know that individual feature vectors are conditionally independent—an assumption that is usually adopted for convenience. Under this assumption, the term that is being maximized in (7.22) becomes a product of terms with disjoint optimization parameters, for which the maximum is attained when each individual factor is maximal, as determined in (7.23). Using this maximum likelihood data association, the correctness of the algorithm follows now directly from the correctness of the EKF localization algorithm with known correspondences.

7.6 Multi-Hypothesis Tracking

There exist a number of extensions of the basic EKF to accommodate situations where the correct data association cannot be determined with sufficient reliability. Several of those techniques will be discussed later in this book, hence our exposition at this point will be brief.

A classical technique that overcomes difficulties in data association is the *multi-hypothesis tracking* filter, or *MHT*. The MHT can represent a belief by

7.6 Multi-Hypothesis Tracking

multiple Gaussians. It represents the posterior by the mixture

$$(7.24) \quad bel(x_t) = \frac{1}{\sum_l \psi_{t,l}} \sum_l \psi_{t,l} \det(2\pi \Sigma_{t,l})^{-\frac{1}{2}} \exp\left\{-\tfrac{1}{2}(x_t - \mu_{t,l})^T \Sigma_{t,l}^{-1}(x_t - \mu_{t,l})\right\}$$

MIXTURE WEIGHT

Here l is the index of the mixture component. Each such component, or "track" in MHT jargon, is itself a Gaussian with mean $\mu_{t,l}$ and covariance $\Sigma_{t,l}$. The scalar $\psi_{t,l} \geq 0$ is a *mixture weight*. It determines the weight of the l-th mixture component in the posterior. Since the posterior is normalized by $\sum_l \psi_{t,l}$, each $\psi_{t,l}$ is a relative weight, and the contribution of the l-th mixture component depends on the magnitude of all other mixture weights.

As we shall see below when we describe the MHT algorithm, each mixture component relies on a unique sequence of data association decisions. Hence, it makes sense to write $c_{t,l}$ for the data association vector associated with the l-th track, and $c_{1:t,l}$ for all past and present data associations associated with the l-th mixture component. With this notation, we can now think of mixture components as contributing local belief functions conditioned on a unique sequence of data associations:

$$(7.25) \quad bel_l(x_t) = p(x_t \mid z_{1:t}, u_{1:t}, c_{1:t,l})$$

Here $c_{1:t,l} = \{c_{1,l}, c_{2,l}, \ldots, c_{t,l}\}$ denotes the sequence of correspondence vectors associated with the l-th track.

Before describing the MHT, it makes sense to discuss a completely intractable algorithm from which the MHT is derived. This algorithm is the full Bayesian implementation of the EKF under unknown data association. It is amazingly simple: Instead of selecting the most likely data association vector, our fictitious algorithm maintains them all. More specifically, at time t each mixture is split into many new mixtures, each conditioned on a unique correspondence vector c_t. Let m be the index of one of the new Gaussians, and l be the index from which this new Gaussian is derived, for the correspondence $c_{t,l}$. The weight of this new mixture is then set to

$$(7.26) \quad \psi_{t,m} = \psi_{t,l}\, p(z_t \mid c_{1:t-1,l}, c_{t,m}, z_{1:t-1}, u_{1:t})$$

This is the product of the mixture weight $\psi_{t,l}$ from which the new component was derived, times the likelihood of the measurement z_t under the specific correspondence vector that led to the new mixture component. In other words, we treat correspondences as latent variable and calculate the posterior likelihood that a mixture component is correct. A nice aspect of this approach is that we already know how to compute the measurement likelihood $p(z_t \mid c_{1:t-1,l}, c_{t,m}, z_{1:t-1}, u_{1:t})$ in Equation (7.26): It is simply the likelihood of the measurement computed in line 21 of the EKF localization algorithm for known data associations (Table 7.2). Thus, we can incrementally calculate the mixture weights for each new component. The only downside of this algorithm is the fact that the number of mixture components, or *tracks*, grows exponentially over time.

PRUNING

The MHT algorithm approximates this algorithm by keeping the number of mixture components small. This process is called *pruning*. Pruning terminates every component whose relative mixture weight

$$\frac{\psi_{t,l}}{\sum_m \psi_{t,m}} \tag{7.27}$$

is smaller than a threshold $\psi_{\min}$. It is easy to see that the number of mixture components is always at most $\psi_{\min}^{-1}$. Thus, the MHT maintains a compact posterior that can be updated efficiently. It is approximate in that it maintains a very small number of Gaussians, but in practice the number of plausible robot locations is usually very small.

We omit a formal description of the MHT algorithm at this point, and instead refer the reader to a large number of related algorithms in this book. We note than when implementing an MHT, it is useful to devise strategies for identifying low-likelihood tracks before instantiating them.

7.7 UKF Localization

UKF localization is a feature-based robot localization algorithm using the unscented Kalman filter. As described in Chapter 3.4, the UKF uses the unscented transform to linearize the motion and measurement models. Instead of computing derivatives of these models, the unscented transform represents Gaussians by sigma points and passes these through the models. Table 7.4 summarizes the UKF algorithm for landmark based robot localization. It assumes that only one landmark detection is contained in the observation z_t and that the identity of the landmark is known.

7.7.1 Mathematical Derivation of UKF Localization

The main difference between the localization version and the general UKF given in Table 3.4 is in the handling of prediction and measurement noise. Recall that the UKF in Table 3.4 is based on the assumption that prediction and measurement noise are additive. This made it possible to consider the

1: **Algorithm UKF_localization**($\mu_{t-1}, \Sigma_{t-1}, u_t, z_t, m$):

Generate augmented mean and covariance

2: $\quad M_t = \begin{pmatrix} \alpha_1 v_t^2 + \alpha_2 \omega_t^2 & 0 \\ 0 & \alpha_3 v_t^2 + \alpha_4 \omega_t^2 \end{pmatrix}$

3: $\quad Q_t = \begin{pmatrix} \sigma_r^2 & 0 \\ 0 & \sigma_\phi^2 \end{pmatrix}$

4: $\quad \mu_{t-1}^a = (\mu_{t-1}^T \; (0\;0)^T \; (0\;0)^T)^T$

5: $\quad \Sigma_{t-1}^a = \begin{pmatrix} \Sigma_{t-1} & 0 & 0 \\ 0 & M_t & 0 \\ 0 & 0 & Q_t \end{pmatrix}$

Generate sigma points

6: $\quad \mathcal{X}_{t-1}^a = (\mu_{t-1}^a \quad \mu_{t-1}^a + \gamma\sqrt{\Sigma_{t-1}^a} \quad \mu_{t-1}^a - \gamma\sqrt{\Sigma_{t-1}^a})$

Pass sigma points through motion model and compute Gaussian statistics

7: $\quad \bar{\mathcal{X}}_t^x = g(u_t + \mathcal{X}_t^u, \mathcal{X}_{t-1}^x)$

8: $\quad \bar{\mu}_t = \sum_{i=0}^{2L} w_i^{(m)} \bar{\mathcal{X}}_{i,t}^x$

9: $\quad \bar{\Sigma}_t = \sum_{i=0}^{2L} w_i^{(c)} (\bar{\mathcal{X}}_{i,t}^x - \bar{\mu}_t)(\bar{\mathcal{X}}_{i,t}^x - \bar{\mu}_t)^T$

Predict observations at sigma points and compute Gaussian statistics

10: $\quad \bar{\mathcal{Z}}_t = h(\bar{\mathcal{X}}_t^x) + \mathcal{X}_t^z$

11: $\quad \hat{z}_t = \sum_{i=0}^{2L} w_i^{(m)} \bar{\mathcal{Z}}_{i,t}$

12: $\quad S_t = \sum_{i=0}^{2L} w_i^{(c)} (\bar{\mathcal{Z}}_{i,t} - \hat{z}_t)(\bar{\mathcal{Z}}_{i,t} - \hat{z}_t)^T$

13: $\quad \Sigma_t^{x,z} = \sum_{i=0}^{2L} w_i^{(c)} (\bar{\mathcal{X}}_{i,t}^x - \bar{\mu}_t)(\bar{\mathcal{Z}}_{i,t} - \hat{z}_t)^T$

Update mean and covariance

14: $\quad K_t = \Sigma_t^{x,z} S_t^{-1}$

15: $\quad \mu_t = \bar{\mu}_t + K_t(z_t - \hat{z}_t)$

16: $\quad \Sigma_t = \bar{\Sigma}_t - K_t S_t K_t^T$

17: $\quad p_{z_t} = \det(2\pi S_t)^{-\frac{1}{2}} \exp\left\{-\frac{1}{2}(z_t - \hat{z}_t)^T S_t^{-1}(z_t - \hat{z}_t)\right\}$

18: $\quad$ return μ_t, Σ_t, p_{z_t}

Table 7.4 The unscented Kalman filter (UKF) localization algorithm, formulated here for a feature-based map and a robot equipped with sensors for measuring range and bearing. This version handles single feature observations only and assumes knowledge of the exact correspondence. L is the dimensionality of the augmented state vector, given by the sum of state, control, and measurement dimensions.

noise terms by simply adding their covariances R_t and Q_t to the predicted state and measurement uncertainty, respectively (lines 5 and 9 in Table 3.4).

UKF_localization provides an alternative, more accurate approach to considering the impact of noise on the estimation process. The key "trick" is to augment the state with additional components representing control and measurement noise. The dimensionality L of the augmented state is given by the sum of the state, control, and measurement dimensions, which is $3 + 2 + 2 = 7$ in this case (the signature of feature measurements is ignored for simplicity). Since we assume zero-mean Gaussian noise, the mean μ_{t-1}^a of the augmented state estimate is given by the mean of the location estimate, μ_{t-1}, and zero vectors for the control and measurement noise (line 4). The covariance Σ_{t-1}^a of the augmented state estimate is given by combining the location covariance, Σ_{t-1}, the control noise covariance, M_t, and the measurement noise covariance, Q_t, as done in line 5.

The sigma point representation of the augmented state estimate is generated in line 6, using Equation (3.66) of the unscented transform. In this example, $\mathcal{X}_{t-1}^a$ contains $2L + 1 = 15$ sigma points, each having components in state, control, and measurement space:

$$(7.28) \quad \mathcal{X}_{t-1}^a = \begin{pmatrix} \mathcal{X}_{t-1}^{x}{}^T \\ \mathcal{X}_t^{u}{}^T \\ \mathcal{X}_t^{z}{}^T \end{pmatrix}$$

We choose mixed time indices to make clear that $\mathcal{X}_{t-1}^x$ refers to x_{t-1} and the control and measurement components refer to u_t and z_t, respectively.

The location components $\mathcal{X}_{t-1}^x$ of these sigma points are then passed through the velocity motion model g, defined in Equation (5.9). Line 7 performs this prediction step by applying the velocity motion model defined in Equation (5.13), using the control u_t with the added control noise component $\mathcal{X}_{i,t}^u$ of each sigma point:

$$(7.29) \quad \bar{\mathcal{X}}_{i,t}^x = \mathcal{X}_{i,t-1}^x + \begin{pmatrix} -\frac{v_{i,t}}{\omega_{i,t}} \sin\theta_{i,t-1} + \frac{v_{i,t}}{\omega_{i,t}} \sin(\theta_{i,t-1} + \omega_{i,t}\Delta t) \\ \frac{v_{i,t}}{\omega_{i,t}} \cos\theta_{i,t-1} - \frac{v_{i,t}}{\omega_{i,t}} \cos(\theta_{i,t-1} + \omega_{i,t}\Delta t) \\ \omega_{i,t}\Delta t \end{pmatrix}$$

where

$$(7.30) \quad v_{i,t} = v_t + \mathcal{X}_{i,t}^{u[v]}$$

$$(7.31) \quad \omega_{i,t} = \omega_t + \mathcal{X}_{i,t}^{u[\omega]}$$

$$(7.32) \quad \theta_{i,t-1} = \mathcal{X}_{i,t-1}^{x[\theta]}$$

are generated from the control $u_t = (v_t \; \omega_t)^T$ and the individual components of the sigma points. For example, $\mathcal{X}_{i,t}^{u[v]}$ represents the translational velocity v_t of the i-th sigma point. The predicted sigma points, $\bar{\mathcal{X}}_t^x$, are thus a set of robot locations, each resulting from a different combination of previous location and control.

Lines 8 and 9 compute the mean and covariance of the predicted robot location, using the unscented transform technique. Line 9 does not require the addition of a motion noise term, which was necessary in the algorithm described in Table 3.4. This is due to the state augmentation, which results in predicted sigma points that already incorporate the motion noise. This fact additionally makes the redrawing of sigma points from the predicted Gaussian obsolete (see line 6 in Table 3.4).

In line 10, the predicted sigma points are then used to generate measurement sigma points based on the measurement model defined in Equation (6.40) in Chapter 6.6:

$$(7.33) \quad \bar{\mathcal{Z}}_{i,t} = \begin{pmatrix} \sqrt{(m_x - \bar{\mathcal{X}}_{i,t}^{x[x]})^2 + (m_y - \bar{\mathcal{X}}_{i,t}^{x[y]})^2} \\ \mathrm{atan2}(m_y - \bar{\mathcal{X}}_{i,t}^{x[y]}, m_x - \bar{\mathcal{X}}_{i,t}^{x[y]}) - \bar{\mathcal{X}}_{i,t}^{x[\theta]} \end{pmatrix} + \begin{pmatrix} \mathcal{X}_{i,t}^{z[r]} \\ \mathcal{X}_{i,t}^{z[\phi]} \end{pmatrix}$$

Observation noise is assumed to be additive in this case.

The remaining updated steps are identical to the general UKF algorithm stated in Table 3.4. Lines 11 and 12 compute the mean and covariance of the predicted measurement. The cross-covariance between robot location and observation is determined in line 13. Lines 14 through 16 update the location estimate. The likelihood of the measurement is computed from the innovation and the predicted measurement uncertainty, just like in the EKF localization algorithm given in Table 7.2.

7.7.2 Illustration

We now illustrate the UKF localization algorithm using the same examples as were used for the EKF localization algorithm. The reader is encouraged to compare the following figures to the ones shown in Chapter 7.4.4.

Prediction Step (Lines 2–9) Figure 7.12 illustrates the UKF prediction step for different motion noise parameters. The location components $\mathcal{X}_{t-1}^x$ of the sigma points generated from the previous belief are indicated by the cross marks located symmetrically around μ_{t-1}. The 15 sigma points have seven different robot locations, only five of which are visible in this x-y-projection.

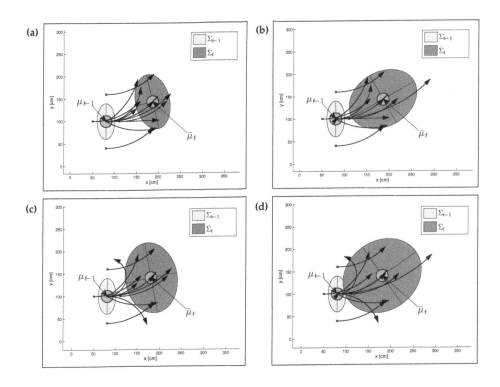

Figure 7.12 Prediction step of the UKF algorithm. The graphs were generated with different motion noise parameters. The robot's initial estimate is represented by the ellipse centered at μ_{t-1}. The robot moves on a circular arc of 90cm length while turning 45 degrees to the left. In panel (a), the motion noise is relatively small in both translation and rotation. The other panels represent (b) high translational noise, (c) high rotational noise, and (d) high noise in both translation and rotation.

The additional two points are located "on top" and "below" the mean sigma point, representing different robot orientations. The arcs indicate the motion prediction performed in line 7. As can be seen, 11 different predictions are generated, resulting from different combinations of previous location and motion noise. The panels illustrate the impact of the motion noise on these updates. The mean $\bar{\mu}_t$ and uncertainty ellipse $\bar{\Sigma}_t$ of the predicted robot location is generated from the predicted sigma points.

Measurement Prediction (Lines 10–12) In the measurement prediction step, the predicted robot locations $\bar{\mathcal{X}}_t^x$ are used to generate the measurement

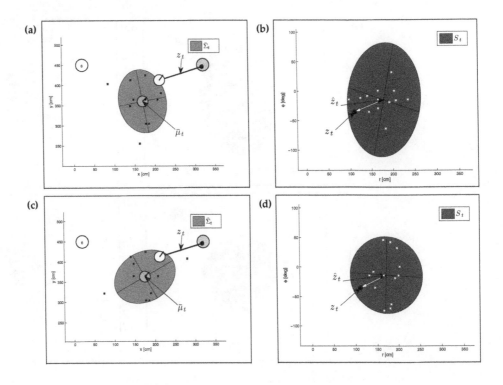

Figure 7.13 Measurement prediction. The left plots show the sigma points predicted from two motion updates along with the resulting uncertainty ellipses. The true robot and the observation are indicated by the white circle and the bold line, respectively. The panels on the right show the resulting measurement prediction sigma points. The white arrows indicate the innovations, the differences between observed and predicted measurements.

sigma points $\bar{\mathcal{Z}}_t$ (line 10). The black cross marks in the left plots of Figure 7.13 represent the location sigma points, and the white cross marks in the right plots indicate the resulting measurement sigma points. Note that the 11 different location sigma points generate 15 different measurements, which is due to different measurement noise components $\mathcal{X}_t^z$ being added in line 10. The panels also show the mean $\hat{z}_t$ and uncertainty ellipse S_t of the predicted measurement, extracted in lines 11 and 12.

Correction Step: Estimation Update (Lines 14–16) The correction step of the UKF localization algorithm is virtually identical to EKF correction step.

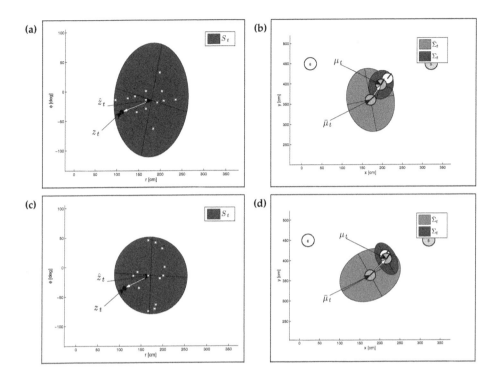

Figure 7.14 Correction step of the UKF algorithm. The panels on the left show the measurement prediction, and the panels on the right the resulting corrections, which update the mean estimate and reduce the position uncertainty ellipses.

The innovation vector and the measurement prediction uncertainty are used to update the estimate, as indicated by the white arrow in Figure 7.14.

Example Figure 7.15 shows a sequence of location estimates generated by a particle filter (upper right), the EKF (lower left), and the UKF (lower right). The upper left graph shows the robot's trajectory according to the motion control (dashed line) and the resulting true trajectory (solid line). Landmark detections are indicated by thin lines. The dashed lines in the other three panels show the paths estimated with the different techniques. The covariances of the particle filter estimates are extracted from the sample sets of a particle filter before and after the measurement update (see Table 8.2). The particle filter estimates are shown here as reference since the particle filter does not

7.7 UKF Localization

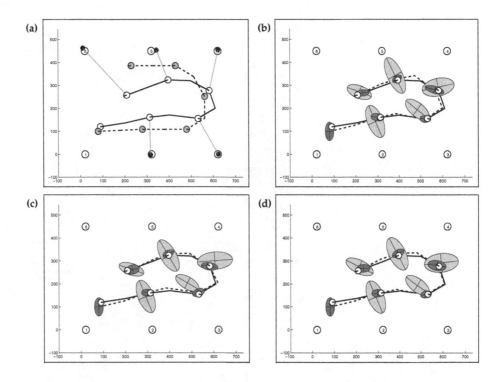

Figure 7.15 Comparison of UKF and EKF estimates: (a) Robot trajectory according to the motion control (dashed lines) and the resulting true trajectory (solid lines). Landmark detections are indicated by thin lines. (b) Reference estimates, generated by a particle filter. (c) EKF and (d) UKF estimates.

make any linearization approximations. As can be seen, the estimates of the EKF and the UKF are extremely close to these reference estimates, with the UKF being slightly closer.

The impact of the improved linearization applied by the UKF is more prominent in the example shown in Figure 7.16. Here, a robot performs two motion controls along the circle indicated by the thin line. The panels show the uncertainty ellipses after the two motions (the robot makes no observation). Again, the covariances extracted from exact, sample-based motion updates are shown as reference. The reference samples were generated using algorithm **sample_motion_model_velocity** in Table 5.3. While the EKF linearization incurs significant errors both in the location of the mean and in the "shape" of the covariance, the UKF estimates are almost identical to

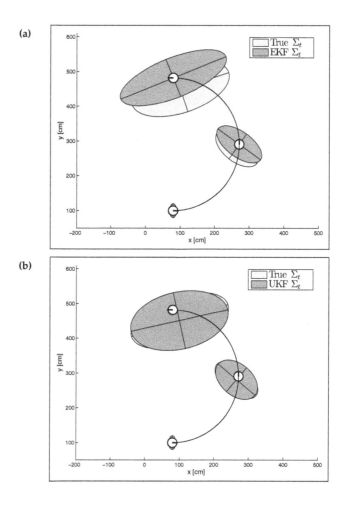

Figure 7.16 Approximation error due to linearization. The robot moves on a circle. Estimates based on (a) EKF prediction and (b) UKF prediction. The reference covariances are extracted from an accurate, sample-based prediction.

the reference estimates. This example also shows a subtle difference between the EKF and the UKF prediction. The mean predicted by the EKF is always exactly on the location predicted from the control (line 6 in Table 7.2). The UKF mean, on the other hand, is extracted from the sigma points and can therefore deviate from the mean predicted by the control (line 7 in Table 7.4).

7.8 Practical Considerations

The EKF localization algorithm and its close relative, MHT localization, are popular techniques for position tracking. There exist a large number of variations of these algorithms that enhance their efficiency and robustness.

- **Efficient search.** First, it is often impractical to loop through all landmarks k in the map, as is done by our EKF localization algorithm for unknown correspondences. Often, there exist simple tests to identify plausible candidate landmarks (e.g., by simply projecting the measurement into x-y-space), enabling one to rule out all but a constant number of candidates. Such algorithms can be orders of magnitude faster than our naive implementation.

- **Mutual exclusion.** A key limitation of our implementation arises from our assumed independence of feature noise in the EKF (and, by inheritance, the MHT). The reader may recall condition (7.16), which enabled us to process individual features sequentially, thereby avoiding a potential exponential search through the space of all correspondence vectors. Unfortunately, such an approach allows for assigning multiple observed features, say z_t^i and z_t^j with $i \neq j$, to be assigned to the same landmark in the map: $\hat{c}_t^i = \hat{c}_t^j$. For many sensors, such a correspondence assignment is wrong by default. For example, if the feature vector is extracted from a single camera image, we know by default that two different regions in the image space must correspond to different locations in the physical world. Put differently, we usually know that $i \neq j \longrightarrow \hat{c}_t^i \neq \hat{c}_t^i$. This (hard!) constraint is called *mutual exclusion principle in data association*. It reduces the space of all possible correspondence vectors. Advanced implementations consider this constraint. For example, one might first search for each correspondence separately—as in our version of the EKF localizer—followed by a "repair" phase in which violations of the mutual exclusion principle are resolved by changing correspondence values accordingly.

MUTUAL EXCLUSION PRINCIPLE IN DATA ASSOCIATION

- **Outlier rejection.** Further, our implementation does not address the issue of outliers. The reader may recall from Chapter 6.6 that we allow for a correspondence $c = N + 1$, with N being the number of landmarks in the map. Such an outlier test is quite easily added to the EKF localization algorithms. In particular, if we set π_{N+1} to be the a prior probability of an outlier, the argmax-step in line 16 of EKF localization (Table 7.3) can default to $N + 1$ if an outlier is the most likely explanation of the measurement vector. Clearly, an outlier does not provide any information on the robot's pose; hence, the pose-related terms are simply omitted in lines 18 and 19 in Table 7.3.

EKF and UKF localization are only applicable to position tracking problems. In general, linearized Gaussian techniques tend to work well only if the position uncertainty is small. There are several complimentary reasons for this observation:

- A unimodal Gaussian is usually a good representation of uncertainty in tracking whereas it is not in more general global localization problems.

- Even during tracking, unimodal Gaussians are not well suited to represent hard spatial constraints such as "the robot is close to a wall but can not be inside the wall". The severity of this limitation increases with the uncertainty in the robot location.

- A narrow Gaussian reduces the danger of erroneous correspondence decisions. This is important particularly for the EKF, since a single false correspondence can derail the tracker by inducing an entire stream of localization and correspondence errors.

- Linearization is usually only good in a close proximity to the linearization point. As a rule of thumb, if the standard deviation for the orientation θ is larger than ± 20 degrees, linearization effects are likely to make both the EKF and the UKF algorithms fail.

The MHT algorithm overcomes most of these problems, at the cost of increased computational complexity.

- It can solve the global localization problem by initializing the belief with multiple Gaussian hypotheses. The hypotheses can be initialized according to the first measurements.

- The kidnapped robot problem can be addressed by injecting additional hypotheses into the mixture.

- Hard spatial constraints are still hard to model, but can be approximated better using multiple Gaussians.

- The MHT is more robust to the problem of erroneous correspondence, though it can fail equally when the correct correspondence is not among those maintained in the Gaussian mixture.

- The MHT discussed here applies the same linearization as the EKF and suffers thus from similar approximation effects. The MHT can also be implemented using a UKF for each hypothesis.

The design of the appropriate features for Gaussian localization algorithms is a bit of an art. This is because multiple competing objectives have to be met. On the one hand, one wants sufficiently many features in the environment, so that the uncertainty in the robot's pose estimate can be kept small. Small uncertainty is absolutely vital for reasons already discussed. On the other hand, one wants to minimize chances that landmarks are confused with each other, or that the landmark detector detects spurious features. Many environments do not possess too many point landmarks that can be detected with high reliability, hence many implementations rely on relatively sparsely distributed landmarks. Here the MHT has a clear advantage, in that it is more robust to data association errors. As a rule of thumb, large numbers of landmarks tend to work better than small numbers even for the EKF and the UKF. When landmarks are dense, however, it is critical to apply the mutual exclusion principle in data association.

Finally, we note that EKF and UKF localization process only a subset of all information in the sensor measurement. By going from raw measurements to features, the amount of information that is being processed is already drastically reduced. Further, EKF and UKF localization are unable to process *negative information*. Negative information pertains to the absence of a feature. Clearly, not seeing a feature when one expects to see it carries relevant information. For example, not seeing the Eiffel Tower in Paris implies that it is unlikely that we are right next to it. The problem with negative information is that it induces non-Gaussian beliefs, which cannot be represented by the mean and variance. For this reason, EKF and UKF implementations simply ignore the issue of negative information, and instead integrate only information from observed features. The standard MHT also avoids negative information. However, it is possible to fold negative information into the mixture weight, by decaying mixture components that failed to observe a landmark.

NEGATIVE INFORMATION

With all these limitations, does this mean that Gaussian techniques are brittle localization techniques? The answer is no. The EKF, the UKF, and especially the MHT are surprisingly robust to violations of the linear system assumptions. In fact, the key to successful localization lies in successful data association. Later in this book, we will encounter more sophisticated techniques for handling correspondences than the ones discussed thus far.

Many of these techniques are applicable (and will be applied) to Gaussian representations, and the resulting algorithms are often among the best ones known.

7.9 Summary

In this chapter, we introduced the mobile robot localization problem and devised a first practical algorithm for solving it.

- The localization problem is the problem of estimating a robot's pose relative to a known map of its environment.

- Position tracking addresses the problem of accommodating the local uncertainty of a robot whose initial pose is known; global localization is the more general problem of localizing a robot from scratch. Kidnapping is a localization problem in which a well-localized robot is secretly teleported somewhere else without being told—it is the hardest of the three localization problems.

- The hardness of the localization problem is also a function of the degree to which the environment changes over time. All algorithms discussed thus far assume a static environment.

- Passive localization approaches are filters: they process data acquired by the robot but do not control the robot. Active techniques control the robot during localization, with the purpose of minimizing the robot's uncertainty. So far, we have only studied passive algorithms. Active algorithms will be discussed in Chapter 17.

- Markov localization is just a different name for the Bayes filter applied to the mobile robot localization problem.

- EKF localization applies the extended Kalman filter to the localization problem. EKF localization is primarily applied to feature-based maps.

- The most common technique for dealing with correspondence problems is the maximum likelihood technique. This approach simply assumes that at each point in time, the most likely correspondence is correct.

- The multi hypothesis tracking algorithm (MHT) pursues multiple correspondences, using a Gaussian mixture to represent the posterior. Mixture components are created dynamically, and terminated if their total likelihood sinks below a user-specified threshold.

- The MHT is more robust to data association problems than the EKF, at an increased computational cost. The MHT can also be implemented using UKF's for the individual hypotheses.

- UKF localization uses the unscented transform to linearize the motion and measurement models in the localization problem.

- All Gaussian filters are well-suited for local position tracking problems with limited uncertainty and in environments with distinct features. The EKF and UKF are less applicable to global localization or in environments where most objects look alike.

- Selecting features for Gaussian filters requires skill. Features must be sufficiently unambiguous to minimize the likelihood of confusing them, and there must be enough of them that the robot frequently encounters features.

- The performance of Gaussian localization algorithms can be improved by a number of measures, such as enforcing mutual exclusion in data association.

In the next chapter, we will discuss alternative localization techniques that aim at dealing with the limitations of the EKF by using different representations of the robot's belief.

7.10 Bibliographical Remarks

Localization has been dubbed "the most fundamental problem to providing a mobile robot with autonomous capabilities" (Cox 1991). The use of EKF for state estimation in outdoor robotics was pioneered by Dickmanns and Graefe (1988), who used EKFs to estimate highway curvature from camera images. Much of the early work on indoor mobile robot localization is surveyed in Borenstein et al. (1996) (see also (Feng et al. 1994)). Cox and Wilfong (1990) provides an early text on the state of the art in mobile robotics, which also covers localization. Many of the early techniques required environmental modifications, such as through artificial beacons. For example, Leonard and Durrant-Whyte (1991) used EKFs when matching geometric beacons extracted from sonar scans with beacons predicted from a geometric map of the environment. The practice of using artificial markers continues to the present day (Salichs et al. 1999), since often the modification of the environment is both feasible and economical. Other early researchers used lasers to scan unmodified environments (Hinkel and Knieriemen 1988).

Moving away from point features, a number of researchers developed more geometric techniques for localization. For example, Cox (1991) developed an algorithm for matching distances measured by infrared sensors, and line segment descriptions of the environment. An approach by Weiss et al. (1994) correlated range measurements for localization. The idea of *map matching*—specifically the comparison of a local occupancy grid map with a global environment map—is due to Moravec (1988). A gradient descent localizer based on this idea was described in Thrun (1993), and used at the first AAAI competition in 1992 (Simmons et al. 1992). Schiele and Crowley (1994) systematically compared different strategies to track the robot's position based on occupancy grid maps and ultrasonic sensors. They showed that matching local occupancy grid maps with a global grid map results in a similar localization performance as if the matching is based on features that are extracted from both maps. Shaffer et al. (1992) compare the robustness of map-matching and feature-based techniques, showing that combinations of both yielded the best empirical results. Yamauchi and Langley (1997) showed the robustness of map matching to environmental change. The idea of using *scan matching* for localization in robotics goes back to Lu and Milios (1994); Gutmann and Schlegel (1996); Lu and Milios (1998), although the basic principle had been popular in other fields (Besl and McKay 1992). A similar technique was proposed by Arras and Vestli (1998), who showed that scan matching made it possible to localize a robot with remarkable accuracy. Ortin et al. (2004) found that using camera data along a laser stripe increases the robustness of range scan matching.

A different strain of research investigated geometric techniques for localization (Betke and Gurvits 1994). The term "kidnapped robot problem" goes back to Engelson and McDermott (1992). The name "Markov localization" is due to Simmons and Koenig (1995), whose localizer used a grid to represent posterior probabilities. However, the intellectual roots of this work goes back to Nourbakhsh et al. (1995), who developed the idea of "certainty factors" for mobile robot localizations. While the update rules for certainty factors did not exactly follow the laws of probability, they captured the essential idea of multi-hypothesis estimation. A seminal paper by Cox and Leonard (1994) also developed this idea, by dynamically maintaining trees of hypotheses for a localizing robot. Fuzzy logic has been proposed for mobile localization by Saffiotti (1997); see also Driankov and Saffiotti (2001).

7.11 Exercises

1. Suppose a robot is equipped with a sensor for measuring range and bearing to a landmark; and for simplicity suppose that the robot can also sense the landmark identity (the identity sensor is noise-free). We want to perform global localization with EKFs. When seeing a single landmark, the posterior is usually poorly approximated by a Gaussian. However, when sensing two or more landmarks at the same time, the posterior is often well-approximated with a Gaussian.

    (a) Explain why.

    (b) Given k simultaneous measurements of ranges and bearings of k identifiably landmarks, devise a procedure for calculating a Gaussian pose estimate for the robot under uniform initial prior. You should start with the range/bearing measurement model provided in Chapter 6.6.

2. In this question we seek to design *hard* environments for global localization. Suppose we can compose a planar environment out of n non-intersecting straight line segments. The free space in the environment

has to be confined; however, there might be island of occupied terrain inside the map. For the sake of this exercise, we assume that the robot is equipped with a circular array of 360 range finders, and that these finders never err.

(a) What is the maximum number of distinct modes that a globally localizing robot might encounter in its belief function? For $n = 3, \ldots, 8$, draw examples of worst-case environments, along with a plausible belief that maximizes the number of modes.

(b) Does your analysis change if the range finders are allowed to err? In particular, can you give an example for $n = 4$ in which the number of plausible modes is larger than the ones derived above? Show such an environment along with the (erroneous) range scan and the posterior.

3. You are requested to derive an EKF localization algorithm for a simplistic underwater robot. The robot lives in a 3-D space and is equipped with a perfect compass (it always knows its orientation). For simplicity, we assume the robot move independently in all three Cartesian directions (x, y, and z), by setting velocities $\dot{x}, \dot{y}, \dot{z}$. Its motion noise is Gaussian and independent for all directions.

The robot is surrounded by a number of beacons that emit acoustic signals. The emission time of each signal is known, and the robot can determine from each signal the identity of the emitting beacon (hence there is no correspondence problem). The robot also knows the location of all beacons, and it is given an accurate clock to measure the arrival time of each signal. However, the robot cannot sense the direction from which it received a signal.

(a) You are asked to devise an EKF localization algorithm for this robot. This involves a mathematical derivation of the motion and the measurement model, along with the Taylor approximation. It also involves the statement of the final EKF algorithm, assuming known correspondence.

(b) Implement your EKF algorithm and a simulation of the environment. Investigate the accuracy and the failure modes of the EKF localizer in the context of the three localization problems: global localization, position tracking, and the kidnapped robot problem.

4. Consider a simplified global localization in any of the following six grid-style environments:

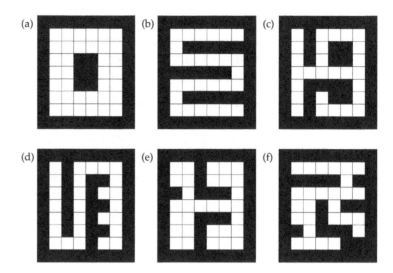

In each environment, the robot will be placed at the random position facing North. You are devised to come up with an open-loop localization strategy that contains a sequence of the following commands

Action L: Turn left 90 degrees.
Action R: Turn right 90 degrees.
Action M: Move forward until you hit an obstacle.

At the end of this strategy, the robot must be at a predictable location. For each such environment, provide a *shortest* such sequence (only "M" actions count). State where the robot will be after executing your action sequence. If no such sequence exists, explain why.

5. Now assume the robot can sense the number of steps it takes while executing an "M" action in the previous exercise. What will be the shortest open-loop sequence for the robot to determine its location? Explain your answer.

Notice: For this question, it might happen that the final location of the robot is a function of its starting location. All we ask here is that the robot localize itself.

8 Mobile Robot Localization: Grid And Monte Carlo

8.1 Introduction

This chapter describes two localization algorithms that are capable of solving global localization problems. The algorithms discussed here possess a number of differences to the unimodal Gaussian techniques discussed in the previous chapter.

- They can process raw sensor measurements. There is no need to extract features from sensor values. As a direct implication, they can also process negative information.

- They are non-parametric. In particular, they are not bound to a unimodal distribution as was the case with the EKF localizer.

- They can solve global localization and—in some instances—kidnapped robot problems. The EKF algorithm is not able to solve such problems— although the MHT (multi-hypothesis tracking) can be modified so as to solve global localization problems.

The techniques presented here have exhibited excellent performance in a number of fielded robotic systems.

The first approach is called *grid localization*. It uses a histogram filter to represent the posterior belief. A number of issues arise when implementing grid localization: with a fine-grained grid, the computation required for a naive implementation may make the algorithm intolerably slow. With a coarse grid, the additional information loss through the discretization negatively affects the filter and—if not properly treated—may even prevent the filter from working.

1:    **Algorithm Grid_localization**($\{p_{k,t-1}\}, u_t, z_t, m$):
2:        for all k do
3:            $\bar{p}_{k,t} = \sum_i p_{i,t-1}$ **motion_model**$(\text{mean}(\mathbf{x}_k), u_t, \text{mean}(\mathbf{x}_i))$
4:            $p_{k,t} = \eta \, \bar{p}_{k,t}$ **measurement_model**$(z_t, \text{mean}(\mathbf{x}_k), m)$
5:        endfor
6:        return $\{p_{k,t}\}$

Table 8.1 Grid localization, a variant of the discrete Bayes filter. The function **motion_model** implements one of the motion models, and **measurement_model** a sensor model. The function "mean" returns the center-of-mass of a grid cell $\mathbf{x}_k$.

The second approach is the Monte Carlo localization (MCL) algorithm, arguably the most popular localization algorithm to date. It uses particle filters to estimate posteriors over robot poses. A number of shortcomings of the MCL are discussed, and techniques for applying it to the kidnapped robot problem and to dynamic environments are presented.

8.2 Grid Localization

8.2.1 Basic Algorithm

Grid localization approximates the posterior using a *histogram filter* over a grid decomposition of the pose space. The discrete Bayes filter was already extensively discussed in Chapter 4.1 and is depicted in Table 4.1. It maintains as posterior a collection of discrete probability values

(8.1) $bel(x_t) = \{p_{k,t}\}$

where each probability $p_{k,t}$ is defined over a grid cell $\mathbf{x}_k$. The set of all grid cells forms a partition of the space of all legitimate poses:

(8.2) $\text{domain}(X_t) = \mathbf{x}_{1,t} \cup \mathbf{x}_{2,t} \cup \ldots \mathbf{x}_{K,t}$

In the most basic version of grid localization, the partitioning of the space of all poses is time-invariant, and each grid cell is of the same size. A common granularity used in many of the indoor environments is 15 centimeters for the x- and y-dimensions, and 5 degrees for the rotational dimension. A finer representation yields better results, but at the expense of increased computation.

Grid localization is largely identical to the basic histogram filter from which it is derived. Table 8.1 provides pseudo-code for the most basic implementation. It requires as input the discrete probability values $\{p_{t-1,k}\}$, along with the most recent measurement, control, and the map. Its inner loop iterates through all grid cells. Line 3 implements the motion model update, and line 4 the measurement update. The final probabilities are normalized through the normalizer η in line 4. The functions **motion_model**, and **measurement_model**, may be implemented by any of the motion models in Chapter 5, and measurement models in Chapter 6, respectively. The algorithm in Table 8.1 assumes that each cell possesses the same volume.

Figure 8.1 illustrates grid localization in our one-dimensional hallway example. This diagram is equivalent to that of the general Bayes filter, except for the discrete nature of the representation. As before, the robot starts out with global uncertainty, represented by a uniform histogram. As it senses, the corresponding grid cells raise their probability values. The example highlights the ability to represent multi-modal distributions with grid localization.

8.2.2 Grid Resolutions

A key variable of the grid localizer is the resolution of the grid. On the surface, this might appear to be a minor detail; however, the type of sensor model that is applicable, the computation involved in updating the belief, and the type results to expect all depend on the grid resolution.

At the extreme end are two types of representations, both of which have been brought to bear successfully in fielded robotics systems.

TOPOLOGICAL GRID REPRESENTATION

A common approach to defining a grid is *topological*; the resulting grids tend to be extremely coarse, and their resolution tends to be influenced by the structure of the environment. Topological representations decompose the space of all poses into regions that correspond to *significant places* in the environment. Such places may be defined by the presence (or absence) of specific landmarks, such as doors and windows. In hallway environments, places may correspond to intersections, T-junctions, dead ends, and so on. Topological representations tend to be coarse, and their environment decomposition depends on the structure of the environment. Figure 8.5 shows such a coarse representation for the one-dimensional hallway example.

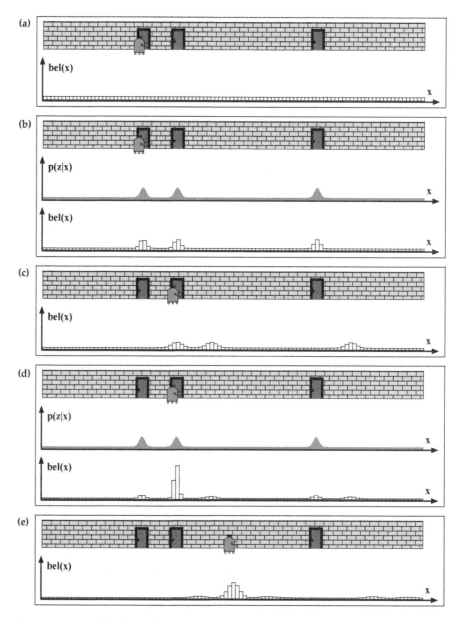

Figure 8.1 Grid localization using a fine-grained metric decomposition. Each picture depicts the position of the robot in the hallway along with its belief $bel(x_t)$, represented by a histogram over a grid.

8.2 Grid Localization

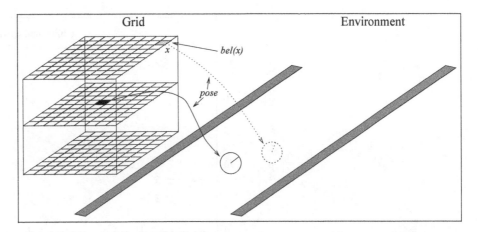

Figure 8.2 Example of a fixed-resolution grid over the robot pose variables x, y, and θ. Each grid cell represents a robot pose in the environment. Different orientations of the robot correspond to different planes in the grid (shown are only three orientations).

METRIC
REPRESENTATIONS

A much finer grained representation is commonly found through *metric representations*. Such representations decompose the state space into fine-grained cells of uniform size. The resolution of such decompositions is usually much higher than that of topological grids. For example, some of the examples in Chapter 7 use grid decompositions with cell sizes of 15 centimeters or less. Hence, they are more accurate, but at the expense of increased computational costs. Figure 8.2 illustrates such a fixed-resolution grid. Fine resolution like these are commonly associated with metric representation of space.

When implementing grid localization for coarse resolutions, it is important to compensate for the coarseness in the resolution in the sensor and motion models. In particular, for a high-resolution sensor like a laser range finder, the value of the measurement model $p(z_t \mid x_t)$ may vary drastically inside each grid cell $\mathbf{x}_{k,t}$. If this is the case, just evaluating it at the center-of-mass will generally yield poor results. Similarly, predicting robot motion from the center-of-mass may yield poor results: If the motion is updated in 1-second intervals for a robot moving at 10cm/sec, and the grid resolution is 1 meter, the naive implementation will never result in a state transition! This is because any location that is approximately 10cm away from the center-of-mass of a grid cell still falls into the same grid cell.

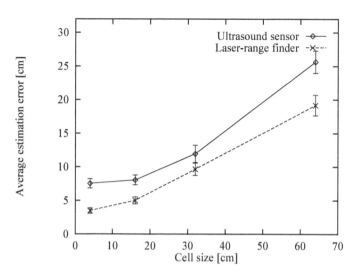

Figure 8.3 Average localization error as a function of grid cell size, for ultrasound sensors and laser range-finders.

A common way to compensate this effect is to modify both the measurement and the motion model by inflating the amount of noise. For example, the variance of a range finder model's main Gaussian cone may be enlarged by half the diameter of the grid cell. In doing so, the new model is much smoother, and its interpretation will be less susceptible to the exact location of the sample point relative to the correct robot location. However, this modified measurement model reduces the information extracted from the sensor measurements,

Similarly a motion model may predict a random transition to a nearby cell with a probability that is proportional to the length of the motion arc, divided by the diameter of a cell. The result of such an inflated motion model is that the robot can indeed move from one cell to another, even if its motion between consecutive updates is small relative to the size of a grid cell. However, the resulting posteriors are wrong in that an unreasonably large probability will be placed on the hypothesis that the robot changes cell at each motion update—and hence moves much faster than commanded.

Figures 8.3 and 8.4 plot the performance of grid localization as a function of the resolution, for two different types of range sensors. As to be expected, the localization error increases as the resolution decreases. The total time

8.2 Grid Localization

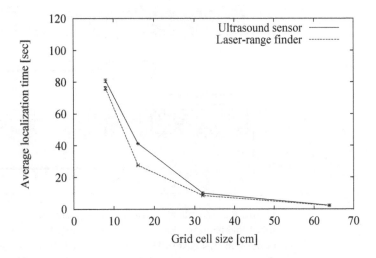

Figure 8.4 Average CPU-time needed for global localization as a function of grid resolution, shown for both ultrasound sensors and laser range-finders.

necessary to localize a robot decreases as the grid becomes coarser, as shown in Figure 8.4.

8.2.3 Computational Considerations

When using a fine-grained grid such as some of the metric grids described in the previous section, the basic algorithm cannot be executed in real-time. At fault are both the motion and the measurement update. The motion update requires a convolution, which for a 3-D grid is a 6-D operation. The measurement update is a 3-D operation, but calculating the likelihood of a full scan is a costly operation.

MODEL PRE-CACHING

There exist a number of techniques to reduce the computational complexity of grid localization. *Model pre-caching* pays tribute to the fact that certain measurement models are expensive to compute. For example, the calculation of a measurement model may require ray casting, which can be pre-cached for any fixed map. As motivated in Chapter 6.3.4, a common strategy is to calculate for each grid cell essential statistics that facilitate the measurement update. In particular, when using a beam model, it is common to cache away the correct range for each grid cell. Further, the sensor model can be pre-

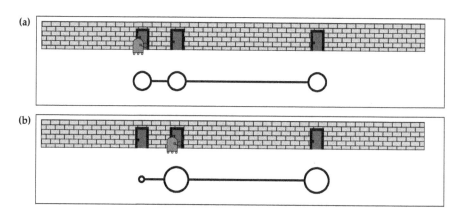

Figure 8.5 Application of a coarse-grained, topological representation to mobile robot localization. Each state corresponds to a distinctive place in the environment (a door in this case). The robot's belief $bel(x_t)$ of being in a state is represented by the size of the circles. (a) The initial belief is uniform over all poses. (b) shows the belief after the robot made one state transition and detected a door. At this point, it is unlikely that the robot is still in the left position.

calculated for a fine-grained array of possible ranges. The calculation of the measurement model reduces then to two table lookups, which is much faster.

SENSOR SUBSAMPLING

Sensor subsampling achieves further speed-ups by evaluating the measurement model only for a subset of all ranges. In some of our systems, we use only 8 of our 360 laser range measurement and still achieve excellent results. Subsampling can take place spatially and in time.

DELAYED MOTION UPDATES

Delayed motion updates applies the motion update at lower frequency than the control or measurement frequency of the robot. This is achieved by geometrically integrating the controls or odometry readings over a short time period. A good delayed motion update technique can easily speed up the algorithm by an order of magnitude.

SELECTIVE UPDATING

Selective updating was already described in Chapter 4.1.4. When updating the grid, selective techniques update a fraction of all grid cells only. A common implementation of this idea updates only those grid cells whose posterior probability exceeds a user-specified threshold. Selective updating techniques can reduce the computational effort involved in updating beliefs by many orders of magnitude. Special care has to be taken to reactivate low-likelihood grid cells when one seeks to apply this approach to the kidnapped robot problem.

With these modifications, grid localization can in fact become quite efficient; even 10 years ago, low-end PCs were fast enough to generate the results shown in this chapter. However, our modifications place an additional burden on the programmer and make a final implementation more complex than the short algorithm in Table 8.1 suggests.

8.2.4 Illustration

Figure 8.6 shows an example of Markov localization with metric grids, at a spatial resolution of 15 centimeters and an angular resolution of 5 degrees. Shown there is a global localization run where a mobile robot equipped with two laser range-finders localizes from scratch. The probabilistic model of the range-finders is computed by the beam model described in Chapter 6.3 and depicted in Table 8.1.

Initially, the robot's belief is uniformly distributed over the pose space. Figure 8.6a depicts a scan of the laser range-finders taken at the start position of the robot. Here, max range measurements are omitted and the relevant part of the map is shaded in gray. After incorporating this sensor scan, the robot's location is focused on just a few regions in the (highly asymmetric) space, as shown by the gray-scale in Figure 8.6b. Notice that beliefs are projected into x-y space; the true belief is defined over a third dimension, the robot's orientation θ, which is omitted in this and the following diagrams. Figure 8.6d shows the belief after the robot moved 2m, and incorporated the second range scan shown in Figure 8.6c. The certainty in the position estimation increases and the global maximum of the belief already corresponds to the true location of the robot. After integrating another scan into the belief the robot finally perceives the sensor scan shown in Figure 8.6e. Virtually all probability mass is now centered at the actual robot pose (see Figure 8.6f). Intuitively, we say that the robot successfully localized itself. This example illustrates that grid localization is capable to globally localize a robot efficiently.

A second example is shown in Figure 8.7; see the figure caption for an explanation. Here the environment is partially symmetric, which causes symmetric modes to appear in the localization process.

Of course, global localization usually requires more than just a few sensor scans to succeed. This is particularly the case in symmetric environments, and if the sensors are less accurate than laser sensors. Figures 8.8 to 8.10 illustrate global localization using a mobile robot equipped with sonar sensors only, and in an environment that possesses many corridors of approximately

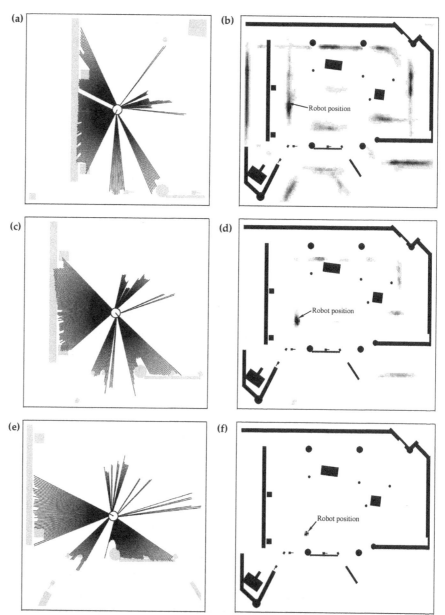

Figure 8.6 Global localization in a map using laser range-finder data. (a) Scan of the laser range-finders taken at the start position of the robot (max range readings are omitted). Figure (b) shows the situation after incorporating this laser scan, starting with the uniform distribution. (c) Second scan and (d) resulting belief. After integrating the final scan shown in (e), the robot's belief is centered at its actual location (see (f)).

8.2 Grid Localization

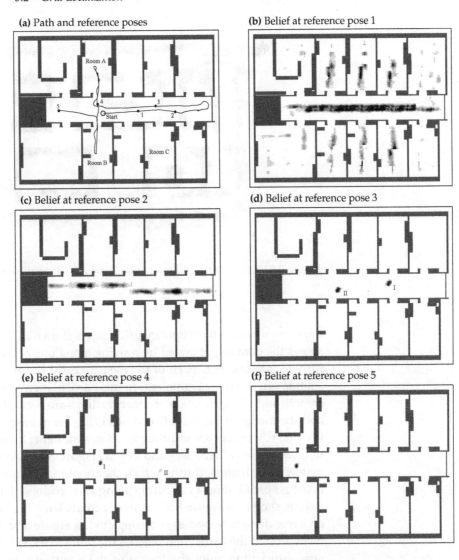

Figure 8.7 Global localization in an office environment using sonar data. (a) Path of the robot. (b) Belief as the robot passes position 1. (c) After some meters of robot motion, the robot knows that it is in the corridor. (d) As the robot reaches position 3 it has scanned the end of the corridor with its sonar sensors and hence the distribution is concentrated on two local maxima. While the maximum labeled I represents the true location of the robot, the second maximum arises due to the symmetry of the corridor (position II is rotated by 180° relative to position I). (e) After moving through Room A, the probability of being at the correct position I is now higher than the probability of being at position II. (f) Finally the robot's belief is centered on the correct pose.

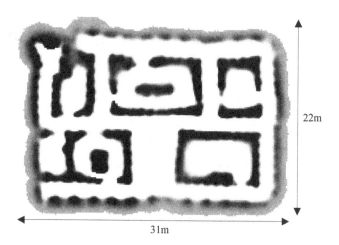

Figure 8.8 Occupancy grid map of the 1994 AAAI mobile robot competition arena.

the same width. An occupancy grid map is shown in Figure 8.8. Figure 8.9a shows the data set, obtained by moving along one of the corridors and then turning into another. Each of the measurement beams in Figure 8.9a corresponds to a sonar measurement. In this particular environment, the walls are smooth and a large fraction of sonar readings are corrupted. Again, the probabilistic model of the sensor readings is the beam-based model described in Chapter 6.3. Figure 8.9 additionally shows the belief for three different points in time, marked "A," "B," and "C" in Figure 8.9a. After moving approximately three meters, during which the robot incorporates 5 sonar scans, the belief is spread almost uniformly along all corridors of approximately equal size, as shown in Figure 8.9b. A few seconds later, the belief is now focused on a few distinct hypotheses, as depicted in Figure 8.9c. Finally, as the robot turns around the corner and reaches the point marked "C," the sensor data is now sufficient to uniquely determine the robot's position. The belief shown in Figure 8.9d is now closely centered around the actual robot pose. This example illustrates that the grid representation works well for high-noise sonar data and in symmetric environments, where multiple hypotheses have to be maintained during global localization.

Figure 8.10 illustrates the ability of the grid approach to correct accumulated dead-reckoning errors by matching sonar data with occupancy grid maps. Figure 8.10a shows the raw odometry data of a 240m long trajectory. Obviously, the rotational error of the odometry quickly increases. After trav-

8.2 Grid Localization

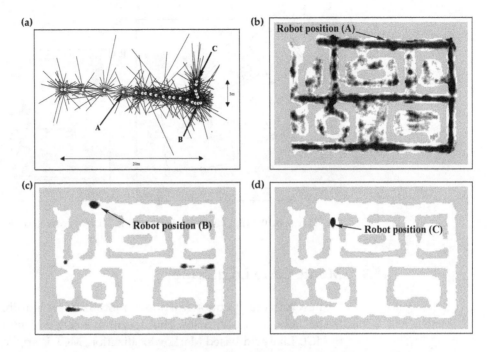

Figure 8.9 (a) Data set (odometry and sonar range scans) collected in the environment shown in Figure 8.8. This data set is sufficient for global localization using the grid localization. The beliefs at the points marked "A," "B" and "C" are shown in (b), (c), and (d).

eling only 40m, the accumulated error in the orientation (raw odometry) is about 50 degrees. Figure 8.10b shows the path of the robot estimated by the grid localizer.

Obviously, the resolution of the discrete representation is a key parameter for grid Markov localization. Given sufficient computing and memory resources, fine-grained approaches are generally preferable over coarse-grained ones. In particular, fine-grained approaches are superior to coarse-grained approaches, assuming that sufficient computing time and memory is available. As we already discussed in Chapter 2.4.4, the histogram representation causes systematic error that may violate the Markov assumption in Bayes filters. The finer the resolution, the less error is introduced, and the better the results. Fine-grained approximations also tend to suffer less from CATASTROPHIC *catastrophic failures* where the robot's belief differs significantly from its actual FAILURE position.

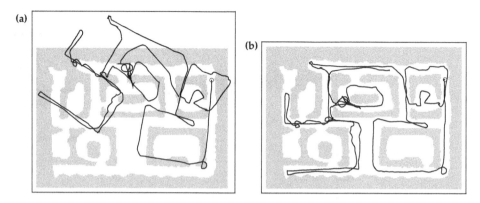

Figure 8.10 (a) Odometry information and (b) corrected path of the robot.

8.3 Monte Carlo Localization

We now turn our attention to a popular localization algorithm that represents the belief $bel(x_t)$ by particles. The algorithm is called *Monte Carlo Localization*, or *MCL*. Like grid-based Markov localization, MCL is applicable to both local and global localization problems. Despite its relatively young age, MCL has already become one of the most popular localization algorithms in robotics. It is easy to implement and tends to work well across a broad range of localization problems.

8.3.1 Illustration

Figure 8.11 illustrates MCL using the one-dimensional hallway example. The initial global uncertainty is achieved through a set of pose particles drawn at random and uniformly over the entire pose space, as shown in Figure 8.11a. As the robot senses the door, MCL assigns importance factors to each particle. The resulting particle set is shown in Figure 8.11b. The height of each particle in this figure shows its importance weight. It is important to notice that this set of particles is identical to the one in Figure 8.11a—the only thing modified by the measurement update are the importance weights.

Figure 8.11c shows the particle set after resampling and after incorporating the robot motion. This leads to a new particle set with uniform importance weights, but with an increased number of particles near the three likely places. The new measurement assigns non-uniform importance weights to the particle set, as shown in Figure 8.11d. At this point, most of the cu-

8.3 Monte Carlo Localization

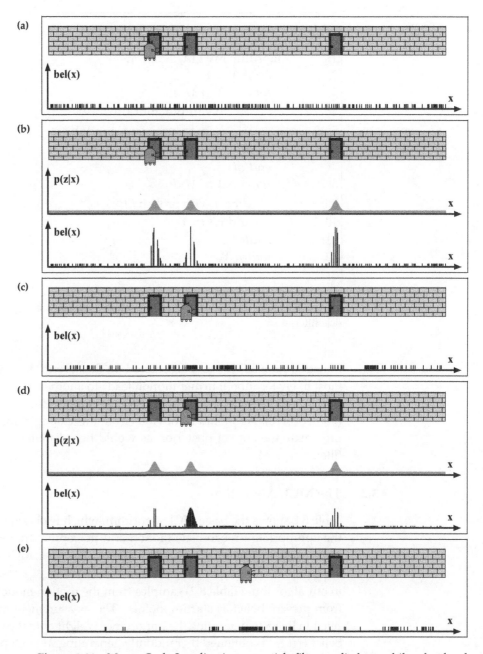

Figure 8.11 Monte Carlo Localization, a particle filter applied to mobile robot localization.

1:   **Algorithm MCL**($\mathcal{X}_{t-1}, u_t, z_t, m$):
2:       $\bar{\mathcal{X}}_t = \mathcal{X}_t = \emptyset$
3:       for $m = 1$ to M do
4:           $x_t^{[m]} = $ **sample_motion_model**$(u_t, x_{t-1}^{[m]})$
5:           $w_t^{[m]} = $ **measurement_model**$(z_t, x_t^{[m]}, m)$
6:           $\bar{\mathcal{X}}_t = \bar{\mathcal{X}}_t + \langle x_t^{[m]}, w_t^{[m]} \rangle$
7:       endfor
8:       for $m = 1$ to M do
9:           draw i with probability $\propto w_t^{[i]}$
10:          add $x_t^{[i]}$ to $\mathcal{X}_t$
11:      endfor
12:      return $\mathcal{X}_t$

Table 8.2 MCL, or Monte Carlo Localization, a localization algorithm based on particle filters.

mulative probability mass is centered on the second door, which is also the most likely location. Further motion leads to another resampling step, and a step in which a new particle set is generated according to the motion model (Figure 8.11e). As should be obvious from this example, the particle sets approximate the correct posterior, as would be calculated by an exact Bayes filter.

8.3.2 The MCL Algorithm

Table 8.2 shows the basic MCL algorithm, which is obtained by substituting the appropriate probabilistic motion and perceptual models into the algorithm **particle_filters** (Table 4.3 on page 98). The basic MCL algorithm represents the belief $bel(x_t)$ by a set of M particles $\mathcal{X}_t = \{x_t^{[1]}, x_t^{[2]}, \ldots, x_t^{[M]}\}$. Line 4 in our algorithm (Table 8.2) samples from the motion model, using particles from present belief as starting points. The measurement model is then applied in line 5 to determine the importance weight of that particle. The initial belief $bel(x_0)$ is obtained by randomly generating M such particles from the prior distribution $p(x_0)$, and assigning the uniform importance factor M^{-1} to each particle. As in grid localization, the functions **motion_model**, and

measurement_model, may be implemented by any of the motion models in Chapter 5, and measurement models in Chapter 6, respectively.

8.3.3 Physical Implementations

It is straightforward to implement the MCL algorithm for the landmark-based localization scenario of Chapter 7. To do so, the sampling procedure in line 4 is implemented using the algorithm **sample_motion_model_velocity** given in Table 5.3. The algorithm **landmark_model_known_correspondence** given in Table 6.4 provides the likelihood model used in line 5 to weigh the predicted samples.

Figure 8.12 illustrates this version of the MCL algorithm. The scenario is identical to the one shown in Figure 7.15. For convenience, the illustration of the robot path and measurements is shown again in the upper left figure. The lower plot shows a sequence of sample sets generated by the MCL algorithm. The solid line represents the true path of the robot, the dotted line represents the path based on the control information, and the dashed line represents the mean path estimated by the MCL algorithm. Predicted sample sets $\bar{\mathcal{X}}_t$ at different points in time are shown in dark, the samples $\mathcal{X}_t$ after the resampling steps are shown in lighter gray. Each particle set is defined over the 3-dimensional pose space, although only the x- and y-coordinates of each particle are shown. The means and covariances extracted from these sets are shown in the upper right figure.

Figure 8.13 shows the result of applying MCL in an actual office environment, for a robot equipped with an array of sonar range finders. This version of MCL computes the likelihood of measurements using algorithm **beam_range_finder_model** given in Table 6.1. The figure depicts particle sets after 5, 28, and 55, meters of robot motion, respectively. A third illustration is provided in Figure 8.14, here using a camera pointed towards the ceiling, and a measurement model that relates the brightness in the center of the image to a previously acquired ceiling map.

8.3.4 Properties of MCL

MCL can approximate almost any distribution of practical importance. It is not bound to a limited parametric subset of distributions, as was the case for EKF localization. Increasing the total number of particles increases the accuracy of the approximation. The number of particles M is a parameter that enables the user to trade off the accuracy of the computation and the compu-

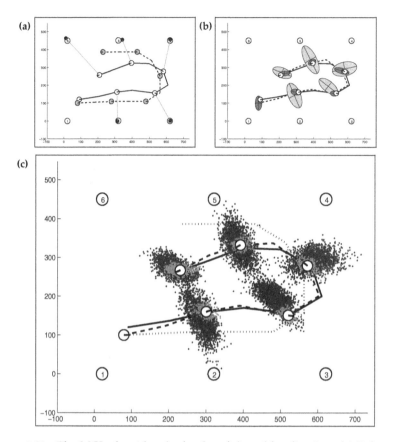

Figure 8.12 The MCL algorithm for landmark-based localization. (a) Robot trajectory according to the motion control (dashed lines) and the resulting true trajectory (solid lines). Landmark detections are indicated by thin lines. (b) Covariances of sample sets before and after resampling. (c) Sample sets before and after resampling.

tational resources necessary to run MCL. A common strategy for setting M is to keep sampling until the next pair u_t and z_t has arrived. In this way, the implementation is adaptive with regards to the computational resources: the faster the underlying processor, the better the localization algorithm. However, as we will see in Chapter 8.3.7, care has to be taken that the number of particles remains high enough to avoid filter divergence.

A final advantage of MCL pertains to the non-parametric nature of the approximation. As our illustrative results suggest, MCL can represent complex multi-modal probability distributions, and blend them seamlessly with

8.3 Monte Carlo Localization

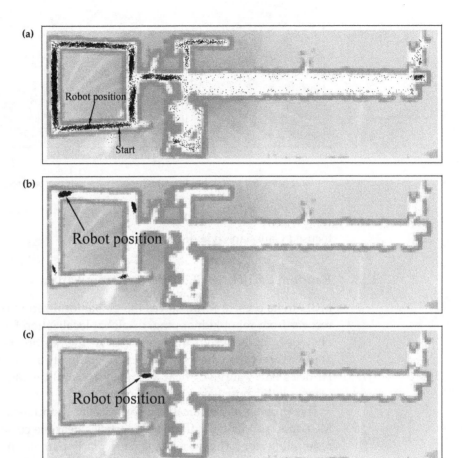

Figure 8.13 Illustration of Monte Carlo localization: Shown here is a robot operating in an office environment of size 54m × 18m. (a) After moving 5m, the robot is still globally uncertain about its position and the particles are spread through major parts of the free-space. (b) Even as the robot reaches the upper left corner of the map, its belief is still concentrated around four possible locations. (c) Finally, after moving approximately 55m, the ambiguity is resolved and the robot knows where it is. All computation is carried out in real-time on a low-end PC.

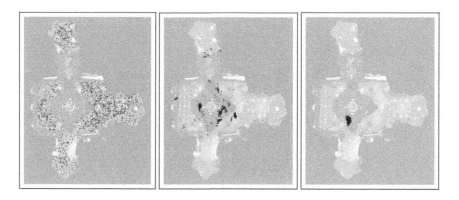

Figure 8.14 Global localization using a camera pointed at the ceiling.

focused Gaussian-style distributions.

8.3.5 Random Particle MCL: Recovery from Failures

MCL, in its present form, solves the global localization problem but cannot recover from robot kidnapping, or global localization failures. This is quite obvious from the results in Figure 8.13: As the position is acquired, particles at places other than the most likely pose gradually disappear. At some point, particles only "survive" near a single pose, and the algorithm is unable to recover if this pose happens to be incorrect.

This problem is significant. In practice, any stochastic algorithm such as MCL may accidentally discard all particles near the correct pose during the resampling step. This problem is particularly paramount when the number of particles is small (e.g., $M = 50$), and when the particles are spread over a large volume (e.g., during global localization).

Fortunately, this problem can be solved by a rather simple heuristic. The idea of this heuristic is to add random particles to the particle sets, as already discussed in Chapter 4.3.4. Such an *injection of random particles* can be justified mathematically by assuming that the robot might get kidnapped with a small probability, thereby generating a fraction of random states in the motion model. Even if the robot does not get kidnapped, however, the random particles add an additional level of robustness.

The approach of adding particles raises two questions. First, how many particles should be added at each iteration of the algorithm and, second, from

INJECTION OF RANDOM PARTICLES

which distribution should we generate these particles? One might add a fixed number of random particles at each iteration. A better idea is to add particles based on some estimate of the localization performance.

One way to implement this idea is to monitor the probability of sensor measurements

$$(8.3) \quad p(z_t \mid z_{1:t-1}, u_{1:t}, m)$$

and relate it to the average measurement probability (which is easily learned from data). In particle filters, an approximation to this quantity is easily obtained from the importance factor, since, by definition, an importance weight is a stochastic estimate of this probability. The average value

$$(8.4) \quad \frac{1}{M} \sum_{m=1}^{M} w_t^{[m]} \approx p(z_t \mid z_{1:t-1}, u_{1:t}, m)$$

approximates the desired probability as stated. It is usually a good idea to smooth this estimate by averaging it over multiple time steps. There exist multiple reasons why the measurement probability may be low, besides a localization failure. The amount of sensor noise might be unnaturally high, or the particles may still be spread out during a global localization phase. For these reasons, it is a good idea to maintain a short-term average of the measurement likelihood, and relate it to the long-term average when determining the number of random samples.

The second problem of determining which sample distribution to use can be addressed in two ways. One can draw particles according to a uniform distribution over the pose space and weight them with the current observation.

For some sensor models, however, it is possible to generate particles directly in accordance to the measurement distribution. One example of such a sensor model is the landmark detection model discussed in Chapter 6.6. In this case the additional particles can be placed directly at locations distributed according to the observation likelihood (see Table 6.5).

Table 8.3 shows a variant of the MCL algorithm that adds random particles. This algorithm is adaptive, in that it tracks the short-term and the long-term average of the likelihood $p(z_t \mid z_{1:t-1}, u_{1:t}, m)$. Its first part is identical to the algorithm **MCL** in Table 8.2: New poses are sampled from old particles using the motion model (line 5), and their importance weight is set in accordance to the measurement model (line 6).

Augmented_MCL calculates the empirical measurement likelihood in line 8, and maintains short-term and long-term averages of this likelihood in lines

1:   **Algorithm Augmented_MCL($\mathcal{X}_{t-1}, u_t, z_t, m$):**
2:       static w_{slow}, w_{fast}
3:       $\bar{\mathcal{X}}_t = \mathcal{X}_t = \emptyset$
4:       for $m = 1$ to M do
5:           $x_t^{[m]} = $ **sample_motion_model**$(u_t, x_{t-1}^{[m]})$
6:           $w_t^{[m]} = $ **measurement_model**$(z_t, x_t^{[m]}, m)$
7:           $\bar{\mathcal{X}}_t = \bar{\mathcal{X}}_t + \langle x_t^{[m]}, w_t^{[m]} \rangle$
8:           $w_{\text{avg}} = w_{\text{avg}} + \frac{1}{M} w_t^{[m]}$
9:       endfor
10:      $w_{\text{slow}} = w_{\text{slow}} + \alpha_{\text{slow}}(w_{\text{avg}} - w_{\text{slow}})$
11:      $w_{\text{fast}} = w_{\text{fast}} + \alpha_{\text{fast}}(w_{\text{avg}} - w_{\text{fast}})$
12:      for $m = 1$ to M do
13:          with probability $\max\{0.0, 1.0 - w_{\text{fast}}/w_{\text{slow}}\}$ do
14:              add random pose to $\mathcal{X}_t$
15:          else
16:              draw $i \in \{1, \ldots, N\}$ with probability $\propto w_t^{[i]}$
17:              add $x_t^{[i]}$ to $\mathcal{X}_t$
18:          endwith
19:      endfor
20:      return $\mathcal{X}_t$

Table 8.3 An adaptive variant of MCL that adds random samples. The number of random samples is determined by comparing the short-term with the long-term likelihood of sensor measurements.

10 and 11. The algorithm requires that $0 \leq \alpha_{\text{slow}} \ll \alpha_{\text{fast}}$. The parameters α_{slow}, and α_{fast}, are decay rates for the exponential filters that estimate the long-term, and short-term, averages, respectively. The crux of this algorithm can be found in line 13: During the resampling process, a random sample is added with probability

(8.5) $\max\{0.0, 1.0 - w_{\text{fast}}/w_{\text{slow}}\}$

Otherwise, resampling proceeds in the familiar way. The probability of adding a random sample takes into consideration the divergence between the short- and the long-term average of the measurement likelihood. If the

8.3 Monte Carlo Localization

short-term likelihood is better or equal to the long-term likelihood, no random sample is added. However, if the short-term likelihood is worse than the long-term one, random samples are added in proportion to the quotient of these values. In this way, a sudden decay in measurement likelihood induces an increased number of random samples. The exponential smoothing counteracts the danger of mistaking momentary sensor noise for a poor localization result.

Figure 8.16 illustrates our augmented MCL algorithm in practice. Shown there is a sequence of particle sets during global localization and relocalization of a legged robot equipped with a color camera, and operating on a 3×2m field as it was used in RoboCup soccer competitions. Sensor measurements correspond to the detection and relative localization of six visual markers placed around the field, as shown in Figure 7.7 on page 210. The algorithm described in Table 6.4 is used to determine the likelihood of detections. Step 14 in Figure 8.3 is replaced by an algorithm for sampling according to the most recent sensor measurement, which is easily implemented using algorithm **sample_landmark_model_known_correspondence** in Table 6.5.

Panels (a) through (d) in Figure 8.16 illustrate global localization. At the first marker detection, virtually all particles are drawn according to this detection (Figure 8.16b). This step corresponds to a situation in which the short-term average of the measurement probability is much worse than its long-term correspondent. After several more detections, the particles are clustered around the true robot position (Figure 8.16d), and both the short- and long-term average of the measurement likelihood increases. At this stage of localization, the robot is merely tracking its position, the observation likelihoods are rather high, and only a very small number of random particles are occasionally added.

As the robot is physically placed at a different location by a referee—a common event in robotic soccer tournaments—the measurement probability drops. The first marker detection at this new location does not yet trigger any additional particles, since the smoothed estimate w_{fast} is still high (see Figure 8.16e). After several marker detections observed at the new location, w_{fast} decreases much faster than w_{slow} and more random particles are added (Figure 8.16f&g). Finally, the robot successfully relocalizes itself as shown in Figure 8.16h, demonstrating that our augmented MCL algorithm is indeed capable of "surviving" the kidnapping.

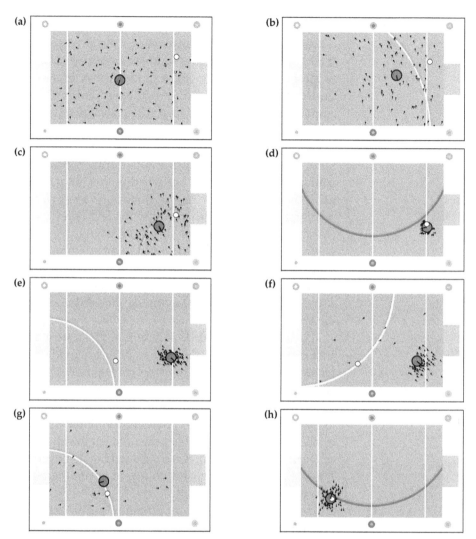

Figure 8.16 Monte Carlo localization with random particles. Each picture shows a particle set representing the robot's position estimate (small lines indicate the orientation of the particles). The large circle depicts the mean of the particles, and the true robot position is indicated by the small, white circle. Marker detections are illustrated by arcs centered at the detected marker. The pictures illustrate global localization (a)–(d) and relocalization (e)–(h).

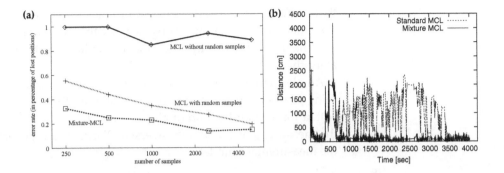

Figure 8.17 (a) plain MCL (top curve), MCL with random samples (center curve), and *Mixture MCL* with mixture proposal distribution (bottom curve). The error rate is measured in percentage of time during which the robot lost track of its position, for a data set acquired by a robot operating in a crowded museum. (b) Error as a function of time for standard MCL and mixture MCL, using a ceiling map for localization.

8.3.6 Modifying the Proposal Distribution

The MCL proposal mechanism is another source that can render MCL inefficient. As discussed in Chapter 4.3.4, the particle filter uses the motion model as proposal distribution, but it seeks to approximate a product of this distribution and the perceptual likelihood. The larger the difference between the proposal and the target distribution, the more samples are needed.

In MCL, this induces a surprising failure mode: If we were to acquire a perfect sensor that—without any noise—always informs the robot of its correct pose, MCL would fail. This is even true for noise-free sensors that do not carry sufficient information for localization. An example of the latter would be a 1-D noise-free range sensor: When receiving such a range measurement, the space of valid pose hypotheses will be a 2-D subspace of the 3-D pose space. We already discussed in length in Chapter 4.3.4 that chances to sample into this 2-D submanifold are zero when sampling from the robot motion model. Thus, we face the strange situation that under certain circumstances, a less accurate sensor would be preferable to a more accurate sensor when using MCL for localization. This is *not* the case for EKF localization, since the EKF update takes the measurements into account when calculating the new mean—instead of generating mean(s) from the motion model alone.

Luckily, a simple trick provides remedy: Simply use a measurement model that artificially inflates the amount of noise in the sensor. One can think of this inflation as accommodating not just the measurement uncertainty, but also the uncertainty induced by the approximate nature of the particle filter algorithm.

An alternative, more sound solution involves a modification of the sampling process which we already discussed briefly in Chapter 4.3.4. The idea is that for a small fraction of all particles, the role of the motion model and the measurement model are reversed: Particles are generated according to the measurement model

$$x_t^{[m]} \sim p(z_t \mid x_t) \tag{8.6}$$

and the importance weight is calculated in proportion to

$$w_t^{[m]} = \int p(x_t^{[m]} \mid u_t, x_{t-1}) \, bel(x_{t-1}) \, dx_{t-1} \tag{8.7}$$

This new sampling process is a legitimate alternative to the plain particle filter. It alone will be inefficient since it entirely ignores the history when generating particles. However, it is equally legitimate to generate a fraction of the particles with either of those two mechanisms and merge the two particle sets. The resulting algorithm is called *MCL with mixture proposal distribution*, or *Mixture MCL*. In practice, it tends to suffice to generate a small fraction of particles (e.g., 5%) through the new process.

MIXTURE MCL

Unfortunately, our idea does not come without challenges. The two main steps—sampling from $p(z_t \mid x_t)$ and calculating the importance weights $w_t^{[m]}$—can be difficult to realize. Sampling from the measurement model is only easy if its inverse possesses a closed form solution from which it is easy to sample. This is usually not the case: imagine sampling from the space of all poses that fit a given laser range scan! Calculating the importance weights is complicated by the integral in (8.7), and by the fact that $bel(x_{t-1})$ is itself represented by a set of particles.

Without delving into too much detail, we note that both steps can be implemented, but only with additional approximations. Figure 8.17 shows comparative results for MCL, MCL augmented with random samples, and Mixture MCL for two real-world data sets. In both cases, $p(z_t \mid x_t)$ was itself learned from data and represented by a density tree—an elaborate procedure whose description is beyond the scope of this book. For calculating the importance weights, the integral was replaced by a stochastic integration, and the prior belief was continued into a space-filling density by convolving each particle with a narrow Gaussian. Details aside, these results illustrate that the mixture idea yields superior results, but it can be challenging to implement.

We also note that the Mixture MCL provides a sound solution to the kidnapped robot problem. By seed-starting particles using the most recent measurement only, we constantly generate particles at locations that are plausible given the momentary sensor input, regardless of past measurements and controls. There exist ample evidence in the literature that such approaches can cope well with total localization failure (Figure 8.17b happens to show one such failure for regular MCL), hence provides improved robustness in practical implementations.

8.3.7 KLD-Sampling: Adapting the Size of Sample Sets

The size of the sample sets used to represent beliefs is an important parameter for the efficiency of particle filters. So far we only discussed particle filters that use sample sets of fixed size. Unfortunately, to avoid divergence due to sample depletion in MCL, one has to choose large sample sets so as to allow a mobile robot to address both the global localization and the position tracking problem. This can be a waste of computational resources, as Figure 8.13 reveals. In this example, all sample sets contain 100,000 particles. While such a high number of particles might be necessary to accurately represent the belief during early stages of localization (cf. Figure 8.13a), it is obvious that only a small fraction of this number suffices to track the position of the robot once it knows where it is (Figure 8.13c).

KLD-SAMPLING

KULLBACK-LEIBLER DIVERGENCE

KLD-sampling is a variant of MCL that adapts the number of particles over time. We do not provide a mathematical derivation of KLD-sampling, but only state the algorithm and show some experimental results. The name *KLD-sampling* is derived from the *Kullback-Leibler divergence*, which is a measure of the difference between two probability distributions. The idea behind KLD-sampling is to determine the number of particles based on a statistical bound on the sample-based approximation quality. More specifically, at each iteration of the particle filter, KLD-sampling determines the number of samples such that, with probability $1 - \delta$, the error between the true posterior and the sample-based approximation is less than ε. Several assumptions not stated here make it possible to derive an efficient implementation of this idea.

The KLD-sampling algorithm is shown in Table 8.4. The algorithm takes as input the previous sample set along with the map and the most recent control and measurement. In contrast to MCL, KLD-sampling takes a weighted sample set as input. That is, the samples in $\mathcal{X}_{t-1}$ are not resampled. Additionally, the algorithm requires the statistical error bounds ε and δ.

In a nutshell, KLD-sampling generates particles until the statistical bound

1:         Algorithm KLD_Sampling_MCL($\mathcal{X}_{t-1}, u_t, z_t, m, \varepsilon, \delta$):
2:         $\mathcal{X}_t = \emptyset$
3:         $M = 0, M_\chi = 0, k = 0$
4:         for all b in H do
5:             $b = $ empty
6:         endfor
7:         do
8:             draw i with probability $\propto w_{t-1}^{[i]}$
9:             $x_t^{[M]} = $ **sample_motion_model**$(u_t, x_{t-1}^{[i]})$
10:            $w_t^{[M]} = $ **measurement_model**$(z_t, x_t^{[M]}, m)$
11:            $\mathcal{X}_t = \mathcal{X}_t + \langle x_t^{[M]}, w_t^{[M]} \rangle$
12:            if $x_t^{[M]}$ falls into empty bin b then
13:                $k = k + 1$
14:                $b = $ non-empty
15:                if $k > 1$ then
16:                    $M_\chi := \frac{k-1}{2\varepsilon} \left\{ 1 - \frac{2}{9(k-1)} + \sqrt{\frac{2}{9(k-1)}} z_{1-\delta} \right\}^3$
17:            endif
18:            $M = M + 1$
19:         while $M < M_\chi$ or $M < M_{\chi min}$
20:         return $\mathcal{X}_t$

Table 8.4 KLD-sampling MCL with adaptive sample set size. The algorithm generates samples until a statistical bound on the approximation error is reached.

in line 16 is satisfied. This bound is based on the "volume" of the state space that is covered by particles. The volume covered by particles is measured by a histogram, or grid, overlayed over the 3-dimensional state space. Each bin in the histogram H is either empty or occupied by at least one particle. Initially, each bin is set to empty (Lines 4 through 6). In line 8, a particle is drawn from the previous sample set. Based on this particle, a new particle is predicted, weighted, and inserted into the new sample set (Lines 9–11, just like in MCL).

Lines 12 through 19 implement the key idea of KLD-sampling. If the newly generated particle falls into an empty bin of the histogram, then the number k of non-empty bins is incremented and the bin is marked as non-empty. Thus, k measures the number of histogram bins filled with at least one particle. This number plays a crucial role in the statistical bound determined in line 16. The quantity M_χ gives the number of particles needed to reach this bound. Note that for a given ε, M_χ is mostly linear in the number k of non-empty bins; the second, nonlinear term becomes negligible as k increases. The term $z_{1-\delta}$ is based on the parameter δ. It represents the upper $1-\delta$ quantile of the standard normal distribution. The values of $z_{1-\delta}$ for typical values of δ are readily available in standard statistical tables.

The algorithm generates new particles until their number M exceeds M_χ and a user-defined minimum $M_{\chi_{\min}}$. As can be seen, the threshold M_χ serves as a moving target for M. The more samples M are generated, the more bins k in the histogram are non-empty, and the higher the threshold M_χ.

In practice, the algorithm terminates based on the following reasoning. In the early stages of sampling, k increases with almost every new sample since virtually all bins are empty. This increase in k results in an increase in the threshold M_χ. However, over time, more and more bins are non-empty and M_χ increases only occasionally. Since M increases with each new sample, M will finally reach M_χ and sampling is stopped. When this happens depends on the belief. The more widespread the particles, the more bins are filled and the higher the threshold M_χ. During tracking, KLD-sampling generates less samples since the particles are concentrated on a small number of different bins. It should be noted that the histogram has no impact on the particle distribution itself. Its only purpose is to measure the complexity, or volume, of the belief. The grid is discarded at the end of each particle filter iteration.

Figure 8.18 shows the sample set sizes during a typical global localization run using KLD-sampling. The figure shows graphs when using a robot's laser range-finder (solid line) or ultrasound sensors (dashed line). In both cases, the algorithm chooses a large number of samples during the initial phase of global localization. Once the robot is localized, the number of particles drops to a much lower level (less than 1% of the initial number of particles). When and how fast this transition from global localization to position tracking happens depends on the type of the environment and the accuracy of the sensors. In this example, the higher accuracy of the laser range-finder is reflected by an earlier transition to a lower level.

Figure 8.19 shows a comparison between the approximation error of KLD-sampling and MCL with fixed sample sets. The approximation error is measured by the Kullback-Leibler distance between the beliefs (sample sets) gen-

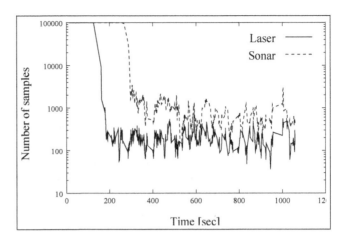

Figure 8.18 KLD-sampling: Typical evolution of number of samples for a global localization run, plotted against time (number of samples is shown on a log scale). The solid line shows the number of samples when using the robot's laser range-finder, the dashed graph is based on sonar sensor data.

erated with varying numbers of samples and the "optimal" beliefs. These "optimal" beliefs were generated by running MCL with sample sets of size 200,000, which is far more than actually needed for position estimation. As expected, the more samples are used by the two approaches, the smaller the approximation error. The dashed graph shows the results achieved by MCL with different sample set sizes. As can be seen, the fixed approach requires about 50,000 samples before it converges to a KL-distance below 0.25. Larger errors typically indicate that the particle filter diverges and the robot is not able to localize. The solid line shows the results when using KLD-sampling. Here the sample set sizes are averages over the global localization runs. The different data points were obtained by varying the error bound ε between 0.4 and 0.015, decreasing from left to right. KLD-sampling converges to a small error level using only 3,000 samples on average. The graph also shows that KLD-sampling is not guaranteed to accurately track the optimal belief. The leftmost data points on the solid line indicate that KLD-sampling diverges due to too loose error bounds.

KLD-sampling can be used by any particle filter, not just MCL. The histogram can be implemented either as a fixed, multi-dimensional grid, or more compactly as tree structures. In the context of robot localization, KLD-

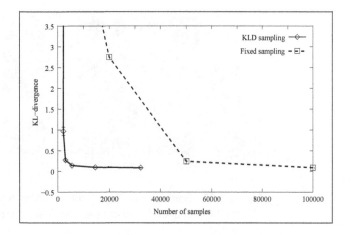

Figure 8.19 Comparison of KLD-sampling and MCL with fixed sample set sizes. The x-axis represents the average sample set size. The y-axis plots the KL-distance between the reference beliefs and the sample sets generated by the two approaches.

sampling has been shown to consistently outperform MCL with fixed sample set sizes. The advantage of this technique is most significant for a combination of global localization and tracking problems. In practice, good results are achieved with error bound values around 0.99 for $(1 - \delta)$ and 0.05 for ε in combination with histogram bin sizes of 50cm × 50cm × 15deg.

8.4 Localization in Dynamic Environments

A key limitation of all localization algorithms discussed thus far arises from the static world assumption, or Markov assumption. Most interesting environments are populated by people, and hence exhibit dynamics not modeled by the state x_t. To some extent, probabilistic approaches are robust to such unmodeled dynamics, due to their ability to accommodate sensor noise. However, as previously noted, the type of sensor noise accommodated in the probabilistic filtering framework must be independent at each time step, whereas unmodeled dynamics induce effects on the sensor measurements over multiple time steps. When such effects are paramount, probabilistic localization algorithms that rely on the static world assumption may fail.

A good example of such a failure situation is shown in Figure 8.20. This example involves a mobile tour-guide robot, navigating in museums full of

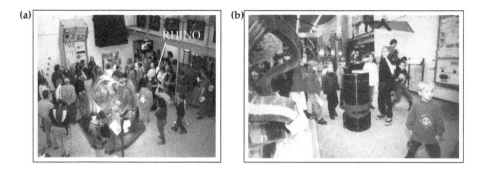

Figure 8.20 Scenes from the *"Deutsches Museum Bonn,"* where the mobile robot "Rhino" was frequently surrounded by dozens of people.

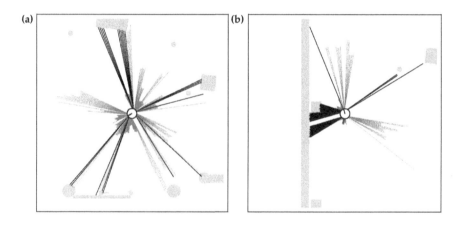

Figure 8.21 Laser range scans are often heavily corrupted when people surround the robot. How can a robot maintain accurate localization under such circumstances?

people. The people—their locations, velocities, intentions etc.—are hidden state relative to the localization algorithm that is not captured in the algorithms discussed thus far. Why is this problematic? Imagine people lining up in a way that suggests the robot is facing a wall. With each single sensor measurement the robot increases its belief of being next to a wall. Since information is treated as independent, the robot will ultimately assign high likelihood to poses near walls. Such an effect is possible with independent sensor noise, but its likelihood is vanishingly small.

8.4 Localization in Dynamic Environments

STATE AUGMENTATION

There exist two fundamental techniques for dealing with dynamic environments. The first technique, *state augmentation* includes the hidden state into the state estimated by the filter; the other, *outlier rejection* pre-processes sensor measurements to eliminate measurements affected by hidden state. The former methodology is mathematically the more general one: Instead of just estimating the robot's pose, one can define a filter that also estimates people's positions, their velocities, etc. In fact, we will later on discuss such an approach, as an extension to a mobile robot mapping algorithm.

The principle disadvantage of estimating the hidden state variables lies in its computational complexity: Instead of estimating 3 variables, the robot must now calculate posteriors over a much larger number of variables. In fact, the number of variables itself is a variable, as the number of people may vary over time. Thus, the resulting algorithm will be substantially more involved than the localization algorithms discussed thus far.

OUTLIER REJECTION

The alternative, *outlier rejection*, works well in certain limited situations, which includes situations where people's presence may affect range finders or (to a lesser extent) camera images. Here we develop it for the beam-based range finder model from Chapter 6.3.

The idea is to investigate the *cause* of a sensor measurement, and to reject those likely to be affected by unmodeled environment dynamics. The sensor models discussed thus far all address different, alternative ways by which a measurement can come into existence. If we manage to associate specific ways with the presence of unwanted dynamic effects—such as people—all we have to do is to discard those measurements that are with high likelihood caused by such an unmodeled entity.

This idea is surprisingly general; and in fact, the mathematics are essentially the same as in the *EM learning algorithm* in Chapter 6.3, but applied in an online fashion. In Equation(6.12), Chapter 6.3, we defined the beam-based measurement model for range finders as a mixture of four terms:

$$(8.8) \quad p(z_t^k \mid x_t, m) = \begin{pmatrix} z_{\text{hit}} \\ z_{\text{short}} \\ z_{\text{max}} \\ z_{\text{rand}} \end{pmatrix}^T \cdot \begin{pmatrix} p_{\text{hit}}(z_t^k \mid x_t, m) \\ p_{\text{short}}(z_t^k \mid x_t, m) \\ p_{\text{max}}(z_t^k \mid x_t, m) \\ p_{\text{rand}}(z_t^k \mid x_t, m) \end{pmatrix}$$

As our derivation of the model clearly states, one of those terms, the one involving z_{short} and p_{short}, corresponds to unexpected objects. To calculate the probability that a measurement z_t^k corresponds to an unexpected object, we have to introduce a new correspondence variable, $\bar{c}_t^k$ which can take on one of the four values {hit, short, max, rand}.

The posterior probability that the range measurement z_t^k corresponds to a "short" reading—our mnemonic from Chapter 6.3 for unexpected obstacle—is then obtained by applying Bayes rule and subsequently dropping irrelevant conditioning variables:

$$\begin{aligned}
(8.9) \quad & p(\bar{c}_t^k = \text{short} \mid z_t^k, z_{1:t-1}, u_{1:t}, m) \\
& = \frac{p(z_t^k \mid \bar{c}_t^k = \text{short}, z_{1:t-1}, u_{1:t}, m) \, p(\bar{c}_t^k = \text{short} \mid z_{1:t-1}, u_{1:t}, m)}{\sum_c p(z_t^k \mid \bar{c}_t^k = c, z_{1:t-1}, u_{1:t}, m) \, p(\bar{c}_t^k = c \mid z_{1:t-1}, u_{1:t}, m)} \\
& = \frac{p(z_t^k \mid \bar{c}_t^k = \text{short}, z_{1:t-1}, u_{1:t}, m) \, p(\bar{c}_t^k = \text{short})}{\sum_c p(z_t^k \mid \bar{c}_t^k = c, z_{1:t-1}, u_{1:t}, m) \, p(\bar{c}_t^k = c)}
\end{aligned}$$

Here the variable c in the denominator takes on any of the four values $\{\text{hit}, \text{short}, \text{max}, \text{rand}\}$. Using the notation in Equation (8.8), the prior $p(\bar{c}_t^k = c)$ is given by the variables z_{hit}, z_{short}, z_{max}, and z_{rand}, for the four values of c. The remaining probability in (8.9) is obtained by integrating out x_t:

$$\begin{aligned}
(8.10) \quad & p(z_t^k \mid \bar{c}_t^k = c, z_{1:t-1}, u_{1:t}, m) \\
& = \int p(z_t^k \mid x_t, \bar{c}_t^k = c, z_{1:t-1}, u_{1:t}, m) \, p(x_t \mid \bar{c}_t^k = c, z_{1:t-1}, u_{1:t}, m) \, dx_t \\
& = \int p(z_t^k \mid x_t, \bar{c}_t^k = c, m) \, p(x_t \mid z_{1:t-1}, u_{1:t}, m) \, dx_t \\
& = \int p(z_t^k \mid x_t, \bar{c}_t^k = c, m) \, \overline{bel}(x_t) \, dx_t
\end{aligned}$$

Probabilities of the form $p(z_t^k \mid x_t, \bar{c}_t^k = c, m)$ were abbreviated as p_{hit}, p_{short}, p_{max}, and p_{rand} in Chapter 6.3. This gives us the expression for desired probability (8.9):

$$(8.11) \quad p(\bar{c}_t^k = \text{short} \mid z_t^k, z_{1:t-1}, u_{1:t}, m) = \frac{\int p_{\text{short}}(z_t^k \mid x_t, m) \, z_{\text{short}} \, \overline{bel}(x_t) \, dx_t}{\int \sum_c p_c(z_t^k \mid x_t, m) \, z_c \, \overline{bel}(x_t) \, dx_t}$$

In general, the integrals in (8.11) do not possess closed-form solutions. To evaluate them, it suffices to approximate them with a representative sample of the posterior $\overline{bel}(x_t)$ over the state x_t. Those samples might be high-likelihood grid cells in grid localizer, or particles in a MCL algorithm. The measurement is then rejected if its probability of being caused by an unexpected obstacle exceeds a user-selected threshold χ.

```
1:          Algorithm test_range_measurement($z_t^k, \bar{\mathcal{X}}_t, m$):
2:              $p = q = 0$
3:              for $m = 1$ to $M$ do
4:                  $p = p + z_{\text{short}} \cdot p_{\text{short}}(z_t^k \mid x_t^{[m]}, m)$
5:                  $q = q + z_{\text{hit}} \cdot p_{\text{hit}}(z_t^k \mid x_t^{[m]}, m) + z_{\text{short}} \cdot p_{\text{short}}(z_t^k \mid x_t^{[m]}, m)$
6:                      $+ z_{\text{max}} \cdot p_{\text{max}}(z_t^k \mid x_t^{[m]}, m) + z_{\text{rand}} \cdot p_{\text{rand}}(z_t^k \mid x_t^{[m]}, m)$
7:              endfor
8:              if $p/q \leq \chi$ then
9:                  return accept
10:             else
11:                 return reject
12:             endif
```

Table 8.5 Algorithm for testing range measurements in dynamic environment.

Table 8.5 depicts an implementation of this technique in the context of particle filters. It requires as input a particle set $\bar{\mathcal{X}}_t$ representative of the belief $\overline{bel}(x_t)$, along with a range measurement z_t^k and a map. It returns "reject" if with probability larger than χ the measurement corresponds to an unexpected object; otherwise it returns "accept." This routine precedes the measurement integration step in MCL.

Figure 8.22 illustrates the effect of the filter. Shown in both panels are a range scan, for a different alignment of the robot pose. The lightly shaded scans are above threshold and rejected. A key property of our rejection mechanism is that it tends to filter out measurements that are "surprisingly" short, but leaves others in place that are "surprisingly" long. This asymmetry reflects the fact that people's presence tends to cause shorter-than-expected measurements. By accepting surprisingly long measurements, the approach maintains its ability to recover from global localization failures.

Figure 8.23 depicts an episode during which a robot navigates through an environment that is densely populated with people (see Figure 8.21). Shown there is the robot's estimated path along with the endpoints of all scans incorporated into the localizer. This figure shows the effectiveness of removing measurements that do not correspond to physical objects in the map: there are very few "surviving" range measurements in the freespace for the right diagram, in which measurements are accepted only if they surpass the

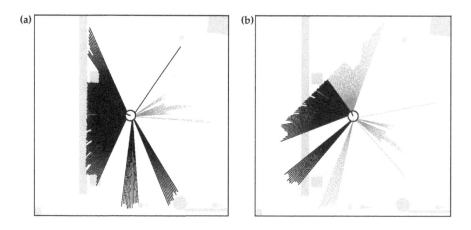

Figure 8.22 Illustration of our measurement rejection algorithm: Shown in both diagrams are range scans (no max-range readings). Lightly shaded readings are filtered out.

threshold test.

As a rule of thumb, outlier rejection of measurements is generally a good idea. There exist almost no static environments; even in office environments furniture is moved, doors are opened/closed, etc. Our specific implementation here benefits from the asymmetry of range measurements: people make measurements shorter, not longer. When applying the same idea to other data (e.g., vision data) or other types of environment modifications (e.g., the removal of a physical obstacle), such an asymmetry might not exist. Nevertheless, the same probabilistic analysis is usually applicable. The disadvantage of the lack of such a symmetry might be that it becomes impossible to recover from global localization failures, as every surprising measurement is rejected. In such cases, it may make sense to impose additional constraints, such as a limit on the fraction of measurements that may be corrupted.

We note that the rejection test has found successful application even in highly static environments, for reasons that are quite subtle. The beam-based sensor model is discontinuous: Small changes of pose can drastically alter the posterior probability of a sensor measurement. This is because the result of ray casting is not a continuous function in pose parameters such as the robot orientation. In environment with cluttered objects, this discontinuity increases the number of particles necessary for successful localization. By manually removing clutter from the map—and instead letting the filter man-

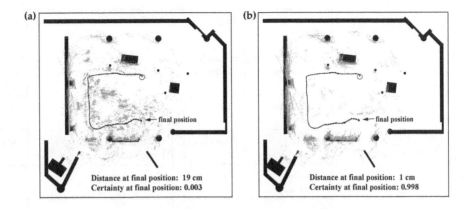

Figure 8.23 Comparison of (a) standard MCL and (b) MCL with the removal of sensor measurements likely caused by unexpected obstacles. Both diagrams show the robot path and the end-points of the scans used for localization.

age the resulting "surprisingly" short measurements—the number of particles can be reduced drastically. The same strategy does not apply to the likelihood field model, since this model is smooth in the pose parameters.

8.5 Practical Considerations

Table 8.6 summarizes and compares the main localization techniques discussed in this and the previous chapter. When choosing a technique, a number of requirements have to be traded off. A first question will always be whether it is preferable to extract features from sensor measurements. Extracting features may be beneficial from a computational perspective, but it comes at the price of reduced accuracy and robustness.

While in this chapter, we discussed techniques for handling dynamic environments in the context of the MCL algorithm, similar ideas can be brought to bear with other localization techniques as well. In fact, the techniques discussed here are only representative of a much richer body of approaches.

When implementing a localization algorithm, it is worthwhile to play with the various parameter settings. For example, the conditional probabilities are often inflated when integrating nearby measurements, so as to accommodate unmodeled dependencies that always exist in robotics. A good strategy is to collect reference data sets, and tune the algorithm until the overall re-

	EKF	MHT	Coarse (topological) grid	fine (metric) grid	MCL
Measurements	landmarks	landmarks	landmarks	raw measurements	raw measurements
Measurement noise	Gaussian	Gaussian	any	any	any
Posterior	Gaussian	mixture of Gaussians	histogram	histogram	particles
Efficiency (memory)	++	++	+	−	+
Efficiency (time)	++	+	+	−	+
Ease of implementation	+	−	+	−	++
Resolution	++	++	−	+	+
Robustness	−	+	+	++	++
Global localization	no	yes	yes	yes	yes

Table 8.6 Comparison of different implementations of Markov localization.

sult is satisfactory. This is necessary because no matter how sophisticated the mathematical model, there will always remain unmodeled dependencies and sources of systematic noise that affect the overall result.

8.6 Summary

In this chapter, we discussed two families of probabilistic localization algorithms, grid techniques and Monte Carlo localization (MCL).

- Grid techniques represent posteriors through histograms.

- The coarseness of the grid trades off accuracy and computational efficiency. For coarse grids, it is usually necessary to adjust the sensor and motion models to account for effects that arise from the coarseness of the representation. For fine grids, it may be necessary to update grid cells selectively to reduce the overall computation.

- The Monte Carlo localization algorithm represents the posterior using particles. The accuracy-computational costs trade-off is achieved through the size of the particle set.

- Both grid localization and MCL can globally localize robots.

- By adding random particles, MCL also solves the kidnapped robot problem.

- Mixture MCL is an extension that inverts the particle generation process for a fraction of all particles. This yields improved performance specifically for robots with low-noise sensors, but at the expense of a more complex implementation.

- KLD-sampling increases the efficiency of particle filters by adapting the size of sample sets over time. The advantage of this approach is maximal if the complexity of the beliefs varies drastically over time.

- Unmodeled environment dynamics can be accommodated by filtering sensor data, rejecting those that with high likelihood correspond to an unmodeled object. When using range sensors, the robot tends to reject measurements that are surprisingly short.

The popularity of MCL is probably due to two facts: MCL is just about the easiest localization algorithm to implement, and it is also one of the most potent ones, in that it can approximate nearly any distribution.

8.7 Bibliographical Remarks

Grid-based Monte Carlo localization was introduced by Simmons and Koenig, based on the related method of maintaining certainty factors by Nourbakhsh et al. (1995). Since Simmons and Koenig's (1995) seminal paper, a number of techniques emerged that maintained histograms for localization (Kaelbling et al. 1996). While the initial work used relatively coarse grids to accommodate the enormous computational overhead of updating grids, Burgard et al. (1996) introduced selective update techniques that could cope with grids of much higher resolution. This development was often seen as a shift from coarse, topological Markov localization to detailed, metric localization. Overview articles about this body of work can be found in Koenig and Simmons (1998); Fox et al. (1999c).

MAP MATCHING

For a number of years, grid-based techniques were considered state of the art in mobile robot localization. Different successful applications of grid-based Markov localization can be found. For example, Hertzberg and Kirchner (1996) applied this technique to robots operating in sewage pipes, Simmons et al. (2000b) used this approach to localize a robot in an office environment, and Burgard et al. (1999a) applied the algorithm to estimate the position of a robot operating in museums. Konolige and Chou (1999) introduced the idea of *map matching* into

Markov localization, by using fast convolution techniques for computing pose probabilities. An extension that combined both global localization and high-accuracy tracking was described in Burgard et al. (1998), who coined their technique *dynamic Markov localization*. A machine learning technique for learning to recognize places was introduced by Oore et al. (1997). Thrun (1998a) extended the approach by a learning component for identifying suitable landmarks in the environment, based on related work by Greiner and Isukapalli (1994). The mathematical framework was extended by Mahadevan and Khaleeli (1999) to a framework known as *semi Markov decision process*, which made it possible to reason about the exact time when a transition from one cell to another occurred. An experimental comparison between grid-approaches and Kalman filtering techniques was carried out by Gutmann et al. (1998). Active localization was introduced for the grid-based paradigm in Burgard et al. (1997), and since extended to multi-hypothesis tracking by Austin and Jensfelt (2000); Jensfelt and Christensen (2001a). Fox et al. (2000) and Howard et al. (2003) extended this approach to the multi-robot localization problem. Moving away from the grid-based paradigm, Jensfelt and Christensen (2001a); Roumeliotis and Bekey (2000); Reuter (2000) showed that multi-hypothesis EKFs were equally suited for global localization problem.

CONDENSATION ALGORITHM

Motivated by the famous *condensation algorithm* in computer vision (Isard and Blake 1998), Dellaert et al. (1999); Fox et al. (1999a) were the first to develop particle filters for mobile robot localization. They also coined the term *Monte Carlo Localization* which has become the common name of this technique in robotics. The idea of adding random samples can be found in Fox et al. (1999a). An improved technique to deal with the kidnapped robot problem was Lenser and Veloso's (2000) *sensor resetting* technique, in which a number of particles were jump-started using only the most recent measurement. Fox built on this technique and introduced the Augmented MCL algorithm for determining the number of particles to be added (Gutmann and Fox 2002). Mixture MCL algorithm is due to Thrun et al. (2000c); see also van der Merwe et al. (2001). It provided a mathematical basis for generating samples from measurements. KLD-sampling, the adaptive version of particle filters, was introduced by Fox (2003). Jensfelt et al. (2000) and Jensfelt and Christensen (2001b) applied MCL to feature-based maps, and Kwok et al. (2004) introduced a real-time version of MCL that adapts the number of particles. Finally, a number of papers have applied MCL to robots with cameras (Lenser and Veloso 2000; Schulz and Fox 2004; Wolf et al. 2005), including omnidirectional cameras (Kröse et al. 2002; Vlassis et al. 2002).

Particle filters have also been used for a number of related tracking and localization problems. Montemerlo et al. (2002b) studied the problem of simultaneous localization and people tracking, using a nested particle filter. A particle filter for tracking variable number of people is described in Schulz et al. (2001b), who demonstrated how moving people can be tracked reliably with a moving robot in an unknown environment (Schulz et al. 2001a).

8.8 Exercises

1. Consider a robot with d state variables. For example, the kinematic state of a free-flying rigid robot is usually $d = 6$; when velocities are included in the state vector, the dimension increases to $d = 12$. How does the complexity (update time and memory) of the following localization algorithms increase with d: EKF localization, grid localization, and Monte Carlo localization. Use the O( ) notation, and argue why your answer is correct.

2. Provide a mathematical derivation of the additive form of the multi-feature information integration in lines 14 and 15 in Table 7.2.

3. Prove the correctness of Equation (8.4), page 257, in the limit $\uparrow \infty$.

4. As noted in the text, Monte Carlo localization is *biased* for any finite sample size—i.e., the expected value of the location computed by the algorithm differs from the true expected value. In this question, you are asked to quantify this bias.

   To simplify, consider a world with four possible robot locations: $X = \{x_1, x_2, x_3, x_4\}$. Initially, we draw $N \geq 1$ samples uniformly from among those locations. As usual, it is perfectly acceptable if more than one sample is generated for any of the locations X. Let Z be a Boolean sensor variable characterized by the following conditional probabilities:

   $$p(z \mid x_1) = 0.8 \qquad p(\neg z \mid x_1) = 0.2$$
   $$p(z \mid x_2) = 0.4 \qquad p(\neg z \mid x_2) = 0.6$$
   $$p(z \mid x_3) = 0.1 \qquad p(\neg z \mid x_3) = 0.9$$
   $$p(z \mid x_4) = 0.1 \qquad p(\neg z \mid x_4) = 0.9$$

   MCL uses these probabilities to generate particle weights, which are subsequently normalized and used in the resampling process. For simplicity, let us assume we only generate one new sample in the resampling process, regardless of N. This sample might correspond to any of the four locations in X. Thus, the sampling process defines a probability distribution over X.

   (a) What is the resulting probability distribution over X for this new sample? Answer this question separately for $N = 1, \ldots, 10$, and for $N = \infty$.

   (b) The difference between two probability distributions p and q can be measured by the KL divergence, which is defined as

   $$KL(p, q) = \sum_i p(x_i) \log \frac{p(x_i)}{q(x_i)}$$

   What are the KL divergences between the distributions in (a) and the true posterior?

   (c) What modification of the problem formulation (not the algorithm!) would guarantee that the specific estimator above is unbiased even for finite values of N? Provide at least two such modifications (each of which should be sufficient).

5. Consider a robot equipped with a range/bearing sensor of the type discussed in Chapter 6.6. In this question, you are asked to devise an *efficient* sampling procedure that can incorporate k simultaneous measurements of identifiable landmarks. To illustrate that your routine works, you might generate plots of different landmark configurations, using $k = 1, \ldots, 5$ adjacent landmarks. Argue what makes your routine efficient.

6. Exercise 3 on page 235 described a simplistic underwater robot that can listen to acoustic beacons for localization. Here you are asked to implement a grid localization algorithm for this robot. Analyze the accuracy and the failure modes in the context of the three localization problems: global localization, position tracking, and the kidnapped robot problem.

Part III

Mapping

9 Occupancy Grid Mapping

9.1 Introduction

The previous two chapters discussed the application of probabilistic techniques to a low-dimensional perceptual problem, that of estimating a robot's pose. We assumed that the robot was given a map in advance. This assumption is legitimate in quite a few real-world applications, as maps are often available a priori or can be constructed by hand. Some application domains, however, do not provide the luxury of coming with an a priori map. Surprisingly enough, most buildings do not comply with the blueprints generated by their architects. And even if blueprints were accurate, they would not contain furniture and other items that, from a robot's perspective, determine the shape of the environment just as much as walls and doors. Being able to learn a map from scratch can greatly reduce the efforts involved in installing a mobile robot, and enable robots to adapt to changes without human supervision. In fact, mapping is one of the core competencies of truly autonomous robots.

Acquiring maps with mobile robots is a challenging problem for a number of reasons:

- The hypothesis space, which is the space of all possible maps, is huge. Since maps are defined over a continuous space, the space of all maps has infinitely many dimensions. Even under discrete approximations, such as the grid approximation that shall be used in this chapter, maps can easily be described 10^5 or more variables. The sheer size of this high-dimensional space makes it challenging to calculate full posteriors over maps; hence, the Bayes filtering approach that worked well for localization is inapplicable to the problem of learning maps, at least in its naive form discussed thus far.

- Learning maps is a "chicken-and-egg" problem, for which reason it is often referred to as the *simultaneous localization and mapping* (SLAM) or *concurrent mapping and localization problem*. First, there is a localization problem. When the robot moves through its environment, it accumulates errors in odometry, making it gradually less certain as to where it is. Methods exist for determining the robot's pose when a map is available, as we have seen in the previous chapter. Second, there is a mapping problem. Constructing a map when the robot's poses are known is also relatively easy—a claim that will be substantiated in this chapter and subsequent chapters. In the absence of both an initial map and exact pose information, however, the robot has to do both: estimating the map and localizing itself relative to this map.

Of course, not all mapping problems are equally hard. The hardness of the mapping problem is the result of a collection of factors, the most important of which are:

- **Size.** The larger the environment relative to the robot's perceptual range, the more difficult it is to acquire a map.

- **Noise in perception and actuation.** If robot sensors and actuators were noise-free, mapping would be a simple problem. The larger the noise, the more difficult the problem.

- **Perceptual ambiguity.** The more frequently different places look alike, the more difficult it is to establish correspondence between different locations traversed at different points in time.

- **Cycles.** Cycles in the environment are particularly difficult to map. If a robot just goes up and down a corridor, it can correct odometry errors incrementally when coming back. Cycles make robots return via different paths, and when closing a cycle the accumulated odometric error can be huge!

To fully appreciate the difficulty of the mapping problem, consider Figure 9.1. Shown there is a data set, collected in a large indoor environment. Figure 9.1a was generated using the robot's raw odometry information. Each black dot in this figure corresponds to an obstacle detected by the robot's laser range finder. Figure 9.1b shows the result of applying mapping algorithms to this data, one of which was the techniques described in this chapter. This example gives a good flavor of the problem at stake.

9.1 Introduction

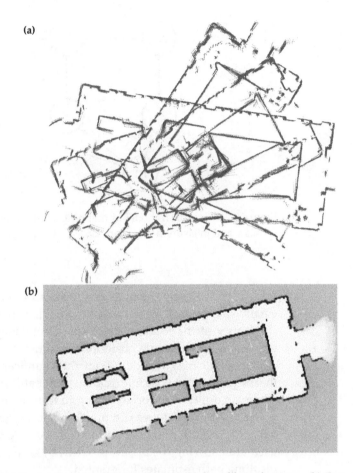

Figure 9.1 (a) Raw range data, position indexed by odometry. (b) Occupancy grid map.

In this chapter, we first study the mapping problem under the restrictive assumption that the robot poses are known. Put differently, we side-step the hardness of the SLAM problem by assuming some oracle informs us of the exact robot path during mapping. This problem, whose graphical model is depicted in Figure 9.2, is also known as *mapping with known poses*. We will discuss a popular family of algorithms, collectively called *occupancy grid mapping*. Occupancy grid mapping addresses the problem of generating consistent maps from noisy and uncertain measurement data, under the assumption that the robot pose is known. The basic idea of the occupancy

MAPPING WITH
KNOWN POSES

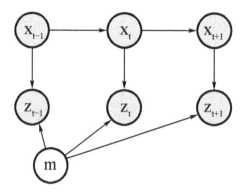

Figure 9.2 Graphical model of mapping with known poses. The shaded variables (poses x and measurements z) are known. The goal of mapping is to recover the map m.

grids is to represent the map as a field of random variables, arranged in an evenly spaced grid. Each random variable is binary and corresponds to the occupancy of the location it covers. Occupancy grid mapping algorithms implement approximate posterior estimation for those random variables.

The reader may wonder about the significance of a mapping technique that requires exact pose information. After all, no robot's odometry is perfect! The main utility of the occupancy grid technique is in post-processing: Many of the SLAM techniques discussed in subsequent chapters do not generate maps fit for path planning and navigation. Occupancy grid maps are often used after solving the SLAM problem by some other means, and taking the resulting path estimates for granted.

9.2 The Occupancy Grid Mapping Algorithm

The gold standard of any occupancy grid mapping algorithm is to calculate the posterior over maps given the data

(9.1) $\quad p(m \mid z_{1:t}, x_{1:t})$

As usual, m is the map, $z_{1:t}$ the set of all measurements up to time t, and $x_{1:t}$ is the path of the robot defined through the sequence of all poses. The controls $u_{1:t}$ play no role in occupancy grid maps, since the path is already known. Hence, they will be omitted throughout this chapter.

The types of maps considered by occupancy grid maps are fine-grained grids defined over the continuous space of locations. By far the most common domain of occupancy grid maps are 2-D floor plan maps, which describe a 2-D slice of the 3-D world. 2-D maps are usually the representation of choice when a robot navigates on a flat surface, and the sensors are mounted so that they capture only a slice of the world. Occupancy grid techniques generalize to 3-D representations but at significant computational expenses.

Let $\mathbf{m}_i$ denote the grid cell with index i. An occupancy grid map partitions the space into finitely many grid cells:

(9.2) $\quad m = \{\mathbf{m}_i\}$

Each $\mathbf{m}_i$ has attached to it a binary occupancy value, which specifies whether a cell is occupied or free. We will write "1" for occupied and "0" for free. The notation $p(\mathbf{m}_i = 1)$ or $p(\mathbf{m}_i)$ refers to the probability that a grid cell is occupied.

The problem with the posterior in Equation (9.1) is its dimensionality: the number of grid cells in maps like the one shown in Figure 9.1 are in the tens of thousands. For a map with 10,000 grid cells, the number of maps that can be represented by this map is $2^{10,000}$. Calculating a posterior probability for each single map is therefore intractable.

The standard occupancy grid approach breaks down the problem of estimating the map into a collection of separate problems, namely that of estimating

(9.3) $\quad p(\mathbf{m}_i \mid z_{1:t}, x_{1:t})$

for all grid cell $\mathbf{m}_i$. Each of these estimation problems is now a binary problem with static state. This decomposition is convenient but not without problems. In particular, it does not enable us to represent dependencies among neighboring cells; instead, the posterior over maps is approximated as the product of its marginals:

(9.4) $\quad p(m \mid z_{1:t}, x_{1:t}) = \prod_i p(\mathbf{m}_i \mid z_{1:t}, x_{1:t})$

We will return to this issue in Chapter 9.4 below, when we discuss more advanced mapping algorithms. For now, we will adopt this factorization for convenience.

Thanks to our factorization, the estimation of the occupancy probability for each grid cell is now a binary estimation problem with static state. A filter for this problem was already discussed in Chapter 4.2: the *binary Bayes filter*. The corresponding algorithm was depicted in Table 4.2 on page 94.

```
1:      Algorithm occupancy_grid_mapping({l_{t-1,i}}, x_t, z_t):
2:          for all cells m_i do
3:              if m_i in perceptual field of z_t then
4:                  l_{t,i} = l_{t-1,i} + inverse_sensor_model(m_i, x_t, z_t) - l_0
5:              else
6:                  l_{t,i} = l_{t-1,i}
7:              endif
8:          endfor
9:          return {l_{t,i}}
```

Table 9.1 The occupancy grid algorithm, a version of the binary Bayes filter in Table 4.2.

The algorithm in Table 9.1 applies this filter to the occupancy grid mapping problem. As in the original filter, our occupancy grid mapping algorithm uses the *log odds* representation of occupancy:

$$(9.5) \quad l_{t,i} = \log \frac{p(\mathbf{m}_i \mid z_{1:t}, x_{1:t})}{1 - p(\mathbf{m}_i \mid z_{1:t}, x_{1:t})}$$

This representation is already familiar from Chapter 4.2. The advantage of the log odds over the probability representation is that we can avoid numerical instabilities for probabilities near zero or one. The probabilities are easily recovered from the log odds ratio:

$$(9.6) \quad p(\mathbf{m}_i \mid z_{1:t}, x_{1:t}) = 1 - \frac{1}{1 + \exp\{l_{t,i}\}}$$

The algorithm **occupancy_grid_mapping** in Table 9.1 loops through all grid cells i, and updates those that fall into the sensor cone of the measurement z_t. For those where it does, it updates the occupancy value by virtue of the function **inverse_sensor_model** in line 4 of the algorithm. Otherwise, the occupancy value remains unchanged, as indicated in line 6. The constant l_0 is the prior of occupancy represented as a log odds ratio:

$$(9.7) \quad l_0 = \log \frac{p(\mathbf{m}_i = 1)}{p(\mathbf{m}_i = 0)} = \log \frac{p(\mathbf{m}_i)}{1 - p(\mathbf{m}_i)}$$

9.2 The Occupancy Grid Mapping Algorithm

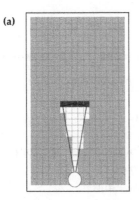 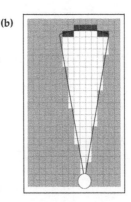

Figure 9.3 Two examples of an inverse measurement model **inverse_range_sensor_model** for two different measurement ranges. The darkness of each grid cell corresponds to the likelihood of occupancy. This model is somewhat simplistic; in contemporary implementations the occupancy probabilities are usually weaker at the border of the measurement cone.

The function **inverse_sensor_model** implements the inverse measurement model $p(\mathbf{m}_i \mid z_t, x_t)$ in its log odds form:

(9.8) $\quad$ **inverse_sensor_model**$(\mathbf{m}_i, x_t, z_t) \;=\; \log \dfrac{p(\mathbf{m}_i \mid z_t, x_t)}{1 - p(\mathbf{m}_i \mid z_t, x_t)}$

A somewhat simplistic example of such a function for range finders is given in Table 9.2 and illustrated in Figure 9.3a&b. This model assigns to all cells within the sensor cone whose range is close to the measured range an occupancy value of l_{occ}. In Table 9.2, the width of this region is controlled by the parameter α, and the opening angle of the beam is given by β. We note that this model is somewhat simplistic; in contemporary implementations the occupancy probabilities are usually weaker at the border of the measurement cone.

The algorithm **inverse_sensor_model** calculates the inverse model by first determining the beam index k and the range r for the center-of-mass of the cell $\mathbf{m}_i$. This calculation is carried out in lines 2 through 5 in Table 9.2. As usual, we assume that the robot pose is given by $x_t = (x \;\; y \;\; \theta)^T$. In line 7, it returns the prior for occupancy in log odds form whenever the cell is outside the measurement range of this sensor beam, or if it lies more than $\alpha/2$ behind the detected range z_t^k. In line 9, it returns $l_{\text{occ}} > l_0$ if the range of the cell is within $\pm \alpha/2$ of the detected range z_t^k. It returns $l_{\text{free}} < l_0$ if the range to the

1: Algorithm inverse_range_sensor_model(m_i, x_t, z_t):
2:      Let x_i, y_i be the center-of-mass of $\mathbf{m}_i$
3:      $r = \sqrt{(x_i - x)^2 + (y_i - y)^2}$
4:      $\phi = \text{atan2}(y_i - y, x_i - x) - \theta$
5:      $k = \text{argmin}_j |\phi - \theta_{j,\text{sens}}|$
6:      if $r > \min(z_{\max}, z_t^k + \alpha/2)$ or $|\phi - \theta_{k,\text{sens}}| > \beta/2$ then
7:          return l_0
8:      if $z_t^k < z_{\max}$ and $|r - z_t^k| < \alpha/2$
9:          return l_{occ}
10:     if $r \leq z_t^k$
11:         return l_{free}
12:     endif

Table 9.2 A simple inverse measurement model for robots equipped with range finders. Here α is the thickness of obstacles, and β the width of a sensor beam. The values l_{occ} and l_{free} in lines 9 and 11 denote the amount of evidence a reading carries for the two different cases.

cell is shorter than the measured range by more than $\alpha/2$. The left and center panel of Figure 9.3 illustrates this calculation for the main cone of a sonar beam.

A typical application of an inverse sensor model for ultrasound sensors is shown in Figure 9.4. Starting with an initial map the robot successively extends the map by incorporating local maps generated using the inverse model. A larger occupancy grid map obtained with this model for the same environment is depicted in Figure 9.5.

Figures 9.6 shows an example map next to a blueprint of a large open exhibit hall, and relates it to the occupancy map acquired by a robot. The map was generated using laser range data gathered in a few minutes. The gray-level in the occupancy map indicates the posterior of occupancy over an evenly spaced grid: The darker a grid cell, the more likely it is to be occupied. While occupancy maps are inherently probabilistic, they tend to quickly converge to estimates that are close to the two extreme posteriors, 1 and 0. In comparison between the learned map and the blueprint, the occupancy grid map shows all major structural elements, and obstacles as they were visi-

9.2 The Occupancy Grid Mapping Algorithm

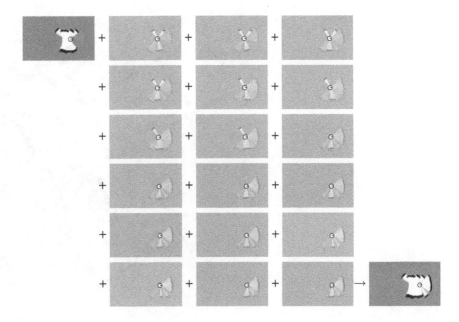

Figure 9.4 Incremental learning of an occupancy grid map using ultra-sound data in a corridor environment. The upper left image shows the initial map and the lower right image contains the resulting map. The maps in columns 2 to 4 are the local maps built from an inverse sensor model. Measurements beyond a 2.5m radius have not been considered. Each cone has an opening angle of 15 degrees. Images courtesy of Cyrill Stachniss, University of Freiburg.

Figure 9.5 Occupancy probability map of an office environment built from sonar measurements. Courtesy of Cyrill Stachniss, University of Freiburg.

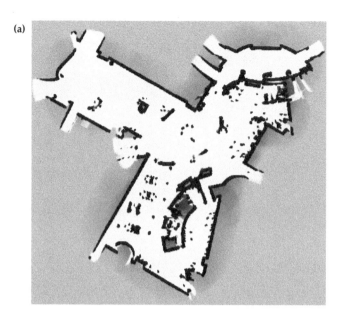

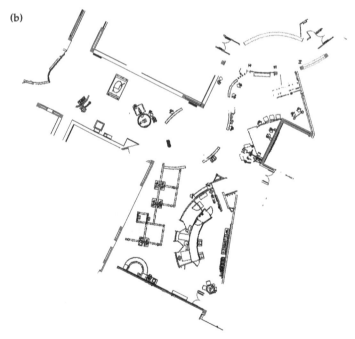

Figure 9.6 (a) Occupancy grid map and (b) architectural blue-print of a large open exhibit space. Notice that the blue-print is inaccurate in certain places.

9.2 The Occupancy Grid Mapping Algorithm

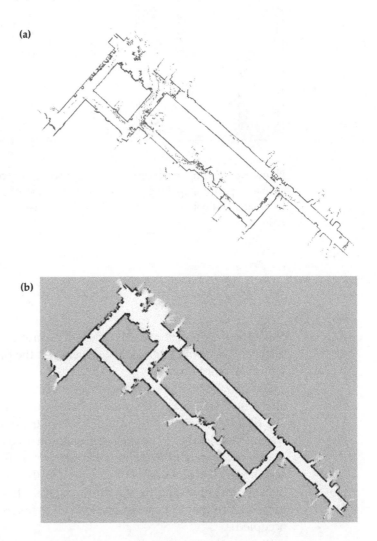

Figure 9.7 (a) Raw laser range data with corrected pose information. Each dot corresponds to a detection of an obstacle. Most obstacles are static (walls etc.), but some were dynamic, since people walked near the robot during data acquisition. (b) Occupancy grid map built from the data. The gray-scale indicates the posterior probability: Black corresponds to occupied with high certainty, and white to free with high certainty. The gray background color represents the prior. Figure (a) courtesy of Steffen Gutmann.

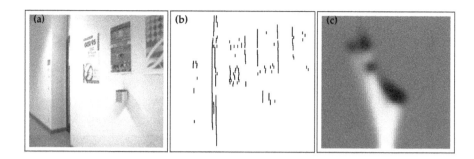

Figure 9.8 Estimation of occupancy maps using stereo vision: (a) camera image, (b) sparse disparity map, (c) occupancy map by projecting the disparity image onto the 2-D plane and convolving the result with a Gaussian. Images courtesy of Thorsten Fröhlinghaus.

ble at the height of the laser. Through a careful inspection, the reader may uncover some small discrepancies between the blueprint and the actual environment configuration.

Figure 9.7 compares a raw dataset with the occupancy grid maps generated from this data. The data in Panel (a) was preprocessed by a SLAM algorithm, so that the poses align. Some of the data is corrupted by the presence of people; the occupancy grid map filters out people quite nicely. This makes occupancy grid maps much better suited for robot navigation than sets of scan endpoint data: A planner fed the raw sensor endpoints would have a hard time finding a path through such scattered obstacles, even if the evidence that the corresponding cell is free outweighed that of it being occupied.

We note that our algorithm makes occupancy decisions exclusively based on sensor measurements. An alternative source of information is the space claimed by the robot itself: When the robot's pose is x_t, the region surrounding x_t must be navigable. Our inverse measurement algorithm in Table 9.2 can easily be modified to incorporate this information, by returning a large negative number for all grid cells occupied by a robot when at x_t. In practice, it is a good idea to incorporate the robot's volume when generating maps, especially if the environment is populated during mapping.

9.2.1 Multi-Sensor Fusion

Robots are often equipped with more than one type of sensor. Hence, a natural objective is to integrate information from more than one sensor into a single map. This question as to how to best integrate data from multiple sensors is particularly interesting if the sensors have different characteristics. For example, Figure 9.8 shows occupancy maps built with a stereo vision system, in which disparities are projected onto the plane and convolved with a Gaussian. Clearly, the characteristics of stereo are different from that of a sonar-based range finder. They are sensitive to different types of obstacles.

Unfortunately, fusing data from multiple sensors with Bayes filters is not an easy endeavor. A naive solution is to execute algorithm **occupancy_grid_mapping** in Table 9.1 with different sensor modalities. However, such an approach has a clear drawback. If different sensors detect different types of obstacles, the result of Bayes filtering is ill-defined. Consider, for example, an obstacle that can be recognized by one sensor type but not by another. Then these two sensor types will generate conflicting information, and the resulting map will depend on the amount of evidence brought by every sensor system. This is generally undesirable, since whether or not a cell is considered occupied depends on the relative frequency at which different sensors are polled.

A popular approach to integrating information from multiple sensors is to build separate maps for each sensor type, and integrate them using an appropriate combination function. Let $m^k = \{\mathbf{m}_i^k\}$ denote the map built by the k-th sensor type. If the measurements of the sensors are independent of each other we can directly combine them using *De Morgan's law*

$$(9.9) \quad p(\mathbf{m}_i) = 1 - \prod_k (1 - p(\mathbf{m}_i^k))$$

Alternatively, one can compute the maximum

$$(9.10) \quad p(\mathbf{m}_i) = \max_k p(\mathbf{m}_i^k)$$

of all maps, which yields the most pessimistic estimates of its components. If any of the sensor-specific maps show that a grid cell is occupied, so will the combined map.

9.3 Learning Inverse Measurement Models

9.3.1 Inverting the Measurement Model

The occupancy grid mapping algorithm requires a marginalized *inverse measurement model*, $p(\mathbf{m}_i \mid x, z)$. This probability is called "inverse" since it reasons from effects to causes: it provides information about the world conditioned on a measurement caused by this world. It is marginalized for the i-th grid cell; a full inverse would be of the type $p(m \mid x, z)$. In our exposition of the basic algorithm, we already provided an ad hoc procedure in Table 9.2 for implementing such an inverse model. This raises the question as to whether we can obtain an inverse model in a more principled manner, starting at the conventional measurement model.

The answer is positive but less straightforward than one might assume at first glance. Bayes rule suggests

$$(9.11) \quad p(m \mid x, z) = \frac{p(z \mid x, m)\, p(m \mid x)}{p(z \mid x)}$$
$$= \eta\, p(z \mid x, m)\, p(m)$$

Here we silently assume $p(m \mid x) = p(m)$, hence the pose of the robot tells us nothing about the map—an assumption that we will adopt for sheer convenience. If our goal was to calculate the inverse model for the entire map at-a-time, we would now be done. However, our occupancy grid mapping algorithm approximates the posterior over maps by its marginals, one for each grid cell $\mathbf{m}_i$. The inverse model for the i-th grid cell is obtained by selecting the marginal for the i-th grid cell:

$$(9.12) \quad p(\mathbf{m}_i \mid x, z) = \eta \sum_{m:m(i)=\mathbf{m}_i} p(z \mid x, m)\, p(m)$$

This expression sums over all maps m for which the occupancy value of grid cell i equals $\mathbf{m}_i$. Clearly, this sum cannot be computed, since the space of all maps is too large.

We will now describe an algorithm for approximating this expression. The algorithm involves generating samples from the measurement model, and approximating the inverse using a *supervised learning algorithm*, such as *logistic regression* or a *neural network*.

SUPERVISED LEARNING ALGORITHM

9.3.2 Sampling from the Forward Model

The basic idea is simple and quite universal: If we can generate random triplets of poses $x_t^{[k]}$, measurements $z_t^{[k]}$, and map occupancy values $\mathbf{m}_i^{[k]}$ for any grid cell $\mathbf{m}_i$, we can learn a function that accepts a pose x and measurement z as an input, and outputs the probability of occupancy for $\mathbf{m}_i$.

A sample of the form $(x_t^{[k]}\ z_t^{[k]}\ \mathbf{m}_i^{[k]})$ can be generated by the following procedure.

1. Sample a random map $m^{[k]} \sim p(m)$. For example, one might already have a database of maps that represents $p(m)$ and randomly draws a map from the database.

2. Sample a pose $x_t^{[k]}$ inside the map. One may safely assume that poses are uniformly distributed.

3. Sample a measurement $z_t^{[k]} \sim p(z \mid x_t^{[k]}, m^{[k]})$. This sampling step is reminiscent of a robot simulator that stochastically simulates a sensor measurement.

4. Extract the desired "true" occupancy value $\mathbf{m}_i$ for the target grid cell from the map m.

The result is a sampled pose $x_t^{[k]}$, a measurement $z_t^{[k]}$, and the occupancy value of the grid cell $\mathbf{m}_i$. Repeated application of this sampling step yields a data set

$$
\begin{array}{ccc}
x_t^{[1]} & z_t^{[1]} & \longrightarrow \quad \mathrm{occ}(\mathbf{m}_i)^{[1]} \\
x_t^{[2]} & z_t^{[2]} & \longrightarrow \quad \mathrm{occ}(\mathbf{m}_i)^{[2]} \\
x_t^{[3]} & z_t^{[3]} & \longrightarrow \quad \mathrm{occ}(\mathbf{m}_i)^{[3]} \\
\vdots & \vdots & \vdots
\end{array}
$$

TRAINING EXAMPLES These triplets may serve as *training examples* for the supervised learning algorithm, which approximates the desired conditional probability $p(\mathbf{m}_i \mid z, x)$. Here the measurements z and the pose x are input variables, and the occupancy value $\mathrm{occ}(\mathbf{m}_i)$ is a target for the output of the learning algorithm.

This approach is somewhat inefficient, since it fails to exploit a number of properties that we know to be the case for the inverse sensor model.

- Measurements should carry no information about grid cells far outside their perceptual range. This observation has two implications: First, we can focus our sample generation process on triplets where the cell $\mathbf{m}_i$ is indeed inside the measurement cone. And second, when making a prediction for this cell, we only have to include a subset of the data in a measurement z (e.g., nearby beams) as input to the learning algorithm.

- The characteristics of a sensor are invariant with respect to the absolute coordinates of the robot or the grid cell when taking a measurement. Only the relative coordinates matter. If we denote the robot pose by $x_t = (x \ y \ \theta)^T$ and the coordinates of the grid cell by $\mathbf{m}_i = (x_{\mathbf{m}_i} \ y_{\mathbf{m}_i})^T$, the coordinates of the grid cell are mapped into the robot's local reference frame via the following translation and rotation:

$$\begin{pmatrix} \cos\theta & -\sin\theta \\ \sin\theta & \cos\theta \end{pmatrix} \begin{pmatrix} x_{\mathbf{m}_i} - x \\ y_{\mathbf{m}_i} - y \end{pmatrix}$$

In robots with circular arrays of range finders, it makes sense to encode the relative location of a grid cell using the familiar polar coordinates (range and bearing).

- Nearby grid cells should have a similar interpretation under the inverse sensor model. This smoothness suggests that it may be beneficial to learn a single function in which the coordinates of the grid cell function as an input, rather than learning a separate function for each grid cell.

- If the robot possesses functionally identical sensors, the inverse sensor model should be interchangeable for different sensors. For robots equipped with a circular array of range sensors, any of the resulting sensor beams are characterized by the same inverse sensor model.

The most basic way to enforce these invariances is to constrain the learning algorithm by choosing appropriate input variables. A good choice is to use relative pose information, so that the learning algorithm cannot base its decision on absolute coordinates. It is also a good idea to omit sensor measurements known to be irrelevant to occupancy predictions, and to confine the prediction to grid cells inside the perceptual field of a sensor. By exploiting these invariances, the training set size can be reduced significantly.

9.3.3 The Error Function

BACKPROPAGATION

To train the learning algorithm, we need an approximate error function. A popular example are artificial neural networks trained with the Backpropagation algorithm. Backpropagation trains *neural networks* by *gradient descent* in parameter space. Given an error function that measures the "mismatch" between the network's actual and desired output, Backpropagation calculates the first derivative of the target function and the parameters of the neural network, and then adapts the parameters in opposite direction of the gradient so as to diminish the mismatch. This raises the question as to what error function to use.

A common approach is to train the learning algorithm so as to maximize the log-likelihood of the training data. More specifically we are given a training set of the form

$$(9.13) \quad \begin{array}{rcl} \text{input}^{[1]} & \longrightarrow & \text{occ}(\mathbf{m}_i)^{[1]} \\ \text{input}^{[2]} & \longrightarrow & \text{occ}(\mathbf{m}_i)^{[2]} \\ \text{input}^{[3]} & \longrightarrow & \text{occ}(\mathbf{m}_i)^{[3]} \\ \vdots & & \vdots \end{array}$$

$\text{occ}(\mathbf{m}_i)^{[k]}$ is the k-th sample of the desired conditional probability, and $\text{input}^{[k]}$ is the corresponding input to the learning algorithm. Clearly, the exact form of the input may vary as a result of the encoding known invariances, but the exact nature of this vector will play no role in the form of the error function.

Let us denote the parameters of the learning algorithm by W. Assuming that each individual item in the training data has been generated independently, the likelihood of the training data is now

$$(9.14) \quad \prod_i p(\mathbf{m}_i^{[k]} \mid \text{input}^{[k]}, W)$$

and its negative logarithm is

$$(9.15) \quad J(W) \;\; = \;\; -\sum_i \log p(\mathbf{m}_i^{[k]} \mid \text{input}^{[k]}, W)$$

Here J defines the function we seek to minimize during training.

Let us denote the learning algorithm by $f(\text{input}^{[k]}, W)$. The output of this function is a value in the interval $[0; 1]$. After training, we want the learning algorithm to output the probability of occupancy:

$$(9.16) \quad p(\mathbf{m}_i^{[k]} \mid \text{input}^{[k]}, W) \;\; = \;\; \begin{cases} f(\text{input}^{[k]}, W) & \text{if } \mathbf{m}_i^{[k]} = 1 \\ 1 - f(\text{input}^{[k]}, W) & \text{if } \mathbf{m}_i^{[k]} = 0 \end{cases}$$

Thus, we seek an error function that adjusts W so as to minimize the deviation of this predicted probability and the one communicated by the training example. To find such an error function, we re-write (9.16) as follows:

$$(9.17) \quad p(\mathbf{m}_i^{[k]} \mid \text{input}^{[k]}, W) = f(\text{input}^{[k]}, W)^{\mathbf{m}_i^{[k]}} (1 - f(\text{input}^{[k]}, W))^{1-\mathbf{m}_i^{[k]}}$$

It is easy to see that this product and Expression (9.16) are identical. In the product, one of the terms is always 1, since its exponent is zero. Substituting the product into (9.15) and multiplying the result by minus one gives us the following function:

$$(9.18) \quad J(W) = -\sum_i \log \left[f(\text{input}^{[k]}, W)^{\mathbf{m}_i^{[k]}} (1 - f(\text{input}^{[k]}, W))^{1-\mathbf{m}_i^{[k]}} \right]$$
$$= -\sum_i \mathbf{m}_i^{[k]} \log f(\text{input}^{[k]}, W) + (1 - \mathbf{m}_i^{[k]}) \log(1 - f(\text{input}^{[k]}, W))$$

$J(W)$ is the error function to minimize when training the learning algorithm. It is easily folded into any algorithm that uses gradient descent to adjusts its parameters.

9.3.4 Examples and Further Considerations

Figure 9.9 shows the result of an artificial neural network trained to mimic the inverse sensor model. The robot in this example is equipped with a circular array of sonar range sensors mounted at approximate table height. The input to the network are the relative range and bearing of a target cell, along with the set of five adjacent range measurements. The output is a probability of occupancy: the darker a cell, the more likely it is occupied. As this example illustrates, the approach correctly learns to distinguish freespace from occupied space. The uniformly gray area behind obstacles matches the prior probability of occupancy, which leads to no change when used in the occupancy grid mapping algorithm. Figure 9.9b contains a faulty short reading on the bottom left. Here a single reading seems to be insufficient to predict an obstacle with high probability.

We note that there exists a number of ways to train a function approximator using actual data collected by a robot, instead of the simulated data from the forward model. In general, this is the most accurate data one can use for learning, since the measurement model is necessarily just an approximation. One such way involves a robot operating in a known environment with a known map. With Markov localization, we can localize the robot, and then use its actual recorded measurements and the known map occupancy to assemble training examples. It is even possible to start with an approxi-

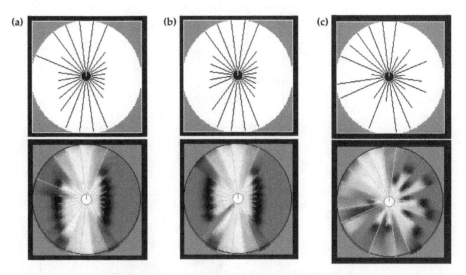

Figure 9.9 Inverse sensor model learned from data: Three sample sonar scans (top row) and local occupancy maps (bottom row), as generated by the neural network. Bright regions indicate free-space, and dark regions indicate walls and obstacles (enlarged by a robot diameter).

mate map, use the learned sensor model to generate a better map, and use the procedure just outlined to improve the inverse measurement model.

9.4 Maximum A Posteriori Occupancy Mapping

9.4.1 The Case for Maintaining Dependencies

In the remainder of this chapter, we will return to one of the very basic assumptions of the occupancy grid mapping algorithm. In Chapter 9.2, we assumed that we can safely decompose the map inference problem defined over high-dimensional space of all maps, into a collection of single-cell mapping problems. This assumption culminated into the factorization in (9.4):

$$(9.19) \quad p(m \mid z_{1:t}, x_{1:t}) = \prod_i p(\mathbf{m}_i \mid z_{1:t}, x_{1:t})$$

This raises the question as to how faithful we should be in the result of any algorithm that relies on such a strong decomposition.

Figure 9.10 illustrates a problem that arises directly as a result of this factor-

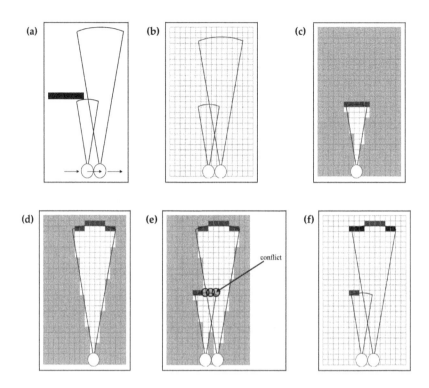

Figure 9.10 The problem with the standard occupancy grid mapping algorithm in Chapter 9.2: For the environment shown in Figure (a), a passing robot might receive the (noise-free) measurement shown in (b). The factorial approach maps these beams into probabilistic maps separately for each grid cell and each beam, as shown in (c) and (d). Combining both interpretations yields the map shown in (e). Obviously, there is a conflict in the overlap region, indicated by the circles in (e). The interesting insight is: There exist maps, such as the one in diagram (f), that perfectly explain the sensor measurement without any such conflict. For a sensor reading to be explained, it suffices to assume an obstacle *somewhere* in the cone of a measurement, and not everywhere.

ization. Shown there is a situation in which the robot facing a wall receives two noise-free sonar range measurements. Because the factored approach predicts an object along the entire arc at the measured range, the occupancy values of all grid cells along this arc are increased. When combining the two different measurements shown in Figure 9.10c&d, a conflict is created, as shown in Figure 9.10e. The standard occupancy grid mapping algorithm

9.4 Maximum A Posteriori Occupancy Mapping

1:      **Algorithm MAP_occupancy_grid_mapping**($x_{1:t}, z_{1:t}$):
2:          set $m = \{0\}$
3:          repeat until convergence
4:              for all cells $\mathbf{m}_i$ do
5:                  $m_i = \underset{k=0,1}{\operatorname{argmax}}\ k\, l_0 + \sum_t \log$
                        measurement_model(z_t, x_t, m with $\mathbf{m}_i = k$)
6:              endfor
7:          endrepeat
8:          return m

Table 9.3 The maximum a posteriori occupancy grid algorithm, which uses conventional measurement models instead of inverse models.

"resolves" this conflict by summing up positive and negative evidence for occupancy; however, the result will reflect the relative frequencies of the two types of measurements, which is undesirable.

However, there exist maps, such as the one in Figure 9.10f, that perfectly explain the sensor measurements without any such conflict. This is because for a sensor reading to be explained, it suffices to assume an obstacle *somewhere* in its measurement cone. Put differently, the fact that cones sweep over multiple grid cells induces important dependencies between neighboring grid cells. When decomposing the mapping into thousands of individual grid cell estimation problems, we lose the ability to consider these dependencies.

9.4.2 Occupancy Grid Mapping with Forward Models

These dependencies are incorporated by an algorithm that outputs the mode of the posterior, instead of the full posterior. The mode is defined as the maximum of the logarithm of the map posterior:

$$(9.20) \quad m^* = \underset{m}{\operatorname{argmax}}\ \log p(m \mid z_{1:t}, x_{1:t})$$

The map posterior factors into a map prior and a measurement likelihood (c.f., Equation (9.11)):

$$\log p(m \mid z_{1:t}, x_{1:t}) = \text{const.} + \log p(z_{1:t} \mid x_{1:t}, m) + \log p(m) \quad (9.21)$$

The log-likelihood $\log p(z_{1:t} \mid x_{1:t}, m)$ decomposes into a sum of individual measurement log-likelihoods:

$$\log p(z_{1:t} \mid x_{1:t}, m) = \sum \log p(z_t \mid x_t, m) \quad (9.22)$$

Further, the log-prior also decomposes. To see, we note that the prior probability of any map m is given by the following product:

$$p(m) = \prod_i p(\mathbf{m})^{\mathbf{m}_i} (1 - p(\mathbf{m}))^{1-\mathbf{m}_i} \quad (9.23)$$

$$= (1 - p(\mathbf{m}))^N \prod_i p(\mathbf{m})^{\mathbf{m}_i} (1 - p(\mathbf{m}))^{-\mathbf{m}_i}$$

$$= \eta \prod_i p(\mathbf{m})^{\mathbf{m}_i} (1 - p(\mathbf{m}))^{-\mathbf{m}_i}$$

Here $p(\mathbf{m})$ is the prior probability of occupancy (e.g., $p(\mathbf{m}) = 0.5$), and N is the number of grid cells in the map. The expression $(1 - p(\mathbf{m}))^N$ is simply a constant, which is replaced by our generic symbol η as usual.

This implies for the log version of the prior:

$$\log p(m) = \text{const.} + \sum_i \mathbf{m}_i \log p(\mathbf{m}) - \mathbf{m}_i \log(1 - p(\mathbf{m})) \quad (9.24)$$

$$= \text{const.} + \sum_i \mathbf{m}_i \log \frac{p(\mathbf{m})}{1 - p(\mathbf{m})}$$

$$= \text{const.} + \sum_i \mathbf{m}_i l_0$$

The constant l_0 is adopted from (9.7). The term $M \log(1 - p(\mathbf{m}_i))$ is obviously independent of the map. Hence it suffices to optimize the remaining expression and the data log-likelihood:

$$m^* = \operatorname*{argmax}_m \sum_t \log p(z_t \mid x_t, m) + l_0 \sum_i \mathbf{m}_i \quad (9.25)$$

A hill-climbing algorithm for maximizing this log-probability is provided in Table 9.3. This algorithm starts with the all-free map (line 2). It "flips" the occupancy value of a grid cell when such a flip increases the likelihood of the data (lines 4-6). For this algorithm it is essential that the prior of occupancy

9.4 Maximum A Posteriori Occupancy Mapping

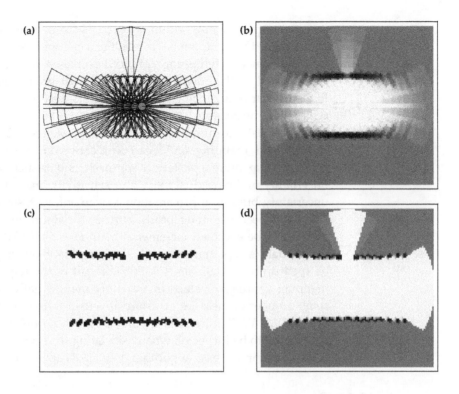

Figure 9.11 (a) Sonar range measurements from a noise-free simulation. (b) Results of the standard occupancy mapper, lacking the open door. (c) A maximum a posterior map. (d) The residual uncertainty in this map, obtained by measuring the sensitivity of the map likelihood function with respect to individual grid cells. This map clearly shows the door, and it also contains flatter walls at both ends.

$p(\mathbf{m}_i)$ is not too close to 1; otherwise it might return an all-occupied map. As any hill climbing algorithm, this approach is only guaranteed to find a local maximum. In practice, there are usually very few, if any, local maxima.

Figure 9.11 illustrates the effect of the MAP occupancy grid algorithm. Figure 9.11a depicts a noise-free data set of a robot passing by an open door. Some of the sonar measurements detect the open door, while others are reflected at the door post. The standard occupancy mapping algorithm with inverse models fails to capture the opening, as shown in Figure 9.11b. The mode of the posterior is shown in Figure 9.11c. This map models the open door correctly, hence it is better suited for robot navigation than the standard occupancy grid map algorithm. Figure 9.11d shows the residual uncertainty of this map. This diagram is the result of a cell-wise sensitivity analysis: The magnitude by which flipping a grid cell decreases the log-likelihood function is illustrated by the grayness of a cell. This diagram, similar in appearance to the regular occupancy grid map, suggests maximum uncertainty for grid cells behind obstacles.

There exists a number of limitations of the algorithm **MAP_occupancy_grid_mapping**, and it can be improved in multiple ways. The algorithm is a maximum a posteriori approach, and as such returns no notion of uncertainty in the residual map. Our sensitivity analysis approximates this uncertainty, but this approximation is overconfident, since sensitivity analysis only inspects the mode locally. Further, the algorithm is a batch algorithm and cannot be executed incrementally. In fact, the MAP algorithm requires that all data is kept in memory. At the computational end, the algorithm can be sped up by initializing it with the result of the regular occupancy grid mapping approach, instead of an empty map. Finally, we note that only a small number of measurements are affected by flipping a grid cell in line 5 of Table 9.3. While each sum is potentially huge, only a small number of elements has to be inspected when calculating the argmax. This property can be exploited in the basic algorithm, to increase its computational efficiency.

9.5 Summary

This chapter introduced algorithms for learning occupancy grids. All algorithms in this chapter require exact pose estimates for the robot, hence they do not solve the general mapping problem.

- The standard occupancy mapping algorithm estimates for each grid cell individually the posterior probability of occupancy. It is an adaptation of the binary Bayes filter for static environments.

- Data from multiple sensors can be fused into a single map in two ways: By maintaining a single map using Bayes filters, and by maintaining multiple maps, one for each sensor modality. The latter approach is usually preferable when different sensors are sensitive to different types of obstacles.

- The standard occupancy grid mapping algorithm relies on inverse measurement models, which reason from effects (measurements) to causes (occupancy). This differs from previous applications of Bayes filters in the context of localization, where the Bayes filter was based on a conventional measurement model that reasons from causes to effects.

- It is possible to learn inverse sensor models from the conventional measurement model, which models the sensor from causes to effects. To do so, one has to generate samples and learn an inverse model using a supervised learning algorithm.

- The standard occupancy grid mapping algorithm does not maintain dependencies in the estimate of occupancy. This is a result of decomposing the map posterior estimation problem into a large number of single-cell posterior estimation problem.

- The full map posterior is generally not computable, due to the large number of maps that can be defined over a grid. However, it can be maximized. Maximizing it leads to maps that are more consistent with the data than the occupancy grid algorithm using Bayes filters. However, the maximization requires the availability of all data, and the resulting maximum a posterior map does not capture the residual uncertainty in the map.

Without a doubt, occupancy grid maps and their various extensions are vastly popular in robotics. This is because they are extremely easy to acquire, and they capture important elements for robot navigation.

9.6 Bibliographical Remarks

Occupancy grid maps are due to Elfes (1987), whose Ph.D. thesis (1989) defined the field. A well-referenced article by Moravec (1988) provides a highly accessible introduction into this topic, and lays out the basic probabilistic approach that forms the core of this chapter. In unpublished work, Moravec and Martin (1994) extended occupancy grid maps to 3-D, using stereo as the primary sensor. Multi-sensor fusion in occupancy grids were introduced in Thrun et al. (1998a). The results for learning inverse sensor models described in this chapter can be found in Thrun (1998b). The forward modeling approach, also described in this chapter, is based on a similar algorithm in Thrun (2003).

Occupancy maps have been used for a number of different purposes. Borenstein and Koren (1991) were the first to adopt occupancy grid maps for collision avoidance. A number of authors have used occupancy grid maps for localization, by cross-matching two occupancy grid maps. Such "map matching" algorithms are discussed in detail in Chapter 7. Biswas et al. (2002) used occupancy grid maps to learn shape models of movable objects in dynamic environments. This approach was later extended to learning hierarchical class models of dynamic objects, all represented with occupancy grid maps (Anguelov et al. 2002). Occupancy grid maps have also extensively been used in the context of the simultaneous localization and mapping problem. Those applications will be discussed in later chapters.

Figure 9.12 Mobile indoor robot of the type RWI B21, with 24 sonar sensors mounted on a circular array around the robot.

The idea of representing space by grids is only one out of many ideas explored in the mobile robotics literature. Classical work on motion planning often assumes that the environment is represented by polygons, but leaves open as to how those models are being acquired from data (Schwartz et al. 1987). An early proposal on learning polygonal maps is due to Chatila and Laumond (1985). A first implementation using Kalman filters for fitting lines from sonar data was done by Crowley (1989). In more recent work, Anguelov et al. (2004) devised techniques for identifying straight-line doors from raw sensor data, and learn visual attributes for improving the door detection rate.

An early paradigm in spatial representation is the topological paradigm, in which space is represented by a set of local relations, often corresponding to specific actions a robot may have to take to navigate between adjacent locations. Examples of topological mapping algorithm include Kuipers and Levitt's (1988) work on their *Spatial Semantic hierarchy* (see also Kuipers et al. (2004)); Matarić's (1990) M.Sc. thesis work, Kortenkamp and Weymouth's (1994) work on topological graphs contracted from sonar and vision data, and Shatkay and Kaelbling's (1997) approach on spatial HMMs with arc length information. Occupancy grid maps are members of the complimentary paradigm: metric representations. Metric representations directly describe the robot environment, in some absolute coordinate system. A second example of a metric approach is the EKF SLAM algorithm, which will be discussed in the subsequent chapter.

SPATIAL SEMANTIC HIERARCHY

There is a history of attempts to generate mapping algorithms that harvest the best of both paradigms, topological and metric. Tomatis et al. (2002) uses topological representations to close loops consistently, then converts to metric maps. Thrun (1998b) first builds a metric occupancy grid map, then extracts a topological skeleton to facilitate fast motion planning. In Chapter 11, we will study techniques that bridge both paradigms, metric and topological.

9.7 Exercises

1. Change the basic occupancy grid algorithm in Table 9.1 to include a provision for the change of occupancy over time. To accommodate such change, evidence collected Δt time steps in the past should be decayed by a factor of $\alpha^{\Delta t}$, for some value of $\alpha < 1$ (e.g., $\alpha = 0.99$). Such a rule is called *exponential decay*. State the exponential decay occupancy grid mapping algorithm in log odds form and argue its correctness. If you cannot find an exact algorithm, state an approximation and argue why it is a suitable approximation. For simplicity, you might want to assume a prior $p(\mathbf{m}_i) = 0.5$ for occupancy.

2. The binary Bayes filter assumes that a cell is either occupied or unoccupied, and the sensor provides noisy evidence for the correct hypothesis. In this question, you will be asked to build an alternative estimator for a grid cell: Suppose the sensor can only measure "0 = unoccupied" or "1 = occupied", and it receives a sequence

    0, 0, 1, 0, 1, 1, 1, 0, 1, 0.

    What is the maximum likelihood probability p for the next reading to be 1? Provide an incremental formula for a general maximum likelihood estimator for this probability p. Discuss the difference of this estimator to the binary Bayes filter (all for a single cell only).

3. We study a common sensor configuration in indoor robotics. Suppose an indoor robot uses sonar sensors with a 15 degree opening cone, mounted at a fixed height so that they point out horizontally and parallel to the ground. Figure 9.12 shows such a robot. Discuss, what happens when the robot faces an obstacle whose height is just below the height of the sensor (for example, 15 cm below). Specifically, answer the following questions.

    (a) Under what conditions will the robot detect the obstacle? Under what conditions will it fail to detect it? Be concise.

    (b) What implications does this all have for the binary Bayes filter and the underlying Markov assumption? How can you make the occupancy grid algorithm fail?

    (c) Based on your answer to the previous question, can you provide an improved occupancy grid mapping algorithm that will detect the obstacle more reliably than the plain occupancy grid mapping algorithm?

EXPONENTIAL DECAY

4. In this question, you are being asked to design a simple sensor model. Suppose you are given binary occupancy measurements for the following four cells:

cell number	type	measurement sequence						
cell 1	occupied	1	1	0	1	0	1	1
cell 2	occupied	0	1	1	1	0	0	1
cell 3	free	0	0	0	0	0	0	0
cell 4	free	1	0	0	1	0	0	0

What is the maximum likelihood measurement model $p(z \mid \mathbf{m}_i)$? (Hint: $\mathbf{m}_i$ is a binary occupancy variable, and z is a binary measurement variable.)

5. For the table in Exercise 4, implement the basic occupancy grid algorithm.

   (a) What is the posterior $p(\mathbf{m}_i \mid z_{1:7})$ for the four different cases, assuming a prior $p(\mathbf{m}_i) = 0.5$?

   (b) Devise a tuning algorithm for your sensor model that makes the output of your occupancy grid mapping algorithm as close as possible to the ground truth, for the four cases in Exercise 4. What do you find? (For this question, you will have to come up with a suitable closeness measure.)

6. The standard occupancy grid mapping algorithm is implemented using the log odds form, even though it would have equally been implementable using probabilities.

   (a) Derive an update rule that represented occupancy probabilities directly, without the detour of the log odds representation.

   (b) For an implementation in a common programming language such as C++, give an example in which the probability implementation yields *different* results form the log odds implementation, due to numerical truncation. Explain your example, and argue whether you judge this to be a problem in practice.

10 Simultaneous Localization and Mapping

10.1 Introduction

This and the following chapters address one of the most fundamental problems in robotics, the *simultaneous localization and mapping problem*. This problem is commonly abbreviated as *SLAM*, and is also known as *Concurrent Mapping and Localization*, or *CML*. SLAM problems arise when the robot does not have access to a map of the environment, nor does it know its own pose. Instead, all it is given are measurements $z_{1:t}$ and controls $u_{1:t}$. The term "simultaneous localization and mapping" describes the resulting problem: In SLAM, the robot acquires a map of its environment while simultaneously localizing itself relative to this map. SLAM is significantly more difficult than all robotics problems discussed thus far. It is more difficult than localization in that the map is unknown and has to be estimated along the way. It is more difficult than mapping with known poses, since the poses are unknown and have to be estimated along the way.

From a probabilistic perspective, there are two main forms of the SLAM problem, which are both of equal practical importance. One is known as the *online SLAM problem*: It involves estimating the posterior over the momentary pose along with the map:

ONLINE SLAM
PROBLEM

$$(10.1) \quad p(x_t, m \mid z_{1:t}, u_{1:t})$$

Here x_t is the pose at time t, m is the map, and $z_{1:t}$ and $u_{1:t}$ are the measurements and controls, respectively. This problem is called the online SLAM problem since it only involves the estimation of variables that persist at time t. Many algorithms for the online SLAM problem are incremental: they discard past measurements and controls once they have been processed. The graphical model of online SLAM is depicted in Figure 10.1.

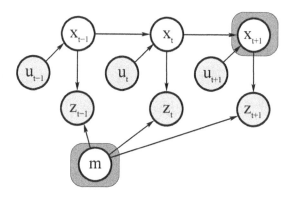

Figure 10.1 Graphical model of the online SLAM problem. The goal of online SLAM is to estimate a posterior over the current robot pose along with the map.

FULL SLAM PROBLEM

The second SLAM problem is called the *full SLAM problem*. In full SLAM, we seek to calculate a posterior over the entire path $x_{1:t}$ along with the map, instead of just the current pose x_t (see also Figure 10.2):

(10.2) $\quad p(x_{1:t}, m \mid z_{1:t}, u_{1:t})$

This subtle difference between online and full SLAM has ramifications in the type of algorithms that can be brought to bear. In particular, the online SLAM problem is the result of integrating out past poses from the full SLAM problem:

(10.3) $\quad p(x_t, m \mid z_{1:t}, u_{1:t}) \;=\; \int \int \cdots \int p(x_{1:t}, m \mid z_{1:t}, u_{1:t}) \, dx_1 \, dx_2 \ldots dx_{t-1}$

In online SLAM, these integrations are typically performed one-at-a-time. They cause interesting changes of the dependency structures in SLAM that we will fully explore in the next chapter.

A second key characteristic of the SLAM problem has to do with the nature of the estimation problem. SLAM problems possess a continuous and a discrete component. The continuous estimation problem pertains to the location of the objects in the map and the robot's own pose variables. Objects may be landmarks in feature-based representation, or they might be object patches detected by range finders. The discrete nature has to do with correspondence: When an object is detected, a SLAM algorithm must reason about the relation of this object to previously detected objects. This reasoning is typically discrete: Either the object is the same as a previously detected one, or it is not.

10.1 Introduction

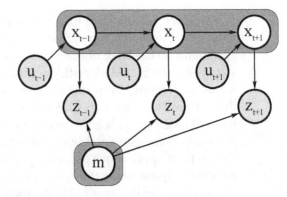

Figure 10.2 Graphical model of the full SLAM problem. Here, we compute a joint posterior over the whole path of the robot and the map.

We already encountered similar continuous-discrete estimation problems in previous chapters. For example, EKF localization in Chapter 7.4 estimates the robot pose, which is continuous. But to do so it also estimates the correspondences of measurements and landmarks in the map, which are discrete. In this and the subsequent chapters, we will discuss a number of different techniques to deal with the continuous and the discrete aspects of the SLAM problem.

At times, it will be useful to make the correspondence variables explicit, as we did in Chapter 7 on localization. The online SLAM posterior is then given by

(10.4) $\quad p(x_t, m, c_t \mid z_{1:t}, u_{1:t})$

and the full SLAM posterior by

(10.5) $\quad p(x_{1:t}, m, c_{1:t} \mid z_{1:t}, u_{1:t})$

The online posterior is obtained from the full posterior by integrating out past robot poses and summing over all past correspondences:

(10.6) $\quad p(x_t, m, c_t \mid z_{1:t}, u_{1:t})$
$$= \int\int \cdots \int \sum_{c_1}\sum_{c_2}\cdots\sum_{c_{t-1}} p(x_{1:t}, m, c_{1:t} \mid z_{1:t}, u_{1:t})\, dx_1\, dx_2 \ldots dx_{t-1}$$

In both versions of the SLAM problems—online and full—estimating the full posterior (10.4) or (10.5) is the gold standard of SLAM. The full posterior captures all there is to be known about the map and the pose or the path.

In practice, calculating a full posterior is usually infeasible. Problems arise from two sources: (1) the high dimensionality of the continuous parameter space, and (2) the large number of discrete correspondence variables. Many state-of-the-art SLAM algorithms construct maps with tens of thousands of features, or more. Even under known correspondence, the posterior over those maps alone involves probability distributions over spaces with 10^5 or more dimensions. This is in stark contrast to localization problems, in which posteriors were estimated over three-dimensional continuous spaces. Further, in most applications the correspondences are unknown. The number of possible assignments to the vector of all correspondence variables $c_{1:t}$ grows exponentially in the time t. Thus, practical SLAM algorithms that can cope with the correspondence problem must rely on approximations.

The SLAM problem will be discussed in a number of subsequent chapters. The remainder of this chapter develops an EKF algorithm for the online SLAM problem. Much of this material builds on Chapter 3.3, where the EKF was introduced, and Chapter 7.4, where we applied the EKF to the mobile robot localization problem. We will derive a progression of EKF algorithms that first apply EKFs to SLAM with known correspondences, and then progress to the more general case with unknown correspondences.

10.2 SLAM with Extended Kalman Filters

10.2.1 Setup and Assumptions

Historically the earliest—and perhaps the most influential—SLAM algorithm is based on the extended Kalman filter, or EKF. In a nutshell, the EKF SLAM algorithm applies the EKF to online SLAM using maximum likelihood data association. In doing so, EKF SLAM is subject to a number of approximations and limiting assumptions:

FEATURE-BASED MAPS

Maps, in EKF SLAM, are *feature-based*. They are composed of point landmarks. For computational reasons, the number of point landmarks is usually small (e.g., smaller than 1,000). Further, the EKF approach tends to work well the less ambiguous the landmarks are. For this reason, EKF SLAM requires significant engineering of feature detectors, sometimes using artificial beacons as features.

GAUSSIAN NOISE ASSUMPTION

As any EKF algorithm, EKF SLAM makes a *Gaussian noise assumption* for robot motion and perception. The amount of uncertainty in the posterior must be relatively small, since otherwise the linearization in EKFs tend to introduce intolerable errors.

The EKF SLAM algorithm, just like the EKF localizer discussed in Chapter 7.4, can only process *positive* sightings of landmarks. It cannot process negative information that arises from the absence of landmarks in sensor measurements. This is a direct consequence of the Gaussian belief representation and was already discussed in Chapter 7.4.

POSITIVE
INFORMATION

10.2.2 SLAM with Known Correspondence

The SLAM algorithm for the case with known correspondence addresses the continuous portion of the SLAM problem only. Its development is in many ways parallel to the derivation of the EKF localization algorithm in Chapter 7.4, but with one key difference: In addition to estimating the robot pose x_t, the EKF SLAM algorithm also estimates the coordinates of all landmarks encountered along the way. This makes it necessary to include the landmark coordinates into the state vector.

For convenience, let us call the state vector comprising robot pose and the map the *combined state vector*, and denote this vector y_t. The combined vector is given by

COMBINED STATE
VECTOR

(10.7)
$$y_t = \begin{pmatrix} x_t \\ m \end{pmatrix}$$
$$= \begin{pmatrix} x & y & \theta & m_{1,x} & m_{1,y} & s_1 & m_{2,x} & m_{2,y} & s_2 & \ldots & m_{N,x} & m_{N,y} & s_N \end{pmatrix}^T$$

Here x, y, and θ denote the robot's coordinates at time t (not to be confused with the state variables x_t and y_t), $m_{i,x}, m_{i,y}$ are the coordinates of the i-th landmark, for $i = 1, \ldots, N$, and s_i is its signature. The dimension of this state vector is $3N + 3$, where N denotes the number of landmarks in the map. Clearly, for any reasonable number of N, this vector is significantly larger than the pose vector that is being estimated in Chapter 7.4, which introduced the EKF localization algorithm. EKF SLAM calculates the online posterior $p(y_t \mid z_{1:t}, u_{1:t})$.

The *EKF SLAM algorithm* is depicted in Table 10.1—notice the similarity to the EKF localization algorithm in Table 7.2. Lines 2 through 5 apply the motion update, whereas lines 6 through 20 incorporate the measurement vector.

Lines 3 and 5 manipulate the mean and covariance of the belief in accordance to the motion model. This manipulation only affects those elements of the belief distribution concerned with the robot pose. It leaves all mean and covariance variables for the map unchanged, along with the pose-map covariances. Lines 7 through 20 iterate through all measurements. The test

1:   **Algorithm EKF_SLAM_known_correspondences**($\mu_{t-1}, \Sigma_{t-1}, u_t, z_t, c_t$):

2: $$F_x = \begin{pmatrix} 1 & 0 & 0 & 0\cdots 0 \\ 0 & 1 & 0 & 0\cdots 0 \\ 0 & 0 & 1 & \underbrace{0\cdots 0}_{3N} \end{pmatrix}$$

3: $$\bar{\mu}_t = \mu_{t-1} + F_x^T \begin{pmatrix} -\frac{v_t}{\omega_t}\sin\mu_{t-1,\theta} + \frac{v_t}{\omega_t}\sin(\mu_{t-1,\theta} + \omega_t\Delta t) \\ \frac{v_t}{\omega_t}\cos\mu_{t-1,\theta} - \frac{v_t}{\omega_t}\cos(\mu_{t-1,\theta} + \omega_t\Delta t) \\ \omega_t \Delta t \end{pmatrix}$$

4: $$G_t = I + F_x^T \begin{pmatrix} 0 & 0 & -\frac{v_t}{\omega_t}\cos\mu_{t-1,\theta} + \frac{v_t}{\omega_t}\cos(\mu_{t-1,\theta} + \omega_t\Delta t) \\ 0 & 0 & -\frac{v_t}{\omega_t}\sin\mu_{t-1,\theta} + \frac{v_t}{\omega_t}\sin(\mu_{t-1,\theta} + \omega_t\Delta t) \\ 0 & 0 & 0 \end{pmatrix} F_x$$

5: $\bar{\Sigma}_t = G_t \Sigma_{t-1} G_t^T + F_x^T R_t F_x$

6: $$Q_t = \begin{pmatrix} \sigma_r^2 & 0 & 0 \\ 0 & \sigma_\phi^2 & 0 \\ 0 & 0 & \sigma_s^2 \end{pmatrix}$$

7:   for all observed features $z_t^i = (r_t^i \; \phi_t^i \; s_t^i)^T$ do
8:      $j = c_t^i$
9:      if landmark j never seen before

10: $$\begin{pmatrix} \bar{\mu}_{j,x} \\ \bar{\mu}_{j,y} \\ \bar{\mu}_{j,s} \end{pmatrix} = \begin{pmatrix} \bar{\mu}_{t,x} \\ \bar{\mu}_{t,y} \\ s_t^i \end{pmatrix} + \begin{pmatrix} r_t^i \cos(\phi_t^i + \bar{\mu}_{t,\theta}) \\ r_t^i \sin(\phi_t^i + \bar{\mu}_{t,\theta}) \\ 0 \end{pmatrix}$$

11:      endif

12: $$\delta = \begin{pmatrix} \delta_x \\ \delta_y \end{pmatrix} = \begin{pmatrix} \bar{\mu}_{j,x} - \bar{\mu}_{t,x} \\ \bar{\mu}_{j,y} - \bar{\mu}_{t,y} \end{pmatrix}$$

13:     $q = \delta^T \delta$

14: $$\hat{z}_t^i = \begin{pmatrix} \sqrt{q} \\ \mathrm{atan2}(\delta_y, \delta_x) - \bar{\mu}_{t,\theta} \\ \bar{\mu}_{j,s} \end{pmatrix}$$

15: $$F_{x,j} = \begin{pmatrix} 1 & 0 & 0 & 0\cdots 0 & 0 & 0 & 0 & 0\cdots 0 \\ 0 & 1 & 0 & 0\cdots 0 & 0 & 0 & 0 & 0\cdots 0 \\ 0 & 0 & 1 & 0\cdots 0 & 0 & 0 & 0 & 0\cdots 0 \\ 0 & 0 & 0 & 0\cdots 0 & 1 & 0 & 0 & 0\cdots 0 \\ 0 & 0 & 0 & 0\cdots 0 & 0 & 1 & 0 & 0\cdots 0 \\ 0 & 0 & 0 & \underbrace{0\cdots 0}_{3j-3} & 0 & 0 & 1 & \underbrace{0\cdots 0}_{3N-3j} \end{pmatrix}$$

16: $$H_t^i = \frac{1}{q} \begin{pmatrix} -\sqrt{q}\delta_x & -\sqrt{q}\delta_y & 0 & +\sqrt{q}\delta_x & \sqrt{q}\delta_y & 0 \\ \delta_y & -\delta_x & -q & -\delta_y & +\delta_x & 0 \\ 0 & 0 & 0 & 0 & 0 & q \end{pmatrix} F_{x,j}$$

17:     $K_t^i = \bar{\Sigma}_t H_t^{iT}(H_t^i \bar{\Sigma}_t H_t^{iT} + Q_t)^{-1}$
18:     $\bar{\mu}_t = \bar{\mu}_t + K_t^i(z_t^i - \hat{z}_t^i)$
19:     $\bar{\Sigma}_t = (I - K_t^i H_t^i) \bar{\Sigma}_t$
20:   endfor
21:   $\mu_t = \bar{\mu}_t$
22:   $\Sigma_t = \bar{\Sigma}_t$
23:   return μ_t, Σ_t

Table 10.1 The EKF algorithm for the SLAM problem (with known correspondences).

10.2 SLAM with Extended Kalman Filters

in line 9 returns true only for landmarks for which we have no initial location estimate. For those, line 10 initializes the location of such a landmark by the projected location obtained from the corresponding range and bearing measurement. As we shall discuss below, this step is important for the linearization in EKFs; it would not be needed in linear Kalman filters. For each measurement, an "expected" measurement is computed in line 14, and the corresponding *Kalman gain* is computed in line 17. Notice that the Kalman gain is a matrix of size 3 by $3N + 3$. This matrix is usually not sparse. Information is propagated through the entire state estimate. The filter update then occurs in lines 18 and 19, where the innovation is folded back into the robot's belief.

The fact that the Kalman gain is fully populated for all state variables—and not just the observed landmark and the robot pose—is important. In SLAM, observing a landmark does not just improve the position estimate of this very landmark, but that of other landmarks as well. This effect is mediated by the robot pose: Observing a landmark improves the robot pose estimate, and as a result it eliminates some of the uncertainty of landmarks previously seen by the same robot. The amazing effect here is that we do not have to model past poses explicitly—which would put us into the realm of the full SLAM problem and make the EKF a non-realtime algorithm. Instead, this dependence is captured in the Gaussian posterior, more specifically, in the off-diagonal covariance elements of the matrix Σ_t.

Figure 10.3 illustrates the EKF SLAM algorithm for an artificial example. The robot navigates from a start pose that serves as the origin of its coordinate system. As it moves, its own pose uncertainty increases, as indicated by uncertainty ellipses of growing diameter. It also senses nearby landmarks and maps them with an uncertainty that combines the fixed measurement uncertainty with the increasing pose uncertainty. As a result, the uncertainty in the landmark locations grows over time. In fact, it parallels that of the pose uncertainty at the time a landmark is observed. The interesting transition happens in Figure 10.3d: Here the robot observes the landmark it saw in the very beginning of mapping, and whose location is relatively well known. Through this observation, the robot's pose error is reduced, as indicated in Figure 10.3d—notice the very small error ellipse for the final robot pose! This observation also reduces the uncertainty for other landmarks in the map. This phenomenon arises from a correlation that is expressed in the covariance matrix of the Gaussian posterior. Since most of the uncertainty in earlier landmarks is caused by the robot pose, and since this very uncertainty persists over time, the location estimates of those landmarks are

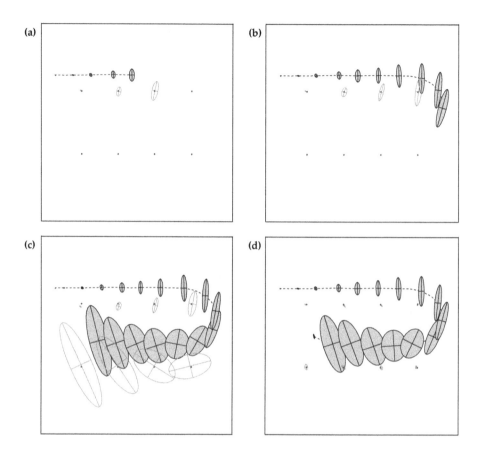

Figure 10.3 EKF applied to the online SLAM problem. The robot's path is a dotted line, and its estimates of its own position are shaded ellipses. Eight distinguishable landmarks of unknown location are shown as small dots, and their location estimates are shown as white ellipses. In (a)–(c) the robot's positional uncertainty is increasing, as is its uncertainty about the landmarks it encounters. In (d) the robot senses the first landmark again, and the uncertainty of *all* landmarks decreases, as does the uncertainty of its current pose. Image courtesy of Michael Montemerlo, Stanford University.

correlated. When gaining information on the robot's pose, this information spreads to previously observed landmarks. This effect is probably the most important characteristic of the SLAM posterior. Information that helps localize the robot is propagated through map, and as a result improves the

10.2 SLAM with Extended Kalman Filters

localization of other landmarks in the map.

10.2.3 Mathematical Derivation of EKF SLAM

The derivation of the EKF SLAM algorithm for the case with known correspondences largely parallels that of the EKF localizer in Chapter 7.4. The key difference is the augmented state vector, which now includes the locations of all landmarks in addition to the robot pose.

In SLAM, the initial pose is taken to be the origin of the coordinate system. This definition is somewhat arbitrary, in that it can be replaced by any coordinate. None of the landmark locations are known initially. The following initial mean and covariance express this belief:

$$\mu_0 = (\;0\;0\;0\;\ldots\;0\;)^T \tag{10.8}$$

$$\Sigma_0 = \begin{pmatrix} 0 & 0 & 0 & 0 & \cdots & 0 \\ 0 & 0 & 0 & 0 & \cdots & 0 \\ 0 & 0 & 0 & 0 & \cdots & 0 \\ 0 & 0 & 0 & \infty & \cdots & 0 \\ \vdots & \vdots & \vdots & \vdots & \ddots & \vdots \\ 0 & 0 & 0 & 0 & \cdots & \infty \end{pmatrix} \tag{10.9}$$

The covariance matrix is of size $(3N+3) \times (3N+3)$. It is composed of a small 3×3 matrix of zeros for the robot pose variables. All other covariance values are infinite.

As the robot moves, the state vector changes according to the standard noise-free velocity model (see Equations (5.13) and (7.4)). In SLAM, this motion model is extended to the augmented state vector:

$$y_t = y_{t-1} + \begin{pmatrix} -\frac{v_t}{\omega_t}\sin\theta + \frac{v_t}{\omega_t}\sin(\theta + \omega_t\Delta t) \\ \frac{v_t}{\omega_t}\cos\theta - \frac{v_t}{\omega_t}\cos(\theta + \omega_t\Delta t) \\ \omega_t\Delta t + \gamma_t\Delta t \\ 0 \\ \vdots \\ 0 \end{pmatrix} \tag{10.10}$$

The variables x, y, and θ denote the robot pose in y_{t-1}. Because the motion only affects the robot's pose and all landmarks remain where they are, only the first three elements in the update are non-zero. This enables us to write the same equation more compactly:

$$(10.11) \quad y_t = y_{t-1} + F_x^T \begin{pmatrix} -\frac{v_t}{\omega_t} \sin\theta + \frac{v_t}{\omega_t} \sin(\theta + \omega_t \Delta t) \\ \frac{v_t}{\omega_t} \cos\theta - \frac{v_t}{\omega_t} \cos(\theta + \omega_t \Delta t) \\ \omega_t \Delta t + \gamma_t \Delta t \end{pmatrix}$$

Here F_x is a matrix that maps the 3-dimensional state vector into a vector of dimension $3N+3$.

$$(10.12) \quad F_x = \begin{pmatrix} 1 & 0 & 0 & 0 \cdots 0 \\ 0 & 1 & 0 & 0 \cdots 0 \\ 0 & 0 & 1 & \underbrace{0 \cdots 0}_{3N \text{ columns}} \end{pmatrix}$$

The full motion model with noise is then as follows

$$(10.13) \quad y_t = \underbrace{y_{t-1} + F_x^T \begin{pmatrix} -\frac{v_t}{\omega_t} \sin\theta + \frac{v_t}{\omega_t} \sin(\theta + \omega_t \Delta t) \\ \frac{v_t}{\omega_t} \cos\theta - \frac{v_t}{\omega_t} \cos(\theta + \omega_t \Delta t) \\ \omega \Delta t \end{pmatrix}}_{g(u_t, y_{t-1})} + \mathcal{N}(0, F_x^T R_t F_x)$$

where $F_x^T R_t F_x$ extends the covariance matrix to the dimension of the full state vector squared.

As usual in EKFs, the motion function g is approximated using a first degree Taylor expansion

$$(10.14) \quad g(u_t, y_{t-1}) \approx g(u_t, \mu_{t-1}) + G_t (y_{t-1} - \mu_{t-1})$$

where the Jacobian $G_t = g'(u_t, \mu_{t-1})$ is the derivative of g with respect to y_{t-1} at u_t and μ_{t-1}, as in Equation (7.7).

Obviously, the additive form in (10.13) enables us to decompose this Jacobian into an identity matrix of dimension $(3N+3) \times (3N+3)$ (the derivative of y_{t-1}) plus a low-dimensional Jacobian g_t that characterizes the change of the robot pose:

$$(10.15) \quad G_t = I + F_x^T g_t F_x$$

with

$$(10.16) \quad g_t = \begin{pmatrix} 0 & 0 & -\frac{v_t}{\omega_t} \cos\mu_{t-1,\theta} + \frac{v_t}{\omega_t} \cos(\mu_{t-1,\theta} + \omega_t \Delta t) \\ 0 & 0 & -\frac{v_t}{\omega_t} \sin\mu_{t-1,\theta} + \frac{v_t}{\omega_t} \sin(\mu_{t-1,\theta} + \omega_t \Delta t) \\ 0 & 0 & 0 \end{pmatrix}$$

Plugging these approximations into the standard EKF algorithm gives us lines 2 through 5 of Table 10.1. Obviously, several of the matrices multiplied in line 5 are sparse, which should be exploited when implementing this algorithm. The result of this update are the mean $\bar{\mu}_t$ and the covariance $\bar{\Sigma}_t$ of the estimate at time t after updating the filter with the control u_t, but before integrating the measurement z_t.

The derivation of the measurement update is similar to the one in Chapter 7.4. In particular, we are given the following measurement model

$$(10.17) \quad z_t^i = \underbrace{\begin{pmatrix} \sqrt{(m_{j,x} - x)^2 + (m_{j,y} - y)^2} \\ \mathrm{atan2}(m_{j,y} - y, m_{j,x} - x) - \theta \\ m_{j,s} \end{pmatrix}}_{h(y_t, j)} + \mathcal{N}(0, \underbrace{\begin{pmatrix} \sigma_r & 0 & 0 \\ 0 & \sigma_\phi & 0 \\ 0 & 0 & \sigma_s \end{pmatrix}}_{Q_t})$$

Here x, y, and θ denotes the pose of the robot, i is the index of an individual landmark observation in z_t, and $j = c_t^i$ is the index of the observed landmark at time t. The variable r denotes the range to a landmark, ϕ is the bearing to a landmark, and s the landmark signature; the terms σ_r, σ_ϕ, and σ_s are the corresponding measurement noise covariances.

This expression is approximated by the linear function

$$(10.18) \quad h(y_t, j) \approx h(\bar{\mu}_t, j) + H_t^i \, (y_t - \bar{\mu}_t)$$

Here H_t^i is the derivative of h with respect to the full state vector y_t. Since h depends only on two elements of that state vector, the robot pose x_t and the location of the j-th landmark m_j, the derivative factors into a low-dimensional Jacobian h_t^i and a matrix $F_{x,j}$, which maps h_t^i into a matrix of the dimension of the full state vector:

$$(10.19) \quad H_t^i = h_t^i \, F_{x,j}$$

Here h_t^i is the Jacobian of the function $h(y_t, j)$ at $\bar{\mu}_t$, calculated with respect to the state variables x_t and m_j:

$$(10.20) \quad h_t^i = \begin{pmatrix} \frac{\bar{\mu}_{t,x} - \bar{\mu}_{j,x}}{\sqrt{q_t}} & \frac{\bar{\mu}_{t,y} - \bar{\mu}_{j,y}}{\sqrt{q_t}} & 0 & \frac{\bar{\mu}_{j,x} - \bar{\mu}_{t,x}}{\sqrt{q_t}} & \frac{\bar{\mu}_{j,y} - \bar{\mu}_{t,y}}{\sqrt{q_t}} & 0 \\ \frac{\bar{\mu}_{j,y} - \bar{\mu}_{t,y}}{q_t} & \frac{\bar{\mu}_{t,x} - \bar{\mu}_{j,x}}{q_t} & -1 & \frac{\bar{\mu}_{t,y} - \bar{\mu}_{j,y}}{q_t} & \frac{\bar{\mu}_{j,x} - \bar{\mu}_{t,x}}{q_t} & 0 \\ 0 & 0 & 0 & 0 & 0 & 1 \end{pmatrix}$$

The scalar $q_t = (\bar{\mu}_{j,x} - \bar{\mu}_{t,x})^2 + (\bar{\mu}_{j,y} - \bar{\mu}_{t,y})^2$, and as before, $j = c_t^i$ is the landmark that corresponds to the measurement z_t^i. The matrix $F_{x,j}$ is of dimension $6 \times (3N + 3)$. It maps the low-dimensional matrix h_t^i into a matrix of dimension $3 \times (3N+3)$:

$$(10.21) \quad F_{x,j} = \begin{pmatrix} 1 & 0 & 0 & 0\cdots 0 & 0 & 0 & 0 & 0\cdots 0 \\ 0 & 1 & 0 & 0\cdots 0 & 0 & 0 & 0 & 0\cdots 0 \\ 0 & 0 & 1 & 0\cdots 0 & 0 & 0 & 0 & 0\cdots 0 \\ 0 & 0 & 0 & 0\cdots 0 & 1 & 0 & 0 & 0\cdots 0 \\ 0 & 0 & 0 & 0\cdots 0 & 0 & 1 & 0 & 0\cdots 0 \\ 0 & 0 & 0 & \underbrace{0\cdots 0}_{3j-3} & 0 & 0 & 1 & \underbrace{0\cdots 0}_{3N-3j} \end{pmatrix}$$

These expressions make up for the gist of the Kalman gain calculation in lines 8 through 17 in our EKF SLAM algorithm in Table 10.1, with one important extension. When a landmark is observed for the first time, its initial pose estimate in Equation (10.8) leads to a poor linearization. This is because with the default initialization in (10.8), the point about which h is being linearized is $(\hat{\mu}_{j,x} \ \hat{\mu}_{j,y} \ \hat{\mu}_{j,s})^T = (0 \ 0 \ 0)^T$, which is a poor estimator of the actual landmark location. A better landmark estimator is given in line 10 of Table 10.1. Here we initialize the landmark estimate $(\hat{\mu}_{j,x} \ \hat{\mu}_{j,y} \ \hat{\mu}_{j,s})^T$ with the expected position. This expected position is derived from the expected robot pose and the measurement variables for this landmark

$$(10.22) \quad \begin{pmatrix} \bar{\mu}_{j,x} \\ \bar{\mu}_{j,y} \\ \bar{\mu}_{j,s} \end{pmatrix} = \begin{pmatrix} \bar{\mu}_{t,x} \\ \bar{\mu}_{t,y} \\ s_t^i \end{pmatrix} + \begin{pmatrix} r_t^i \cos(\phi_t^i + \bar{\mu}_{t,\theta}) \\ r_t^i \sin(\phi_t^i + \bar{\mu}_{t,\theta}) \\ 0 \end{pmatrix}$$

We note that this initialization is only possible because the measurement function h is bijective. Measurements are two-dimensional, as are landmark locations. In cases where a measurement is of lower dimensionality than the coordinates of a landmark, h is a true projection and it is impossible to calculate a meaningful expectation for $(\bar{\mu}_{j,x} \ \bar{\mu}_{j,y} \ \bar{\mu}_{j,s})^T$ from a single measurement only. This is, for example, the case in computer vision implementations of SLAM, since cameras often calculate the angle to a landmark but not the range. SLAM is then usually performed by integrating multiple sightings and applying triangulation to determine an appropriate initial location estimate. In the SLAM literature, such a problem is known as *bearing only SLAM* and will be further discussed in one of the exercises (page 334).

BEARING ONLY SLAM

Finally, we note that the EKF algorithm requires memory that is quadratic in N, the number of landmarks in the map. Its update time is also quadratic in N. The quadratic update complexity stems from the matrix multiplications that take place at various locations in the EKF.

10.2 SLAM with Extended Kalman Filters

1: **Algorithm EKF_SLAM**($\mu_{t-1}, \Sigma_{t-1}, u_t, z_t, N_{t-1}$):

2:     $N_t = N_{t-1}$

3:     $F_x = \begin{pmatrix} 1 & 0 & 0 & 0 \cdots 0 \\ 0 & 1 & 0 & 0 \cdots 0 \\ 0 & 0 & 1 & 0 \cdots 0 \end{pmatrix}$

4:     $\bar{\mu}_t = \mu_{t-1} + F_x^T \begin{pmatrix} -\frac{v_t}{\omega_t} \sin \mu_{t-1,\theta} + \frac{v_t}{\omega_t} \sin(\mu_{t-1,\theta} + \omega_t \Delta t) \\ \frac{v_t}{\omega_t} \cos \mu_{t-1,\theta} - \frac{v_t}{\omega_t} \cos(\mu_{t-1,\theta} + \omega_t \Delta t) \\ \omega_t \Delta t \end{pmatrix}$

5:     $G_t = I + F_x^T \begin{pmatrix} 0 & 0 & -\frac{v_t}{\omega_t} \cos \mu_{t-1,\theta} + \frac{v_t}{\omega_t} \cos(\mu_{t-1,\theta} + \omega_t \Delta t) \\ 0 & 0 & -\frac{v_t}{\omega_t} \sin \mu_{t-1,\theta} + \frac{v_t}{\omega_t} \sin(\mu_{t-1,\theta} + \omega_t \Delta t) \\ 0 & 0 & 0 \end{pmatrix} F_x$

6:     $\bar{\Sigma}_t = G_t \, \Sigma_{t-1} \, G_t^T + F_x^T \, R_t \, F_x$

7:     $Q_t = \begin{pmatrix} \sigma_r & 0 & 0 \\ 0 & \sigma_\phi & 0 \\ 0 & 0 & \sigma_s \end{pmatrix}$

8:     for all observed features $z_t^i = (r_t^i \;\; \phi_t^i \;\; s_t^i)^T$ do

9:        $\begin{pmatrix} \bar{\mu}_{N_t+1,x} \\ \bar{\mu}_{N_t+1,y} \\ \bar{\mu}_{N_t+1,s} \end{pmatrix} = \begin{pmatrix} \bar{\mu}_{t,x} \\ \bar{\mu}_{t,y} \\ s_t^i \end{pmatrix} + r_t^i \begin{pmatrix} \cos(\phi_t^i + \bar{\mu}_{t,\theta}) \\ \sin(\phi_t^i + \bar{\mu}_{t,\theta}) \\ 0 \end{pmatrix}$

10:       for $k = 1$ to N_t+1 do

11:         $\delta_k = \begin{pmatrix} \delta_{k,x} \\ \delta_{k,y} \end{pmatrix} = \begin{pmatrix} \bar{\mu}_{k,x} - \bar{\mu}_{t,x} \\ \bar{\mu}_{k,y} - \bar{\mu}_{t,y} \end{pmatrix}$

12:         $q_k = \delta_k^T \delta_k$

see next page for continuation

continued from the previous page

13: $\quad \hat{z}_t^k = \begin{pmatrix} \sqrt{q_k} \\ \operatorname{atan2}(\delta_{k,y}, \delta_{k,x}) - \bar{\mu}_{t,\theta} \\ \bar{\mu}_{k,s} \end{pmatrix}$

14: $\quad F_{x,k} = \begin{pmatrix} 1 & 0 & 0 & 0\cdots 0 & 0 & 0 & 0 & 0\cdots 0 \\ 0 & 1 & 0 & 0\cdots 0 & 0 & 0 & 0 & 0\cdots 0 \\ 0 & 0 & 1 & 0\cdots 0 & 0 & 0 & 0 & 0\cdots 0 \\ 0 & 0 & 0 & 0\cdots 0 & 1 & 0 & 0 & 0\cdots 0 \\ 0 & 0 & 0 & 0\cdots 0 & 0 & 1 & 0 & 0\cdots 0 \\ 0 & 0 & 0 & 0\cdots 0 & 0 & 0 & 1 & 0\cdots 0 \end{pmatrix}$

15: $\quad H_t^k = \dfrac{1}{q_k} \begin{pmatrix} -\sqrt{q_k}\delta_{k,x} & -\sqrt{q_k}\delta_{k,y} & 0 & \sqrt{q_k}\delta_{k,x} & \sqrt{q_k}\delta_{k,y} & 0 \\ \delta_{k,y} & -\delta_{k,x} & -1 & -\delta_{k,y} & \delta_{k,x} & 0 \\ 0 & 0 & 0 & 0 & 0 & 1 \end{pmatrix} F_{x,k}$

16: $\quad \Psi_k = H_t^k \, \bar{\Sigma}_t \, [H_t^k]^T + Q_t$

17: $\quad \pi_k = (z_t^i - \hat{z}_t^k)^T \, \Psi_k^{-1} \, (z_t^i - \hat{z}_t^k)$

18: endfor

19: $\quad \pi_{N_t+1} = \alpha$

20: $\quad j(i) = \operatorname*{argmin}_{k} \pi_k$

21: $\quad N_t = \max\{N_t, j(i)\}$

22: $\quad K_t^i = \bar{\Sigma}_t \, [H_t^{j(i)}]^T \, \Psi_{j(i)}^{-1}$

23: $\quad \bar{\mu}_t = \bar{\mu}_t + K_t^i \, (z_t^i - \hat{z}_t^{j(i)})$

24: $\quad \bar{\Sigma}_t = (I - K_t^i \, H_t^{j(i)}) \, \bar{\Sigma}_t$

25: endfor

26: $\mu_t = \bar{\mu}_t$

27: $\Sigma_t = \bar{\Sigma}_t$

28: return μ_t, Σ_t

Table 10.2 The EKF SLAM algorithm with ML correspondences, shown here with outlier rejection.

10.3 EKF SLAM with Unknown Correspondences

10.3.1 The General EKF SLAM Algorithm

MAXIMUM LIKELIHOOD CORRESPONDENCE

The EKF SLAM algorithm with known correspondences is now extended into the general EKF SLAM algorithm, which uses an incremental *maximum likelihood* (ML) estimator to determine correspondences. Table 10.2 depicts the algorithm for unknown correspondences.

Since the correspondence is unknown, the input to the algorithm **EKF_SLAM** lacks a correspondence variable c_t. Instead, it includes the momentary size of the map, N_{t-1}. The motion update in lines 3 through 6 is equivalent to the one in **EKF_SLAM_known_correspondences** in Table 10.1. The measurement update loop, however, is different. Starting in line 8, it first creates the hypothesis of a new landmark with index $N_t + 1$; this index is one larger than the landmarks in the map at this point in time. The new landmark's location is initialized in line 9, by calculating its expected location given the estimate of the robot pose and the range and bearing in the measurement. Line 9 also assigns the observed signature value to this new landmark. Next, various update quantities are then computed in lines 10 through 18 for all $N_t + 1$ possible landmarks, including the "new" landmark. Line 19 sets the threshold for the creation of a new landmark: A new landmark is created if the Mahalanobis distance to all existing landmarks in the map exceeds the value α. The ML correspondence is then selected in line 20. If the measurement is associated with a previously unseen landmark, the landmark counter is incremented in line 21, and various vectors and matrices are enlarged accordingly—this somewhat tedious step is not made explicit in Table 10.2. The update of the EKF finally takes place in lines 23 and 24. The algorithm **EKF_SLAM** returns the new number of landmarks N_t along with the mean μ_t and the covariance Σ_t.

The derivation of this EKF SLAM follows directly from previous derivations. In particular, the initialization in line 9 is identical to the one in line 10 in **EKF_SLAM_known_correspondences**, Table 10.1. Lines 10 through 18 parallel lines 12 through 17 in **EKF_SLAM_known_correspondences**, with the added variable π_k needed for calculating the ML correspondence. The selection of the ML correspondence in line 20, and the definition of the Mahalanobis distance in line 17, is analogous to the ML correspondence discussed in Chapter 7.5; in particular, the algorithm **EKF_localization** in Table 7.3 on page 217 used an analogous equation to determine the most likely landmark (line 16). The measurement updates in lines 23 and 24 of Table 10.2 are also analogous to those in the EKF algorithm with known correspondences, assuming that the participating vectors and matrices are of the appropriate dimension in case the map has just been extended.

Our example implementation of **EKF_SLAM** can be made more efficient by restricting the landmarks considered in lines 10 through 18 to those that are near the robot. Further, many of the values and matrices calculated in this inner loop can safely be cached away when looping through more than one feature measurement vector z_t^i. In practice, a good management of features in the map and a tight optimization of this loop can greatly reduce the running time.

10.3.2 Examples

Figure 10.4 shows the EKF SLAM algorithm—here with known correspondence—applied in simulation. The left panel of each of the three diagrams plots the posterior distributions, marginalized for the individual landmarks and the robot pose. The right side depicts the correlation matrix for the augmented state vector y_t; the correlation is the normalized covariance. As is easily seen from the result in Figure 10.4c, over time all x- and y-coordinate estimates become fully correlated. This means the map becomes known in relative terms, up to an uncertain global location that cannot be reconciled. This highlights an important characteristic of the SLAM problem: The absolute coordinates of a map relative to the coordinate system defined by the initial robot pose can only be determined in approximation, whereas the relative coordinates can be determined asymptotically with certainty.

In practice, EKF SLAM has been applied successfully to a large range of navigation problems, involving airborne, underwater, indoor, and various other vehicles. Figure 10.5 shows an example result obtained using the underwater robot Oberon, developed at the University of Sydney, Australia, and shown in Figure 10.6. This vehicle is equipped with a pencil sonar, a sonar that can scan at very high resolutions and detect obstacles up to 50 meters away. To facilitate the mapping problem, researchers have deposited long, small vertical objects in the water, which can be extracted from the sonar scans with relative ease. In this specific experiment, there is a row of such objects, spaced approximately 10 meters apart. In addition, a more distant cliff offers additional point features that can be detected using the pencil sonar.

In the experiment shown in Figure 10.5, the robot moves by these land-

10.3 EKF SLAM with Unknown Correspondences

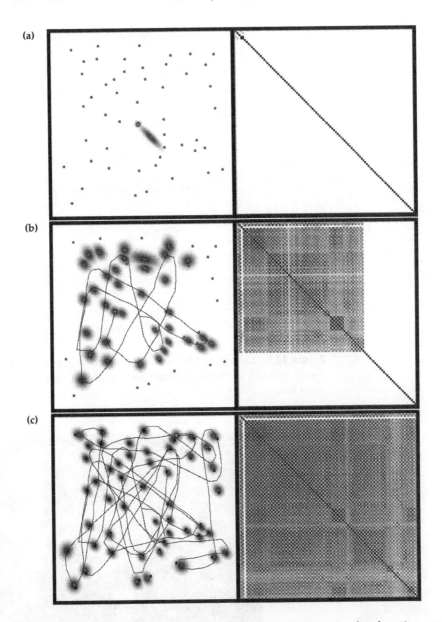

Figure 10.4 EKF SLAM with known data association in a simulated environment. The map is shown on the left, with the gray-level corresponding to the uncertainty of each landmark. The matrix on the right is the correlation matrix, which is the normalized covariance matrix of the posterior estimate. After some time, all x- and all y-coordinate estimates become fully correlated.

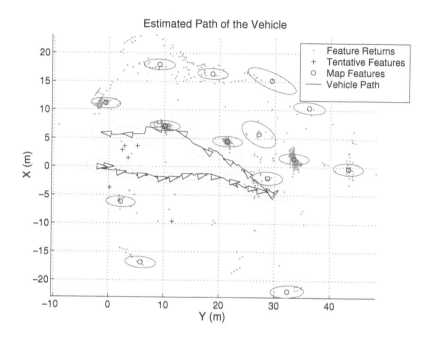

Figure 10.5 Example of Kalman filter estimation of the map and the vehicle pose. Image courtesy of Stefan Williams and Hugh Durrant-Whyte, Australian Centre for Field Robotics.

Figure 10.6 Underwater vehicle Oberon, developed at the University of Sydney. Image courtesy of Stefan Williams and Hugh Durrant-Whyte, Australian Centre for Field Robotics.

10.3 EKF SLAM with Unknown Correspondences

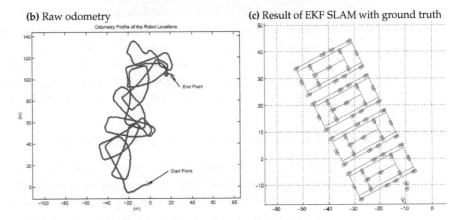

Figure 10.7 (a) The MIT B21 mobile robot in a calibrated testing facility. (b) Raw odometry of the robot, as it is manually driven through the environment. (c) The result of EKF SLAM is a highly accurate map. The image shows the estimated map overlayed on a manually constructed map. All images and results are courtesy of John Leonard and Matthew Walter, MIT.

marks, then turns around and moves back. While doing so, it measures and integrates landmarks into its map using the EKF SLAM algorithm described in this chapter.

The map shown in Figure 10.5 shows the robot's path, marked by the triangles connected by a line. Around each triangle one can see an ellipse, which corresponds to the covariance matrix of the Kalman filter estimate projected into the robot's x-y position. The ellipse shows the variance; the larger it is, the less certain the robot is about its current pose. Various small dots in

Figure 10.5 show landmark sightings, obtained by searching the sonar scan for small and highly reflective objects. The majority of these sightings are rejected, using a mechanism described in the next section. However, some are believed to correspond to a landmark and are added to the map. At the end of the run, the robot has classified 14 such objects as landmarks, each of which is plotted with the projected uncertainty ellipse in Figure 10.5. These landmarks include the artificial landmarks put out by the researchers, but they also include various other terrain features in the vicinity of the robot. The residual pose uncertainty is small.

Figure 10.7 shows the result of another EKF SLAM implementation. Panel (a) shows MIT's RWI B21 mobile robot, situated in a testing environment. The testing environment is a tennis court; obstacles are hurdles whose position was measured manually with centimeter accuracy for evaluation purposes. Panel (b) of Figure 10.7 shows the raw odometry path. The result of EKF SLAM is shown in Panel (c), overlayed with the manually constructed map. The reader should verify that this is indeed an accurate map.

10.3.3 Feature Selection and Map Management

Making EKF SLAM robust in practice often requires additional techniques for *map management*. Many of them pertain to the fact that the Gaussian noise assumption is unrealistic, and many spurious measurements occur in the far tail end of the noise distribution. Such spurious measurements can cause the creation of fake landmarks in the map which, in turn, negatively affect the localization of the robot.

OUTLIERS

PROVISIONAL
LANDMARK LIST

Many state-of-the-art techniques possess mechanisms to deal with *outliers* in the measurement space. Such outliers are defined as spurious landmark sightings outside the uncertainty range of any landmark in the map. The most simple technique to reject such outliers is to maintain a *provisional landmark list*. Instead of augmenting the map by a new landmark once a measurement indicates the existence of a new landmark, such a new landmark is first added to a provisional list of landmarks. This list is just like the map, but landmarks on this list are *not* used to adjust the robot pose (the corresponding gradients in the measurement equations are set to zero). Once a landmark has consistently been observed and its uncertainty ellipse has shrunk, it is transitioned into the regular map.

In practical implementations, this mechanism tends to reduce the number of landmarks in the map by a significant factor, while still retaining all physical landmarks with high probability. A further step, also commonly found

LANDMARK EXISTENCE PROBABILITY

in state-of-the-art implementations, is to maintain a *landmark existence probability*. Such a posterior probability may be implemented as log odds ratio and be denoted o_j for the j-th landmark in the map. Whenever the j-th landmark m_j is observed, o_j is incremented by a fixed value. Not observing m_j when it would be in the perceptual range of the robot's sensors leads to a decrement of o_j. Since it can never be known with certainty whether a landmark is within a robot's perceptual range, the decrement may factor in the probability of such an event. Landmarks are removed from the map when the value o_j drops below a threshold. Such techniques lead to much leaner maps in the face of non-Gaussian measurement noise.

NUMERICAL INSTABILITY OF EKF SLAM

When initializing the estimate for a new landmark starting with a covariance with very large elements—as suggested in Equation (10.9)—may induce *numerical instabilities*. This is because the very first covariance update step will change this value by several orders of magnitude, too many perhaps for generating a matrix that is still positive definite. A better strategy involves an explicit initialization step for any feature that has not been observed before. In particular, such a step would initialize the covariance Σ_t directly with the actual landmark uncertainty, instead of executing line 24 in Table 10.2 (same with the mean in line 23).

As noted previously, the maximum likelihood approach to data association has a clear limitation, which arises from the fact that the maximum likelihood approach deviates from the idea of full posterior estimation in probabilistic robotics. Instead of maintaining a joint posterior over augmented states and data associations, it reduces the data association problem to a deterministic determination, which is treated as if the maximum likelihood association was always correct. This limitation makes EKF brittle with regards to landmark confusion, which in turn can lead to wrong results. In practice, researchers often remedy the problem by choosing one of the following two methods, both of which reduce the chances of confusing landmarks:

- **Spatial arrangement.** The further apart landmarks are, the smaller the chance to accidentally confuse them. It is therefore common practice to choose landmarks that are sufficiently far away from each other so that the probability of confusing one with another becomes small. This introduces an interesting trade-off: a large number of landmarks increases the danger of confusing them. Too few landmarks makes it more difficult to localize the robot, which in turn also increases the chances of confusing landmarks. Little is currently known about the optimal density of landmarks, and researchers often use intuition when selecting specific landmarks.

- **Signatures.** When selecting appropriate landmarks, it is essential to maximize the perceptual distinctiveness of landmarks. For example, doors might possess different colors, or corridors might have different widths. The resulting signatures are essential for successful SLAM.

With these additions, the EKF SLAM algorithm has indeed been applied successfully to a wide range of practical mapping problems, involving robotic vehicles in the air, on the ground, and underwater.

A key limitation of EKF SLAM lies in the necessity to select appropriate landmarks. By reducing the sensor stream to the presence and absence of landmarks, a lot of sensor data is usually discarded. This results in an information loss relative to a SLAM algorithm that can exploit sensors without extensive pre-filtering. Further, the quadratic update time of the EKF limits this algorithm to relatively scarce maps with less than 1,000 features. In practice, one often seeks maps with 10^6 features or more, in which case the EKF ceases to be applicable.

The relatively low dimensionality of the map tends to create a harder data association problem. This is easily verified: When you open your eyes and look at the full room you are in, you probably have no difficulty in recognizing where you are! However, if you are only told the location of a small number of landmarks—e.g., the location of all light sources—the decision is much harder. As a result, data association in EKF SLAM is more difficult than in some of the SLAM algorithms discussed in subsequent chapters, and capable of handling orders of magnitude more features. This culminates into the *fundamental dilemma of EKF SLAM*: While incremental maximum likelihood data association might work well with dense maps with hundreds of millions of features, it tends to be brittle with scarce maps. However, EKFs require sparse maps because of the quadratic update complexity. In subsequent chapters, we will discuss SLAM algorithms that are more efficient and can handle much larger maps. We will also discuss more robust data association techniques. For its many limitations, the value of the EKF SLAM algorithm presented in this chapter is mostly historical.

FUNDAMENTAL DILEMMA EKF SLAM

10.4 Summary

This chapter described the general SLAM problem and introduced the EKF approach.

10.4 Summary

- The SLAM problem is defined as a concurrent localization and mapping problem, in which a robot seeks to acquire a map of the environment while simultaneously seeking to localize itself relative to this map.

- The SLAM problem comes in two versions: online and global. Both problems involve the estimation of the map. The online problem seeks to estimate the momentary robot pose, whereas the global problem seeks to determine all poses. Both problems are of equal importance in practice, and have found equal coverage in the literature.

- The EKF SLAM algorithm is arguably the earliest SLAM algorithm. It applies the extended Kalman filter to the online SLAM problem. With known correspondences, the resulting algorithm is incremental. Updates require time quadratic in the number of landmarks in the map.

- When correspondences are unknown, the EKF SLAM algorithm applies an incremental maximum likelihood estimator to the correspondence problem. The resulting algorithm works well when landmarks are sufficiently distinct.

- Additional techniques were discussed for managing maps. Two common strategies for identifying outliers include a provisional list for landmarks that are not yet observed sufficiently often, and a landmark evidence counter that calculates the posterior evidence of the existence of a landmark.

- EKF SLAM has been applied with considerable success in a number of robotic mapping problems. Its main drawback is the need for sufficiently distinct landmarks, and the computational complexity required for updating the filter.

In practice, EKF SLAM has been applied with some success. When landmarks are sufficiently distinct, the approach approximates the posterior well. The advantage of calculating a full posterior are manifold: It captures all residual uncertainty and enables the robot to reason about its control taking its true uncertainty into account. However, the EKF SLAM algorithm suffers from its enormous update complexity, and the limitation to sparse maps. This, in turn, makes the data association problem a difficult one, and EKF SLAM tends to work poorly in situations where landmarks are highly ambiguous. Further brittleness is due to the fact that the EKF SLAM algorithm relies on an incremental maximum likelihood data association technique. This technique makes it impossible to revise past data associations, and can induce failure when the ML data association is incorrect.

The EKF SLAM algorithm applies to the online SLAM problem; it is inapplicable to the full SLAM problem. In the full SLAM problem, the addition of a new pose to the state vector at each time step would make both the state vector and the covariance grow without bounds. Updating the covariance would therefore require an ever-increasing amount of time, and the approach would quickly run out of computational time no matter how fast the processor.

10.5 Bibliographical Remarks

The problem of SLAM predates the invention of modern robots by many centuries. The problem of modeling a physical structure from a moving sensor platform is at the core of a number of fields, such as *geosciences, photogrammetry*, and *computer vision*. Many of the mathematical techniques that form the core of the SLAM work today were first developed for calculating planetary orbits. For example, the least squares method can be traced back to Johann Carl Friedrich Gauss (1809). SLAM is essentially a geographic surveying problem. Teleported to a robot it creates challenges that human surveyors rarely face, such as the correspondence problem and the problem of finding appropriate features.

In robotics, the EKF to the SLAM problem was introduced through a series of seminal papers by Cheeseman and Smith (1986); Smith and Cheeseman (1986); Smith et al. (1990). These papers were the first to describe the EKF approach discussed in this chapter. Just like us in this book, Smith et al. discussed the EKF in the context of feature-based mapping with point landmarks, and known data association. The first implementations of EKF SLAM were due to Moutarlier and Chatila (1989a,b) and Leonard and Durrant-Whyte (1991), some using artificial beacons as landmarks. The EKF became fashionable at a time when many authors investigated alternative techniques for maintaining accurate pose estimates during mapping (Cox 1991). Early work by Dickmanns and Graefe (1988) on estimation road curvature in autonomous cars is highly related; see (Dickmanns 2002) for a survey.

SLAM is a highly active field of research, as a recent workshop indicates (Leonard et al. 2002b). An extensive literature for the field of SLAM—or CML for *concurrent mapping and localization* as Leonard and Durrant-Whyte (1991) call it—can be found in Thrun (2002). The importance of maintaining correlations in the map was pointed out by Csorba (1997), who in his Ph.D. thesis, who also established some basic convergence results. Since then, a number of authors have extended the basic paradigm in many different ways. The feature management techniques described in this chapter are due to Dissanayake et al. (2001, 2002); see also Bailey (2002). Williams et al. (2001) developed the idea of provisional feature lists in SLAM, to reduce the effect of feature detection errors. Feature initialization is discussed in Leonard et al. (2002a), who explicitly maintains an estimate of previous poses to accommodate sensors that provide incomplete data on feature coordinates. A representation that avoids singularities by explicitly factoring "perturbations" out of the posterior was devised by Castellanos et al. (1999), who reports improved numerical stability over the vanilla EKF. Jensfelt et al. (2002) found significant improvements in indoor SLAM when utilizing basic geometric constraints, such as the fact that most walls are parallel or orthogonal. Early work on SLAM with sonars goes back to Rencken

10.5 Bibliographical Remarks

(1993); a state-of-the-art system for SLAM with sonar sensors can be found in Tardós et al. (2002). Castellanos et al. (2004) provided a critical consistency analysis for the EKF. An empirical comparison of multiple algorithms can be found in Vaganay et al. (2004). A few open questions are discussed by Salichs and Moreno (2000). Research on the important data association problem will be reviewed in a later chapter (see page 481).

As noted, a key limitation of the EKF solution to the SLAM problem lies in the quadratic nature of the covariance matrix. This "flaw" has not remained unnoticed. In the past few years, a number of researchers have proposed EKF SLAM algorithms that gain remarkable scalability through decomposing the map into submaps, for which covariances are maintained separately. Some of the original work in this field is by Leonard and Feder (1999), Guivant and Nebot (2001), and Williams (2001). Leonard and Feder's (1999) *decoupled stochastic mapping* algorithm decomposes the map into collections of smaller, more manageable submaps. This approach is computationally efficient, but does not provide a mechanism to propagate information through the network of local maps (Leonard and Feder 2001). Guivant and Nebot (2001, 2002), in contrast, provided an approximate factorization of the covariance matrix which reduced the actual complexity of EKF updating by a significant factor. Williams (2001) and Williams et al. (2002) proposed the *constrained local submap filter (CLSF)*, which relies on creating independent local submaps of the features in the immediate vicinity of the vehicle. Williams et al. (2002) provides results for underwater mapping (see Figure 10.5 for some of his early work). The *sequential map joining techniques* described in Tardós et al. (2002) is a related decomposition. Bailey (2002) devises a similar technique for representing SLAM maps hierarchically. Folkesson and Christensen (2003) describes a technique by which frequent updates are limited to a small region near the robot, whereas the remainder of the map is updated at much lower frequencies. All these techniques achieve the same rate of convergence as the full EKF solution, but incur an $O(n^2)$ computational burden. However, they scale much better to large problems with tens of thousands of features.

A number of researchers have developed hybrid SLAM techniques, which combine EKF-style SLAM techniques with volumetric techniques, such as occupancy grid maps. The *hybrid metric map (HYMM)* by Guivant et al. (2004) and Nieto et al. (2004) decomposes maps into triangular regions (LTRs) using volumetric maps such as occupancy grid maps as a basic representation for those regions. These local maps are combined using EKFs. Burgard et al. (1999b) also decomposes maps into local occupancy grid maps, but uses the *expectation maximization* (EM) algorithm (see Dempster et al. (1977)) for combining local maps into a joint global map. The work by Betgé-Brezetz et al. (1995, 1996) integrated two types of representations into a SLAM framework: bitmaps for representing outdoor terrain, and object representations for sparse outdoor objects.

Extensions of SLAM to dynamic environments can be found in Wang et al. (2003), Hähnel et al. (2003c), and Wolf and Sukhatme (2004). Wang et al. (2003) developed an algorithm called *SLAM with DATMO*, short for SLAM with the detection and tracking of moving objects. Their approach is based on the EKF, but it allows for the possible motion of features. Hähnel et al. (2003c) studied the problem of performing SLAM in environments with many moving objects. They successfully employed the EM algorithm for filtering out measurements that likely correspond to moving objects. By doing so, they were able to acquire maps in environments where conventional SLAM techniques failed. The approach in Wolf and Sukhatme (2004) maintains two coupled occupancy grids of the environment, one for the static map, and one for moving objects. SLAM-style localization is achieved by a regular landmark-based SLAM algorithm.

SLAM systems have been brought to bear in a number of deployed systems. Rikoski et al. (2004) applied SLAM to sonar odometry of submarine, providing a new approach for "auditory odometry." SLAM in abandoned mines is described in Nüchter et al. (2004), who extended the paradigm to full 6-D pose estimation. Extensions to the multi-robot SLAM problem have been proposed by a number of researchers. Some of the earlier work is by Nettleton et al. (2000), who developed a technique by which vehicles maintain local EKF maps, but fuse them using the information representation of the posterior. An alternative technique is due to Rekleitis et al. (2001a), who use a team of stationary and moving robots to reduce the localization error when performing SLAM. Fenwick et al. (2002) provides a comprehensive theoretical investigation of multi vehicle map fusion, specifically for landmark based SLAM. Techniques for fusing scans were developed in Konolige et al. (1999); Thrun et al. (2000b); Thrun (2001).

A number of researchers have developed SLAM systems for specific sensor types. An important sensor is a camera; however, cameras only provide bearing to features. This problem is well-studied in the computer vision literature as *structure from motion* (SFM) (Tomasi and Kanade 1992; Soatto and Brockett 1998; Dellaert et al. 2003), and in the field of photogrammetry (Konecny 2002). Within SLAM, the seminal work on bearing only SLAM is due to Deans and Hebert (2000, 2002). Their approach recursively estimates features of the environment that are invariant to the robot pose, so as to decouple the pose error from the map error. A great number of researchers has applied SLAM using cameras as the primary sensor (Neira et al. 1997; Cid et al. 2002; Davison 2003). Davison (1998) provides active vision techniques in the context of SLAM. Dudek and Jegessur's (2000) work relies on place recognition based on appearance, whereas Hayet et al. (2002) and Bouguet and Perona (1995) use visual landmarks. Diebel et al. (2004) developed a filter for SLAM with an active stereo sensor that accounts for the nonlinear noise distribution of a stereo range finder. Sensor fusion techniques for SLAM were developed by Devy and Bulata (1996). Castellanos et al. (2001) found empirically that fusing laser and camera outperformed each sensor modality in isolation.

SLAM has also been extended to the problem of building dense 3-D models. Early systems for acquiring 3-D models with indoor mobile robots can be found in Reed and Allen (1997); Iocchi et al. (2000); Thrun et al. (2004b). Devy and Parra (1998) acquire 3-D models using parametric curves. Zhao and Shibasaki (2001), Teller et al. (2001), and Frueh and Zakhor (2003) have developed impressive systems for building large textured 3-D maps of urban environments. Neither of these systems addresses the full SLAM problem due to the availability of outdoor GPS, but they are highly related to the mathematical basis of SLAM. These techniques blend smoothly with a rich body of work on aerial reconstruction of urban environments (Jung and Lacroix 2003; Thrun et al. 2003).

The following chapters discuss alternatives to the plain EKFs. The techniques described there share many of the intuitions with the extensions discussed here, as the boundary between different types of filters has become almost impossible to draw. The literature review will be continued after the next chapter, when discussing a SLAM algorithm using information-theoretic representations.

10.6 Exercises

1. What is the computational complexity of the motion update in EKF SLAM? Use the $O(\ )$ notation. Compare this with the worst case complexity for EKFs over a feature vector of the same size.

2. *Bearing only SLAM* refers to the SLAM problem when the sensors can only measure the bearing of a landmark but not its range. As noted, bearing

only SLAM is closely related to Structure from Motion (SFM) in Computer Vision. One problem in bearing only SLAM with EKFs concerns the initialization of landmark location estimates, even if the correspondences are known. Discuss why, and devise a technique for initializing the landmark location estimates (means and covariances) that can be applied in bearing only SLAM.

3. On page 329, we remarked that the EKF algorithm in Table 10.2 can become numerically instable. Devise a method for setting μ_t and Σ_t directly when a new feature is observed for the first time. Such a technique would not require initializing the covariance with very large values. Show that the result is mathematically equivalent to lines 23 and 24 when the covariance is initialized as in Equation (10.9).

4. The text suggests using a binary Bayes filter to compute the probability that a landmark represented in the posterior actually exists in the physical world.

   (a) In the first part of this exercise, you are asked to design such a binary Bayes filter.

   (b) Now extend your filter to a situation in which landmarks sporadically disappear with a probability p^*.

   (c) Picture a situation where for a well-established landmark, no information is received for a long time with regards to its existence (no positive and no negative information). To what value will your filter converge? Prove your answer.

5. The EKF SLAM algorithm, as presented here, is unable to cope with the data association problem in a sound statistical way. Lay out an algorithm (and a statistical framework) for posterior estimation with unknown data association that represents the posterior by mixtures of Gaussians, and characterize its advantages and disadvantages. How does the complexity of the posterior grow over time?

6. Based on the previous problem, develop an approximate method for posterior estimation with unknown data association where the time needed for each incremental update step does not grow over time (assuming a fixed number of landmarks).

7. Develop a Kalman filter algorithm that uses local occupancy grid maps as its basic components, instead of landmarks. Among other things, prob-

lems that have to be solved are how to relate local grids to each other, and how to deal with the ever-growing number of local grids.

11 The GraphSLAM Algorithm

11.1 Introduction

The EKF SLAM algorithm described in the previous chapter is subject to a number of limitations. One of them is its quadratic update complexity; another is the linearization technique in EKFs, which is only performed once in the EKF for each nonlinear term. In this chapter, we introduce an alternative SLAM algorithm, called *GraphSLAM*. In contrast to the EKF, GraphSLAM solves the *full SLAM problem*. It calculates a solution for the offline problem defined over all poses and all features in the map. As we will show in this chapter, the posterior of the *full SLAM problem* naturally forms a *sparse graph*. This graph leads to a sum of nonlinear quadratic constraints. Optimizing these constraints yields a maximum likelihood map and a corresponding set of robot poses. Historically, this idea can be found in a large number of SLAM publications. The name "GraphSLAM" has been chosen because it captures the essence of this approach.

SPARSE GRAPH

Figure 11.1 illustrates the GraphSLAM algorithm. Shown there is the graph that GraphSLAM extracts from five poses labeled $x_0, \ldots, x_4$, and two map features m_1, m_2. Arcs in this graph come in two types: motion arcs and measurement arcs. *Motion arcs* link any two consecutive robot poses, and *measurement arcs* link poses to features that were measured there. Each edge in the graph corresponds to a nonlinear constraint. As we shall see later, these constraints represent the negative log likelihood of the measurement and the motion models, hence are best thought of as *information constraints*. Adding such a constraint to the graph shall prove to be trivial for GraphSLAM; it involves no significant computation. The sum of all constraints results in a nonlinear *least squares problem*, as stated in Figure 11.1.

LEAST SQUARES

To compute a map posterior, GraphSLAM linearizes the set of constraints.

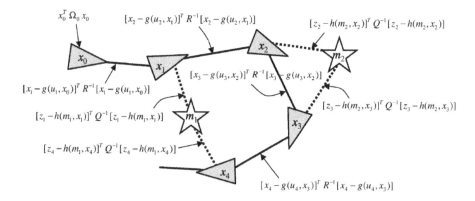

Sum of all constraints:
$$J_{\text{GraphSLAM}} = x_0^T \Omega_0 x_0 + \sum_t [x_t - g(u_t, x_{t-1})]^T R^{-1} [x_t - g(u_t, x_{t-1})] + \sum_t [z_t - h(m_{c_t}, x_t)]^T Q^{-1} [z_t - h(m_{c_t}, x_t)]$$

Figure 11.1 GraphSLAM illustration, with 4 poses and two map features. Nodes in the graphs are robot poses and feature locations. The graph is populated by two types of edges: Solid edges link consecutive robot poses, and dashed edges link poses with features sensed while the robot assumes that pose. Each link in GraphSLAM is a non-linear quadratic constraint. Motion constraints integrate the motion model; measurement constraints the measurement model. The target function of GraphSLAM is the sum of these constraints. Minimizing it yields the most likely map and the most likely robot path.

The result of linearization is an information matrix and an information vector of essentially the same form as already encountered in Chapter 3, when we discussed the information filter. However, the information matrix inherits the sparseness from the graph constructed by GraphSLAM. This sparseness enables GraphSLAM to apply the variable elimination algorithm, thereby transforming the graph into a much smaller one only defined over robot poses. The path posterior map is then calculated using standard inference techniques. GraphSLAM also computes a map and certain marginal posteriors over the map; the full map posterior is of course quadratic in the size of the map and hence is usually not recovered.

In may ways, EKF SLAM and GraphSLAM are extreme ends of the spectrum of SLAM algorithms. A primary difference between EKF SLAM and GraphSLAM pertains to the representation of information. While EKF

11.1 Introduction

SLAM represents information through a covariance matrix and a mean vector, GraphSLAM represents the information as a graph of soft constraints. Updating the covariance in an EKF is computationally expensive; whereas growing the graph is cheap!

Such savings come at a price. GraphSLAM requires additional inference when recovering the map and the path, whereas EKF maintains its best estimate of the map and the robot pose at all times. The build-up of the graph is followed by a separate computational phase in which this information is transformed into an estimate of the state. No such phase is required for EKF SLAM.

PROACTIVE SLAM

Consequently, one may think of EKF as a *proactive SLAM* algorithm, in the sense that it resolves any new piece of information immediately into an improved estimate of the state of the world. GraphSLAM, in contrast, is

LAZY SLAM

more like a *lazy SLAM* technique, which simply accumulates information into its graph without resolving it. This difference is significant. GraphSLAM can acquire maps that are many orders of magnitude larger than EKFs can handle.

There are further differences between EKF SLAM and GraphSLAM. As a solution to the full SLAM problem, GraphSLAM calculates posteriors over robot paths, hence is not an incremental algorithm. This approach is different from EKF SLAM, which, as a filter, only maintains a posterior over the momentary pose of the robot. EKF SLAM enables a robot to update its map forever, whereas GraphSLAM is best suited for problems where one seeks a map from a data set of fixed size. EKF SLAM can maintain a map over the entire lifetime of a robot without having to worry about the total number of time steps elapsed since the beginning of data acquisition.

Because GraphSLAM has access to the full data when building the map, it can apply improved linearization and data association techniques. In EKF SLAM, the linearization and the correspondence for a measurement at time t are calculated based on the data up to time t. In GraphSLAM *all* data can be used to linearize and to calculate correspondences. Put differently, GraphSLAM can revise past data association, and it can linearize more than once. In fact, GraphSLAM iterates the three crucial steps in mapping: the construction of the map, the calculation of correspondence variables, and the linearization of the measurement and motion models—so as to obtain the best estimate of all of those quantities. As a result of all this, GraphSLAM tends to produce maps that are superior in accuracy to maps generated by EKFs.

However, GraphSLAM is not without limitations when compared to the

EKF approach. One was already discussed: The size of the graph grows linearly over time, whereas the EKF shows no such time dependence in the amount of memory allocated to its estimate. Another pertains to data association. Whereas in EKF SLAM data association probabilities can easily be obtained from the posterior's covariance matrix, computing the same probabilities in GraphSLAM requires inference. This difference will be elucidated below, where we define an explicit algorithm for computing correspondences in GraphSLAM. Thus, which method is preferable is very much a question of the application, as there is no single method that would be superior in all critical dimensions.

This chapter first describes the intuition behind GraphSLAM and its basic updates steps. We then derive the various update steps mathematically and prove its correctness relative to specific linear approximations. A technique for data association will also be devised, followed by a discussion of actual implementations of the GraphSLAM algorithm.

11.2 Intuitive Description

The basic intuition behind GraphSLAM is remarkably simple: GraphSLAM extracts from the data a set of soft constraints, represented by a sparse graph. It obtains the map and the robot path by resolving these constraints into a globally consistent estimate. The constraints are generally nonlinear, but in the process of resolving them they are linearized and transformed into an information matrix. Thus, GraphSLAM is essentially an information-theoretic technique. We will describe GraphSLAM both as a technique for building a sparse graph of nonlinear constraints, and as a technique for populating a sparse information matrix of linearized constraints.

11.2.1 Building Up the Graph

Suppose we are given a set of measurements $z_{1:t}$ with associated correspondence variables $c_{1:t}$, and a set of controls $u_{1:t}$. GraphSLAM turns this data into a graph. The nodes in the graph are the robot poses $x_{1:t}$ and the features in the map $m = \{m_j\}$. Each edge in the graph corresponds to an event: a motion event generates an edge between two robot poses, and a measurement event creates a link between a pose and a feature in the map. Edges represent soft constraints between poses and features in GraphSLAM.

For a linear system, these constraints are equivalent to entries in an information matrix and an information vector of a large system of equations. As

11.2 Intuitive Description

usual, we will denote the information matrix by Ω and the information vector by ξ. As we shall see below, each measurement and each control leads to a *local* update of Ω and ξ, which corresponds to a local addition of an edge to the graph in GraphSLAM. In fact, the rule for incorporating a control or a measurement into Ω and ξ is a local addition, paying tribute to the important fact that information is an additive quantity.

Figure 11.2 illustrates the process of constructing the graph along with the corresponding information matrix. First consider a measurement z_t^i. This measurement provides information between the location of the feature $j = c_t^i$ and the robot pose x_t at time t. In GraphSLAM, this information is mapped into a constraint between x_t and m_j. We can think of this edge as a "spring" in a spring-mass model. As we shall see below, the constraint is of the type:

$$(11.1) \quad (z_t^i - h(x_t, m_j))^T \, Q_t^{-1} \, (z_t^i - h(x_t, m_j))$$

Here h is the familiar measurement function, and Q_t is the covariance of the measurement noise. Figure 11.2a shows the addition of such a link into the graph maintained by GraphSLAM.

Now consider robot motion. The control u_t provides information about the relative value of the robot pose at time $t-1$ and the pose at time t. Again, this information induces a constraint in the graph, which will be of the form

$$(11.2) \quad (x_t - g(u_t, x_{t-1}))^T \, R_t^{-1} \, (x_t - g(u_t, x_{t-1}))$$

Here g is the familiar kinematic motion model of the robot, and R_t is the covariance of the motion noise.

Figure 11.2b illustrates the addition of such a link in the graph. It also shows the addition of a new element in the information matrix, between the pose x_t and the measurement z_t^i. This update is again additive. As before, the magnitude of these values reflects the residual uncertainty R_t due to the measurement noise; the less noisy the sensor, the larger the value added to Ω and ξ.

After incorporating all measurements $z_{1:t}$ and controls $u_{1:t}$, we obtain a sparse graph of soft constraints. The number of constraints in the graph is linear in the time elapsed, hence the graph is sparse. The sum of all constraints in the graph will be of the form

$$(11.3) \quad J_{\text{GraphSLAM}} = x_0^T \, \Omega_0 \, x_0 + \sum_t (x_t - g(u_t, x_{t-1}))^T \, R_t^{-1} \, (x_t - g(u_t, x_{t-1}))$$
$$+ \sum_t \sum_i (z_t^i - h(y_t, c_t^i))^T \, Q_t^{-1} \, (z_t^i - h(y_t, c_t^i))$$

342   11 The GraphSLAM Algorithm

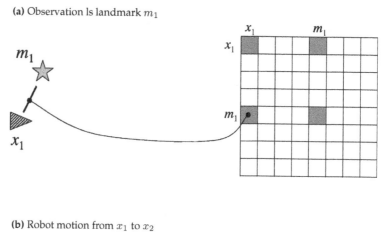

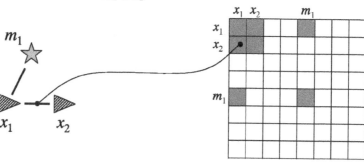

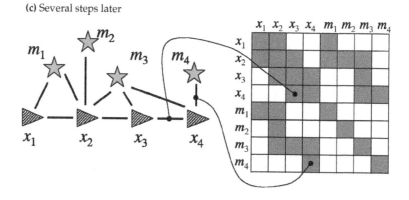

Figure 11.2 Illustration of the acquisition of the information matrix in GraphSLAM. The left diagram shows the dependence graph, the right the information matrix.

ANCHORING CONSTRAINT

It is a function defined over pose variables $x_{1:t}$ and all feature locations in the map m. Notice that this expression also features an *anchoring constraint* of the form $x_0^T \, \Omega_0 \, x_0$. This constraint anchors the absolute coordinates of the map by initializing the very first pose of the robot as $(0 \ 0 \ 0)^T$.

In the associated information matrix Ω, the off-diagonal elements are all zero with two exceptions: Between any two consecutive poses x_{t-1} and x_t will be a non-zero value that represents the information link introduced by the control u_t. Also non-zero will be any element between a map feature m_j and a pose x_t, if m_j was observed when the robot was at x_t. All elements between pairs of different features remain zero. This reflects the fact that we never received information pertaining to their relative location—all we receive in SLAM are measurements that constrain the location of a feature relative to a robot pose. Thus, the information matrix is equally sparse; all but a linear number of its elements are zero.

11.2.2 Inference

Of course, neither the graph representation nor the information matrix representation gives us what we want: the map and the path. In GraphSLAM, the map and the path are obtained from the linearized information matrix, via $\mu = \Omega^{-1} \xi$ (see Equation (3.73) on page 72). This operation requires us to solve a system of linear equations. This raises the question on how efficiently we can recover the map estimate μ and the covariance Σ.

The answer to the complexity question depends on the topology of the world. If each feature is seen only locally in time, the graph represented by the constraints is linear. Thus, Ω can be reordered so that it becomes a band-diagonal matrix, and all non-zero values occur near its diagonal. The equation $\mu = \Omega^{-1} \xi$ can then be computed in linear time. This intuition carries over to cycle-free world that is traversed once, so that each feature is seen for a short, consecutive period of time.

CYCLES

The more common case, however, involves features that are observed multiple times, with large time delays in between. This might be the case because the robot goes back and forth through a corridor, or because the world possesses *cycles*. In either situation, there will exist features m_j that are seen at drastically different time steps x_{t_1} and x_{t_2}, with $t_2 \gg t_1$. In our constraint graph, this introduces a cyclic dependence: x_{t_1} and x_{t_2} are linked through the sequence of controls $u_{t_1+1}, u_{t_1+2}, \ldots, u_{t_2}$ and through the joint observation links between x_{t_1} and m_j, and x_{t_2} and m_j, respectively. Such links make our variable reordering trick inapplicable, and recovering the map becomes more complex. In fact, since the inverse of Ω is multiplied with a vector, the result can be computed with optimization techniques such as conjugate gradient, without explicitly computing the full inverse matrix. Since most worlds possess cycles, this is the case of interest.

FACTORIZATION

The GraphSLAM algorithm now employs an important *factorization trick*, which we can think of as propagating information through the information matrix (in fact, it is a generalization of the well-known *variable elimination algorithm* for matrix inversion). Suppose we would like to remove a feature m_j from the information matrix Ω and the information state ξ. In our spring mass model, this is equivalent to removing the node and all springs attached to this node. As we shall see below, this is possible by a remarkably simple operation: We can remove all those springs between m_j and the poses at which m_j was observed, by introducing new springs between any pair of such poses.

This process is illustrated in Figure 11.3, which shows the removal of two map features, m_1 and m_3 (the removal of m_2 and m_4 is trivial in this example). On both cases, the feature removal modifies the link between any pair of poses from which a feature was originally observed. As illustrated in Figure 11.3b, this operation may lead to the introduction of new links in the graph. In the example shown there, the removal of m_3 leads to a new link between x_2 and x_4.

Formally, let $\tau(j)$ be the set of poses at which m_j was observed (that is: $x_t \in \tau(j) \iff \exists i : c_t^i = j$). Then we already know that the feature m_j is only linked to poses x_t in $\tau(j)$; by construction, m_j is *not* linked to any other pose, or to any feature in the map. We can now set all links between m_j and the poses $\tau(j)$ to zero by introducing a new link between any two poses $x_t, x_{t'} \in \tau(j)$. Similarly, the information vector values for all poses $\tau(j)$ are also updated. An important characteristic of this operation is that it is local: It only involves a small number of constraints. After removing all links to m_j, we can safely remove m_j from the information matrix and vector. The resulting information matrix is smaller—it lacks an entry for m_j. However, it is equivalent for the remaining variables, in the sense that the posterior defined by this information matrix is mathematically equivalent to the original posterior before removing m_j. This equivalence is intuitive: We simply have replaced springs connecting m_j to various poses in our spring mass model by a set of springs directly linking these poses. In doing so, the total force asserted by these springs remains equivalent, with the only exception that m_j is now disconnected.

The virtue of this reduction step is that we can gradually transform our

11.2 Intuitive Description

(a) The removal of m_1 changes the link between x_1 and x_2

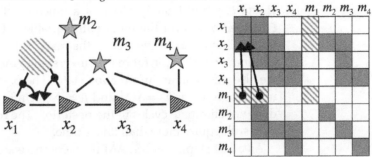

(b) The removal of m_3 introduces a new link between x_2 and x_4

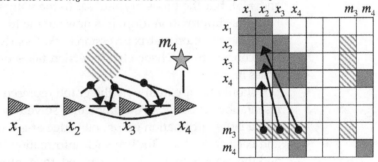

(c) Final Result after removing all map features

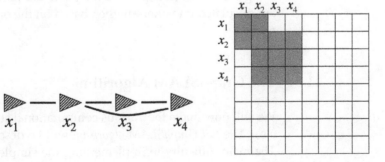

Figure 11.3 Reducing the graph in GraphSLAM: Arcs are removed to yield a network of links that only connect robot poses.

inference problem into a smaller one. By removing each feature m_j from Ω and ξ, we ultimately arrive at a much smaller information form $\tilde{\Omega}$ and $\tilde{\xi}$ defined only over the robot path variables. The reduction can be carried out in time linear in the size of the map; in fact, it generalizes the variable elimination technique for matrix inversion to the information form, in which we also maintain an information state. The posterior over the robot path is now recovered as $\tilde{\Sigma} = \tilde{\Omega}^{-1}$ and $\tilde{\mu} = \tilde{\Sigma}\xi$. Unfortunately, our reduction step does not eliminate cycles in the posterior. The remaining inference problem may still require more than linear time.

As a last step, GraphSLAM recovers the feature locations. Conceptually, this is achieved by building a new information matrix Ω_j and information vector ξ_j for each m_j. Both are defined over the variable m_j and the poses $\tau(j)$ at which m_j were observed. It contains the original links between m_j and $\tau(j)$, but the poses $\tau(j)$ are set to the values in $\tilde{\mu}$, without uncertainty. From this information form, it is now simple to calculate the location of m_j, using the common matrix inversion trick. Clearly, Ω_j contains only elements that connect to m_j; hence the inversion takes time linear in the number of poses in $\tau(j)$.

It should be apparent why the graph representation is such a natural representation. The full SLAM problem is solved by locally adding information into a large information graph, one edge at-a-time for each measurement z_t^i and each control u_t. To turn such information into an estimate of the map and the robot path, it is first linearized, then information between poses and features is gradually shifted to information between pairs of poses. The resulting structure only constraints the robot poses, which are then calculated using matrix inversion. Once the poses are recovered, the feature locations are calculated one-after-another, based on the original feature-to-pose information.

11.3 The GraphSLAM Algorithm

We will now make the various computational steps of the GraphSLAM precise. The *full GraphSLAM algorithm* will be described in a number of steps. The main difficulty in implementing the simple additive information algorithm pertains to the conversion of a conditional probability of the form $p(z_t^i \mid x_t, m)$ and $p(x_t \mid u_t, x_{t-1})$ into a link in the information matrix. The information matrix elements are all linear; hence this step involves linearizing $p(z_t^i \mid x_t, m)$ and $p(x_t \mid u_t, x_{t-1})$. In EKF SLAM, this linearization was

11.3 The GraphSLAM Algorithm

1: **Algorithm GraphSLAM_initialize($u_{1:t}$):**

2: $\begin{pmatrix} \mu_{0,x} \\ \mu_{0,y} \\ \mu_{0,\theta} \end{pmatrix} = \begin{pmatrix} 0 \\ 0 \\ 0 \end{pmatrix}$

3: for all controls $u_t = (v_t \ \omega_t)^T$ do

4: $\begin{pmatrix} \mu_{t,x} \\ \mu_{t,y} \\ \mu_{t,\theta} \end{pmatrix} = \begin{pmatrix} \mu_{t-1,x} \\ \mu_{t-1,y} \\ \mu_{t-1,\theta} \end{pmatrix}$

4: $+ \begin{pmatrix} -\frac{v_t}{\omega_t} \sin \mu_{t-1,\theta} + \frac{v_t}{\omega_t} \sin(\mu_{t-1,\theta} + \omega_t \Delta t) \\ \frac{v_t}{\omega_t} \cos \mu_{t-1,\theta} - \frac{v_t}{\omega_t} \cos(\mu_{t-1,\theta} + \omega_t \Delta t) \\ \omega_t \Delta t \end{pmatrix}$

5: endfor

6: return $\mu_{0:t}$

Table 11.1 Initialization of the mean pose vector $\mu_{1:t}$ in the GraphSLAM algorithm.

1: **Algorithm GraphSLAM_linearize($u_{1:t}, z_{1:t}, c_{1:t}, \mu_{0:t}$):**

2: set $\Omega = 0, \xi = 0$

3: add $\begin{pmatrix} \infty & 0 & 0 \\ 0 & \infty & 0 \\ 0 & 0 & \infty \end{pmatrix}$ to Ω at x_0

4: for all controls $u_t = (v_t \ \omega_t)^T$ do

5: $\hat{x}_t = \mu_{t-1} + \begin{pmatrix} -\frac{v_t}{\omega_t} \sin \mu_{t-1,\theta} + \frac{v_t}{\omega_t} \sin(\mu_{t-1,\theta} + \omega_t \Delta t) \\ \frac{v_t}{\omega_t} \cos \mu_{t-1,\theta} - \frac{v_t}{\omega_t} \cos(\mu_{t-1,\theta} + \omega_t \Delta t) \\ \omega_t \Delta t \end{pmatrix}$

6: $G_t = \begin{pmatrix} 1 & 0 & -\frac{v_t}{\omega_t} \cos \mu_{t-1,\theta} + \frac{v_t}{\omega_t} \cos(\mu_{t-1,\theta} + \omega_t \Delta t) \\ 0 & 1 & -\frac{v_t}{\omega_t} \sin \mu_{t-1,\theta} + \frac{v_t}{\omega_t} \sin(\mu_{t-1,\theta} + \omega_t \Delta t) \\ 0 & 0 & 1 \end{pmatrix}$

see next page for continuation

continued from the previous page

7: $\quad$ add $\begin{pmatrix} -G_t^T \\ 1 \end{pmatrix} R_t^{-1} (-G_t \; 1)$ to Ω at x_t and x_{t-1}

8: $\quad$ add $\begin{pmatrix} -G_t^T \\ 1 \end{pmatrix} R_t^{-1} [\hat{x}_t - G_t \mu_{t-1}]$ to ξ at x_t and x_{t-1}

9: $\quad$ endfor

10: $\quad$ for all measurements z_t do

11: $\quad\quad Q_t = \begin{pmatrix} \sigma_r^2 & 0 & 0 \\ 0 & \sigma_\phi^2 & 0 \\ 0 & 0 & \sigma_s^2 \end{pmatrix}$

12: $\quad\quad$ for all observed features $z_t^i = (r_t^i \; \phi_t^i \; s_t^i)^T$ do

13: $\quad\quad\quad j = c_t^i$

14: $\quad\quad\quad \delta = \begin{pmatrix} \delta_x \\ \delta_y \end{pmatrix} = \begin{pmatrix} \mu_{j,x} - \mu_{t,x} \\ \mu_{j,y} - \mu_{t,y} \end{pmatrix}$

15: $\quad\quad\quad q = \delta^T \delta$

16: $\quad\quad\quad \hat{z}_t^i = \begin{pmatrix} \sqrt{q} \\ \text{atan2}(\delta_y, \delta_x) - \mu_{t,\theta} \\ s_j \end{pmatrix}$

17: $\quad\quad\quad H_t^i = \frac{1}{q} \begin{pmatrix} -\sqrt{q}\delta_x & -\sqrt{q}\delta_y & 0 & +\sqrt{q}\delta_x & \sqrt{q}\delta_y & 0 \\ \delta_y & -\delta_x & -q & -\delta_y & +\delta_x & 0 \\ 0 & 0 & 0 & 0 & 0 & q \end{pmatrix}$

18: $\quad\quad\quad$ add $H_t^{iT} Q_t^{-1} H_t^i$ to Ω at x_t and m_j

19: $\quad\quad\quad$ add $H_t^{iT} Q_t^{-1} [\; z_t^i - \hat{z}_t^i + H_t^i \begin{pmatrix} \mu_{t,x} \\ \mu_{t,y} \\ \mu_{t,\theta} \\ \mu_{j,x} \\ \mu_{j,y} \\ \mu_{j,s} \end{pmatrix} \;]$ to ξ at x_t and m_j

20: $\quad\quad$ endfor

21: $\quad$ endfor

22: $\quad$ return Ω, ξ

Table 11.2 Calculation of Ω and ξ in GraphSLAM.

11.3 The GraphSLAM Algorithm

1: **Algorithm GraphSLAM_reduce**(Ω, ξ):
2: $\tilde{\Omega} = \Omega$
3: $\tilde{\xi} = \xi$
4: for each feature j do
5:      let $\tau(j)$ be the set of all poses x_t at which j was observed
6:      subtract $\tilde{\Omega}_{\tau(j),j} \, \tilde{\Omega}_{j,j}^{-1} \, \xi_j$ from $\tilde{\xi}$ at $x_{\tau(j)}$ and m_j
7:      subtract $\tilde{\Omega}_{\tau(j),j} \, \tilde{\Omega}_{j,j}^{-1} \, \tilde{\Omega}_{j,\tau(j)}$ from $\tilde{\Omega}$ at $x_{\tau(j)}$ and m_j
8:      remove from $\tilde{\Omega}$ and $\tilde{\xi}$ all rows/columns corresponding to j
9: endfor
10: return $\tilde{\Omega}, \tilde{\xi}$

Table 11.3 Algorithm for reducing the size of the information representation of the posterior in GraphSLAM.

1: **Algorithm GraphSLAM_solve**($\tilde{\Omega}, \tilde{\xi}, \Omega, \xi$):
2: $\Sigma_{0:t} = \tilde{\Omega}^{-1}$
3: $\mu_{0:t} = \Sigma_{0:t} \, \tilde{\xi}$
4: for each feature j do
5:      set $\tau(j)$ to the set of all poses x_t at which j was observed
6:      $\mu_j = \Omega_{j,j}^{-1} \, (\xi_j + \Omega_{j,\tau(j)} \, \tilde{\mu}_{\tau(j)})$
7: endfor
8: return $\mu, \Sigma_{0:t}$

Table 11.4 Algorithm for updating the posterior μ.

1:	**Algorithm GraphSLAM_known_correspondence**($u_{1:t}, z_{1:t}, c_{1:t}$):
2:	$\mu_{0:t}$ = **GraphSLAM_initialize**($u_{1:t}$)
3:	repeat
4:	Ω, ξ = **GraphSLAM_linearize**($u_{1:t}, z_{1:t}, c_{1:t}, \mu_{0:t}$)
5:	$\tilde{\Omega}, \tilde{\xi}$ = **GraphSLAM_reduce**(Ω, ξ)
6:	$\mu, \Sigma_{0:t}$ = **GraphSLAM_solve**($\tilde{\Omega}, \tilde{\xi}, \Omega, \xi$)
7:	until convergence
8:	return μ

Table 11.5 The GraphSLAM algorithm for the full SLAM problem with known correspondence.

found by calculating a Jacobian at the estimated mean poses $\mu_{0:t}$. To build our initial information matrix Ω and ξ, we need an initial estimate $\mu_{0:t}$ for all poses $x_{0:t}$.

There exist a number of solutions to the problem of finding an initial mean μ suitable for linearization. For example, we can run an EKF SLAM and use its estimate for linearization. For the sake of this chapter, we will use an even simpler technique: Our initial estimate will simply be provided by chaining together the motion model $p(x_t \mid u_t, x_{t-1})$. Such an algorithm is outlined in Table 11.1, and called there **GraphSLAM_initialize**. This algorithm takes the controls $u_{1:t}$ as input, and outputs sequence of pose estimates $\mu_{0:t}$. It initializes the first pose by zero, and then calculates subsequent poses by recursively applying the velocity motion model. Since we are only interested in the mean poses vector $\mu_{0:t}$, **GraphSLAM_initialize** only uses the deterministic part of the motion model. It also does not consider any measurement in its estimation.

Once an initial $\mu_{0:t}$ is available, the GraphSLAM algorithm constructs the full SLAM information matrix Ω and the corresponding information vector ξ. This is achieved by linearizing the links in the graph. The algorithm **GraphSLAM_linearize** is depicted in Table 11.2. This algorithm contains a good amount of mathematical notation, much of which shall become clear in our derivation of the algorithm further below. **GraphSLAM_linearize** accepts as an input the set of controls, $u_{1:t}$, the measurements $z_{1:t}$ and associated correspondence variables $c_{1:t}$, and the mean pose estimates $\mu_{0:t}$. It then gradually constructs the information matrix Ω and the information vector ξ through linearization, by locally adding submatrices in accordance with the information obtained from each measurement and each control.

In particular, line 2 in **GraphSLAM_linearize** initializes the information elements. The "infinite" information entry in line 3 fixes the initial pose x_0 to $(0 \ 0 \ 0)^T$. It is necessary, since otherwise the resulting matrix becomes singular, reflecting the fact that from relative information alone we cannot recover absolute estimates.

Controls are integrated in lines 4 through 9 of **GraphSLAM_linearize**. The pose $\hat{x}$ and the Jacobian G_t calculated in lines 5 and 6 represent the linear approximation of the non-linear control function g. As obvious from these equations, this linearization step utilizes the pose estimates $\mu_{0:t-1}$, with $\mu_0 = (0 \ 0 \ 0)^T$. This leads to the updates for Ω, and ξ, calculated in lines 7, and 8, respectively. Both terms are added into the corresponding rows and columns of Ω and ξ. This addition realizes the inclusion of a new constraint into the SLAM posterior, very much along the lines of the intuitive description in the previous section.

Measurements are integrated in lines 10 through 21 of **GraphSLAM_linearize**. The matrix Q_t calculated in line 11 is the familiar measurement noise covariance. Lines 13 through 17 compute the Taylor expansion of the measurement function, here stated for the feature-based measurement model defined in Chapter 6.6. Attention has to be payed to the implementation of line 16, since the angular expressions can be shifted arbitrarily by 2π. This calculation culminates into the computation of the measurement update in lines 18 and 19. The matrix that is being added to Ω in line 18 is of dimension 6×6. To add it, we decompose it into a matrix of dimension 3×3 for the pose x_t, a matrix of dimension 3×3 for the feature m_j, and two matrices of dimension 3×3 and 3×3 for the link between x_t and m_j. Those are added to Ω at the corresponding rows and columns. Similarly, the vector added to the information vector ξ is of vertical dimension 5. It is also chopped into two vectors of size 3 and 2, and added to the elements corresponding to x_t and m_j, respectively. The result of **GraphSLAM_linearize** is an information vector ξ and a matrix Ω. We already noted that Ω is sparse. It contains only non-zero submatrices along the main diagonal, between subsequent poses, and between poses and features in the map. The running time of this algorithm is linear in t, the number of time steps at which data was accrued.

The next step of the GraphSLAM algorithm pertains to reducing the dimensionality of the information matrix/vector. This is achieved through the algorithm **GraphSLAM_reduce** in Table 11.3. This algorithm takes as input Ω and ξ defined over the full space of map features and poses, and outputs a reduced matrix $\tilde{\Omega}$ and vectors $\tilde{\xi}$ defined over the space of all poses (but not the map!). This transformation is achieved by removing features m_j one-at-a-time, in lines 4 through 9 of **GraphSLAM_reduce**. The bookkeeping of the exact indexes of each item in $\tilde{\Omega}$ and $\tilde{\xi}$ is a bit tedious, hence Table 11.3 only provides an intuitive account.

Line 5 calculates the set of poses $\tau(j)$ at which the robot observed feature j. It then extracts two submatrices from the present $\tilde{\Omega}$: $\tilde{\Omega}_{j,j}$ and $\tilde{\Omega}_{\tau(j),j}$. $\tilde{\Omega}_{j,j}$ is the quadratic submatrix between m_j and m_j, and $\tilde{\Omega}_{\tau(j),j}$ is composed of the off-diagonal elements between m_j and the pose variables $\tau(j)$. It also extracts from the information state vector $\tilde{\xi}$ the elements corresponding to the j-th feature, denoted here as ξ_j. It then subtracts information from $\tilde{\Omega}$ and $\tilde{\xi}$ as stated in lines 6 and 7. After this operation, the rows and columns for the feature m_j are zero. These rows and columns are then removed, reducing the dimension on $\tilde{\Omega}$ and $\tilde{\xi}$ accordingly. This process is iterated until all features have been removed, and only pose variables remain in $\tilde{\Omega}$ and $\tilde{\xi}$. The complexity of **GraphSLAM_reduce** is once again linear in t.

The last step in the GraphSLAM algorithm computes the mean and covariance for all poses in the robot path, and a mean location estimate for all features in the map. This is achieved through **GraphSLAM_solve** in Table 11.4. Lines 2 and 3 compute the path estimates $\mu_{0:t}$, by inverting the reduced information matrix $\tilde{\Omega}$ and multiplying the resulting covariance with the information vector. Subsequently, **GraphSLAM_solve** computes the location of each feature in lines 4 through 7. The return value of **GraphSLAM_solve** contains the mean for the robot path and all features in the map, but only the covariance for the robot path. We note that there exist other, more efficient ways to compute $\mu_{0:t}$ that bypass the matrix inversion step. Those will be discussed towards the end of this chapter, when applying standard optimization techniques to GraphSLAM.

The quality of the solution calculated by the GraphSLAM algorithm depends on the goodness of the initial mean estimates, calculated by **GraphSLAM_initialize**. The x- and y- components of these estimates affect the respective models in a linear way, hence the linearization does not depend on these values. Not so for the orientation variables in $\mu_{0:t}$. Errors in these initial estimates affect the accuracy of the Taylor approximation, which in turn affects the result.

To reduce potential errors due to the Taylor approximation in the linearization, the procedures **GraphSLAM_linearize**, **GraphSLAM_reduce**, and **GraphSLAM_solve** are run multiple times over the same data set. Each iteration takes as an input an estimated mean vector $\mu_{0:t}$ from the previous iteration, and outputs a new, improved estimate. The iteration of the GraphSLAM optimization are only necessary when the initial pose estimates have high error (e.g., more than 20 degrees orientation error). A small number of iterations (e.g., 3) is usually sufficient.

Table 11.5 summarizes the resulting algorithm. It initializes the means, then repeats the construction step, the reduction step, and the solution step. Typically, two or three iterations suffice for convergence. The resulting mean μ is our best guess of the robot's path and the map.

11.4 Mathematical Derivation of GraphSLAM

The derivation of the GraphSLAM algorithm begins with a derivation of a recursive formula for calculating the full SLAM posterior, represented in information form. We then investigate each term in this posterior, and derive from them the additive SLAM updates through Taylor expansions. From that, we will derive the necessary equations for recovering the path and the map.

11.4.1 The Full SLAM Posterior

As in the discussion of EKF SLAM, it will be beneficial to introduce a variable for the augmented state of the full SLAM problem. We will use y to denote state variables that combine one or more poses x with the map m. In particular, we define $y_{0:t}$ to be a vector composed of the path $x_{0:t}$ and the map m, whereas y_t is composed of the momentary pose at time t and the map m:

$$(11.4) \quad y_{0:t} = \begin{pmatrix} x_0 \\ x_1 \\ \vdots \\ x_t \\ m \end{pmatrix} \quad \text{and} \quad y_t = \begin{pmatrix} x_t \\ m \end{pmatrix}$$

The posterior in the full SLAM problem is $p(y_{0:t} \mid z_{1:t}, u_{1:t}, c_{1:t})$, where $z_{1:t}$ are the familiar measurements with correspondences $c_{1:t}$, and $u_{1:t}$ are the controls. Bayes rule enables us to factor this posterior:

$$(11.5) \quad p(y_{0:t} \mid z_{1:t}, u_{1:t}, c_{1:t})$$

$$= \eta \, p(z_t \mid y_{0:t}, z_{1:t-1}, u_{1:t}, c_{1:t}) \, p(y_{0:t} \mid z_{1:t-1}, u_{1:t}, c_{1:t})$$

where η is the familiar normalizer. The first probability on the right-hand side can be reduced by dropping irrelevant conditioning variables:

(11.6) $\quad p(z_t \mid y_{0:t}, z_{1:t-1}, u_{1:t}, c_{1:t}) \;=\; p(z_t \mid y_t, c_t)$

Similarly, we can factor the second probability by partitioning $y_{0:t}$ into x_t and $y_{0:t-1}$, and obtain

(11.7) $\quad p(y_{0:t} \mid z_{1:t-1}, u_{1:t}, c_{1:t})$
$$= p(x_t \mid y_{0:t-1}, z_{1:t-1}, u_{1:t}, c_{1:t}) \, p(y_{0:t-1} \mid z_{1:t-1}, u_{1:t}, c_{1:t})$$
$$= p(x_t \mid x_{t-1}, u_t) \, p(y_{0:t-1} \mid z_{1:t-1}, u_{1:t-1}, c_{1:t-1})$$

Putting these expressions back into (11.5) gives us the recursive definition of the full SLAM posterior:

(11.8) $\quad p(y_{0:t} \mid z_{1:t}, u_{1:t}, c_{1:t})$
$$= \eta \, p(z_t \mid y_t, c_t) \, p(x_t \mid x_{t-1}, u_t) \, p(y_{0:t-1} \mid z_{1:t-1}, u_{1:t-1}, c_{1:t-1})$$

The closed form expression is obtained through induction over t. Here $p(y_0)$ is the prior over the map m and the initial pose x_0.

(11.9) $\quad p(y_{0:t} \mid z_{1:t}, u_{1:t}, c_{1:t}) \;=\; \eta \, p(y_0) \prod_t p(x_t \mid x_{t-1}, u_t) \, p(z_t \mid y_t, c_t)$
$$= \eta \, p(y_0) \prod_t \left[p(x_t \mid x_{t-1}, u_t) \prod_i p(z_t^i \mid y_t, c_t^i) \right]$$

Here, as before, z_t^i is the i-th measurement in the measurement vector z_t at time t. The prior $p(y_0)$ factors into two independent priors, $p(x_0)$ and $p(m)$. In SLAM, we usually have no prior knowledge about the map m. We simply replace $p(y_0)$ by $p(x_0)$ and subsume the factor $p(m)$ into the normalizer η.

11.4.2 The Negative Log Posterior

The information form represents probabilities in logarithmic form. The log-SLAM posterior follows directly from the previous equation:

(11.10) $\quad \log p(y_{0:t} \mid z_{1:t}, u_{1:t}, c_{1:t})$
$$= \text{const.} + \log p(x_0) + \sum_t \left[\log p(x_t \mid x_{t-1}, u_t) + \sum_i \log p(z_t^i \mid y_t, c_t^i) \right]$$

Just as in Chapter 10, we assume the outcome of robot motion is distributed normally according to $\mathcal{N}(g(u_t, x_{t-1}), R_t)$, where g is the deterministic motion function, and R_t is the covariance of the motion error. Similarly, measurements z_t^i are generated according to $\mathcal{N}(h(y_t, c_t^i), Q_t)$, where h is the familiar measurement function and Q_t is the measurement error covariance. In equations, we have

$$(11.11) \quad p(x_t \mid x_{t-1}, u_t) = \eta \exp\left\{-\tfrac{1}{2}(x_t - g(u_t, x_{t-1}))^T R_t^{-1} (x_t - g(u_t, x_{t-1}))\right\}$$

$$(11.12) \quad p(z_t^i \mid y_t, c_t^i) = \eta \exp\left\{-\tfrac{1}{2}(z_t^i - h(y_t, c_t^i))^T Q_t^{-1} (z_t^i - h(y_t, c_t^i))\right\}$$

The prior $p(x_0)$ in (11.10) is also easily expressed by a Gaussian-type distribution. It *anchors* the initial pose x_0 to the origin of the global coordinate system: $x_0 = (0\ 0\ 0)^T$:

$$(11.13) \quad p(x_0) = \eta \exp\left\{-\tfrac{1}{2} x_0^T \Omega_0 x_0\right\}$$

with

$$(11.14) \quad \Omega_0 = \begin{pmatrix} \infty & 0 & 0 \\ 0 & \infty & 0 \\ 0 & 0 & \infty \end{pmatrix}$$

For now, it shall not concern us that the value of ∞ cannot be implemented, as we can easily substitute ∞ with a large positive number. This leads to the following quadratic form of the negative log-SLAM posterior in (11.10):

$$(11.15) \quad -\log p(y_{0:t} \mid z_{1:t}, u_{1:t}, c_{1:t})$$
$$= \text{const.} + \tfrac{1}{2}\left[x_0^T \Omega_0 x_0 + \sum_t (x_t - g(u_t, x_{t-1}))^T R_t^{-1} (x_t - g(u_t, x_{t-1})) \right.$$
$$\left. + \sum_t \sum_i (z_t^i - h(y_t, c_t^i))^T Q_t^{-1} (z_t^i - h(y_t, c_t^i)) \right]$$

This is essentially the same as $J_{\text{GraphSLAM}}$ in Equation (11.3), with a few differences pertaining to the omission of normalization constants (including a multiplication with -1). Equation (11.15) highlights an essential characteristic of the full SLAM posterior in the information form: It is composed of a number of quadratic terms, one for the prior, and one for each control and each measurement.

11.4.3 Taylor Expansion

The various terms Equation (11.15) are quadratic in the functions g and h, not in the variables we seek to estimate (poses and the map). GraphSLAM alle-

LINEARIZATION viates this problem by *linearizing* g and h via Taylor expansion—completely analogously to Equations (10.14) and (10.18) in the derivation of the EKF. In particular, we have:

(11.16) $\quad g(u_t, x_{t-1}) \approx g(u_t, \mu_{t-1}) + G_t(x_{t-1} - \mu_{t-1})$

(11.17) $\quad h(y_t, c_t^i) \approx h(\mu_t, c_t^i) + H_t^i(y_t - \mu_t)$

Here μ_t is the current estimate of the state vector y_t, and $H_t^i = h_t^i \, F_{x,j}$ as defined already in Equation (10.19).

This linear approximation turns the log-likelihood (11.15) into a function that is quadratic in $y_{0:t}$. In particular, we obtain

(11.18) $\quad \log p(y_{0:t} \mid z_{1:t}, u_{1:t}, c_{1:t}) \;=\; \text{const.} - \tfrac{1}{2}$
$$\Big\{ x_0^T \, \Omega_0 \, x_0 + \sum_t [x_t - g(u_t, \mu_{t-1}) - G_t(x_{t-1} - \mu_{t-1})]^T $$
$$R_t^{-1} \, [x_t - g(u_t, \mu_{t-1}) - G_t(x_{t-1} - \mu_{t-1})] $$
$$+ \sum_i [z_t^i - h(\mu_t, c_t^i) - H_t^i(y_t - \mu_t)]^T \, Q_t^{-1} \, [z_t^i - h(\mu_t, c_t^i) - H_t^i(y_t - \mu_t)] \Big\} $$

This function is indeed a quadratic in $y_{0:t}$, and it shall prove convenient to reorder its terms, omitting several constant terms.

(11.19) $\quad \log p(y_{0:t} \mid z_{1:t}, u_{1:t}, c_{1:t}) \;=\; \text{const.}$
$$-\tfrac{1}{2} \underbrace{x_0^T \, \Omega_0 \, x_0}_{\text{quadratic in } x_0} \;-\; \tfrac{1}{2} \sum_t \underbrace{x_{t-1:t}^T \begin{pmatrix} -G_t^T \\ 1 \end{pmatrix} R_t^{-1} (-G_t \; 1) \, x_{t-1:t}}_{\text{quadratic in } x_{t-1:t}}$$
$$+ \underbrace{x_{t-1:t}^T \begin{pmatrix} -G_t^T \\ 1 \end{pmatrix} R_t^{-1} [g(u_t, \mu_{t-1}) - G_t \, \mu_{t-1}]}_{\text{linear in } x_{t-1:t}}$$
$$-\tfrac{1}{2} \sum_i \underbrace{y_t^T \, H_t^{iT} \, Q_t^{-1} \, H_t^i \, y_t}_{\text{quadratic in } y_t} \;+\; \underbrace{y_t^T \, H_t^{iT} \, Q_t^{-1} \, [z_t^i - h(\mu_t, c_t^i) + H_t^i \mu_t]}_{\text{linear in } y_t}$$

Here $x_{t-1:t}$ denotes the state vector concatenating x_{t-1} and x_t; hence we can write $(x_t - G_t \, x_{t-1})^T = x_{t-1:t}^T \, (-G_t \; 1)^T = x_{t-1:t}^T \begin{pmatrix} -G_t^T \\ 1 \end{pmatrix}$.

If we collect all quadratic terms into the matrix Ω, and all linear terms into a vector ξ, we see that expression (11.19) is of the form

(11.20) $\quad \log p(y_{0:t} \mid z_{1:t}, u_{1:t}, c_{1:t}) \;=\; \text{const.} - \tfrac{1}{2} y_{0:t}^T \, \Omega \, y_{0:t} + y_{0:t}^T \, \xi$

11.4.4 Constructing the Information Form

We can read off these terms directly from (11.19), and verify that they are indeed implemented in the algorithm **GraphSLAM_linearize** in Table 11.2:

- **Prior.** The initial pose prior manifests itself by a quadratic term Ω_0 over the initial pose variable x_0 in the information matrix. Assuming appropriate extension of the matrix Ω_0 to match the dimension of $y_{0:t}$, we have

$$(11.21) \qquad \Omega \longleftarrow \Omega_0$$

This initialization is performed in lines 2 and 3 of the algorithm **GraphSLAM_linearize**.

- **Controls.** From (11.19), we see that each control u_t adds to Ω and ξ the following terms, assuming that the matrices are rearranged so as to be of matching dimensions:

$$(11.22) \qquad \Omega \longleftarrow \Omega + \begin{pmatrix} -G_t^T \\ 1 \end{pmatrix} R_t^{-1} \begin{pmatrix} -G_t & 1 \end{pmatrix}$$

$$(11.23) \qquad \xi \longleftarrow \xi + \begin{pmatrix} -G_t^T \\ 1 \end{pmatrix} R_t^{-1} [g(u_t, \mu_{t-1}) - G_t\, \mu_{t-1}]$$

This is realized in lines 4 through 9 in **GraphSLAM_linearize**.

- **Measurements.** According to Equation (11.19), each measurement z_t^i transforms Ω and ξ by adding the following terms, once again assuming appropriate adjustment of the matrix dimensions:

$$(11.24) \qquad \Omega \longleftarrow \Omega + H_t^{iT}\, Q_t^{-1}\, H_t^i$$

$$(11.25) \qquad \xi \longleftarrow \xi + H_t^{iT}\, Q_t^{-1}\, [z_t^i - h(\mu_t, c_t^i) + H_t^i \mu_t]$$

This update occurs in lines 10 through 21 in **GraphSLAM_linearize**.

This proves the correctness of the construction algorithm **GraphSLAM_linearize**, relative to our Taylor expansion approximation.

We also note that the steps above only affect off-diagonal elements that involve at least one pose. Thus, all between-feature elements are zero in the resulting information matrix.

Marginals of a multivariate Gaussian. Let the probability distribution $p(x, y)$ over the random vectors x and y be a Gaussian represented in the information form

$$\Omega = \begin{pmatrix} \Omega_{xx} & \Omega_{xy} \\ \Omega_{yx} & \Omega_{yy} \end{pmatrix} \quad \text{and} \quad \xi = \begin{pmatrix} \xi_x \\ \xi_y \end{pmatrix}$$

If Ω_{yy} is invertible, the marginal $p(x)$ is a Gaussian whose information representation is

$$\bar{\Omega}_{xx} = \Omega_{xx} - \Omega_{xy} \Omega_{yy}^{-1} \Omega_{yx} \quad \text{and} \quad \bar{\xi}_x = \xi_x - \Omega_{xy} \Omega_{yy}^{-1} \xi_y$$

Proof. The marginal for a Gaussian in its moments parameterization

$$\Sigma = \begin{pmatrix} \Sigma_{xx} & \Sigma_{xy} \\ \Sigma_{yx} & \Sigma_{yy} \end{pmatrix} \quad \text{and} \quad \mu = \begin{pmatrix} \mu_x \\ \mu_y \end{pmatrix}$$

is $\mathcal{N}(\mu_x, \Sigma_{xx})$. By definition, the information matrix of this Gaussian is therefore Σ_{xx}^{-1}, and the information vector is $\Sigma_{xx}^{-1} \mu_x$. We show $\Sigma_{xx}^{-1} = \bar{\Omega}_{xx}$ via the Inversion Lemma from Table 3.2 on page 50. Let $P = (0\ 1)^T$, and let $[\infty]$ be a matrix of the same size as Ω_{yy} but whose entries are all infinite (with $[\infty]^{-1} = 0$). This gives us

$$(\Omega + P[\infty]P^T)^{-1} = \begin{pmatrix} \Omega_{xx} & \Omega_{xy} \\ \Omega_{yx} & [\infty] \end{pmatrix}^{-1} \stackrel{(*)}{=} \begin{pmatrix} \Sigma_{xx}^{-1} & 0 \\ 0 & 0 \end{pmatrix}$$

The same expression can also be expanded by the inversion lemma into:

$(\Omega + P[\infty]P^T)^{-1}$

$= \Omega - \Omega\, P([\infty]^{-1} + P^T\, \Omega\, P)^{-1}\, P^T\, \Omega$

$= \Omega - \Omega\, P(0 + P^T\, \Omega\, P)^{-1}\, P^T\, \Omega$

$= \Omega - \Omega\, P(\Omega_{yy})^{-1}\, P^T\, \Omega$

$= \begin{pmatrix} \Omega_{xx} & \Omega_{xy} \\ \Omega_{yx} & \Omega_{yy} \end{pmatrix} - \begin{pmatrix} \Omega_{xx} & \Omega_{xy} \\ \Omega_{yx} & \Omega_{yy} \end{pmatrix} \begin{pmatrix} 0 & 0 \\ 0 & \Omega_{yy}^{-1} \end{pmatrix} \begin{pmatrix} \Omega_{xx} & \Omega_{xy} \\ \Omega_{yx} & \Omega_{yy} \end{pmatrix}$

$\stackrel{(*)}{=} \begin{pmatrix} \Omega_{xx} & \Omega_{xy} \\ \Omega_{yx} & \Omega_{yy} \end{pmatrix} - \begin{pmatrix} 0 & \Omega_{xy} \Omega_{yy}^{-1} \\ 0 & 1 \end{pmatrix} \begin{pmatrix} \Omega_{xx} & \Omega_{xy} \\ \Omega_{yx} & \Omega_{yy} \end{pmatrix}$

$= \begin{pmatrix} \Omega_{xx} & \Omega_{xy} \\ \Omega_{yx} & \Omega_{yy} \end{pmatrix} - \begin{pmatrix} \Omega_{xy} \Omega_{yy}^{-1} \Omega_{yx} & \Omega_{xy} \\ \Omega_{yx} & \Omega_{yy} \end{pmatrix} = \begin{pmatrix} \bar{\Omega}_{xx} & 0 \\ 0 & 0 \end{pmatrix}$

see next page for continuation

11.4 Mathematical Derivation of GraphSLAM

> *continued from the previous page*

The remaining statement, $\Sigma_{xx}^{-1} \mu_x = \bar{\xi}_x$, is obtained analogously, exploiting the fact that $\mu = \Omega^{-1}\xi$ (see Equation (3.73)) and the equality of the two expressions marked "(*)" above:

$$\begin{pmatrix} \Sigma_{xx}^{-1} \mu_x \\ 0 \end{pmatrix} = \begin{pmatrix} \Sigma_{xx}^{-1} & 0 \\ 0 & 0 \end{pmatrix} \begin{pmatrix} \mu_x \\ \mu_y \end{pmatrix} = \begin{pmatrix} \Sigma_{xx}^{-1} & 0 \\ 0 & 0 \end{pmatrix} \Omega^{-1} \begin{pmatrix} \xi_x \\ \xi_y \end{pmatrix}$$

$$\stackrel{(*)}{=} \left[\Omega - \begin{pmatrix} 0 & \Omega_{xy}\Omega_{yy}^{-1} \\ 0 & 1 \end{pmatrix} \Omega \right] \Omega^{-1} \begin{pmatrix} \xi_x \\ \xi_y \end{pmatrix}$$

$$= \begin{pmatrix} \xi_x \\ \xi_y \end{pmatrix} - \begin{pmatrix} 0 & \Omega_{xy}\Omega_{yy}^{-1} \\ 0 & 1 \end{pmatrix} \begin{pmatrix} \xi_x \\ \xi_y \end{pmatrix} = \begin{pmatrix} \bar{\xi}_x \\ 0 \end{pmatrix}$$

Table 11.6 Lemma for marginalizing Gaussians in information form. The form of the covariance $\bar{\Omega}_{xx}$ in this lemma is also known as *Schur complement*.

Conditionals of a multivariate Gaussian. Let the probability distribution $p(x, y)$ over the random vectors x and y be a Gaussian represented in the information form

$$\Omega = \begin{pmatrix} \Omega_{xx} & \Omega_{xy} \\ \Omega_{yx} & \Omega_{yy} \end{pmatrix} \quad \text{and} \quad \xi = \begin{pmatrix} \xi_x \\ \xi_y \end{pmatrix}$$

The conditional $p(x \mid y)$ is a Gaussian with information matrix Ω_{xx} and information vector $\xi_x - \Omega_{xy} y$.

Proof. The result follows trivially from the definition of a Gaussian in information form:

$$p(x \mid y)$$
$$= \eta \exp\left\{-\frac{1}{2}\begin{pmatrix} x \\ y \end{pmatrix}^T \begin{pmatrix} \Omega_{xx} & \Omega_{xy} \\ \Omega_{yx} & \Omega_{yy} \end{pmatrix} \begin{pmatrix} x \\ y \end{pmatrix} + \begin{pmatrix} x \\ y \end{pmatrix}^T \begin{pmatrix} \xi_x \\ \xi_y \end{pmatrix}\right\}$$
$$= \eta \exp\{-\tfrac{1}{2}x^T \Omega_{xx} x - x^T \Omega_{xy} y - \tfrac{1}{2}y^T \Omega_{yy} y + x^T \xi_x + y^T \xi_y\}$$
$$= \eta \exp\{-\tfrac{1}{2}x^T \Omega_{xx} x + x^T(\xi_x - \Omega_{xy} y) \underbrace{- \tfrac{1}{2}y^T \Omega_{yy} y + y^T \xi_y}_{\text{const.}}\}$$
$$= \eta \exp\{-\tfrac{1}{2}x^T \Omega_{xx} x + x^T(\xi_x - \Omega_{xy} y)\}$$

Table 11.7 Lemma for conditioning Gaussians in information form.

11.4.5 Reducing the Information Form

The reduction step **GraphSLAM_reduce** is based on a factorization of the full SLAM posterior.

$$(11.26) \quad p(y_{0:t} \mid z_{1:t}, u_{1:t}, c_{1:t}) = p(x_{0:t} \mid z_{1:t}, u_{1:t}, c_{1:t})\, p(m \mid x_{0:t}, z_{1:t}, u_{1:t}, c_{1:t})$$

Here $p(x_{0:t} \mid z_{1:t}, u_{1:t}, c_{1:t}) \sim \mathcal{N}(\xi, \Omega)$ is the posterior over paths alone, with the map integrated out:

$$(11.27) \quad p(x_{0:t} \mid z_{1:t}, u_{1:t}, c_{1:t}) = \int p(y_{0:t} \mid z_{1:t}, u_{1:t}, c_{1:t})\, dm$$

As we will show shortly, this probability is indeed calculated by the algorithm **GraphSLAM_reduce** in Table 11.3, since

$$(11.28) \quad p(x_{0:t} \mid z_{1:t}, u_{1:t}, c_{1:t}) \sim \mathcal{N}(\tilde{\xi}, \tilde{\Omega})$$

In general, the integration in (11.27) will be intractable, due to the large number of variables in m. For Gaussians, this integral can be calculated in closed form. The key insight is given in Table 11.6, which states and proves the *marginalization lemma* for Gaussians.

Let us subdivide the matrix Ω and the vector ξ into submatrices, for the robot path $x_{0:t}$ and the map m:

$$(11.29) \quad \Omega = \begin{pmatrix} \Omega_{x_{0:t},x_{0:t}} & \Omega_{x_{0:t},m} \\ \Omega_{m,x_{0:t}} & \Omega_{m,m} \end{pmatrix}$$

$$(11.30) \quad \xi = \begin{pmatrix} \xi_{x_{0:t}} \\ \xi_m \end{pmatrix}$$

According to the *marginalization lemma*, the probability (11.28) is obtained as

$$(11.31) \quad \tilde{\Omega} = \Omega_{x_{0:t},x_{0:t}} - \Omega_{x_{0:t},m}\, \Omega_{m,m}^{-1}\, \Omega_{m,x_{0:t}}$$

$$(11.32) \quad \tilde{\xi} = \xi_{x_{0:t}} - \Omega_{x_{0:t},m}\, \Omega_{m,m}^{-1}\, \xi_m$$

The matrix $\Omega_{m,m}$ is block-diagonal. This follows from the way Ω is constructed, in particular the absence of any links between pairs of features. This makes the inversion efficient:

$$(11.33) \quad \Omega_{m,m}^{-1} = \sum_j F_j^T\, \Omega_{j,j}^{-1}\, F_j$$

where $\Omega_{j,j} = F_j \Omega F_j^T$ is the sub-matrix of Ω that corresponds to the j-th feature in the map:

$$(11.34) \quad F_j = \begin{pmatrix} 0 \cdots 0 & 1\ 0\ 0 & 0 \cdots 0 \\ 0 \cdots 0 & 0\ 1\ 0 & 0 \cdots 0 \\ 0 \cdots 0 & \underbrace{0\ 0\ 1}_{j\text{-th feature}} & 0 \cdots 0 \end{pmatrix}$$

This insight makes it possible to decompose the implement Equations (11.31) and (11.32) into a sequential update:

$$(11.35) \quad \tilde{\Omega} = \Omega_{x_{0:t},x_{0:t}} - \sum_j \Omega_{x_{0:t},j}\, \Omega_{j,j}^{-1}\, \Omega_{j,x_{0:t}}$$

$$(11.36) \quad \tilde{\xi} = \xi_{x_{0:t}} - \sum_j \Omega_{x_{0:t},j}\, \Omega_{j,j}^{-1}\, \xi_j$$

The matrix $\Omega_{x_{0:t},j}$ is non-zero only for elements in $\tau(j)$, the set of poses at which feature j was observed. This essentially proves the correctness of the reduction algorithm **GraphSLAM_reduce** in Table 11.3. The operation performed on Ω in this algorithm can be thought of as the variable elimination algorithm for matrix inversion, applied to the feature variables but not the robot pose variables.

11.4.6 Recovering the Path and the Map

The algorithm **GraphSLAM_solve** in Table 11.4 calculates the mean and variance of the Gaussian $\mathcal{N}(\tilde{\xi}, \tilde{\Omega})$, using the standard equations, see Equations (3.72) and (3.73) on page 72:

$$(11.37) \quad \tilde{\Sigma} = \tilde{\Omega}^{-1}$$
$$(11.38) \quad \tilde{\mu} = \tilde{\Sigma}\, \tilde{\xi}$$

In particular, this operation provides us with the mean of the posterior on the robot path; it does not give us the locations of the features in the map.

It remains to recover the second factor of Equation (11.26):

$$(11.39) \quad p(m \mid x_{0:t}, z_{1:t}, u_{1:t}, c_{1:t})$$

The *conditioning lemma*, stated and proved in Table 11.7, shows that this probability distribution is Gaussian with the parameters

$$(11.40) \quad \Sigma_m = \Omega_{m,m}^{-1}$$
$$(11.41) \quad \mu_m = \Sigma_m(\xi_m + \Omega_{m,x_{0:t}}\tilde{\xi})$$

Here ξ_m and $\Omega_{m,m}$ are the subvector of ξ, and the submatrix of Ω, respectively, restricted to the map variables. The matrix $\Omega_{m,x_{0:t}}$ is the off-diagonal submatrix of Ω that connects the robot path to the map. As noted before, $\Omega_{m,m}$ is block-diagonal, hence we can decompose

(11.42) $\quad p(m \mid x_{0:t}, z_{1:t}, u_{1:t}, c_{1:t}) = \prod_j p(m_j \mid x_{0:t}, z_{1:t}, u_{1:t}, c_{1:t})$

where each $p(m_j \mid x_{0:t}, z_{1:t}, u_{1:t}, c_{1:t})$ is distributed according to

(11.43) $\quad \Sigma_j = \Omega_{j,j}^{-1}$

(11.44) $\quad \mu_j = \Sigma_j(\xi_j + \Omega_{j,x_{0:t}}\tilde{\mu}) = \Sigma_j(\xi_j + \Omega_{j,\tau(j)}\tilde{\mu}_{\tau(j)})$

The last transformation exploited the fact that the submatrix $\Omega_{j,x_{0:t}}$ is zero except for those pose variables $\tau(j)$ from which the j-th feature was observed.

It is important to notice that this is a Gaussian $p(m \mid x_{0:t}, z_{1:t}, u_{1:t}, c_{1:t})$ conditioned on the true path $x_{0:t}$. In practice, we do not know the path, hence one might want to know the posterior $p(m \mid z_{1:t}, u_{1:t}, c_{1:t})$ without the path in the conditioning set. This Gaussian cannot be factored in the moments parameterization, as locations of different features are correlated through the uncertainty over the robot pose. For this reason, **GraphSLAM_solve** returns the mean estimate of the posterior but only the covariance over the robot path. Luckily, we never need the full Gaussian in moments representation—which would involve a fully populated covariance matrix of massive dimensions—as all essential questions pertaining to the SLAM problem can be answered at least in approximation without knowledge of Σ.

11.5 Data Association in GraphSLAM

Data association in GraphSLAM is realized through correspondence variables, just as in EKF SLAM. GraphSLAM searches for a single best correspondence vector, instead of calculating an entire distribution over correspondences. Thus, finding a correspondence vector is a search problem. However, it shall prove convenient to define correspondences slightly differently in GraphSLAM than before: correspondences are defined over pairs of features in the map, rather than associations of measurements to features. Specifically, we say $c(j,k) = 1$ if m_j and m_k correspond to the same physical feature in the world. Otherwise, $c(j,k) = 0$. This feature-correspondence is in fact logically equivalent to the correspondence defined in the previous section, but it simplifies the statement of the basic algorithm.

11.5 Data Association in GraphSLAM

The technique for searching the space of correspondences is greedy, just as in the EKF. Each step in the search of the best correspondence value leads to an improvement, as measured by the appropriate log-likelihood function. However, because GraphSLAM has access to all data at the same time, it is possible to devise correspondence techniques that are considerably more powerful than the incremental approach in the EKF. In particular:

1. At any point in the search, GraphSLAM can consider the correspondence of any set of features. There is no requirement to process the observed features sequentially.

2. Correspondence search can be combined with the calculation of the map. Assuming that two observed features correspond to the same physical feature in the world affects the resulting map. By incorporating such a correspondence hypothesis into the map, other correspondence hypotheses will subsequently look more or less likely.

3. Data association decisions in GraphSLAM can be *undone*. The goodness of a data association depends on the value of other data association decisions. What appears to be a good choice early on in the search may, at some later time in the search, turn out to be inferior. To accommodate such a situation, GraphSLAM can effectively undo a previous data association decision.

We will now describe one specific correspondence search algorithm that exploits the first two properties, but not the third. The data association algorithm will still be greedy, and it will sequentially search the space of possible correspondences to arrive at a plausible map. However, like all greedy algorithms, our approach is subject to local maxima; the true space of correspondences is of course exponential in the number of features in the map. Nevertheless, we will be content with a hill climbing algorithm and postpone the treatment of an exhaustive algorithm to the next chapter.

11.5.1 The GraphSLAM Algorithm with Unknown Correspondence

LIKELIHOOD TEST FOR CORRESPONDENCE

The key component of our algorithm is a *likelihood test for correspondence*. Specifically, GraphSLAM correspondence is based on a simple test: What is the probability that two different features in the map, m_j and m_k, correspond to the same physical feature in the world? If this probability exceeds a threshold, we will accept this hypothesis and merge both features in the map.

1:     **Algorithm GraphSLAM_correspondence_test($\Omega, \xi, \mu, \Sigma_{0:t}, j, k$):**

2:     $\Omega_{[j,k]} = \Omega_{jk,jk} - \Omega_{jk,\tau(j,k)} \Sigma_{\tau(j,k),\tau(j,k)} \Omega_{\tau(j,k),jk}$

3:     $\xi_{[j,k]} = \Omega_{[j,k]} \mu_{j,k}$

4:     $\Omega_{\Delta j,k} = \begin{pmatrix} 1 \\ -1 \end{pmatrix}^T \Omega_{[j,k]} \begin{pmatrix} 1 \\ -1 \end{pmatrix}$

5:     $\xi_{\Delta j,k} = \begin{pmatrix} 1 \\ -1 \end{pmatrix}^T \xi_{[j,k]}$

6:     $\mu_{\Delta j,k} = \Omega_{\Delta j,k}^{-1} \xi_{\Delta j,k}$

7:     **return** $|2\pi \, \Omega_{\Delta j,k}^{-1}|^{-\frac{1}{2}} \exp\left\{-\frac{1}{2} \mu_{\Delta j,k}^T \, \Omega_{\Delta j,k}^{-1} \, \mu_{\Delta j,k}\right\}$

Table 11.8 The GraphSLAM test for correspondence: It accepts as input an information representation of the SLAM posterior, along with the result of the **GraphSLAM_solve** step. It then outputs the posterior probability that m_j corresponds to m_k.

The algorithm for the correspondence test is depicted in Table 11.8: The input to the test are two feature indexes, j and k, for which we seek to compute the probability that those two features correspond to the same feature in the physical world. To calculate this probability, our algorithm utilizes a number of quantities: The information representation of the SLAM posterior, as manifest by Ω and ξ, and the result of the procedure **GraphSLAM_solve**, which is the mean vector μ and the path covariance $\Sigma_{0:t}$.

The correspondence test then proceeds in the following way: First, it computes the marginalized posterior over the two target features. This posterior is represented by the information matrix $\Omega_{[j,k]}$ and vector $\xi_{[j,k]}$ computed in lines 2 and 3 in Table 11.8. This step of the computation utilizes various sub-elements of the information form Ω, ξ, the mean feature locations as specified through μ, and the path covariance $\Sigma_{0:t}$. Next, it calculates the parameters of a new Gaussian random variable, whose value is the difference between m_j and m_k. Denoting the difference variable $\Delta_{j,k} = m_j - m_k$, the information parameters $\Omega_{\Delta j,k}, \xi_{\Delta j,k}$ are calculated in lines 4 and 5, and the correspond-

11.5 Data Association in GraphSLAM

1:     **Algorithm GraphSLAM**($u_{1:t}, z_{1:t}$):
2:         *initialize all c_t^i with a unique value*
3:         $\mu_{0:t} =$ **GraphSLAM_initialize**($u_{1:t}$)
4:         $\Omega, \xi =$ **GraphSLAM_linearize**($u_{1:t}, z_{1:t}, c_{1:t}, \mu_{0:t}$)
5:         $\tilde{\Omega}, \tilde{\xi} =$ **GraphSLAM_reduce**(Ω, ξ)
6:         $\mu, \Sigma_{0:t} =$ **GraphSLAM_solve**($\tilde{\Omega}, \tilde{\xi}, \Omega, \xi$)
7:         *repeat*
8:             *for each pair of non-corresponding features m_j, m_k do*
9:                 $\pi_{j=k} =$ **GraphSLAM_correspondence_test**
                                                  ($\Omega, \xi, \mu, \Sigma_{0:t}, j, k$)
10:             *if $\pi_{j=k} > \chi$ then*
11:                 *for all $c_t^i = k$ set $c_t^i = j$*
12:                 $\Omega, \xi =$ **GraphSLAM_linearize**($u_{1:t}, z_{1:t}, c_{1:t}, \mu_{0:t}$)
13:                 $\tilde{\Omega}, \tilde{\xi} =$ **GraphSLAM_reduce**(Ω, ξ)
14:                 $\mu, \Sigma_{0:t} =$ **GraphSLAM_solve**($\tilde{\Omega}, \tilde{\xi}, \Omega, \xi$)
15:             *endif*
16:         *endfor*
17:         *until no more pair m_j, m_k found with $\pi_{j=k} < \chi$*
18:         *return μ*

Table 11.9 The GraphSLAM algorithm for the full SLAM problem with unknown correspondence. The inner loop of this algorithm can be made more efficient by selective probing feature pairs m_j, m_k, and by collecting multiple correspondences before solving the resulting collapsed set of equations.

ing expectation for the difference is computed in line 6. Line 7 returns the probability that the difference between m_j and m_k is zero.

The correspondence test provides us with an algorithm for performing data association search in GraphSLAM. Table 11.9 shows such an algorithm. It initializes the correspondence variables with unique values. The four steps that follow (lines 3-7) are the same as in our GraphSLAM algorithm with known correspondence, stated in Table 11.5. However, this general SLAM algorithm then engages in the data association search. Specifically, for each pair of different features in the map, it calculates the probability of correspondence (line 9 in Table 11.9). If this probability exceeds a threshold χ, the correspondence vectors are set to the same value (line 11).

The GraphSLAM algorithm iterates the construction, reduction, and solution of the SLAM posterior (lines 12 through 14). As a result, subsequent correspondence tests factor in previous correspondence decisions though a newly constructed map. The map construction is terminated when no further features are found in its inner loop.

Clearly, the algorithm **GraphSLAM** is not particularly efficient. In particular, it tests all feature pairs for correspondence, not just nearby ones. Further, it reconstructs the map whenever a single correspondence is found; rather than processing sets of corresponding features in batch. Such modifications, however, are relatively straightforward. A good implementation of **GraphSLAM** will be more refined than our basic implementation discussed here.

11.5.2 Mathematical Derivation of the Correspondence Test

We essentially restrict our derivation to showing the correctness of the correspondence test in Table 11.8. Our first goal shall be to define a posterior probability distribution over a variable $\Delta_{j,k} = m_j - m_k$, the *difference* between the location of feature m_j and feature m_k. Two features m_j and m_k are equivalent if and only if their location is the same. Hence, by calculating the posterior probability of $\Delta_{j,k}$, we obtain the desired correspondence probability.

We obtain the posterior for $\Delta_{j,k}$ by first calculating the joint over m_j and m_k:

$$(11.45) \quad p(m_j, m_k \mid z_{1:t}, u_{1:t}, c_{1:t})$$
$$= \int p(m_j, m_k \mid x_{1:t}, z_{1:t}, c_{1:t}) \, p(x_{1:t} \mid z_{1:t}, u_{1:t}, c_{1:t}) \, dx_{1:t}$$

We will denote the information form of this marginal posterior by $\xi_{[j,k]}$ and $\Omega_{[j,k]}$. Note the use of the squared brackets, which distinguish these values from the submatrices of the joint information form.

The distribution (11.45) is obtained from the joint posterior over $y_{0:t}$, by applying the marginalization lemma. Specifically, Ω and ξ represent the joint posterior over the full state vector $y_{0:t}$ in information form, and $\tau(j)$ and $\tau(k)$

denote the sets of poses at which the robot observed feature j, and feature k, respectively. GraphSLAM gives us the mean pose vector $\tilde{\mu}$. To apply the marginalization lemma (Table 11.6), we shall leverage the result of the algorithm **GraphSLAM_solve**. Specifically, **GraphSLAM_solve** provides us already with a mean for the features m_j and m_k. We simply restate the computation here for the joint feature pair:

(11.46) $\quad \mu_{[j,k]} \;=\; \Omega_{jk,jk}^{-1} \left(\xi_{jk} + \Omega_{jk,\tau(j,k)} \mu_{\tau(j,k)} \right)$

Here $\tau(j,k) = \tau(j) \cup \tau(k)$ denotes the set of poses at which the robot observed m_j or m_k.

For the joint posterior, we also need a covariance. This covariance is *not* computed in **GraphSLAM_solve**, simply because the joint covariance over multiple features requires space quadratic in the number of features. However, for pairs of features the covariance of the joint is easily recovered.

Let $\Sigma_{\tau(j,k),\tau(j,k)}$ be the submatrix of the covariance $\Sigma_{0:t}$ restricted to all poses in $\tau(j,k)$. Here the covariance $\Sigma_{0:t}$ is calculated in line 2 of the algorithm **GraphSLAM_solve**. Then the marginalization lemma provides us with the marginal information matrix for the posterior over $(m_j \; m_k)^T$:

(11.47) $\quad \Omega_{[j,k]} \;=\; \Omega_{jk,jk} - \Omega_{jk,\tau(j,k)} \, \Sigma_{\tau(j,k),\tau(j,k)} \, \Omega_{\tau(j,k),jk}$

The information form representation for the desired posterior is now completed by the following information vector:

(11.48) $\quad \xi_{[j,k]} \;=\; \Omega_{[j,k]} \, \mu_{[j,k]}$

Hence we have for the joint

(11.49) $\quad p(m_j, m_k \mid z_{1:t}, u_{1:t}, c_{1:t})$

$$= \; \eta \, \exp \left\{ -\tfrac{1}{2} \begin{pmatrix} m_j \\ m_k \end{pmatrix}^T \Omega_{[j,k]} \begin{pmatrix} m_j \\ m_k \end{pmatrix} + \begin{pmatrix} m_j \\ m_k \end{pmatrix}^T \xi_{[j,k]} \right\}$$

These equations are identical to lines 2 and 3 in Table 11.8.

The nice thing about our representation is that it immediately lets us define the desired correspondence probability. For that, let us consider the random variable

(11.50) $\quad \Delta_{j,k} \;=\; m_j - m_k$

$$= \; \begin{pmatrix} 1 \\ -1 \end{pmatrix}^T \begin{pmatrix} m_j \\ m_k \end{pmatrix}$$

$$= \; \begin{pmatrix} m_j \\ m_k \end{pmatrix}^T \begin{pmatrix} 1 \\ -1 \end{pmatrix}$$

Plugging this into the definition of a Gaussian in information representation, we obtain:

$$(11.51) \quad p(\Delta_{j,k} \mid z_{1:t}, u_{1:t}, c_{1:t})$$

$$= \eta \exp\left\{ -\tfrac{1}{2} \Delta_{j,k}^T \underbrace{\begin{pmatrix} 1 \\ -1 \end{pmatrix}^T \Omega_{[j,k]} \begin{pmatrix} 1 \\ -1 \end{pmatrix}}_{=:\, \Omega_{\Delta j,k}} \Delta_{j,k} + \Delta_{j,k}^T \underbrace{\begin{pmatrix} 1 \\ -1 \end{pmatrix}^T \xi_{[j,k]}}_{=:\, \xi_{\Delta j,k}} \right\}$$

$$= \eta \exp\left\{ -\tfrac{1}{2} \Delta_{j,k}^T \Omega_{\Delta j,k} + \Delta_{j,k}^T \xi_{\Delta j,k} \right\}^T$$

which is Gaussian with information matrix $\Omega_{\Delta j,k}$ and information vector $\xi_{\Delta j,k}$ as defined above. To calculate the probability that this Gaussian assumes the value of $\Delta_{j,k} = 0$, it shall be useful to rewrite this Gaussian in moments parameterization:

$$(11.52) \quad p(\Delta_{j,k} \mid z_{1:t}, u_{1:t}, c_{1:t})$$

$$= |2\pi \, \Omega_{\Delta j,k}^{-1}|^{-\tfrac{1}{2}} \exp\left\{ -\tfrac{1}{2} (\Delta_{j,k} - \mu_{\Delta j,k})^T \Omega_{\Delta j,k}^{-1} (\Delta_{j,k} - \mu_{\Delta j,k}) \right\}$$

where the mean is given by the obvious expression:

$$(11.53) \quad \mu_{\Delta j,k} = \Omega_{\Delta j,k}^{-1} \xi_{\Delta j,k}$$

These steps are found in lines 4 through 6 in Table 11.8.

The desired probability for $\Delta_{j,k} = 0$ is the result of plugging 0 into this distribution, and reading off the resulting probability:

$$(11.54) \quad p(\Delta_{j,k} = 0 \mid z_{1:t}, u_{1:t}, c_{1:t}) = |2\pi \, \Omega_{\Delta j,k}^{-1}|^{-\tfrac{1}{2}} \exp\left\{ -\tfrac{1}{2} \mu_{\Delta j,k}^T \Omega_{\Delta j,k}^{-1} \mu_{\Delta j,k} \right\}$$

This expression is the probability that two features in the map, m_j and m_k, correspond to the same features in the map. This calculation is implemented in line 7 in Table 11.8.

11.6 Efficiency Consideration

Practical implementations of GraphSLAM rely on a number of additional insights and techniques for improving efficiency. Possibly the biggest deficiency of GraphSLAM, as discussed thus far, is due to the fact that in the very beginning, we assume that all observed features constitute different features. Our algorithm unifies them one-by-one. For any reasonable number of features, such an approach will be unbearably slow. Further, it will neglect the important constraint that at any point in time, the same feature can only be observed once, but not twice.

Existing implementations of the GraphSLAM idea exploit such opportunities.

Features that are immediately identified to correspond with high likelihood are often unified form the very beginning, before running the full GraphSLAM solution. For example, it is quite common to compile short segments into *local submaps*, e.g., local occupancy grid maps. GraphSLAM inference is then performed only between those local occupancy grid maps, where the match of two maps is taken as a probabilistic constraint between the relative poses of these maps. Such a hierarchical technique reduces the complexity of SLAM by orders of magnitude, while still retaining some of the key elements of GraphSLAM, specifically the ability to perform data association over large data sets.

LOCAL SUBMAPS

Many robots are equipped with sensors that observe large number of features at a time. For example, laser range finders observe dozens of features within a single scan. For any such scan, one commonly assumes that different measurements indeed correspond to different features in the environment, by virtue of the fact that each scan points in a different direction. This is known as the *mutual exclusion principle*, which was already discussed in Chapter 7. It therefore follows that $i \neq j \longrightarrow c_t^i \neq c_t^j$. No two measurements acquired within a single scan correspond to the same feature in the world.

Our pairwise data association technique above is unable to incorporate this constraint. Specifically, it may assign two measurements z_t^i and z_t^j to the same feature z_s^k for some $s \neq t$. To overcome this problem, it is common to associate entire measurement vectors z_t and z_s at the same time. This involves a calculation of a joint over all features in z_t and z_s. Such a calculation generalizes our pairwise calculation and is mathematically straightforward.

The GraphSLAM algorithm stated in this chapter does not make use of its ability to *undo a data association*. Once a data association decision is made, it cannot be reverted further down in the search. Mathematically, it is relatively straightforward to undo past data association decisions in the information framework. One can change the correspondence variables of any two measurements in arbitrary ways in our algorithm above. However, it is more difficult to test whether a data association should be undone, as there is no (obvious) test for testing whether two previously associated features should be distinct. A simple implementation involves undoing a data association in question, rebuilding the map, and testing whether our criterion above still calls for correspondence. Such an approach can be computation-

Figure 11.4 The Groundhog robot is a 1,500 pound custom-built vehicle equipped with onboard computing, laser range sensing, gas and sinkage sensors, and video recording equipment. The robot has been built to map abandoned mines.

ally involved, as it provides no means of detecting which data association to test. Mechanisms for detecting unlikely associations are outside the scope of this book, but should be considered when implementing this approach.

NEGATIVE INFORMATION

Finally, the GraphSLAM algorithm does not consider *negative information*. In practice, not seeing a feature can be as informative as seeing one. However, our simple formulation does not perform the necessary geometric computations.

In practice, whether or not we can exploit negative information depends on the nature of the sensor model, and the model of our features in the world. For example, we might have to compute probabilities of occlusion, which might be tricky for certain type sensors (e.g., range and bearing sensors for landmarks). However, contemporary implementations indeed consider negative information, but often by replacing proper probabilistic calculations through approximations. One such example will be given in the next section.

11.7 Empirical Implementation

We will now highlight empirical results for a GraphSLAM implementation. The vehicle used in our experiment is Figure 11.4; it is a robot designed to map abandoned mines.

The type map collected by the robot is shown in Figure 11.5. This map is an occupancy grid map, using effectively pairwise scan matching for recovering the robot's poses. Pairwise scan matching can be thought of as a version of

11.7 Empirical Implementation

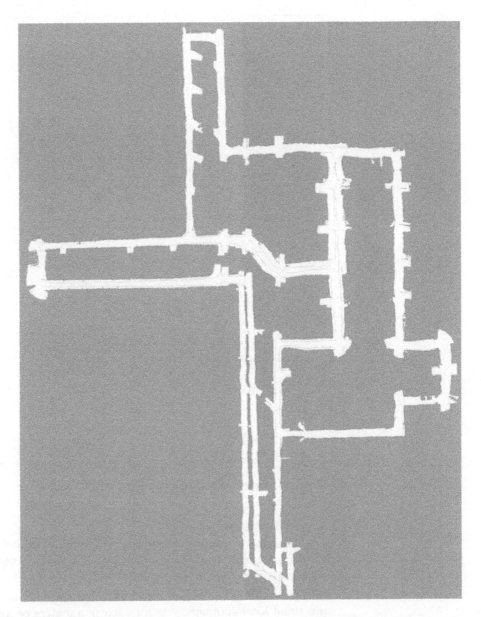

Figure 11.5 Map of a mine, acquired by pairwise scan matching. The diameter of this environment is approximately 250 meters. The map is obviously inconsistent, in that several hallways show up more than once. Image courtesy of Dirk Hähnel, University of Freiburg.

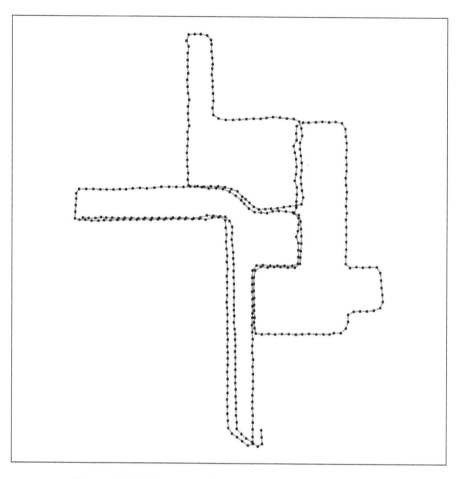

Figure 11.6 Mine map skeleton, visualizing the local maps.

GraphSLAM, but correspondence is only established between immediately consecutive scans. The result of this approach leads to an obvious deficiency of the map shown in Figure 11.5.

To apply the GraphSLAM algorithm, our software decomposes the map into small local submaps, one for each five meters of robot travel. Within these five meters, the maps are sufficiently accurate, as general drift is small and hence scan matching performs essentially flawlessly. Each submap's coordinates become a pose node in the GraphSLAM. Adjacent submaps are linked through the relative motion constraints between them. The resulting

11.7 Empirical Implementation

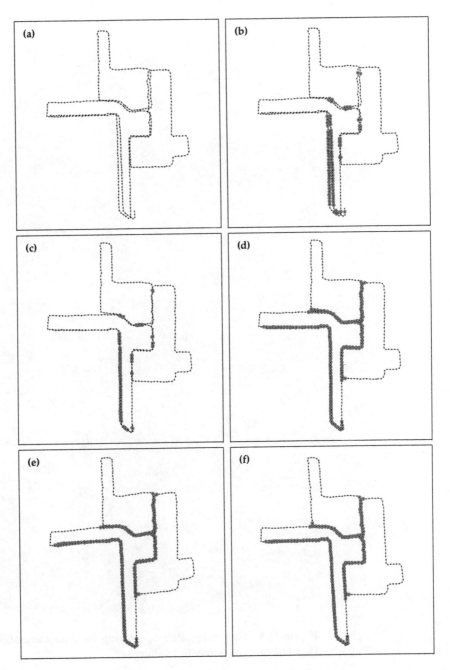

Figure 11.7 Data association search. See text.

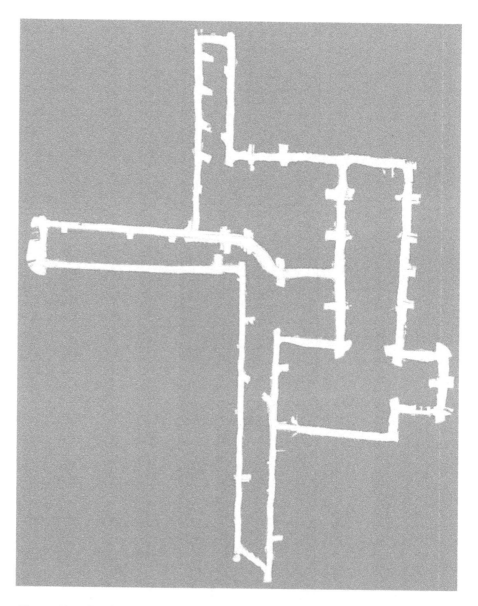

Figure 11.8 Final map, after optimizing for data associations. Image courtesy of Dirk Hähnel, University of Freiburg.

11.7 Empirical Implementation

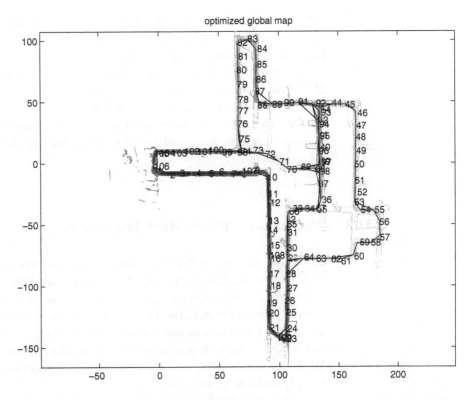

Figure 11.9 Mine map generated by the *Atlas* SLAM algorithm by Bosse et al. (2004). Image courtesy of Michael Bosse, Paul Newman, John Leonard, and Seth Teller, MIT.

structure is shown in Figure 11.6.

Next, we apply the recursive data association search. The correspondence test is now implemented using a correlation analysis for two overlaying maps, and the Gaussian matching constraints are recovered by approximating this match function through a Gaussian. Figure 11.7 illustrates the process of data association: The circles each correspond to a new constraint that is imposed when constructing the information form with GraphSLAM. This figure illustrates the iterative nature of the search: Certain correspondences are only discovered when others have been propagated, and others are dissolved in the process of the search. The final model is stable, in that additional search for new data association induces no further changes. Displayed as a grid map, it yields the 2-D map shown in Figure 11.8. While this map is far from being perfect—largely due to a crude implementation of the local map matching constraints—it nevertheless is superior to the one found through incremental scan matching.

The reader should notice that other information-theoretic techniques for SLAM have produced similar results. Figure 11.9 shows a map of the same data set generated by Bosse et al. (2004), using an algorithm called *Atlas*. This algorithm decomposes maps into submaps whose relation is maintained through information-theoretic relative links. See the bibliographical remarks for more detail.

ATLAS

11.8 Alternative Optimization Techniques

The reader may recall that the central target function $J_{\text{GraphSLAM}}$ in GraphSLAM is the nonlinear quadratic function in Equation (11.3). GraphSLAM minimizes this function through a sequence of linearizations, variable eliminations, and optimizations. The inference technique in **GraphSLAM_solve** in Table 11.4 is generally not very efficient. If all one is interested in is a map and a path without covariances, the calculation of the inverse in line 2 in Table 11.4 can be avoided. The resulting implementation will be computationally much more efficient.

The key to efficient inference lies in the form of the function $J_{\text{GraphSLAM}}$. This function is of a general least squares form, and hence can be minimized by a number of different algorithms in the literature. Examples include *gradient descent techniques*, *Levenberg Marquardt*, and *conjugate gradient*.

CONJUGATE GRADIENT

Figure 11.10a shows results obtained using *conjugate gradient* for minimizing $J_{\text{GraphSLAM}}$. The data is a map of an outdoor environment of the approximate size 600 by 800 meters, collected on Stanford's campus with the robot shown in Figure 11.10b. Figure 11.11 illustrates the alignment process, from a data set that is based on pose data only, to a fully aligned map and robot path. This dataset contains approximately 10^8 features and 10^5 poses. Running an EKF would be infeasible on such a large data set. As would be inverting the matrix Ω in Table 11.4. Conjugate gradient required only a few seconds to minimize $J_{\text{GraphSLAM}}$. For this reason, many contemporary implementations of this approach use modern optimization techniques, instead of the relatively slow algorithm discussed here. The interested reader shall be referred to the bibliographical remarks for more pointers to alternative optimization techniques.

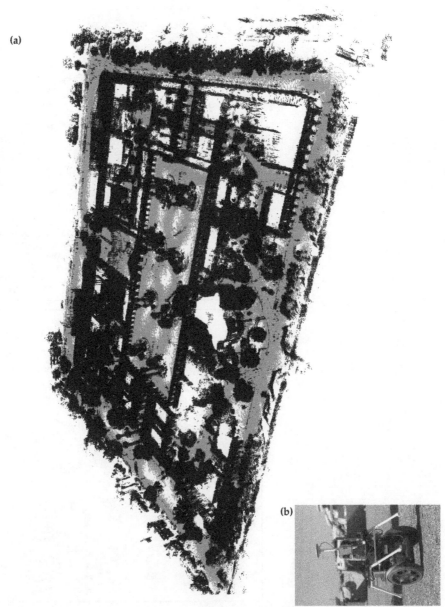

Figure 11.10 (a) A 3-D map of Stanford's campus. (b) The robot used for acquiring this data is based on a Segway RMP platform, whose development was funded by the DARPA MARS program. Image courtesy of Michael Montemerlo, Stanford University. *(Turn this page 90 degrees to the right to view this figure.)*

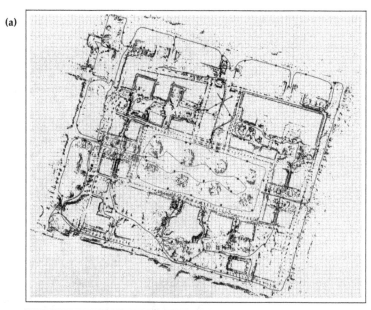

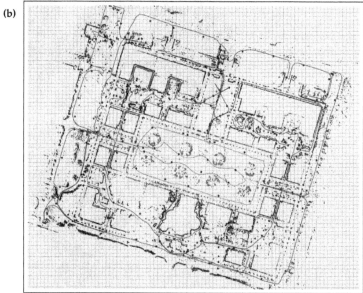

Figure 11.11 2-D slice through the Stanford campus map (a) before and (b) after alignment using conjugate gradient. Such an optimization takes only a few seconds with the conjugate gradient method applied to the least square formulation of Graph-SLAM. Images courtesy of Michael Montemerlo, Stanford University.

11.9 Summary

This chapter introduced the GraphSLAM algorithm to the full SLAM problem.

- GraphSLAM algorithm addresses the full SLAM problem. It calculates posteriors over the full robot path along with the map. Therefore, GraphSLAM is a batch algorithm, not an online algorithm like EKF SLAM.

- GraphSLAM constructs a graph of soft constraints from the data set. In particular, measurements are mapped into edges that represent nonlinear constraints between poses and sensed features, and motion commands map into soft constraints between consecutive poses. The graph is naturally sparse. The number of edges is a linear function of the number of nodes, and each node is only connected to finitely many other nodes, regardless of the size of the graph.

  GraphSLAM simply records all this information in the graph, through links that are defined between poses and features, and pairs of subsequent poses. However, this information representation does not provide estimates of the map or of the robot path.

- The sum of all such constraints is given by a function $J_{\text{GraphSLAM}}$. The maximum likelihood estimates for the robot path and the map can be obtained by minimizing this function $J_{\text{GraphSLAM}}$.

- GraphSLAM performs inference by mapping the graph into an isomorphic information matrix and information vector, defined over all pose variables and the entire map. The key insight of the GraphSLAM algorithm is that the structure of information is sparse. Measurements provide information of a feature relative to the robot's pose at the time of measurement. In information space, they form constraints between these pairs of variables. Similarly, motion provides information between two subsequent poses. In information space, each motion command forms a constraint between subsequent pose variables. This sparseness is inherited from the sparse graph.

- The vanilla GraphSLAM algorithm recovers maps through an iterative procedure that involves three steps: Construction of a linear information form through Taylor expansion, reduction of this form to remove the map, and solving the resulting optimization problem over robot poses. These three steps effectively resolve the information, and produce a consistent probabilistic posterior over the path and the map. Since GraphSLAM is run batch, we can repeat the linearization step to improve the result.

- Alternative implementations perform inference through nonlinear least squares optimization of the function $J_{\text{GraphSLAM}}$. However, such techniques only find the mode of the posterior, not its covariance.

- Data association in GraphSLAM is performed by calculating the probability that two features have identical world coordinates. Since GraphSLAM is a batch algorithm, this can be done for any pair of features, at any time. This led to an iterative greedy search algorithm over all data association variables, which recursively identifies pairs of features in the map that likely correspond.

- Practical implementations of the GraphSLAM often use additional tricks to keep the computation low and to avoid false data associations. Specifically, practical implementations tend to reduce the data complexity by extracting local maps and using each map as the basic entity; they tend to match multiple features at-a-time, and they tend to consider negative information in the data association search.

- We briefly provided results for a variant of GraphSLAM that follows the decomposition idea, but uses occupancy grid maps for representing sets of range scans. Despite these approximations, we find that data association and inference techniques yield favorable results in a large-scale mapping problem.

- Results were also provided for a conjugate gradient implementation to the underlying least square problem. We noted that the general target function of GraphSLAM can be optimized by any least squares technique. Certain techniques, such as conjugate gradient, are significantly faster than the basic optimization technique in GraphSLAM.

As noted in the introduction, EKF SLAM and GraphSLAM are extreme ends of a spectrum of SLAM algorithms. Algorithms that fall in between these extremes will be discussed in the next two chapters. References to further techniques can be found in the bibliographical remarks of the next few chapters.

11.10 Bibliographical Remarks

Graphical inference techniques are well known in computer vision and photogrammetry, and are related to *structure from motion* and *bundle adjustment* (Hartley and Zisserman 2000; B et al. 2000; Mikhail et al. 2001). The first mention of relative, graph-like constraints in the SLAM literature goes back to Cheeseman and Smith (1986) and Durrant-Whyte (1988), but these approaches did not perform any global relaxation, or optimization. The algorithm presented in this chapter is loosely based on a seminal paper by Lu and Milios (1997). They were historically the first to represent the SLAM prior as a set of links between robot poses, and to formulate a global optimization algorithm for generating a map from such constraints. Their original algorithm for globally consistent range scan alignment used the robot pose variables as the frame of reference, which differed from the standard EKF view in which poses were integrated out. Through analyzing odometry and laser range scans, their approach generated relative constraints between poses that can be viewed as the edges in GraphSLAM; however, they did not phrase their method using information representations. Lu and Milios's (1997) algorithm was first successfully implemented by Gutmann and Nebel (1997), who reported numerical instabilities, possibly due to the extensive use of matrix inversion. Golfarelli et al. (1998) were the first to establish the relation of SLAM problems and spring-mass models, and Duckett et al. (2000, 2002) provided a first efficient technique for solving such problems. The relation between covariances and the information matrix is discussed in Frese and Hirzinger (2001). Araneda (2003) developed a more detailed elaborate graphical model.

The Lu and Milios algorithm initiated a development of offline SLAM algorithms that up to the present date runs largely parallel to the EKF work. Gutmann and Konolige combined their implementation with a Markov localization step for establishing correspondence when closing a loop in a cyclic environment. Bosse et al. (2003, 2004) developed Atlas, which is a hierarchical mapping framework based on the decoupled stochastic mapping paradigm, which retains relative information between submaps. It uses an optimization technique similar to the one in Duckett et al. (2000) and GraphSLAM when aligning multiple submaps. Folkesson and Christensen (2004a,b) exploited the optimization perspective of SLAM by applying gradient descent to the log-likelihood version of the SLAM posterior. Their *Graphical SLAM* algorithm reduced the number of variables to the path variables—just like GraphSLAM—when closing the loop. This reduction (which is mathematically an approximation since the map is simply omitted) significantly accelerated gradient descent. Konolige (2004) and Montemerlo and Thrun (2004) introduced *conjugate gradient* into the field of SLAM, which is known to be more efficient than gradient descent. Both also reduced the number of variables when closing large cycles, and report that maps with 10^8 features can be aligned in just a few seconds. The Levenberg Marquardt technique mentioned in the text is due to Levenberg (1944) and Marquardt (1963), who devised it in the context of least squares optimization. Frese et al. (2005) analyzed the efficiency of SLAM in the information form, and developed highly efficient optimization techniques using multi-grid optimization techniques. He reports speed-ups of several orders of magnitude; the resulting optimization techniques are presently the state-of-the-art. Dellaert (2005) developed efficient factorization techniques for the GraphSLAM constraint graph, specifically aimed at transforming the constraint graph into more compact versions while retaining sparseness.

It should be mentioned that the intuition to maintain relative links between local entities is at the core of many of the submapping techniques discussed in the previous section—although it is rarely made explicit. Authors such as Guivant and Nebot (2001); Williams (2001); Tardós et al. (2002); Bailey (2002) report of data structures for minuting the relative displacement between submaps, which are easily mapped to information theoretic concepts. While many of these algorithms are filters, they nevertheless share a good amount of insight with the information form discussed in this chapter.

To our knowledge, the GraphSLAM algorithm presented here has never been published in the present form (an earlier draft of this book referred to this algorithm as *extended information form*). However, GraphSLAM is closely tied to the literature reviewed above, building on Lu and Milios's (1997) seminal algorithm. The name *GraphSLAM* bears resemblance to the name *Graphical SLAM* by Folkesson and Christensen (2004a); we have chosen it for this chapter because graphs of constraints are the essence of this entire line of SLAM research. A number of authors have developed *filters* in information form, which address the online SLAM problem instead of the full SLAM problem. These algorithms will be discussed in the coming chapter, which explicitly addresses the problem of filtering.

The GraphSLAM formulation of the SLAM problem relates to a decades-old discussion in the representation of spatial maps, which was already mentioned in the bibliographical remarks to Chapter 9. Information representations bring together two different paradigms of map representation: topological and metric. Behind this distinction is a decades-old debate about the representation of space, in people and for robots (Chown et al. 1995). Topological approaches were already discussed in the bibliographical remarks to Chapter 9. A key feature of topological representation pertains to the fact that they only specify *relative* information between entities in the map. Hence, they are free of the problem of finding a consistent metric embedding of this relative information. State-of-the-art topological tend to augment links with metric information, such as the distance between two locations.

Just like topological representations, information-theoretic methods accumulate relative information between adjacent objects (landmarks and robots). However, the relative map information is "translated" into metric embeddings by inference. In the linear Gaussian case, this inference step is loss-free and invertible. Computing the full posterior, including the covariance, requires matrix inversion. A second matrix inversion operation leads back to the original form of relative constraints. Thus, both the topological view and the metric view appear to be duals of each other, just like information-theoretic and probabilistic representations (or EKF and GraphSLAM). So maybe there shall be a unifying mathematical framework that embraces both, topological and metric maps? The reader shall be forewarned that this view has not yet been adopted by the mainstream research community.

11.11 Exercises

1. We already encountered *bearing only SLAM* in the exercises to the previous chapter, as a form of SLAM in which sensors can only measure the bearing of landmarks but not the range. We conjecture that GraphSLAM is better suited for this problem than the EKF. Why?

2. In this question, you are asked to prove convergence results for a special class of SLAM problems: *linear Gaussian SLAM*. In linear Gaussian SLAM, the motion equation is of the simple additive type

LINEAR GAUSSIAN SLAM

$$x_t \sim \mathcal{N}(x_{t-1} + u_t, R)$$

and the measurement equation is of the type

$$z_t = \mathcal{N}(m_j - x_t, Q)$$

where R and Q are diagonal covariances matrices, and m_j is the feature observed at time t. You may assume that the number of landmarks is finite, that all landmarks are seen infinitely often and in no specific order, and that the correspondence is known.

(a) Prove that for GraphSLAM the distance between any two landmarks converges to the correct distance with probability 1.

(b) What does this proof entail for EKF SLAM?

(c) Does GraphSLAM converge for the general SLAM problem with known correspondence? If so, argue why; if not, argue why not (without proof).

3. The algorithm **GraphSLAM_reduce** reduces the set of constraints by integrating out the map variables, leaving a constraint system over robot poses only. Is it possible to instead integrate out the pose variables, so that the resulting network of constraints is defined only over map variables? If so, what would the resulting inference problem be sparse? How would cycles in the robot's path affect this new set of constraints?

4. The GraphSLAM algorithm in this chapter ignored the landmark signatures. Extend the basic GraphSLAM algorithm to utilize such signatures in its measurements and in its map.

12 The Sparse Extended Information Filter

12.1 Introduction

The previous two chapters covered two extreme ends of a spectrum of SLAM algorithms. We already noted that EKF SLAM is *proactive*. Every time information is acquired, it resolves this information into a probability distribution—which is computationally expensive. The GraphSLAM algorithm is different: It simply accumulates information. We noted that such an accumulation is *lazy*: at the time of data acquisition, GraphSLAM simply memorizes the information it receives. To turn the accumulated information into a map, GraphSLAM performs inference. This inference is performed after all data is acquired. This makes GraphSLAM an offline algorithm.

This raises the question as to whether we can devise an *online* filter algorithm that inherits the efficiency of the information representation. The answer is yes, but only with a number of approximations. The *sparse extended information filter*, or *SEIF*, implements an information solution to the online SLAM problem. Just like the EKF, the SEIF integrates out past robot poses, and only maintains a posterior over the present robot pose and the map. But like GraphSLAM and unlike EKF SLAM, SEIF maintains an information representation of all knowledge. In doing so, updating the SEIF becomes a lazy information shifting operation, which is superior to the proactive probabilistic update of the EKF. Thus, the SEIF can be seen as the best of both worlds: It runs online, and it is computationally efficient.

As an online algorithm, the SEIF maintains a belief over the very same state vector as the EKF:

$$(12.1) \quad y_t = \begin{pmatrix} x_t \\ m \end{pmatrix}$$

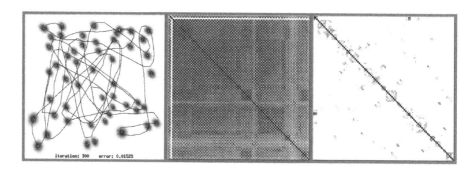

Figure 12.1 Motivation for using an information filter for online SLAM. Left: Simulated robot run with 50 landmarks. Center: The correlation matrix of an EKF, which shows strong correlations between any two landmarks' coordinates. Right: The normalized information matrix of the EKF is naturally sparse. This sparseness leads to a SLAM algorithm that can be updated more efficiently.

Here x_t is the robot state, and m the map. The posterior under known correspondences is given by $p(y_t \mid z_{1:t}, u_{1:t}, c_{1:t})$.

The key insight for turning GraphSLAM into an online SLAM algorithm is illustrated in Figure 12.1. This figure shows the result of the EKF SLAM algorithm in a simulated environment containing 50 landmarks. The left panel displays a moving robot along with its probabilistic estimate of the location of all 50 point features. The central information maintained by the EKF SLAM is a covariance matrix of these different estimates. The correlation, which is the normalized covariance, is visualized in the center panel of this figure. Each of the two axes lists the robot pose (location and orientation) followed by the 2-D locations of all 50 landmarks. Dark entries indicate strong correlations. We already discussed in the EKF SLAM chapter that in the limit, all feature coordinates become fully correlated—hence the checker-board appearance of the correlation matrix.

The right panel of Figure 12.1 shows the information matrix Ω_t, normalized just like the correlation matrix. As in the previous chapter, elements in this normalized information matrix can be thought of as constraints, or links, which constrain the relative locations of pairs of features in the map: The darker an entry in the display, the stronger the link. As this depiction suggests, the normalized information matrix appears to be *sparse*. It is dominated by a small number of strong links; and it possesses a large number of links whose values, when normalized, are practically zero.

12.1 Introduction

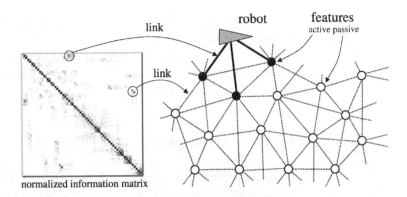

Figure 12.2 Illustration of the network of features generated by our approach. Shown on the left is a sparse information matrix, and on the right a map in which entities are linked whose information matrix element is non-zero. As argued in the text, the fact that not all features are connected is a key structural element of the SLAM problem, and at the heart of our constant time solution.

The strength of each link is related to the distance of the corresponding features: Strong links are found only between nearby features. The more distant two features, the weaker their link.

This sparseness is distinctly different from that in the previous chapter: First, there exist links between pairs of landmarks. In the previous chapter, no such links could exist. Second, the sparseness is only approximate: In fact, all elements of the normalized information matrix are non-zero, but nearly all of them are very close to zero.

SPARSE INFORMATION MATRIX

ACTIVE FEATURES

The SEIF SLAM algorithm exploits this insight by maintaining a sparse information matrix, in which only nearby features are linked through a non-zero element. The resulting network structure is illustrated in the right panel of Figure 12.2, where disks correspond to point features and dashed arcs to links, as specified in the information matrix visualized on the left. This diagram also shows the robot, which is linked to a small subset of all features only. Those features are called *active features* and are drawn in black. Storing a sparse information matrix requires space linear in the number of features in the map. More importantly, all essential updates in SEIF SLAM can be performed in constant time, regardless of the number of features in the map. This result is somewhat surprising, as a naive implementation of motion updates in information filters—as stated in Table 3.6 on page 76—requires inversion of the entire information matrix.

The SEIF is an online SLAM algorithm that maintains such a sparse information matrix, and for which the time required for all update steps is *independent* of the size of the map for the case with known data association, and logarithmic is data association search is involved. This makes SEIF the first efficient online SLAM algorithm encountered in this book.

12.2 Intuitive Description

We begin with an intuitive description of the SEIF update, using graphical illustrations. Specifically, a SEIF update is composed of 4 steps: a motion update step, a measurement update step, a sparsification step, and a state estimation step.

We begin with the measurement update step, depicted in Figure 12.3. Each of the two panels shows the information matrix maintained by the SEIF, along with the graph defined by the information links. Just as in GraphSLAM, sensing a feature m_1 leads the SEIF to update the off-diagonal element of its information matrix, which links the robot pose estimate x_t to the observed feature m_1. This is illustrated in the left panel of Figure 12.3a.

Sensing m_2 leads it to update the elements in the information matrix that link the robot pose x_t and the feature m_2, as illustrated in Figure 12.3b. As we shall see, each of these updates correspond to local additions in the information matrix and the information vector. In both cases (information matrix and vector), this addition touches only elements that link the robot pose variable to the observed feature. As in GraphSLAM, the complexity of incorporating a measurement into a SEIF takes time independent of the size of the map.

The motion update differs from GraphSLAM, since, as a filter, the SEIF eliminates past pose estimates. It is shown in Figure 12.4. Here a robot's pose changes; Figure 12.4a depicts a the information state before, and Figure 12.4b after motion, respectively. The motion affects the information state in multiple ways. First, the links between the robot's pose and the features m_1, m_2 are weakened. This is a result of the fact that robot motion introduces new uncertainty, hence causes us to lose information about where the robot is relative to the map. However, this information is not entirely lost. Some of it is mapped into information links between pairs of features. This shift of information comes about since even though we lost information on the robot pose, we did not lose information on the relative location of features in the map. Whereas previously, those features were linked indirectly through the robot pose, they are now linked also directly after the update step.

12.2 Intuitive Description

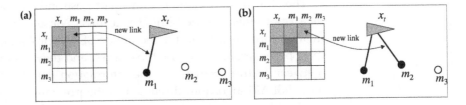

Figure 12.3 The effect of measurements on the information matrix and the associated network of features: (a) Observing m_1 results in a modification of the information matrix elements Ω_{x_t,m_1}. (b) Similarly, observing m_2 affects Ω_{x_t,m_2}.

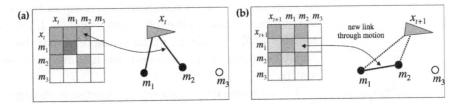

Figure 12.4 The effect of motion on the information matrix and the associated network of features: (a) before motion, and (b) after motion. If motion is non-deterministic, motion updates introduce new links (or reinforce existing links) between any two active features, while weakening the links between the robot and those features. This step introduces links between pairs of features.

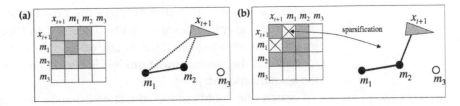

Figure 12.5 Sparsification: A feature is deactivated by eliminating its link to the robot. To compensate for this change in information state, links between active features and/or the robot are also updated. The entire operation can be performed in constant time.

The shift of information from robot pose links to between-feature links is a key element of the SEIF. It is a direct consequence of using the information form as a filter, for the online SLAM problem. By integrating out past pose variables, we lose those links, and they are mapped back into the between-feature elements in the information matrix. This differs from the GraphSLAM algorithm discussed in the previous chapter, which never introduced any links between pairs of features in the map.

For a pair of features to acquire a direct link in this process, both have to be active before the update, hence their corresponding elements linking them to the robot pose in the information matrix have to be non-zero. This is illustrated in Figure 12.4: A between-feature link is only introduced between features m_1 and m_2. Feature m_3, which is not active, remains untouched. This suggests that by controlling the number of active landmarks at any point in time, we can control the computational complexity of the motion update, and the number of links in the information matrix. If the number of active links remains small, so will the update complexity for the motion update, and so will the number of non-zero between-landmark elements in the information matrix.

SPARSIFICATION SEIF therefore employs a *sparsification* step, illustrated in Figure 12.5. The sparsification involves the removal of a link between the robot and an active feature, effectively turning the active feature into a passive one. In SEIFs, this arc removal leads to a redistribution of information into neighboring links, specifically between other active features and the robot pose. The time required for sparsification is independent of the size of the map. However, it is an approximation, one that induces an information loss in the robot's posterior. The benefit of this approximation is that it induces true sparseness, and hence makes it possible to update the filter efficiently.

There exists one final step in the SEIF algorithm, which is not depicted in any of the figures. This step involves the propagation of a mean estimate through the graph. As was already discussed in Chapter 3, the extended information filter requires an estimate of the state μ_t for linearization of the motion and the measurement model. SEIFs also require a state estimate for the sparsification step.

Clearly, one could recover the state estimate through the equation $\mu = \Omega^{-1}\xi$, where Ω is the information vector, and ξ the information state. However, this would require solving an inference problem that is too large to be RELAXATION run at each time step. SEIFs circumvent the step by an iterative *relaxation al-*ALGORITHM *gorithm* that propagates state estimates through the information graph. Each local state estimate is updated based on the best estimates of its neighbors in the information graph. This relaxation algorithm converges to the true mean μ. Since the information form is sparse in SEIFs, each such update requires constant time, though with the caveat that more than a finite number of such updates may be needed to achieve good results. To keep the computation independent of the size of the state space, SEIFs perform a fixed number of such updates at any iteration. The resulting state vector is only an approximation, which is used instead of the correct mean estimate in all updating steps.

12.3 The SEIF SLAM Algorithm

The outer loop of the SEIF update is depicted in Table 12.1. The algorithm accepts as input an information matrix Ω_{t-1}, an information vector ξ_{t-1}, and an estimate of the state μ_{t-1}. It also accepts a measurement z_t, a control u_t, and a correspondence vector c_t. The output of the algorithm **SEIF_SLAM_known_correspondences** is a new state estimate, represented by the information matrix Ω_t and the information vector ξ_t. The algorithm also outputs an improved estimate μ_t.

As stated in Table 12.1, the SEIF update proceeds in four major steps. The motion update in Table 12.2 incorporates the control u_t into the filter estimate. It does so through a number of computationally efficient operations. Specifically, the only components of the information vector/matrix that are modified in this update are those of the robot pose and the active features. The measurement update in Table 12.3 incorporates the measurement vector z_t under known correspondence c_t. This step is also local, just like the motion update step. It only updates the information values of the robot pose and the observed features in the map. The sparsification step, shown in Table 12.4, is an approximate step: It removes active features by transforming the information matrix and the information vector accordingly. This step is again efficient; it only modifies links between the robot and the active landmarks. Finally, the state estimate update in Table 12.5, applies an amortized coordinate descent technique to recover the state estimate μ_t. This step once again exploits the sparseness of the SEIF, through which it only has to consult a small number of other state vector elements in each incremental update.

Together, the entire update loop of the SEIF is constant time, in that the processing time is independent of the size of the map. This is in stark contrast to the only other online SLAM algorithm discussed so far—the EKF—which requires time quadratic the size of the map for each update. However, such

1:         **Algorithm SEIF_SLAM_known_correspondences**$(\xi_{t-1}, \Omega_{t-1}, \mu_{t-1}, u_t, z_t, c_t)$:
2:         $\bar{\xi}_t, \bar{\Omega}_t, \bar{\mu}_t = \textbf{SEIF\_motion\_update}(\xi_{t-1}, \Omega_{t-1}, \mu_{t-1}, u_t)$
3:         $\mu_t = \textbf{SEIF\_update\_state\_estimate}(\bar{\xi}_t, \bar{\Omega}_t, \bar{\mu}_t)$
4:         $\xi_t, \Omega_t = \textbf{SEIF\_measurement\_update}(\bar{\xi}_t, \bar{\Omega}_t, \mu_t, z_t, c_t)$
5:         $\tilde{\xi}_t, \tilde{\Omega}_t = \textbf{SEIF\_sparsification}(\xi_t, \Omega_t)$
6:         return $\tilde{\xi}_t, \tilde{\Omega}_t, \mu_t$

Table 12.1 The Sparse Extended Information Filter algorithm for the SLAM Problem, here with known data association.

1:         **Algorithm SEIF_motion_update**$(\xi_{t-1}, \Omega_{t-1}, \mu_{t-1}, u_t)$:

2:         $F_x = \begin{pmatrix} 1 & 0 & 0 & 0 \cdots 0 \\ 0 & 1 & 0 & 0 \cdots 0 \\ 0 & 0 & 1 & \underbrace{0 \cdots 0}_{3N} \end{pmatrix}$

3:         $\delta = \begin{pmatrix} -\frac{v_t}{\omega_t}\sin\mu_{t-1,\theta} + \frac{v_t}{\omega_t}\sin(\mu_{t-1,\theta} + \omega_t \Delta t) \\ \frac{v_t}{\omega_t}\cos\mu_{t-1,\theta} - \frac{v_t}{\omega_t}\cos(\mu_{t-1,\theta} + \omega_t \Delta t) \\ \omega_t \Delta t \end{pmatrix}$

4:         $\Delta = \begin{pmatrix} 0 & 0 & \frac{v_t}{\omega_t}\cos\mu_{t-1,\theta} - \frac{v_t}{\omega_t}\cos(\mu_{t-1,\theta} + \omega_t \Delta t) \\ 0 & 0 & \frac{v_t}{\omega_t}\sin\mu_{t-1,\theta} - \frac{v_t}{\omega_t}\sin(\mu_{t-1,\theta} + \omega_t \Delta t) \\ 0 & 0 & 0 \end{pmatrix}$

5:         $\Psi_t = F_x^T \left[(I + \Delta)^{-1} - I \right] F_x$
6:         $\lambda_t = \Psi_t^T \, \Omega_{t-1} + \Omega_{t-1} \, \Psi_t + \Psi_t^T \, \Omega_{t-1} \, \Psi_t$
7:         $\Phi_t = \Omega_{t-1} + \lambda_t$
8:         $\kappa_t = \Phi_t \, F_x^T (R_t^{-1} + F_x \, \Phi_t \, F_x^T)^{-1} F_x \, \Phi_t$
9:         $\bar{\Omega}_t = \Phi_t - \kappa_t$
10:      $\bar{\xi}_t = \xi_{t-1} + (\lambda_t - \kappa_t) \, \mu_{t-1} + \bar{\Omega}_t \, F_x^T \, \delta_t$
11:      $\bar{\mu}_t = \mu_{t-1} + F_x^T \, \delta$
12:      return $\bar{\xi}_t, \bar{\Omega}_t, \bar{\mu}_t$

Table 12.2 The motion update in SEIFs.

1: **Algorithm SEIF_measurement_update($\bar{\xi}_t, \bar{\Omega}_t, \mu_t, z_t, c_t$):**

2: $\quad Q_t = \begin{pmatrix} \sigma_r & 0 & 0 \\ 0 & \sigma_\phi & 0 \\ 0 & 0 & \sigma_s \end{pmatrix}$

3: $\quad$ for all observed features $z_t^i = (r_t^i \;\; \phi_t^i \;\; s_t^i)^T$ do

4: $\quad\quad j = c_t^i$

5: $\quad\quad$ if landmark j never seen before

6: $\quad\quad \begin{pmatrix} \mu_{j,x} \\ \mu_{j,y} \\ \mu_{j,s} \end{pmatrix} = \begin{pmatrix} \mu_{t,x} \\ \mu_{t,y} \\ s_t^i \end{pmatrix} + r_t^i \begin{pmatrix} \cos(\phi_t^i + \mu_{t,\theta}) \\ \sin(\phi_t^i + \mu_{t,\theta}) \\ 0 \end{pmatrix}$

7: $\quad\quad$ endif

8: $\quad\quad \delta = \begin{pmatrix} \delta_x \\ \delta_y \end{pmatrix} = \begin{pmatrix} \mu_{j,x} - \mu_{t,x} \\ \mu_{j,y} - \mu_{t,y} \end{pmatrix}$

9: $\quad\quad q = \delta^T \delta$

10: $\quad\quad \hat{z}_t^i = \begin{pmatrix} \sqrt{q} \\ \operatorname{atan2}(\delta_y, \delta_x) - \mu_{t,\theta} \\ \mu_{j,s} \end{pmatrix}$

11: $\quad\quad H_t^i = \frac{1}{q} \begin{pmatrix} \sqrt{q}\delta_x & -\sqrt{q}\delta_y & 0 & 0\cdots 0 & -\sqrt{q}\delta_x & \sqrt{q}\delta_y & 0 & 0\cdots 0 \\ \delta_y & \delta_x & -1 & 0\cdots 0 & -\delta_y & -\delta_x & 0 & 0\cdots 0 \\ 0 & 0 & 0 & \underbrace{0\cdots 0}_{3j-3} & 0 & 0 & 1 & \underbrace{0\cdots 0}_{3j} \end{pmatrix}$

12: $\quad$ endfor

13: $\quad \xi_t = \bar{\xi}_t + \sum_i H_t^{iT} \, Q_t^{-1} \, [z_t^i - \hat{z}_t^i - H_t^i \, \mu_t]$

14: $\quad \Omega_t = \bar{\Omega}_t + \sum_i H_t^{iT} \, Q_t^{-1} \, H_t^i$

15: $\quad$ return ξ_t, Ω_t

Table 12.3 The measurement update step in SEIFs.

1:   **Algorithm SEIF_sparsification(ξ_t, Ω_t):**

2:   define F_{m_0}, F_{x,m_0}, F_x as projection matrices
     from y_t to m_0, $\{x, m_0\}$, and x, respectively

3:   $\tilde{\Omega}_t = \Omega_t - \Omega_t^0 \, F_{m_0} \, (F_{m_0}^T \, \Omega_t^0 \, F_{m_0})^{-1} \, F_{m_0}^T \, \Omega_t^0$
     $\quad + \Omega_t^0 \, F_{x,m_0} \, (F_{x,m_0}^T \, \Omega_t^0 \, F_{x,m_0})^{-1} \, F_{x,m_0}^T \, \Omega_t^0$
     $\quad - \Omega_t \, F_x \, (F_x^T \, \Omega_t F_x)^{-1} \, F_x^T \, \Omega_t$

4:   $\tilde{\xi}_t = \xi_t + \mu_t \, (\tilde{\Omega}_t - \Omega_t)$

5:   return $\tilde{\xi}_t, \tilde{\Omega}_t$

Table 12.4 The sparsification step in SEIFs.

1:   **Algorithm SEIF_update_state_estimate($\bar{\xi}_t, \bar{\Omega}_t, \bar{\mu}_t$):**

2:   for a small set of map features m_i do

3:   $F_i = \begin{pmatrix} 0 \cdots 0 & 1 & 0 & 0 \cdots 0 \\ \underbrace{0 \cdots 0}_{2(N-i)} & 0 & 1 & \underbrace{0 \cdots 0}_{2(i-1)x} \end{pmatrix}$

4:   $\mu_{i,t} = (F_i \, \Omega_t \, F_i^T)^{-1} \, F_i \, [\xi_t - \Omega_t \, \bar{\mu}_t + \Omega_t \, F_i^T \, F_i \, \bar{\mu}_t]$

5:   endfor

6:   for all other map features m_i do

7:   $\mu_{i,t} = \bar{\mu}_{i,t}$

8:   endfor

9:   $F_x = \begin{pmatrix} 1 & 0 & 0 & 0 \cdots 0 \\ 0 & 1 & 0 & 0 \cdots 0 \\ 0 & 0 & 1 & \underbrace{0 \cdots 0}_{3N} \end{pmatrix}$

10: $\mu_{x,t} = (F_x \, \Omega_t \, F_x^T)^{-1} \, F_x \, [\xi_t - \Omega_t \, \bar{\mu}_t + \Omega_t \, F_x^T \, F_x \, \bar{\mu}_t]$

11: return μ_t

Table 12.5 The amortized state update step in SEIFs updates a small number of state estimates.

CONSTANT TIME SLAM a *"constant time SLAM"* conjecture should be taken with a grain of salt: The recovery of state estimates is a computational problem for which no linear-time solution is presently known, if the environment possesses large cycles.

12.4 Mathematical Derivation of the SEIF

12.4.1 Motion Update

The motion update in SEIF processes the control u_t by transforming the information matrix Ω_{t-1} and the information vector ξ_{t-1} into a new matrix $\bar{\Omega}_t$ and vector $\bar{\xi}_t$. As usual, the bar in our notation indicates that this prediction is only based on the control; it does not yet take the measurement into account.

The motion update in SEIFs exploits the sparseness of the information matrix, which makes it possible to perform this update in time independent of the map size n. This derivation is best started with the corresponding formula for the EKF. We begin with the algorithm **EKF_SLAM_known_correspondences** in Table 10.1, page 314. Lines 3 and 5 state the motion update, which we restate here for the reader's convenience:

$$(12.2) \quad \bar{\mu}_t = \mu_{t-1} + F_x^T \, \delta$$
$$(12.3) \quad \bar{\Sigma}_t = G_t \, \Sigma_{t-1} \, G_t^T + F_x^T \, R_t \, F_x$$

The essential elements of this update were defined as follows:

$$(12.4) \quad F_x = \begin{pmatrix} 1 & 0 & 0 & 0 \cdots 0 \\ 0 & 1 & 0 & 0 \cdots 0 \\ 0 & 0 & 1 & 0 \cdots 0 \end{pmatrix}$$

$$(12.5) \quad \delta = \begin{pmatrix} -\frac{v_t}{\omega_t} \sin \mu_{t-1,\theta} + \frac{v_t}{\omega_t} \sin(\mu_{t-1,\theta} + \omega_t \Delta t) \\ \frac{v_t}{\omega_t} \cos \mu_{t-1,\theta} - \frac{v_t}{\omega_t} \cos(\mu_{t-1,\theta} + \omega_t \Delta t) \\ \omega_t \Delta t \end{pmatrix}$$

$$(12.6) \quad \Delta = \begin{pmatrix} 0 & 0 & \frac{v_t}{\omega_t} \cos \mu_{t-1,\theta} - \frac{v_t}{\omega_t} \cos(\mu_{t-1,\theta} + \omega_t \Delta t) \\ 0 & 0 & \frac{v_t}{\omega_t} \sin \mu_{t-1,\theta} - \frac{v_t}{\omega_t} \sin(\mu_{t-1,\theta} + \omega_t \Delta t) \\ 0 & 0 & 0 \end{pmatrix}$$

$$(12.7) \quad G_t = I + F_x^T \, \Delta \, F_x$$

In SEIFs, we have to define the motion update over the information vector ξ and the information matrix Ω. From Equation (12.3), the definition of G_t in

(12.7), and the information matrix equation $\Omega = \Sigma^{-1}$, it follows that

$$(12.8) \quad \bar{\Omega}_t = \left[G_t \, \Omega_{t-1}^{-1} \, G_t^T + F_x^T \, R_t \, F_x \right]^{-1}$$

$$= \left[(I + F_x^T \, \Delta \, F_x) \, \Omega_{t-1}^{-1} \, (I + F_x^T \, \Delta \, F_x)^T + F_x^T \, R_t \, F_x \right]^{-1}$$

A key insight is this update can be implemented in constant time—regardless of the dimension of Ω. The fact that this is possible for sparse matrices Ω_{t-1} is somewhat non-trivial, since Equation (12.8) seems to require two nested inversions of matrices of size $(3N + 3) \times (3N + 3)$. As we shall see, if Ω_{t-1} is sparse, this update step can be carried out efficiently. We define

$$(12.9) \quad \Phi_t = \left[G_t \, \Omega_{t-1}^{-1} \, G_t^T \right]^{-1}$$

$$= [G_t^T]^{-1} \, \Omega_{t-1} \, G_t^{-1}$$

and hence Equation (12.8) can be rewritten as

$$(12.10) \quad \bar{\Omega}_t = \left[\Phi_t^{-1} + F_x^T \, R_t \, F_x \right]^{-1}$$

We now apply the matrix inversion lemma and obtain:

$$(12.11) \quad \bar{\Omega}_t = \left[\Phi_t^{-1} + F_x^T \, R_t \, F_x \right]^{-1}$$

$$= \Phi_t - \underbrace{\Phi_t \, F_x^T (R_t^{-1} + F_x \, \Phi_t \, F_x^T)^{-1} \, F_x \, \Phi_t}_{\kappa_t}$$

$$= \Phi_t - \kappa_t$$

Here κ_t is defined as indicated. This expression can be calculated in constant time *if* we can compute Φ_t in constant time from Ω_{t-1}. To see that this is indeed possible, we note that the argument inside the inverse, $R_t^{-1} + F_x \, \Phi_t \, F_x^T$, is 3-dimensional. Multiplying this inverse with F_x^T and F_x induces a matrix that is of the same size as Ω; however, this matrix is only non-zero for the 3×3 sub-matrix corresponding to the robot pose. Multiplying this matrix with a sparse matrix Ω_{t-1} (left and right) touches only elements for which the off-diagonal element in Ω_{t-1} between the robot pose and a map feature is non-zero. Put differently, the result of this operation only touches rows and columns that correspond to active features in the map. Since sparsity implies that the number of active features in Ω_{t-1} is independent of the size of Ω_{t-1}, the total number of non-zero elements in κ_t is also $O(1)$. Consequently, the subtraction requires $O(1)$ time.

It remains to be shown that we can calculate Φ_t from Ω_{t-1} in constant time. We begin with a consideration of the inverse of G_t, which is efficiently

12.4 Mathematical Derivation of the SEIF

calculated as follows:

$$
\begin{align}
(12.12) \quad G_t^{-1} &= (I + F_x^T \Delta F_x)^{-1} \\
&= (I \underbrace{- F_x^T I F_x + F_x^T I F_x}_{=0} + F_x^T \Delta F_x)^{-1} \\
&= (I - F_x^T I F_x + F_x^T (I + \Delta) F_x)^{-1} \\
&= I - F_x^T I F_x + F_x^T (I + \Delta)^{-1} F_x b \\
&= I + \underbrace{F_x^T [(I + \Delta)^{-1} - I] F_x}_{\Psi_t} \\
&= I + \Psi_t
\end{align}
$$

By analogy, we get for the transpose $[G_t^T]^{-1} = (I + F_x^T \Delta^T F_x)^{-1} = I + \Psi_t^T$. Here the matrix Ψ_t is only non-zero for elements that correspond to the robot pose. It is zero for all features in the map, and hence can be computed in constant time. This gives us for our desired matrix Φ_t the following expression:

$$
\begin{align}
(12.13) \quad \Phi_t &= (I + \Psi_t^T) \, \Omega_{t-1} \, (I + \Psi_t) \\
&= \Omega_{t-1} + \underbrace{\Psi_t^T \Omega_{t-1} + \Omega_{t-1} \Psi_t + \Psi_t^T \Omega_{t-1} \Psi_t}_{\lambda_t} \\
&= \Omega_{t-1} + \lambda_t
\end{align}
$$

where Ψ_t is zero except for the sub-matrix corresponding to the robot pose. Since Ω_{t-1} is sparse, λ_t is zero except for a finite number of elements, which correspond to active map features and the robot pose.

Hence, Φ_t can be computed from Ω_{t-1} in constant time, assuming that Ω_{t-1} is sparse. Equations (12.11) through (12.13) are equivalent to lines 5 through 9 in Table 12.2, which proves the correctness of the information matrix update in **SEIF_motion_update**.

Finally, we show a similar result for the information vector. From (12.2) we obtain

$$
(12.14) \quad \bar{\mu}_t = \mu_{t-1} + F_x^T \delta_t
$$

This implies for the information vector:

$$
\begin{align}
(12.15) \quad \bar{\xi}_t &= \bar{\Omega}_t \, (\Omega_{t-1}^{-1} \xi_{t-1} + F_x^T \delta_t) \\
&= \bar{\Omega}_t \, \Omega_{t-1}^{-1} \xi_{t-1} + \bar{\Omega}_t F_x^T \delta_t \\
&= (\bar{\Omega}_t + \Omega_{t-1} - \Omega_{t-1} + \Phi_t - \Phi_t) \, \Omega_{t-1}^{-1} \xi_{t-1} + \bar{\Omega}_t F_x^T \delta_t \\
&= (\bar{\Omega}_t \underbrace{- \Phi_t + \Phi_t}_{=0} \underbrace{- \Omega_{t-1} + \Omega_{t-1}}_{=0}) \, \Omega_{t-1}^{-1} \xi_{t-1} + \bar{\Omega}_t F_x^T \delta_t
\end{align}
$$

$$\begin{aligned}
&= \underbrace{(\bar{\Omega}_t - \Phi_t}_{=-\kappa_t} + \underbrace{\Phi_t - \Omega_{t-1})}_{=\lambda_t} \underbrace{\Omega_{t-1}^{-1} \xi_{t-1}}_{=\mu_{t-1}} + \underbrace{\Omega_{t-1} \Omega_{t-1}^{-1}}_{=I} \xi_{t-1} + \bar{\Omega}_t F_x^T \delta_t \\
&= \xi_{t-1} + (\lambda_t - \kappa_t) \mu_{t-1} + \bar{\Omega}_t F_x^T \delta_t
\end{aligned}$$

Since λ_t and κ_t are both sparse, the product $(\lambda_t - \kappa_t) \mu_{t-1}$ only contains finitely many non-zero elements and can be calculated in constant time. Further, $F_x^T \delta_t$ is a sparse matrix. The sparseness of the product $\bar{\Omega}_t F_x^T \delta_t$ follows now directly from the fact that $\bar{\Omega}_t$ is sparse as well.

12.4.2 Measurement Updates

The second important step of SLAM concerns the update of the filter in accordance to robot motion. The measurement update in SEIF directly implements the general extended information filter update, as stated in lines 6 and 7 of Table 3.6, page 76:

(12.16) $\quad \Omega_t = \bar{\Omega}_t + H_t^T Q_t^{-1} H_t$

(12.17) $\quad \xi_t = \bar{\xi}_t + H_t^T Q_t^{-1} [z_t - h(\bar{\mu}_t) - H_t \mu_t]$

Writing the prediction $\hat{z}_t = h(\bar{\mu}_t)$ and summing over all individual elements in the measurement vector leads to the form in lines 13 and 14 in Table 12.3:

(12.18) $\quad \Omega_t = \bar{\Omega}_t + \sum_i H_t^{iT} Q_t^{-1} H_t^i$

(12.19) $\quad \xi_t = \bar{\xi}_t + \sum_i H_t^{iT} Q_t^{-1} [z_t^i - \hat{z}_t^i - H_t^i \mu_t]$

Here Q_t, δ, q, and H_t^i are defined as before (e.g., Table 11.2 on page 348).

12.5 Sparsification

12.5.1 General Idea

The key step in SEIFs concerns the sparsification of the information matrix Ω_t. Because sparsification is so essential to SEIFs, let us first discuss it in general terms before we apply it to the information filter. Sparsification is an approximation through which a posterior distribution is approximated by two of its marginals. Suppose a, b, and c are sets of random variables (not to be confused with any other occurrence of these variables in this book!), and suppose we are given a joint distribution $p(a, b, c)$ over these variables. To sparsify this distribution, we have to remove any direct link between the

12.5 Sparsification

variables a and b. In other words, we would like to approximate p by a distribution $\tilde{p}$ for which the following property holds: $\tilde{p}(a \mid b, c) = p(a \mid c)$ and $\tilde{p}(b \mid a, c) = p(b \mid c)$. In multivariate Gaussians, it is easily shown that this conditional independence is equivalent to the absence of a direct link between a and b. The corresponding element in the information matrix is zero.

A good approximation $\tilde{p}$ is obtained by a term proportional to the product of the marginals, $p(a, c)$ and $p(b, c)$. Neither of these marginals retain dependence between the variables a and b, since they both contain only one of those variables. Thus, the product $p(a, c)\, p(b, c)$ does not contain any *direct* dependencies between a and b; instead, a and b are conditionally independent given c. However, $p(a, c)\, p(b, c)$ is not yet a valid probability distribution over a, b, and c. This is because c occurs twice in this expression. However, proper normalization by $p(c)$ yields a probability distribution (assuming $p(c) > 0$):

$$(12.20) \quad \tilde{p}(a, b, c) = \frac{p(a, c)\, p(b, c)}{p(c)}$$

To understand the effect of this approximation, we apply the following transformation:

$$(12.21) \quad \tilde{p}(a, b, c) = \frac{p(a, b, c)}{p(a, b, c)} \frac{p(a, c)\, p(b, c)}{p(c)}$$
$$= p(a, b, c) \frac{p(a, c)}{p(c)} \frac{p(b, c)}{p(a, b, c)}$$
$$= p(a, b, c) \frac{p(a \mid c)}{p(a \mid b, c)}$$

In other words, removing the direct dependence between a and b is equivalent to approximating the conditional $p(a \mid b, c)$ by a conditional $p(a \mid c)$. We also note (without proof) that among all approximations q of p where a and b are conditionally independent given c, the one described here is "closest" to p, where closeness is measured by the Kullback-Leibler divergence, a common asymmetric measure of the "nearness" of one probability distribution to another.

An important observation pertains to the fact that the original $p(a \mid b, c)$ is *at least as informative* as $p(a \mid c)$, the conditional that replaces $p(a \mid b, c)$ in $\tilde{p}$. This is because $p(a \mid b, c)$ is conditioned on a superset of variables of the conditioning variables in $p(a \mid c)$. For Gaussians, this implies that the variances of the approximation $p(a \mid c)$ is equal or larger than the variance of the original conditional, $p(a \mid b, c)$. Further, the variances of the marginals $\tilde{p}(a)$, $\tilde{p}(b)$, and $\tilde{p}(c)$ are also larger than or equal to the corresponding variances of $p(a), p(b)$, and $p(c)$. In other words, it is impossible that the variance shrinks under this approximation.

12.5.2 Sparsification in SEIFs

The SEIF applies the idea of sparsification to the posterior $p(y_t \mid z_{1:t}, u_{1:t}, c_{1:t})$, to maintain an information matrix Ω_t that is sparse at all times. To do so, it suffices to deactivate links between the robot pose and individual features in the map. If done correctly, this also limits the number of links between pairs of features.

To see, let us briefly consider the two circumstances under which a new link may be introduced. First, observing a passive feature activates this feature and hence introduces a new link between the robot pose and the very feature. Second, motion introduces links between any two active features. This consideration suggests that controlling the number of active features can avoid violation of both sparseness bounds. Thus, sparseness is achieved simply by keeping the number of active features small at any point in time.

To define the sparsification step, it will prove useful to partition the set of all features into three disjoint subsets:

$$(12.22) \quad m \;=\; m^+ + m^0 + m^-$$

where m^+ is the set of all active features that shall remain active. The set m^0 are one or more active features that we seek to deactivate. Put differently, we seek to remove the links between m^0 and the robot pose. And finally, m^- are all currently passive features; they shall remain passive in the process of sparsification. Since $m^+ \cup m^0$ contains all currently active features, the posterior can be factored as follows:

$$(12.23) \quad \begin{aligned} p(y_t \mid z_{1:t}, u_{1:t}, c_{1:t}) &= p(x_t, m^0, m^+, m^- \mid z_{1:t}, u_{1:t}, c_{1:t}) \\ &= p(x_t \mid m^0, m^+, m^-, z_{1:t}, u_{1:t}, c_{1:t})\, p(m^0, m^+, m^- \mid z_{1:t}, u_{1:t}, c_{1:t}) \\ &= p(x_t \mid m^0, m^+, m^- = 0, z_{1:t}, u_{1:t}, c_{1:t})\, p(m^0, m^+, m^- \mid z_{1:t}, u_{1:t}, c_{1:t}) \end{aligned}$$

In the last step we exploited the fact that if we know the active features m^0 and m^+, the variable x_t does not depend on the passive features m^-. We can hence set m^- to an arbitrary value without affecting the conditional posterior over x_t, $p(x_t \mid m^0, m^+, m^-, z_{1:t}, u_{1:t}, c_{1:t})$. Here we simply choose $m^- = 0$.

12.5 Sparsification

Following the sparsification idea discussed in general terms in the previous section, we now replace $p(x_t \mid m^0, m^+, m^- = 0)$ by $p(x_t \mid m^+, m^- = 0)$ and thereby drop the dependence on m^0.

$$\tilde{p}(x_t, m \mid z_{1:t}, u_{1:t}, c_{1:t}) \tag{12.24}$$
$$= p(x_t \mid m^+, m^- = 0, z_{1:t}, u_{1:t}, c_{1:t})\, p(m^0, m^+, m^- \mid z_{1:t}, u_{1:t}, c_{1:t})$$

This approximation is obviously equivalent to the following expression:

$$\tilde{p}(x_t, m \mid z_{1:t}, u_{1:t}, c_{1:t}) \tag{12.25}$$
$$= \frac{p(x_t, m^+ \mid m^- = 0, z_{1:t}, u_{1:t}, c_{1:t})}{p(m^+ \mid m^- = 0, z_{1:t}, u_{1:t}, c_{1:t})}\, p(m^0, m^+, m^- \mid z_{1:t}, u_{1:t}, c_{1:t})$$

12.5.3 Mathematical Derivation of the Sparsification

In the remainder of this section, we show that the algorithm **SEIF_sparsification** in Table 12.4 implements this probabilistic calculation, and that it does so in constant time. We begin by calculating the information matrix for the distribution $p(x_t, m^0, m^+ \mid m^- = 0)$ of all variables but m^-, and conditioned on $m^- = 0$. This is obtained by extracting the sub-matrix of all state variables but m^-:

$$\Omega_t^0 = F_{x,m^+,m^0}\, F_{x,m^+,m^0}^T\, \Omega_t\, F_{x,m^+,m^0}\, F_{x,m^+,m^0}^T \tag{12.26}$$

With that, the matrix inversion lemma (Table 3.2 on page 50) leads to the following information matrices for the terms $p(x_t, m^+ \mid m^- = 0, z_{1:t}, u_{1:t}, c_{1:t})$ and $p(m^+ \mid m^- = 0, z_{1:t}, u_{1:t}, c_{1:t})$, denoted Ω_t^1 and Ω_t^2, respectively:

$$\Omega_t^1 = \Omega_t^0 - \Omega_t^0\, F_{m^0}\, (F_{m^0}^T\, \Omega_t^0\, F_{m^0})^{-1}\, F_{m^0}^T\, \Omega_t^0 \tag{12.27}$$
$$\Omega_t^2 = \Omega_t^0 - \Omega_t^0\, F_{x,m^0}\, (F_{x,m^0}^T\, \Omega_t^0\, F_{x,m^0})^{-1}\, F_{x,m^0}^T\, \Omega_t^0 \tag{12.28}$$

Here the various F-matrices are projection matrices that project the full state y_t into the appropriate sub-state containing only a subset of all variables—in analogy to the matrix F_x used in various previous algorithms. The final term in our approximation (12.25), $p(m^0, m^+, m^- \mid z_{1:t}, u_{1:t}, c_{1:t})$, possesses the following information matrix:

$$\Omega_t^3 = \Omega_t - \Omega_t F_x (F_x^T \Omega_t F_x)^{-1} F_x^T \Omega_t \tag{12.29}$$

Putting these expressions together according to Equation (12.25) yields the following information matrix, in which the feature m^0 is now indeed deactivated:

$$\tilde{\Omega}_t = \Omega_t^1 - \Omega_t^2 + \Omega_t^3 \tag{12.30}$$

$$
\begin{aligned}
&= \Omega_t - \Omega_t^0 \, F_{m_0} \, (F_{m_0}^T \, \Omega_t^0 \, F_{m_0})^{-1} \, F_{m_0}^T \, \Omega_t^0 \\
&\quad + \Omega_t^0 \, F_{x,m_0} \, (F_{x,m_0}^T \, \Omega_t^0 \, F_{x,m_0})^{-1} \, F_{x,m_0}^T \, \Omega_t^0 \\
&\quad - \Omega_t \, F_x \, (F_x^T \, \Omega_t \, F_x)^{-1} \, F_x^T \, \Omega_t
\end{aligned}
$$

The resulting information vector is now obtained by the following simple consideration:

$$
\begin{aligned}
(12.31) \quad \tilde{\xi}_t &= \tilde{\Omega}_t \, \mu_t \\
&= (\Omega_t - \Omega_t + \tilde{\Omega}_t) \, \mu_t \\
&= \Omega_t \, \mu_t + (\tilde{\Omega}_t - \Omega_t) \, \mu_t \\
&= \xi_t + (\tilde{\Omega}_t - \Omega_t) \, \mu_t
\end{aligned}
$$

This completes the derivation of lines 3 and 4 in Table 12.4.

12.6 Amortized Approximate Map Recovery

The final update step in SEIFs is concerned with the computation of the mean μ. Throughout this section, we will drop the time index from our notation, since it plays no role in the techniques to be discussed. So we will write μ instead of μ_t.

Before deriving an algorithm for recovering the state estimate μ from the information form, let us briefly consider what parts of μ are needed in SEIFs, and when. SEIFs need the state estimate μ of the robot pose and the active features in the map. These estimates are needed at three different occasions:

1. The mean is used for the linearization of the motion model, which takes place in lines 3, 4, and 10 in Table 12.2.

2. It is also used for linearization of the measurement update, see lines 6, 8, 10, 13 in Table 12.3.

3. Finally, it is used in the sparsification step, specifically in line 4 in Table 12.4.

However, we never need the full vector μ. We only need an estimate of the robot pose, and an estimate of the locations of all active features. This is a small subset of all state variables in μ. Nevertheless, computing these estimates efficiently requires some additional mathematics, as the *exact* approach for recovering the mean via $\mu = \Omega^{-1} \, \xi$ requires matrix inversion or the use of some other optimization technique—even when recovering a subset of variables.

12.6 Amortized Approximate Map Recovery

Once again, the key insight is derived from the sparseness of the matrix Ω. The sparseness enables us do define an iterative algorithm for recovering state variables online, as the data is being gathered and the estimates ξ and Ω are being constructed. To do so, it will prove convenient to reformulate $\mu = \Omega^{-1} \xi$ as an optimization problem. As we will show in just a minute, the state μ is the mode

$$(12.32) \quad \hat{\mu} = \underset{\mu}{\operatorname{argmax}}\ p(\mu)$$

of the following Gaussian distribution, defined over the variable μ:

$$(12.33) \quad p(\mu) = \eta\ \exp\left\{-\tfrac{1}{2}\mu^T \Omega \mu + \xi^T \mu\right\}$$

Here μ is a vector of the same form and dimensionality as μ. To see that this is indeed the case, we note that the derivative of $p(\mu)$ vanishes at $\mu = \Omega^{-1}\xi$:

$$(12.34) \quad \frac{\partial p(\mu)}{\partial \mu} = \eta\ (-\Omega\mu + \xi)\ \exp\left\{-\tfrac{1}{2}\mu^T \Omega \mu + \xi^T \mu\right\} \overset{!}{=} 0$$

which implies $\Omega\mu = \xi$ or, equivalently, $\mu = \Omega^{-1}\xi$.

This transformation suggests that recovering the state vector μ is equivalent to finding the mode of (12.33), which now has become an optimization problem. For this optimization problem, we will now describe an iterative hill climbing algorithm which, thanks to the sparseness of the information matrix.

COORDINATE DESCENT

Our approach is an instantiation of *coordinate descent*. For simplicity, we state it here for a single coordinate only; our implementation iterates a constant number K of such optimizations after each measurement update step. The mode $\hat{\mu}$ of (12.33) is attained at:

$$(12.35) \quad \begin{aligned}\hat{\mu} &= \underset{\mu}{\operatorname{argmax}}\ \exp\left\{-\tfrac{1}{2}\mu^T \Omega \mu + \xi^T \mu\right\} \\ &= \underset{\mu}{\operatorname{argmin}}\ \tfrac{1}{2}\mu^T \Omega \mu - \xi^T \mu\end{aligned}$$

We note that the argument of the min-operator in (12.35) can be written in a form that makes the individual coordinate variables μ_i (for the i-th coordinate of μ_t) explicit:

$$(12.36) \quad \tfrac{1}{2}\mu^T \Omega \mu - \xi^T \mu = \tfrac{1}{2}\sum_i \sum_j \mu_i^T\ \Omega_{i,j}\ \mu_j - \sum_i \xi_i^T\ \mu_i$$

where $\Omega_{i,j}$ is the element with coordinates (i,j) in the matrix Ω, and ξ_i if the i-th component of the vector ξ. Taking the derivative of this expression with respect to an arbitrary coordinate variable μ_i gives us

$$(12.37) \quad \frac{\partial}{\partial \mu_i} \left\{ \frac{1}{2} \sum_i \sum_j \mu_i^T \, \Omega_{i,j} \, \mu_j - \sum_i \xi_i^T \, \mu_i \right\} \;=\; \sum_j \Omega_{i,j} \, \mu_j - \xi_i$$

Setting this to zero leads to the optimum of the i-th coordinate variable μ_i given all other estimates μ_j:

$$(12.38) \quad \mu_i \;=\; \Omega_{i,i}^{-1} \left[\xi_i - \sum_{j \neq i} \Omega_{i,j} \, \mu_j \right]$$

The same expression can conveniently be written in matrix notation. Here we define $F_i = (\, 0 \ldots 0 \; 1 \; 0 \ldots 0 \,)$ to be a projection matrix for extracting the i-th component from the matrix Ω:

$$(12.39) \quad \mu_i \;=\; (F_i \, \Omega \, F_i^T)^{-1} \, F_i \, [\xi - \Omega \, \mu + \Omega \, F_i^T \, F_i \, \mu]$$

This consideration derives our incremental update algorithm. Repeatedly updating

$$(12.40) \quad \mu_i \;\longleftarrow\; (F_i \, \Omega \, F_i^T)^{-1} \, F_i \, [\xi - \Omega \, \mu + \Omega \, F_i^T \, F_i \, \mu]$$

for some element of the state vector μ_i reduces the error between the left-hand side and the right-hand side of Equation (12.39). Repeating this update indefinitely for all elements of the state vector converges to the correct mean (without proof).

As is easily seen, the number of elements in the summation in (12.38), and hence the vector multiplication in the update rule (12.40), is constant if Ω is sparse. Hence, each update requires constant time. To maintain the constant-time property of our SLAM algorithm, we can afford a constant number of updates K per time step. This usually leads to convergence over many updates.

However, a note of caution is in order. The quality of this approximation depends on a number of factors, among them the size of the largest cyclic structure in the map. In general, a constant number of K updates per time step might be insufficient to yield good results. Also, there exists a number of optimization techniques that are more efficient than the coordinate descent algorithm described here. A "classical" example is the conjugate gradient discussed in the context of GraphSLAM. In practical implementations it is advisable to rely on efficient optimization techniques to recover μ.

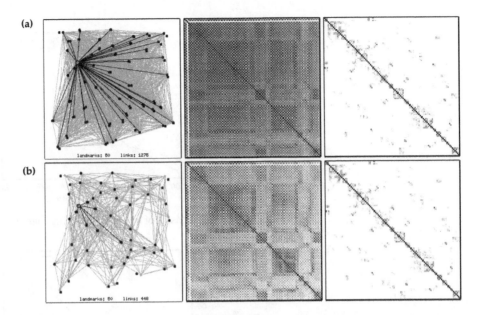

Figure 12.6 Comparison of (a) SEIF without sparsification with (b) SEIF using the sparsification step with 4 active landmarks. The comparison is carried out in a simulated environment with 50 landmarks. In each row, the left panel shows the set of links in the filter, the center panel the correlation matrix, and the right panel the normalized information matrix. Obviously, the sparsified SEIF maintains many fewer links, but its result is less confident as indicated by its less-expressed correlation matrix.

12.7 How Sparse Should SEIFs Be?

A key question pertains to the degree of sparseness one should enforce in a SEIF. In particular, the number of active features in SEIFs determines the degree of sparseness. The sparseness trades off two factors: the computational efficiency of the SEIF, and the accuracy of the result. When implementing a SEIF algorithm, it is therefore advisable to get a feeling for this trade-off.

The "gold standard" for a SEIF is the EKF, which avoids sparsification and also does not rely on relaxation techniques for recovering the state estimate. The following comparison characterizes the three key performance measures that set sparse SEIFs apart from EKFs. Our comparison is based on a simulated robot world, in which the robot senses the range, proximity, and identity of nearby landmarks.

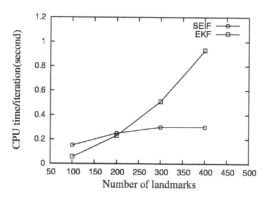

Figure 12.7 The comparison of average CPU time between SEIF and EKF.

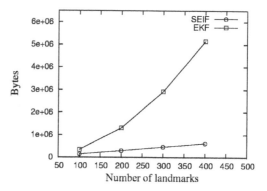

Figure 12.8 The comparison of average memory usage between SEIF and EKF.

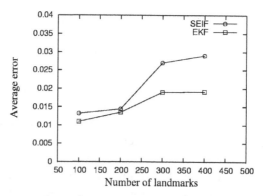

Figure 12.9 The comparison of root mean square distance error between SEIF and EKF.

1. **Computation.** Figures 12.7 compares the computation per update in SEIFs with that in EKFs; in both cases the implementation is optimized. This graph illustrates the major computational ramification of the probabilistic versus information representation in the filter. While EKFs indeed require time quadratic in the map size, SEIFs level off and require constant time.

2. **Memory.** Figure 12.8 compares the memory use of EKFs with that of SEIFs. Here once again, EKFs scale quadratically, whereas SEIFs scale linearly, due to the sparseness of its information representation.

3. **Accuracy.** Here EKFs outperform SEIFs, due to the fact that SEIFS require approximation for maintaining sparseness, and when recovering the state estimate μ_t. This is shown in Figure 12.9, which plots the error of both methods as a function of map size.

One way to get a feeling for the effect of the degree of sparseness can be obtained via simulation. Figure 12.10 plots the update time and the approximation error as a function of the number of active landmarks in the SEIF update, for a map consisting of 50 landmarks. The update time falls monotonically with the number of active features. Figure 12.11 shows the corresponding plot for the error, comparing the EKF with the SEIF at different degrees of sparseness. The solid line is the SEIF as described, whereas the dashed line corresponds to a SEIF that recovers the mean μ_t exactly. As this plot suggests, 6 active features seem to provide competitive results, at significant computational savings over the EKF. For smaller numbers of active features, the error increases drastically. A careful implementation of SEIFs will require the experimenter to vary this important parameter, and graph its effect on key factors as done here.

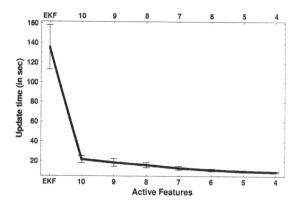

Figure 12.10 The update time of the EKF (leftmost data point only) and the SEIF, for different degrees of sparseness, as induced by a bound on the number of active features as indicated.

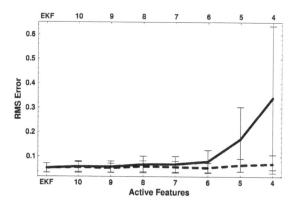

Figure 12.11 The approximation error EKF (leftmost data point only) and SEIF for different degrees of sparseness. In both figures, the map consists of 50 landmarks.

12.8 Incremental Data Association

We will now turn our attention to the problem of data association in SEIFs. Our first technique will be the familiar incremental approach, which greedily identifies the most likely correspondence, and then treats this value as if it was ground truth. We already encountered an instance of such a greedy data association technique in Chapter 10.3, where we discussed data association in the EKF. In fact, the only difference between greedy incremental data association in SEIFs and EKFs pertains to the calculation of the data association probability. As a rule of thumb, computing this probability is generally more difficult in an information filter than in a probabilistic filter such as the EKF, since the information filter does not keep track of covariances.

12.8.1 Computing Incremental Data Association Probabilities

As before, the data association vector at time t will be denoted c_t. The greedy incremental technique maintains a set of data association guesses, denoted $\hat{c}_{1:t}$. In the incremental regime, we are given the estimated correspondences $\hat{c}_{1:t-1}$ from previous updates when computing $\hat{c}_t$. The data associating step then pertains to the estimation of the most likely value for the data association variable $\hat{c}_t$ at time t. This is achieved via the following maximum likelihood estimator:

$$
\begin{aligned}
(12.41) \quad \hat{c}_t &= \operatorname*{argmax}_{c_t} p(z_t \mid z_{1:t-1}, u_{1:t}, \hat{c}_{1:t-1}, c_t) \\
&= \operatorname*{argmax}_{c_t} \int p(z_t \mid y_t, c_t) \underbrace{p(y_t \mid z_{1:t-1}, u_{1:t}, \hat{c}_{1:t-1})}_{\bar{\Omega}_t, \bar{\xi}_t} \, dy_t \\
&= \operatorname*{argmax}_{c_t} \int\!\!\int p(z_t \mid x_t, y_{c_t}, c_t) \, p(x_t, y_{c_t} \mid z_{1:t-1}, u_{1:t}, \hat{c}_{1:t-1}) \, dx_t \, dy_{c_t}
\end{aligned}
$$

Our notation $p(z_t \mid x_t, y_{c_t}, c_t)$ of the sensor model makes the correspondence variable c_t explicit. Calculating this probability exactly is not possible in constant time, since it involves marginalizing out almost all variables in the map. However, the same type of approximation that was essential for the efficient sparsification can also be applied here as well.

In particular, let us denote by $m_{c_t}^+$ the combined Markov blanket of the robot pose x_t and the landmark y_{c_t}. This Markov blanket is the set of all features in the map that are linked to the robot of landmark y_{c_t}. Figure 12.12 illustrates this set. Notice that $m_{c_t}^+$ includes by definition all active landmarks. The spareness of $\bar{\Omega}_t$ ensures that $m_{c_t}^+$ contains only a fixed number

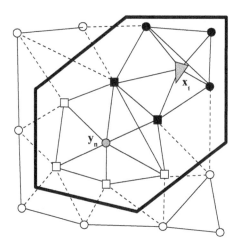

Figure 12.12 The combined Markov blanket of feature y_n and the observed features is usually sufficient for approximating the posterior probability of the feature locations, conditioning away all other features.

of features, regardless of the size of the map N. If the Markov blankets of x_t and of y_{c_t} do not intersect, further features are added that represent the shortest path in the information graph between x_t and of y_{c_t}.

All remaining features will now be collectively referred to as $m_{c_t}^-$:

(12.42) $\quad m_{c_t}^- = m - m_{c_t}^+ - \{y_{c_t}\}$

The set $m_{c_t}^-$ contains only features that have only a minor impact on the target variables, x_t and y_{c_t}. The SEIF approximates the probability $p(x_t, y_{c_t} \mid z_{1:t-1}, u_{1:t}, \hat{c}_{1:t-1})$ in Equation (12.41) by essentially ignoring these indirect influences:

(12.43) $\quad p(x_t, y_{c_t} \mid z_{1:t-1}, u_{1:t}, \hat{c}_{1:t-1})$
$\qquad = \iint p(x_t, y_{c_t}, m_{c_t}^+, m_{c_t}^- \mid z_{1:t-1}, u_{1:t}, \hat{c}_{1:t-1}) \, dm_{c_t}^+ \, dm_{c_t}^-$
$\qquad = \iint p(x_t, y_{c_t} \mid m_{c_t}^+, m_{c_t}^-, z_{1:t-1}, u_{1:t}, \hat{c}_{1:t-1})$
$\qquad \qquad p(m_{c_t}^+ \mid m_{c_t}^-, z_{1:t-1}, u_{1:t}, \hat{c}_{1:t-1}) \, p(m_{c_t}^- \mid z_{1:t-1}, u_{1:t}, \hat{c}_{1:t-1}) \, dm_{c_t}^+ \, dm_{c_t}^-$
$\qquad \approx \int p(x_t, y_{c_t} \mid m_{c_t}^+, m_{c_t}^- = \mu_{c_t}^-, z_{1:t-1}, u_{1:t}, \hat{c}_{1:t-1})$
$\qquad \qquad p(m_{c_t}^+ \mid m_{c_t}^- = \mu_{c_t}^-, z_{1:t-1}, u_{1:t}, \hat{c}_{1:t-1}) \, dm_{c_t}^+$

This probability can be computed in constant time if the set of features considered in this calculation is independent of the map size (which it generally is). In complete analogy to various derivations above, we note that the approximation of the posterior is simply obtained by carving out the submatrix corresponding to the two target variables:

$$\Sigma_{t:c_t} = F^T_{x_t,y_{c_t}} (F^T_{x_t,y_{c_t},m^+_{c_t}} \Omega_t F_{x_t,y_{c_t},m^+_{c_t}})^{-1} F_{x_t,y_{c_t}} \qquad (12.44)$$

$$\mu_{t:c_t} = \mu_t F_{x_t,y_{c_t}} \qquad (12.45)$$

This calculation is constant time, since it involves a matrix whose size is independent of N. From this Gaussian, the desired measurement probability in Equation (12.41) is now easily recovered.

As in our EKF SLAM algorithm, features are labeled as new when the likelihood $p(z_t \mid z_{1:t-1}, u_{1:t}, \hat{c}_{1:t-1}, c_t)$ remains below a threshold α. We then simply set $\hat{c}_t = N_{t-1} + 1$ and $N_t = N_{t-1} + 1$. Otherwise the size of the map remains unchanged, hence $N_t = N_{t-1}$. The value $\hat{c}_t$ is chosen that maximizes the data association probability.

As a last caveat, sometimes the combines Markov blanket is insufficient, in that it does not contain a path between the robot pose and the landmark that is being tested for correspondence. This will usually be the case when closing a large cycle in the environment. Here we need to augment the set of features $m^+_{c_t}$ by a set of landmarks along at least one path between m_{c_t} and the robot pose x_t. Depending on the size of this cycle, the numbers of landmarks contained in the resulting set may now depend on N, the size of the map. We leave the details of such an extension as an exercise.

12.8.2 Practical Considerations

In general, the incremental greedy data association technique is brittle, as discussed in the chapters EKF SLAM. Spurious measurements can easily cause false associations and induce significant errors into the SLAM estimate. The standard remedy for this brittleness—in EKFs and SEIFs alike—pertains to the creation of a *provisional landmark list*. We already discussed this approach in depth in Chapter 10.3.3, in the context of EKF SLAM. A provisional list adds any new feature that has not been previously observed into a candidate list, which is maintained separately from the SEIF. In the measurement steps that follow, the newly arrived candidates are checked against all candidates in the waiting list, reasonable matches increase the weight of corresponding candidates, and not seeing a nearby feature decreases its weight.

Figure 12.13 The vehicle used in our experiments is equipped with a 2-D laser range finder and a differential GPS system. The vehicle's ego-motion is measured by a linear variable differential transformer sensor for the steering, and a wheel-mounted velocity encoder. In the background, the Victoria Park test environment can be seen. Image courtesy of José Guivant and Eduardo Nebot, Australian Centre for Field Robotics.

When a candidate's weight is above a certain threshold, it joins the SEIF network of features.

We notice that data association violates the constant time property of SEIFs. This is because when calculating data associations, multiple features have to be tested. If we can ensure that all plausible features are already connected in the SEIF by a short path to the set of active features, it would be feasible to perform data association in constant time. In this way, the SEIF structure naturally facilitates the search of the most likely feature given a measurement. However, this is not the case when closing a cycle for the first time, in which case the correct association might be far away in the SEIF adjacency graph.

We will now briefly turn our attention to an implementation of the SEIF algorithm using a physical vehicle. The data used here is a common benchmark in the SLAM field. This data set was collected with an instrumented outdoor vehicle driven through a park in Sydney, Australia.

The vehicle and its environment are shown in Figures 12.13 and 12.14, respectively. The robot is equipped with a SICK laser range finder and a system for measuring steering angle and forward velocity. The laser is used to detect trees in the environment, but it also picks up hundreds of spurious features

12.8 Incremental Data Association

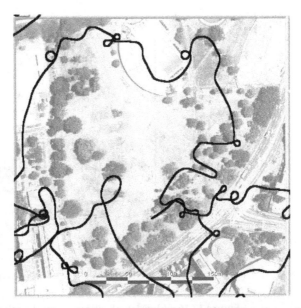

Figure 12.14 The testing environment: A 350 meter by 350 meter patch in Victoria Park in Sydney. Overlayed is the integrated path from odometry readings. Data and aerial image courtesy of José Guivant and Eduardo Nebot, Australian Centre for Field Robotics; results courtesy of Michael Montemerlo, Stanford University.

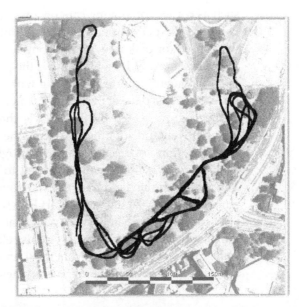

Figure 12.15 The path recovered by the SEIF, is correct within ±1m. Courtesy of Michael Montemerlo, Stanford University.

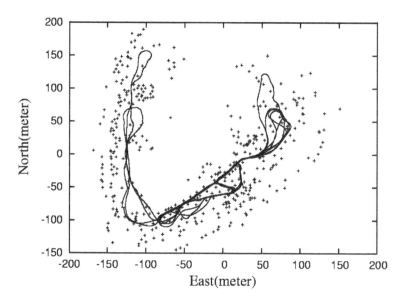

Figure 12.16 Overlay of estimated landmark positions and robot path. Images courtesy of Michael Montemerlo, Stanford University

such as corners of moving cars on a nearby highway. The raw odometry, as used in our experiments, is poor, resulting in several hundred meters of error when used for path integration along the vehicle's 3.5km path. This is illustrated in Figure 12.14, which shows the path of the vehicle. The poor quality of the odometry information along with the presence of many spurious features make this dataset particularly amenable for testing SLAM algorithms.

The path recovered by the SEIF is shown in Figure 12.15. This path is quantitatively indistinguishable from the one produced by the EKF. The average position error, as measured through differential GPS, is smaller than 0.50 meters, which is small compared to the overall path length of 3.5 km. The corresponding landmark map is shown in Figure 12.16. It, too, is of comparable accuracy as state-of-the-art EKF results. Compared with the EKF, the SEIF runs approximately twice as fast and consumes less than a quarter of the memory EKF uses. This saving is relatively small, but it a result of the small map size, and the fact that most time is spent preprocessing the sensor data. For larger maps, the relative savings are larger.

12.9 Branch-and-Bound Data Association

SEIFs make it possible to define a radically different data association approach, which can be proven to yield the optimal results (although possibly in exponential time). The technique is built on three key insights:

SOFT DATA ASSOCIATION CONSTRAINTS

- Just like GraphSLAM, SEIFs make it possible to add *soft data association constraints*. Given two features m_i and m_j, a soft data association constraint is nothing else but an information link that forces the distance between m_i and m_j to be small. We already encountered examples of such soft links in the previous chapter. In sparse extended information filters, introducing such a link is a simple, local addition of values in the information matrix.

- We can also easily *remove* soft association constraints. Just as introducing a new constraint amounts to a local addition in the information matrix, removing it is nothing else but a local subtraction. Such an "undo" operation can be applied to arbitrary data association links, regardless when they were added, or when the respective feature was last observed. This makes it possible to revise past data association decisions.

- The ability to freely add and subtract data associations arbitrarily enables us to search the tree of possible data associations in a way that is both efficient and complete—as will be shown below.

To develop a *branch-and-bound data association* algorithm, it shall prove useful to consider the data association tree that defines the sequence of data association decisions over time. At each point in time, each observed feature can be associated with a number of other features, or considered a new, previously unobserved feature. The resulting tree of data association choices, starting at time $t = 1$ all the way to the present time, is illustrated in Figure 12.17a. Of course, the tree grows exponentially over time, hence searching it exhaustively is impossible. The incremental greedy algorithm described in the previous section, in contrast, follows one path through this tree, defined by the locally most likely data associations. Such a path is visualized in Figure 12.17a as the thick gray path.

Obviously, if the incremental greedy approach succeeds, the resulting path is optimal. However, the incremental greedy technique may fail. Once a wrong choice has been made, the incremental approach cannot recover. Moreover, wrong data association decisions introduce errors in the map which, subsequently, can induce more errors in the data association.

12.9.1 Recursive Search

FRONTIER

The approach discussed in the remainder of this chapter generalizes the incremental greedy algorithm into a full-blown search algorithm for the tree that is provably optimal. Of course, searching all branches in the tree is intractable. However, if we maintain the log-likelihood of all nodes on the *frontier* of the tree expanded thus far, we can guarantee optimality. Figure 12.17b illustrates the idea: The branch-and-bound SEIF maintains not just a single path through the data association tree, but an entire frontier. Every time a node is expanded (e.g., through incremental ML), all alternative outcomes are also assessed and the corresponding likelihoods are memorized. This is illustrated in Figure 12.17b, which depicts the log-likelihood for an entire frontier of the tree.

Finding the maximum in Equation (12.41) implies that the log-likelihood of the chosen leaf is greater or equal to that of any other leaf at the same depth. Since the log-likelihood decreases monotonically with the depth of the tree, we can guarantee that we have indeed found the optimal data association values when the log-likelihood of the chosen leaf is greater or equal to the log-likelihood of any other node on the frontier. Put differently, when a frontier node assumes a log-likelihood greater than the one of the chosen leaf, there might be an opportunity to further increase the likelihood of the data by revising past data association decisions. Our approach then simply expands such frontier nodes. If an expansion reaches a leaf and its value is larger than the one of the present best leaf, this leaf is chosen as the new data association. Otherwise the search is terminated when the entire frontier possesses values that are all smaller or equal to the one of the chosen leaf. This approach is guaranteed to always maintain the best set of values for the data association variables; however, occasionally it might require substantial search.

12.9.2 Computing Arbitrary Data Association Probabilities

To test whether or not two features in the map should be linked, we now need a technique that, for any two features in the map, calculates the probability of equality. This test is essentially the same as GraphSLAM's correspondence test, stated in Table 11.8 on page 364. However, in SEIFs this test is approximate, in that the exact calculation of the log-likelihood would require additional computation.

Table 12.6 lists an algorithm that tests the probability that two features in

12.9 Branch-and-Bound Data Association

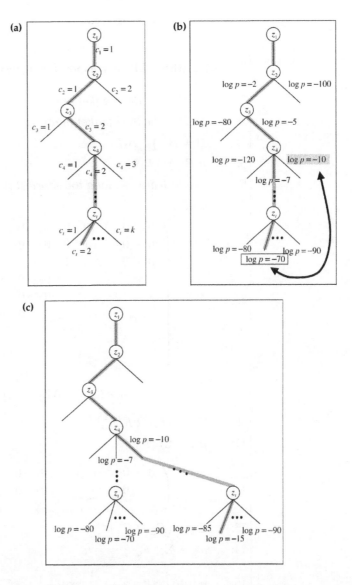

Figure 12.17 (a) The data association tree, whose branching factor grows with the number of landmarks in the map. (b) The tree-based SEIF maintains the log-likelihood for the entire frontier of expanded nodes, enabling it to find alternative paths. (c) Improved path.

1: **Algorithm SEIF_correspondence_test($\Omega, \xi, \mu, m_j, m_k$):**

2:     let $B(j)$ be the blanket of m_j

3:     let $B(k)$ be the blanket of m_k

4:     $B = B(j) \cup B(k)$

5:     if $B(j) \cap B(k) = \emptyset$

5:         add features along the shortest path between m_i and m_j to B

7:     endif

8:     $F_B = \begin{pmatrix} 0\cdots0 & 1 & 0 & 0 & 0\cdots0 & \cdots & & & & & \\ 0\cdots0 & 0 & 1 & 0 & 0\cdots0 & \cdots & & & & & \\ 0\cdots0 & 0 & 0 & 1 & 0\cdots0 & \cdots & & & & & \\ & \cdots & & & & 0\cdots0 & 1 & 0 & 0 & 0\cdots0 \\ & \cdots & & & & 0\cdots0 & 0 & 1 & 0 & 0\cdots0 \\ & \cdots & & & & 0\cdots0 & 0 & 0 & 1 & 0\cdots0 \\ & & & & & & & & & & \ddots \\ & \cdots & & & & & & & & & 0\cdots0 \\ & \cdots & & & & & & & & & 0\cdots0 \end{pmatrix}$

9:         (size $(3N+3)$ by $3|B|$)

10:    $\Sigma_B = (F_B \, \Omega \, F_B^T)^{-1}$

11:    $\mu_B = \Sigma_B \, F_B \, \xi$

12:    $F_\Delta = \begin{pmatrix} 0\cdots0 & 1 & 0 & 0\cdots0 & -1 & 0 & 0\cdots0 \\ 0\cdots0 & \underbrace{0 \quad 1}_{\text{feature } m_j} & 0\cdots0 & \underbrace{0 \quad -1}_{\text{feature } m_j} & 0\cdots0 \end{pmatrix}$

13:    $\Sigma_\Delta = (F_\Delta \, \Omega \, F_\Delta^T)^{-1}$

14:    $\mu_\Delta = \Sigma_\Delta \, F_\Delta \, \xi$

15:    **return** $\det(2\pi \, \Sigma_\Delta)^{-\frac{1}{2}} \exp\{-\frac{1}{2} \mu_\Delta^T \, \Sigma_\Delta^{-1} \, \mu_\Delta\}$

Table 12.6 The SEIF SLAM test for correspondence.

the map are one and the same—this test is sufficient to implement the greedy data association technique. The key calculation here pertains to the recovery of a joint covariance and mean vector over a small set of map features B. To determine whether two features in the map are identical, SEIFs must consider the information links between them. Technically, the more links included in this consideration, the more accurate the result, but at the expense of increased computation. In practice, it usually suffices to identify the two *Markov blankets* of the features in question. A Markov blanket of a feature is the feature itself, and all other features that are connected via a non-zero element in the information matrix. In most cases, both Markov blankets intersect; if they do not, the algorithm in Table 12.6 identifies a path between the landmarks (which must exist if both were observed by the same robot).

MARKOV BLANKET

The algorithm in Table 12.6 then proceeds by cutting out a local information matrix and information vector, employing the very same mathematical "trick" that led to an efficient sparsification step: SEIFs condition away features outside the Markov blankets. As a result, SEIFs obtain an efficient technique for calculating the desired probability, one that is approximate (because of the conditioning), but works very well in practice.

This result is interesting in that it not only enables SEIFs to make a data association decision, but it provides a way for calculating the log-likelihood of such a decision. The logarithm of the result of this procedure corresponds to the log-likelihood of this specific data item, and summing those up along the path in the data association tree becomes the total data log-likelihood under a specific association.

12.9.3 Equivalence Constraints

Once two features in the map have determined to be equivalent in the data association search, SEIFs add a soft link to the information matrix. Suppose the first feature is m_i and the second is m_j. The soft link constrains their position to be equal through the following exponential-quadratic constraint

$$(12.46) \quad \exp\left\{-\tfrac{1}{2}(m_i - m_j)^T C (m_i - m_j)\right\}$$

Here C is a diagonal penalty matrix of the type

$$(12.47) \quad C = \begin{pmatrix} \infty & 0 & 0 \\ 0 & \infty & 0 \\ 0 & 0 & \infty \end{pmatrix}$$

In practice, the diagonal elements of C are replaced by large positive values; the larger those values, the stronger the constraint.

It is easily seen that the non-normalized Gaussian (12.46) can be written as a link between m_i and m_j in the information matrix. Simply define the projection matrix

$$(12.48) \quad F_{m_i - m_j} = \begin{pmatrix} 0 \cdots 0 & 1\ 0\ 0 & 0 \cdots 0 & -1\ 0\ 0 & 0 \cdots 0 \\ 0 \cdots 0 & 0\ 1\ 0 & 0 \cdots 0 & 0\ -1\ 0 & 0 \cdots 0 \\ 0 \cdots 0 & \underbrace{0\ 0\ 1}_{m_i} & 0 \cdots 0 & \underbrace{0\ 0\ -1}_{m_j} & 0 \cdots 0 \end{pmatrix}$$

This matrix maps the state y_t to the difference $m_i - m_j$. Thus, the expression (12.46) becomes

$$(12.49) \quad \exp\left\{-\tfrac{1}{2} (F_{m_i - m_j}\, y_t)^T\, C\, (F_{m_i - m_j}\, y_t)\right\}$$
$$= \exp\left\{-\tfrac{1}{2} y_t^T\, [F_{m_i - m_j}^T\, C\, F_{m_i - m_j}]\, y_t\right\}$$

Thus, to implement this soft constraint, SEIFs have to add $F_{m_i - m_j}^T\, C\, F_{m_i - m_j}$ to the information matrix, while leaving the information vector unchanged:

$$(12.50) \quad \Omega_t \longleftarrow \Omega_t + F_{m_i - m_j}^T\, C\, F_{m_i - m_j}$$

Clearly, the additive term is sparse: it only contains non-zero off-diagonal elements between the features m_i and m_j. Once a soft link has been added, it can be removed by the inverse operation

$$(12.51) \quad \Omega_t \longleftarrow \Omega_t - F_{m_i - m_j}^T\, C\, F_{m_i - m_j}$$

This removal can occur even regardless of the time that elapsed since a constraint was introduced in the filter. However, careful bookkeeping is necessary to guarantee that SEIFs never remove a non-existent data association constraint—otherwise the information matrix may no longer be positive semidefinite, and the resulting belief might not correspond to a valid probability distribution.

12.10 Practical Considerations

In any competitive implementation of this approach, there will usually only exist a small number of data association paths that are plausible at any point in time. When closing a loop in an indoor environment, for example, there are usually at most three plausible hypotheses: a closure, a continuation on the left, and a continuation on the right. But all should quickly become unlikely, so the number of times in which the tree is searched recursively should be small.

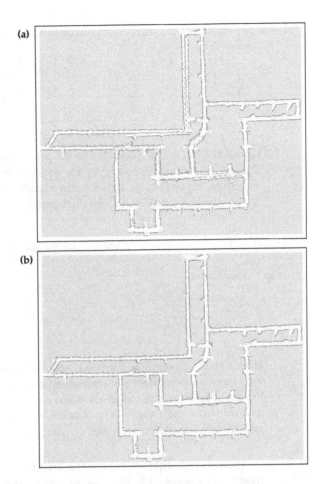

Figure 12.18 (a) Map with incremental ML scan matching and (b) full recursive branch-and-bound data association. Images courtesy of Dirk Hähnel, University of Freiburg.

One way to make the data association succeed more often is to incorporate *negative measurement information*. Range sensors, which are brought to bear in our implementation, return positive and negative information with regards to the presence of objects in the world. The positive information are object detections. The negative information applies to the space between the detection and the sensor. The fact that the robot failed to detect an object closer than its actual reading provides information about the *absence* of an object within the measurement range.

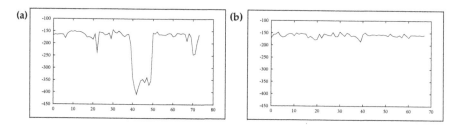

Figure 12.19 (a) Log-likelihood of the actual measurement, as a function of time. The lower likelihood is caused by the wrong assignment. (b) Log-likelihood, when recursively fixing false data association hypotheses through the tree search. The success is manifested by the lack of a distinct dip.

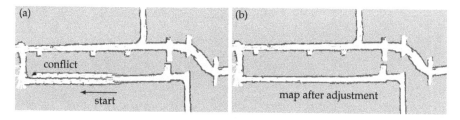

Figure 12.20 Example of the tree-based data association technique: (a) When closing a large loop, the robot first erroneously assumes the existence of a second, parallel hallway. However, this model leads to a gross inconsistency as the robot encounters a corridor at a right angle. At this point, the approach recursively searches for improved data association decisions, arriving on the map shown in diagram (b).

An approach that evaluates the effect of a new constraint on the overall likelihood considers both types of information: positive and negative. Both are obtained by calculating the pairwise (mis)match of two scans under their pose estimate. When using range scanners, one way to obtain a combination of positive and negative information is by superimposing a scan onto a local occupancy grid map built by another scan. In doing so, it is straightforward to determine an approximate matching probability for two local maps in a way that incorporates both the positive and the negative information.

The remainder of this section highlights practical results achieved using SEIFs with tree-based data association. The left panel of Figure 12.18a depicts the result of incremental ML data association, which is equivalent to regular incremental scan matching. Clearly, certain corridors are represented doubly in this map, illustrating the shortcomings of the ML approach. The right panel, in comparison, shows the result. Clearly, this map is more accurate

12.10 Practical Considerations

(a) Robot path

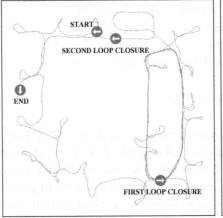

(b) Incremental ML (map inconsistent on left)

(c) FastSLAM (see next Chapter)

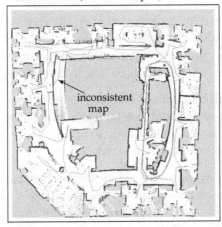

(d) SEIFs with branch-and-bound data association

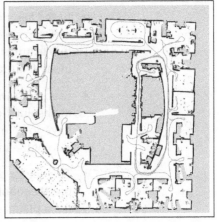

Figure 12.21 (a) Path of the robot. (b) Incremental ML (scan matching) (c) Fast-SLAM. (d) SEIFs with lazy data association. Image courtesy of Dirk Hähnel, University of Freiburg.

than the one generated by the incremental ML approach.

Figure 12.19a illustrates the log-likelihood of the most recent measurement (not the entire path), which drops significantly as the map becomes inconsistent. At this point, the SEIF engages in searching alternative data association values. It quickly finds the "correct" one and produces the map shown in Figure 12.18b. The area in question is shown in Figure 12.20, illustrating the moment at which the likelihood takes its dip. The log-likelihood of the measurement is shown in Figure 12.19b.

Finally, Figure 12.21 compares various techniques in the context of mapping a large building with multiple cycles.

12.11 Multi-Robot SLAM

The SEIF is also applicable to *multi-robot SLAM problems*. The multi-robot SLAM problem involves several robots that independently explore and map an environment, with the eventual goal of integrating their maps into a single, monolithic map. In many ways, the multi-robot SLAM problem is reminiscent of the single-mapping problem, in that data needs to be integrated into a single posterior over time. However, the multi-robot problem is significantly more difficult in a number of dimensions:

- In the absence of prior information on the relative location of two robots, the correspondence problem becomes a *global* problem. In principal, any two features in the map may correspond, and only through comparison of many features will the robots be able to determine good correspondences.

- Each map will be acquired in a local coordinate frame, which may differ in absolute locations and orientations. Before integrating two maps, they have to be oriented and shifted. In SEIFs, this requires a re-linearization step of the information matrix and vector.

- The degree of overlap between the different maps is unknown. For example, the robots might operate at different floors of a building with identical floor plans. In such a situation, the ability to differentiate the different maps may rely on small environmental features that differ, such as furniture that might be arranged slightly differently in the different floors.

In this section, we will only sketch some of the main ideas necessary to implement an algorithm for multi-robot mapping. We will present an algorithm for integrating two maps once correspondence has been established. We will also discuss, but not prove, techniques for establishing global correspondence in multi-robot SLAM.

12.11.1 Integrating Maps

The critical subroutine for fusing maps under known correspondence is shown in Table 12.7. This algorithm accepts as an input two local robot pos-

1:     **Algorithm SEIF_map_fusion**($\Omega^j, \xi^j, \Omega^k, \xi^k, d, \alpha, \mathcal{C}$):

2:     $\Delta = (d_x \ d_y \ \alpha \ d_x \ d_y \ 0 \ \cdots \ d_x \ d_y \ 0)^T$

3:     $\mathcal{A} = \begin{pmatrix} \cos\alpha & \sin\alpha & 0 & & & & & 0 \\ -\sin\alpha & \cos\alpha & 0 & & & & & \vdots \\ 0 & 0 & 1 & & & & & \\ & & & \ddots & & & & \\ & & & & \cos\alpha & \sin\alpha & 0 \\ \vdots & & & & -\sin\alpha & \cos\alpha & 0 \\ 0 & \cdots & & & 0 & 0 & 1 \end{pmatrix}$

4:     $\Omega^{j \to k} = \mathcal{A} \, \Omega^j \, \mathcal{A}^T$

5:     $\xi^{j \to k} = \mathcal{A} \, (\xi^j - \Omega^{j \to k} \, \Delta)$

6:     $\Omega = \begin{pmatrix} \Omega^k & 0 \\ 0 & \Omega^{j \to k} \end{pmatrix}$

7:     $\xi = \begin{pmatrix} \xi^k \\ \xi^{j \to k} \end{pmatrix}$

8:     for any pair $(m_j, m_k) \in \mathcal{C}^{j,k}$ do

9:     $F = \begin{pmatrix} 0\cdots 0 & 1 & 0 & 0 & 0\cdots 0 & -1 & 0 & 0 & 0\cdots 0 \\ 0\cdots 0 & 0 & 1 & 0 & 0\cdots 0 & 0 & -1 & 0 & 0\cdots 0 \\ 0\cdots 0 & \underbrace{0 \ 0 \ 1}_{m_j} & & 0\cdots 0 & \underbrace{0 \ 0 \ -1}_{m_k} & & 0\cdots 0 \end{pmatrix}$

10:     $\Omega \longleftarrow \Omega + F^T \begin{pmatrix} \infty & 0 & 0 \\ 0 & \infty & 0 \\ 0 & 0 & \infty \end{pmatrix} F$

11:     endfor

12:     return Ω, ξ

Table 12.7 The map fusion loop in multi-robot mapping with SEIFs.

teriors, represented by the information form Ω^j, ξ^j and Ω^k, ξ^k, respectively. It also requires three other items

1. A linear displacement tuple d
2. A relative rotation angle α
3. A set of feature correspondences $\mathcal{C}^{j,k}$

The displacement vector $d = (d_x \ d_y)^T$ and rotation α specify the relative orientation of the two robots' coordinate systems. In particular, the j-th robot pose x^j and features in the map of the j-th robot are mapped into the k-th robot's coordinate frame through a rotation by α followed by a translation by d. Here we use "$j \to k$" to denote the coordinates of an item in the j-th robot's map represented in the k-th robot coordinate frame.

1. For the j-th robot pose x_t^j

(12.52)
$$\underbrace{\begin{pmatrix} x^{j \to k} \\ y^{j \to k} \\ \theta^{j \to k} \end{pmatrix}}_{x_t^{j \to k}} = \begin{pmatrix} d_x \\ d_y \\ \alpha \end{pmatrix} + \begin{pmatrix} \cos\alpha & \sin\alpha & 0 \\ -\sin\alpha & \cos\alpha & 0 \\ 0 & 0 & 1 \end{pmatrix} \underbrace{\begin{pmatrix} x^j \\ y^j \\ \theta^j \end{pmatrix}}_{x_t^j}$$

2. For each feature in the j-th robot's map m_i^j

(12.53)
$$\underbrace{\begin{pmatrix} m_{i,x}^{j \to k} \\ m_{i,y}^{j \to k} \\ m_{i,s}^{j \to k} \end{pmatrix}}_{m_i^{j \to k}} = \begin{pmatrix} d_x \\ d_y \\ 0 \end{pmatrix} + \begin{pmatrix} \cos\alpha & \sin\alpha & 0 \\ -\sin\alpha & \cos\alpha & 0 \\ 0 & 0 & 1 \end{pmatrix} \underbrace{\begin{pmatrix} m_{i,x}^j \\ m_{i,y}^j \\ m_{i,s}^j \end{pmatrix}}_{m_i^j}$$

These two mappings are performed in lines 2 through 5 of the algorithm **SEIF_map_fusion** in Table 12.7. This step involves a local rotation and shift of the information matrix and the information vector, which preserved the sparseness of the SEIF. Afterwards, the map fusion proceeds by building a single joint posterior map, in lines 6 and 7. The final step in the fusion algorithm pertains to the correspondence list $\mathcal{C}^{j,k}$. This set consists of pairs (m_j, m_k) of features that mutually correspond in the maps of robot j and robot k. The fusion is performed analogously to the soft equivalence constraints considered in Chapter 12.9.3. Specifically, for any two corresponding features, we simply add large terms into the information matrix at the elements linking these two features.

We note that an alternative way to implement the map fusing step *collapses* the corresponding rows and columns of the resulting information matrix and vector. The following example illustrates the operation of collapsing feature 2 and 4 in the filter, which would occur when our correspondence list states that feature 2 and 4 are identical:

$$(12.54) \quad \begin{pmatrix} \Omega_{11} & \Omega_{12} & \Omega_{13} & \Omega_{14} \\ \Omega_{21} & \Omega_{22} & \Omega_{23} & \Omega_{24} \\ \Omega_{31} & \Omega_{32} & \Omega_{33} & \Omega_{34} \\ \Omega_{41} & \Omega_{42} & \Omega_{43} & \Omega_{44} \end{pmatrix} \longrightarrow \begin{pmatrix} \Omega_{11} & \Omega_{12}+\Omega_{14} & \Omega_{13} \\ \Omega_{21}+\Omega_{41} & \Omega_{22}+\Omega_{42}+\Omega_{24}+\Omega_{44} & \Omega_{23}+\Omega_{43} \\ \Omega_{31} & \Omega_{32}+\Omega_{34} & \Omega_{33} \end{pmatrix}$$

$$(12.55) \quad \begin{pmatrix} \xi_1 \\ \xi_2 \\ \xi_3 \\ \xi_4 \end{pmatrix} \longrightarrow \begin{pmatrix} \xi_1 \\ \xi_2+\xi_4 \\ \xi_3 \end{pmatrix}$$

Collapsing the information state exploits the additivity of the information state.

12.11.2 Mathematical Derivation of Map Integration

For the derivation, it shall prove successful to give the rotation matrix and the shift vectors in (12.52) and (12.53) names. Let us define the variables δ_x, δ_m, and A as follows:

$$(12.56) \quad \delta_x = (d_x \ d_y \ \alpha)^T$$

$$(12.57) \quad \delta_m = (d_x \ d_y \ 0)^T$$

$$(12.58) \quad A = \begin{pmatrix} \cos\alpha & \sin\alpha & 0 \\ -\sin\alpha & \cos\alpha & 0 \\ 0 & 0 & 1 \end{pmatrix}$$

We can then rewrite (12.52) and (12.53) as

$$(12.59) \quad x_t^{j \to k} = \delta_x + A\, x_t^j$$
$$(12.60) \quad m_i^{j \to k} = \delta_m + A\, m_i^j$$

For the full state vector, we now obtain

$$(12.61) \quad y_t^{j \to k} = \Delta + \mathcal{A}\, y_t^j$$

with

$$(12.62) \quad \Delta = (\delta_r \ \delta_m \ \delta_m \ \cdots \ \delta_m)^T$$

(12.63) $$\mathcal{A} = \begin{pmatrix} A_r & 0 & \cdots & 0 \\ 0 & A_m & \cdots & 0 \\ \vdots & \vdots & \ddots & \vdots \\ 0 & 0 & \cdots & A_m \end{pmatrix}$$

The coordinate transformations needed in information space are similar. To see, let the posterior of the j-th robot at time t be defined through the information matrix Ω^j and the information vector ξ^j. The following transformation applies the shift and rotation:

$$
\begin{aligned}
(12.64) \quad p(y^{j \to k} \mid z^j_{1:t}, u^j_{1:t}) \\
= \eta \exp\left\{ -\tfrac{1}{2} y^{j \to k, T} \Omega^{j \to k} y^{j \to k} + y^{j \to k, T} \xi^{j \to k} \right\} \\
= \eta \exp\left\{ -\tfrac{1}{2} (\Delta + \mathcal{A} y^j)^T \Omega^{j \to k} (\Delta + \mathcal{A} y^j) + (\Delta + \mathcal{A} y^j)^T \xi^{j \to k} \right\} \\
= \eta \exp\left\{ -\tfrac{1}{2} y^{jT} \mathcal{A}^T \Omega^{j \to k} \mathcal{A} y^j + y^{jT} \Omega^{j \to k} \Delta - \underbrace{\tfrac{1}{2} \Delta^T \Omega^{j \to k} \Delta}_{\text{const.}} \right. \\
\left. + \underbrace{\Delta^T \xi^{j \to k}}_{\text{const.}} + y^{jT} \mathcal{A}^T \xi^{j \to k} \right\} \\
= \eta \exp\left\{ -\tfrac{1}{2} y^{jT} \mathcal{A}^T \Omega^{j \to k} \mathcal{A} y^j + y^{jT} \Omega^{j \to k} \Delta + y^{jT} \mathcal{A}^T \xi^{j \to k} \right\} \\
= \eta \exp\left\{ -\tfrac{1}{2} y^{jT} \underbrace{\mathcal{A}^T \Omega^{j \to k} \mathcal{A}}_{\Omega^j} y^j + y^{jT} \underbrace{(\Omega^{j \to k} \Delta + \mathcal{A}^T \xi^{j \to k})}_{\xi^j} \right\}
\end{aligned}
$$

Thus, we have

(12.65) $\quad \Omega^j = \mathcal{A}^T \Omega^{j \to k} \mathcal{A}$

(12.66) $\quad \xi^j = (\Omega^{j \to k} \Delta + \mathcal{A}^T \xi^{j \to k})$

From $\mathcal{A}^{-1} = \mathcal{A}^T$, it follows that

$$\Omega^{j \to k} = \mathcal{A} \Omega^j \mathcal{A}^T$$

(12.67) $\quad \xi^{j \to k} = \mathcal{A} (\xi^j - \Omega^{j \to k} \Delta)$

This proves the correctness of lines 2 through 7 in Table 12.7. The remaining soft equality constraints follow directly from the deliberations in Chapter 12.9.3.

12.11.3 Establishing Correspondence

The remaining problem pertains to establishing *correspondence* between different maps, and calculating the rotation α and translation δ. There exists a myriad of possible approaches, hence we will only sketch one possible algorithm here. Clearly, the problem lies in the large number of features that can potentially be matched in both local maps.

A canonical algorithm for landmark-based maps might seek to cache away local configurations of sufficiently nearby landmarks so that a comparison of such local configurations yields good candidates for correspondences. For example, one might identify sets of m nearby landmarks (for a small number m), and calculate relative distances or angles between them. Such a vector of distances or angles will then serve as statistics that can be used to compare two maps. Using hash tables or kd-trees, they may be efficiently accessible, so that a query "do the following m landmarks in the j-th robot's map correspond to any m landmarks in the map of robot k?" can be answered efficiently, at least in approximation. Once an initial correspondence has been identified, we can easily calculate d and α by minimizing the quadratic distance between these m features in both maps.

The fusion then proceeds as follows: First, the fusion operator is called using the d, α, and $C^{j,k}$ computed from those m local features in both maps. Subsequently, additional landmarks are identified for which our correspondence test in Table 12.6 generates a probability that falls below a threshold. A simple termination may occur when no such pairs of landmarks can be found.

A comparison of both components of the unified maps—and specifically of nearby landmarks that are not in correspondence—will then provide a criterion for acceptance of the resulting match. Formally, once the search has terminated, a fusion is accepted if the resulting reduction of the overall likelihood (in logarithmic form) is offset by the number of collapsed features times a constant; this effectively implements a Bayesian MAP estimator with an exponential prior over the number of features in the world.

In general, we note that the search for the optimal correspondence is NP-hard. However, hill climbing tends to work extremely well in practice.

12.11.4 Example

Figure 12.22 shows an example of eight local maps. These maps are obtained by partitioning our benchmark data set discussed previously, into 8 disjoint

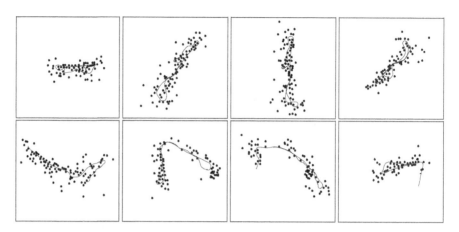

Figure 12.22 Eight local maps obtained by splitting the data into eight sequences.

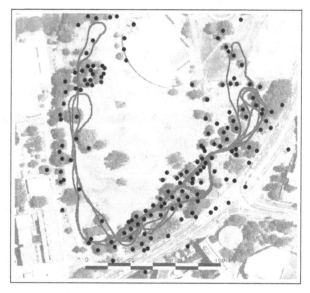

Figure 12.23 A multi-robot SLAM result, obtained using the algorithm described in this chapter. Image courtesy of Yufeng Liu.

12.11 Multi-Robot SLAM

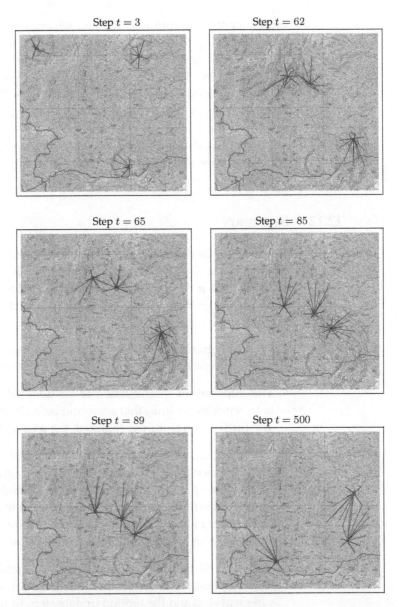

Figure 12.24 Snapshots from our multi-robot SLAM simulation at different points in time. During Steps 62 through 64, vehicle 1 and 2 traverse the same area for the first time; as a result, the uncertainty in their local maps shrinks. Later, in steps 85 through 89, vehicle 2 observes the same landmarks as vehicle 3, with a similar effect on the overall uncertainty. After 500 steps, all landmarks are accurately localized.

subsequences, and running SEIF on each one of those in separation.

By combining those local maps using $m = 4$ local features in a hash table for the correspondence search, the SEIF arrives reliably at the map shown in Figure 12.23. This map, when calculated through $\mu = \Omega^{-1}\xi$, is not just the superposition of the individual local maps that participated. Instead, each local map is slightly bent in the process, which is a result of additively combining the information forms.

Figure 12.24 shows a simulation of three air vehicles. The diagram illustrates that through fusing maps, the uncertainty in each individual map is reduced.

12.12 Summary

This chapter has described an efficient solution to the online SLAM problem: the *sparse extended information filter*, of SEIF. The SEIF is similar to GraphSLAM in that it represents the posterior in information form. However it differs in that past poses are integrated out, which results in an online SLAM algorithm. We learned:

- When integrating out past poses, features that were observed from those poses become linked directly in the information matrix.

- The information matrix tends to be dominated by a small number of between-features links that are found between physically nearby features. The further two features separated, the weaker their link.

- By sparsifying the matrix, which amounts to shifting information through the SEIF in a way that reduces the number of links, the information matrix will remain sparse at all times. Sparseness implies that every element in this matrix is only linked through a non-zero information value to finitely many other elements, regardless of the total map size N. However, sparsification is an approximation, and not an exact operation.

- We observed that for sparse information matrix, *both* essential filtering steps can be carried out in time independent of the map size: the measurement step and the motion update step. In regular information filters, only the measurement update step requires constant time. The motion update step requires more time.

- For a number of steps, the SEIF still requires a state estimate. The SEIF uses an amortized algorithm for recovering these estimates.

- We discussed two techniques for data association. The first is identical to the one discussed for EKF SLAM: Incremental maximum likelihood. This technique associates measurements to the most likely one at each point in time, but never revises a correspondence decision.

- An improved technique recursively searches the tree of all data associations, so as to arrive at a data association vector that maximizes the likelihood of all data associations together. It does this based on an online version of branch-and-bound. This techniques uses a lazy tree expansion technique, in which the log-likelihood values of the data are remembered along the fringe of a partially expanded tree. When the present best leaf arrives at a value that is inferior to a partially expanded value at the fringe, the fringe is expanded until it either becomes inferior itself or a better global solution is found to the data association problem.

- We also discussed the use of SEIF in the context of multi-robot mapping. The algorithm uses as an inner loop a technique for rotating and shifting maps represented in information form, without ever computing the underlying map itself. This operation maintains the sparseness of the information matrix.

- An algorithm was sketched that makes it possible to efficiently carry out the global correspondence between two maps in the multi-robot mapping problem. This algorithm hashes away local feature configurations and uses fast search techniques to establish correspondence. It then fuses maps recursively, and accepts a merge if the resulting maps fit well.

The SEIF is our first efficient online SLAM algorithm in this book. It marries the elegance of the information representation with the idea of integrating out past poses. It is the "lazy" sister of the EKF: Whereas EKFs proactively spread the information of each new measurement through the network of features so as to calculate a correct joint covariance, the SEIF merely accumulates this information, and resolves it slowly over time. The tree-based data association in SEIF is also lazy: It only considered alternative paths to the best known one when necessary. This will be in contrast to the technique described in the coming chapter, which applies particle filters to the problem of data association.

To attain efficient online, however, the SEIF has to make a number of approximations, which make its result less accurate than that of GraphSLAM or the EKF. In particular, the SEIF has two limitations: First, it linearizes only once, just like the EKF. GraphSLAM can re-linearize, which generally improves the accuracy of the result. Second, the SEIF uses an approximation step to maintain sparsity of its information matrix. This sparsity was naturally given for the GraphSLAM algorithm, by nature of the information that was being integrated.

While each of the basic SEIF steps (with known correspondence) can be implemented in "constant time," a final note of caution is in order. If SEIFs were applied to a linear system (meaning, we do not need Taylor series approximations, and the data association is known), the update would be truly constant time. However, because of the need to linearize, we need an estimate of the mean μ_t along with the information state. This estimate is not maintained in the traditional information filter, and recovering it requires a certain amount of time. Our SEIF implementation only approximates it, and the quality of the posterior estimate depends on the quality of this approximation.

12.13 Bibliographical Remarks

The literature on information-theoretic representations in SLAM was already discussed in the previous chapter, insofar it pertains to offline optimization. Information filters have a relatively young history in the SLAM field. In 1997, Csorba developed an information filter that maintained relative information between triplets of three landmarks. He was possibly the first to observe that such information links maintained global correlation information implicitly, paving the way for algorithms with quadratic to linear memory requirements. Newman (2000); Newman and Durrant-Whyte (2001) developed a similar information filter, but left open the question how the landmark-landmark information links are actually acquired. Under the ambitious name *"consistent, convergent, and constant-time SLAM,"* Leonard and Newman further developed this approach into an efficient alignment algorithm, which was successfully applied to an autonomous underwater vehicle using synthetic aperture sonar (Newman and Rikoski 2003). Another seminal algorithm in the field is Paskin's (2003) *thin junction filter* algorithm, which represents the SLAM posterior in a sparse network known as thin junction trees (Pearl 1988; Cowell et al. 1999). The same idea was exploited by Frese (2004), who developed a similar tree factorization of the information matrix for efficient inference. Julier and Uhlmann developed a scalable technique called *covariance intersection*, which sparsely approximates the posterior in a way that provably prevents overconfidence. Their algorithm was successfully implemented on NASA's MARS Rover fleet (Uhlmann et al. 1999). The information filter perspective is also related to early work by Bulata and Devy (1996), whose approach acquired landmark models first in local landmark-centric reference frames, and only later assembles a consistent global map by resolving the relative information between landmarks. Finally, certain "offline" SLAM algorithms that solve the full SLAM problem, such as the ones by Bosse et al. (2004), Gutmann and Konolige (2000), Frese (2004), and Montemerlo and Thrun (2004), have shown to be fast enough to run online on limited-sized data sets.

Multi-robot map merging is discussed in Gutmann and Konolige (2000). Nettleton et al. (2003) was the first to extended the information representation to multi-robot SLAM problems.

They realized that the additivity of information enabled the asynchronous integration of local maps across vehicles. They also realized that the addition of submaps led to efficient communication algorithms, whereby the integration of maps would be possible in time logarithmic in the number of vehicles involved. However, they left open as to how to align such maps, a problem later addressed by Thrun and Liu (2003).

The SEIF algorithm was developed by Thrun et al. (2002); see also Thrun et al. (2004a). To our knowledge, it is the first algorithm that derives the creation of information links between pairs of features from a filtering perspective. A greedy data association algorithm for SEIFs was developed by Liu and Thrun (2003), which was subsequently extended to multi-robot SLAM by Thrun and Liu (2003). The branch-and-bound data association search is due to Hähnel et al. (2003a), based on earlier branch-and-bound methods by Lawler and Wood (1966) and Narendra and Fukunaga (1977). It parallels work by Kuipers et al. (2004), who developed a similar data association technique, albeit not in the context of information theoretic concepts. SEIFs were applied to mapping problems of abandoned mines (Thrun et al. 2004c), involving maps with 10^8 features.

The Victoria Park dataset referenced in this chapter is due to Guivant et al. (2000).

12.14 Exercises

1. Compare the sparseness in GraphSLAM with the sparseness in SEIFs: what are the advantages and disadvantages of each? Provide conditions under which either would be clearly preferable. The more concise your reasoning, the better.

CONSISTENCY

2. An important concept for many SLAM researchers is *consistency*. The SLAM community defines consistency a bit different from the general field of statistics (in which consistency is an asymptotic property).

    Let x be a random vector, and $\mathcal{N}(\mu, \Sigma)$ be a Gaussian estimate of x. The Gaussian is said to be consistent if it meets the following two properties:

    Condition 1: Unbiasedness: The mean μ is an unbiased estimator of x:

    $$E[\mu] \;=\; x$$

    Condition 2: No Overconfidence: The covariance Σ is not overconfident. Let Ξ be the *true* the covariance of the estimator μ:

    $$\Xi \;=\; E[(\mu - E[\mu])\,(\mu - E[\mu])^T]$$

    Then Σ overconfident if there exists a vector $\bar{x}$ for which

    $$\bar{x}^T\, \Sigma^{-1}\, \bar{x} \;>\; \bar{x}^T\, \Xi^{-1}\, \bar{x}$$

    Overconfidence implies that the 95% confidence ellipse of the estimated covariance Σ falls inside or intersects with the true confidence ellipse of the estimator.

Proving consistency is generally difficult for SLAM algorithms. Here we want you to prove or disprove that sparsification retains consistency (see Equation (12.20)). In particular, prove or disprove the following conjecture: *Given a consistent joint $p(a, b, c)$ in Gaussian form, the following approximation will also be consistent:*

$$\tilde{p}(a, b, c) = \frac{p(a, c)\, p(b, c)}{p(c)}$$

3. You are asked to implement the SEIF algorithm for linear Gaussian SLAM. In linear Gaussian SLAM, the motion equation is of the simple additive type

$$x_t \sim \mathcal{N}(x_{t-1} + u_t, R)$$

and the measurement equation is of the type

$$z_t = \mathcal{N}(m_j - x_t, Q)$$

where R and Q are diagonal covariances matrices. The data associations are known in linear Gaussian SLAM.

   (a) Run it on simple simulations and verify the correctness of your implementation.

   (b) Graph the error of the SEIF as a function of the sparseness of the information matrix. What can you learn?

   (c) Graph the computation time of your SEIF implementation as a function of the sparseness of the information matrix. Report any interesting findings.

4. The sparsification rule in SEIFs conditions away all passive features m^-, by assuming $m^- = 0$. Why is this done? What would be the update equation if these features would not be conditioned away? Would the result be more accurate or less accurate? Would the computation be more or less efficient? Be concise.

5. At present, SEIFs linearize as soon as a measurement or a motion command is integrated into the filter. Brainstorm about a SEIF algorithm that allows for retro-actively altering the linearization. How would the posterior of such an algorithm be represented? What would be the representation of the information matrix?

13 The FastSLAM Algorithm

We will now turn our attention to the *particle filter* approach to SLAM. We already encountered particle filters in several chapters of this book. We noted that particle filters are at the core of some of the most effective robotics algorithms. This raises the question as to whether particle filters are applicable to the SLAM problem. Unfortunately, particle filters are subject to the curse of dimensionality: whereas Gaussians scale between linearly and quadratically with the number of dimensions of the estimation problem, particle filters scale exponentially! A straightforward implementation of particle filters for the SLAM problem would be doomed to fail, due to the large number of variables involved in describing a map.

The algorithm in this chapter is based on an important characteristic of the SLAM problem, which has not yet been explicitly discussed in this book. Specifically, the full SLAM problem with known correspondences possesses a conditional independence between any two disjoint sets of features in the map, given the robot pose. Put differently, if an oracle told us the true robot path, we could estimate the location of all features independently of each other. Dependencies in these estimates arise *only* through robot pose uncertainty. This structural observation will make it possible to apply a version of particle filters to SLAM known as *Rao-Blackwellized particle filters*. Rao-Blackwellized particle filters use particles to represent the posterior over some variables, along with Gaussians (or some other parametric PDF) to represent all other variables.

RAO-BLACKWELLIZED PARTICLE FILTER

CONDITIONAL INDEPENDENCE

FastSLAM uses particle filters for estimating the robot path. As we shall see, for each of these particles the individual map errors are *conditionally independent*. Hence the mapping problem can be factored into many separate problems, one for each feature in the map. FastSLAM estimates these map feature locations by EKFs, but using a separate low-dimensional EKF for each individual feature. This is fundamentally different from SLAM algorithms discussed in previous chapters, which all use a single Gaussian to estimate the location of all features jointly.

The basic algorithm can be implemented in time logarithmic in the number of features. Hence, FastSLAM offers computational advantages over plain EKF implementations and many of its descendants. The key advantage of FastSLAM, however, stems from the fact that data association decisions can be made on a per-particle basis. As a result, the filter maintains posteriors over multiple data associations, not just the most likely one. This is in stark contrast to all SLAM algorithms discussed so far, which track only a single data association at any point in time. In fact, by sampling over data associations, FastSLAM approximates the full posterior, not just the maximum likelihood data association. As shown empirically, the ability to pursue multiple data associations simultaneously makes FastSLAM significantly more robust to data association problems than algorithms based on incremental maximum likelihood data association.

Another advantage of FastSLAM over other SLAM algorithms arises from the fact that particle filters can cope with non-linear robot motion models, whereas previous techniques approximate such models via linear functions. This is important when the kinematics are highly non-linear, or when the pose uncertainty is relatively high.

The use of particle filters creates the unusual situation that FastSLAM solves both the *full SLAM problem* and the *online SLAM problem*. As we shall see, FastSLAM is formulated to calculate the full path posterior—only the full path renders feature locations conditionally independent. However, because particle filters estimate one pose at-a-time, FastSLAM is indeed an online algorithm. Hence it also solves the online SLAM problem. Among all SLAM algorithms discussed thus far, FastSLAM is the only algorithm that fits both categories.

This chapter describes several instantiations of the FastSLAM algorithm. FastSLAM 1.0 is the original FastSLAM algorithm, which is conceptually simple and easy to implement. In certain situations, however, the particle filter component of FastSLAM 1.0 generates samples inefficiently. The algorithm FastSLAM 2.0 overcomes this problem through an improved proposal distribution, but at the expense of an implementation that is significantly more involved (as is the mathematical derivation). Both of these FastSLAM algorithms assume the feature-based sensor model discussed earlier. The application of FastSLAM to range sensors results in an algorithm that solves the SLAM problem in the context of occupancy grid maps. For all algorithms,

13.1 The Basic Algorithm

Particles in the basic FastSLAM algorithm are of the form shown in Table 13.1. Each particle contains an estimated robot pose, denoted $x_t^{[k]}$, and a set of Kalman filters with mean $\mu_{j,t}^{[k]}$ and covariance $\Sigma_{j,t}^{[k]}$, one for each feature m_j in the map. Here $[k]$ is the index of the particle. As usual, the total number of particles is denoted M.

The basic FastSLAM update step is depicted in Table 13.2. Barring the many details of the update step, the main loop is in large parts identical to the particle filter, as discussed in Chapter 4 of this book. The initial step involves the retrieval of a particle representing the posterior at time $t-1$, and the sampling of a robot pose for time t using the probabilistic motion model. The step that follows updates the EKFs for the observed features, using the standard EKF update equation. This update is not part of the vanilla particle filter, but it is necessary in FastSLAM to learn a map. The final steps are concerned with the calculation of an importance weight, which are then used to resample the particles.

We will now investigate each of these steps in more detail and derive them from the basic mathematical properties of the SLAM problem. We note that our derivation presupposes that FastSLAM solves the full SLAM problem, not the online problem. However, as shall become apparent further below in this chapter, FastSLAM is a solution to both of these problems: Each particle can be thought of as a sample in path space as required for the full SLAM problem, but the update only requires the most recent pose. For this reason, FastSLAM can be run just like a filter.

13.2 Factoring the SLAM Posterior

FastSLAM's key mathematical insight pertains to the fact that the full SLAM posterior, $p(y_{1:t} \mid z_{1:t}, u_{1:t})$ in Equation (10.2) can be written in the factored form

$$(13.1) \quad p(y_{1:t} \mid z_{1:t}, u_{1:t}, c_{1:t}) = p(x_{1:t} \mid z_{1:t}, u_{1:t}, c_{1:t}) \prod_{n=1}^{N} p(m_n \mid x_{1:t}, z_{1:t}, c_{1:t})$$

This *factorization* states that the calculation of the posterior over paths and maps can be decomposed into $N+1$ probabilities.

	robot path	feature 1	feature 2	...	feature N
Particle $k=1$	$x_{1:t}^{[1]} = \{(x\ y\ \theta)^T\}_{1:t}^{[1]}$	$\mu_1^{[1]}, \Sigma_1^{[1]}$	$\mu_2^{[1]}, \Sigma_2^{[1]}$	...	$\mu_N^{[1]}, \Sigma_N^{[1]}$
Particle $k=2$	$x_{1:t}^{[2]} = \{(x\ y\ \theta)^T\}_{1:t}^{[2]}$	$\mu_1^{[2]}, \Sigma_1^{[2]}$	$\mu_2^{[2]}, \Sigma_2^{[2]}$	...	$\mu_N^{[2]}, \Sigma_N^{[2]}$
$\vdots$					
Particle $k=M$	$x_{1:t}^{[M]} = \{(x\ y\ \theta)^T\}_{1:t}^{[M]}$	$\mu_1^{[M]}, \Sigma_1^{[M]}$	$\mu_2^{[M]}, \Sigma_2^{[M]}$	...	$\mu_N^{[M]}, \Sigma_N^{[M]}$

Figure 13.1 Particles in FastSLAM are composed of a path estimate and a set of estimators of individual feature locations with associated covariances.

- Do the following M times:
  - **Retrieval.** Retrieve a pose $x_{t-1}^{[k]}$ from the particle set Y_{t-1}.
  - **Prediction.** Sample a new pose $x_t^{[k]} \sim p(x_t \mid x_{t-1}^{[k]}, u_t)$.
  - **Measurement update.** For each observed feature z_t^i identify the correspondence j for the measurement z_t^i, and incorporate the measurement z_t^i into the corresponding EKF, by updating the mean $\mu_{j,t}^{[k]}$ and covariance $\Sigma_{j,t}^{[k]}$.
  - **Importance weight.** Calculate the importance weight $w^{[k]}$ for the new particle.
- **Resampling.** Sample, with replacement, M particles, where each particle is sampled with a probability proportional to $w^{[k]}$.

Figure 13.2 The basic steps of the FastSLAM algorithm.

13.2 Factoring the SLAM Posterior

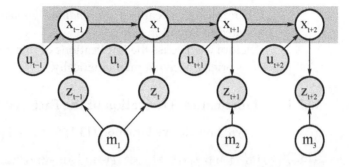

Figure 13.3 The SLAM problem depicted as Bayes network graph. The robot moves from pose x_{t-1} to pose x_{t+2}, driven by a sequence of controls. At each pose x_t it observes a nearby feature in the map $m = \{m_1, m_2, m_3\}$. This graphical network illustrates that the pose variables "separate" the individual features in the map from each other. If the poses are known, there remains no other path involving variables whose value is not known, between any two features in the map. This lack of a path renders the posterior of any two features in the map conditionally independent (given the poses).

FastSLAM uses a particle filter to compute the posterior over robot paths, denoted $p(x_{1:t} \mid z_{1:t}, u_{1:t}, c_{1:t})$. For each feature in the map, FastSLAM now uses a separate estimator over its location $p(m_n \mid x_{1:t}, c_{1:t}, z_{1:t})$ one for each $n = 1, \ldots, N$. Thus, in total there are $N + 1$ posteriors in FastSLAM. The feature estimators are conditioned on the robot path, which means we will have a separate copy of each feature estimator, one for each particle. With M particles, the number of filters will actually be $1 + MN$. The product of these probabilities represents the desired posterior in a factored way. As we shall show below, this factored representation is exact, not just an approximation. It is a generic characteristic of the SLAM problem.

To illustrate the correctness of this factorization, Figure 13.3 depicts the data acquisition process graphically, in the form of a dynamic Bayesian network. As this graph suggests, each measurement $z_1, \ldots, z_t$ is a functions of the position of the corresponding feature, along with the robot pose at the time the measurement was taken. Knowledge of the robot path separates the individual feature estimation problems and renders them independent of one another, in the sense that no direct path exists in this graphical depiction from one feature to another that would *not* involve variables on the robot's path. Knowledge of the exact location of one feature will therefore tell us nothing about the locations of other features, if the robot path is known.

This implies that features are *conditionally independent* given the robot path, as stated in Equation (13.1).

Before we discuss the implications of this property to the SLAM problem, let us briefly derive it mathematically.

13.2.1 Mathematical Derivation of the Factored SLAM Posterior

We will now derive Equation (13.1) from first principles. Clearly, we have

$$(13.2) \quad p(y_{1:t} \mid z_{1:t}, u_{1:t}, c_{1:t}) = p(x_{1:t} \mid z_{1:t}, u_{1:t}, c_{1:t}) \, p(m \mid x_{1:t}, z_{1:t}, c_{1:t})$$

It therefore suffices to show that the second term on the right-hand side factors as follows:

$$(13.3) \quad p(m \mid x_{1:t}, c_{1:t}, z_{1:t}) = \prod_{n=1}^{N} p(m_n \mid x_{1:t}, c_{1:t}, z_{1:t})$$

We will prove this by mathematical induction. Our derivation requires the distinction of two possible cases, depending on whether or not the feature m_n was observed in the most recent measurement. In particular, if $c_t \neq n$, the most recent measurement z_t has no effect on the posterior, and neither has the robot pose x_t or the correspondence c_t. Thus, we obtain:

$$(13.4) \quad p(m_n \mid x_{1:t}, c_{1:t}, z_{1:t}) = p(m_n \mid x_{1:t-1}, c_{1:t-1}, z_{1:t-1})$$

If $c_t = n$ and hence $m_n = m_{c_t}$ was observed by the most recent measurement z_t, the situation calls for applying Bayes rule, followed by some standard simplifications:

$$(13.5) \quad p(m_{c_t} \mid x_{1:t}, c_{1:t}, z_{1:t}) = \frac{p(z_t \mid m_{c_t}, x_{1:t}, c_{1:t}, z_{1:t-1}) \, p(m_{c_t} \mid x_{1:t}, c_{1:t}, z_{1:t-1})}{p(z_t \mid x_{1:t}, c_{1:t}, z_{1:t-1})}$$

$$= \frac{p(z_t \mid x_t, m_{c_t}, c_t) \, p(m_{c_t} \mid x_{1:t-1}, c_{1:t-1}, z_{1:t-1})}{p(z_t \mid x_{1:t}, c_{1:t}, z_{1:t-1})}$$

This gives us the following expression for the probability of the observed feature m_{c_t}:

$$(13.6) \quad p(m_{c_t} \mid x_{1:t-1}, c_{1:t-1}, z_{1:t-1}) = \frac{p(m_{c_t} \mid x_{1:t}, c_{1:t}, z_{1:t}) \, p(z_t \mid x_{1:t}, c_{1:t}, z_{1:t-1})}{p(z_t \mid x_t, m_{c_t}, c_t)}$$

The proof of the correctness of (13.3) is now carried out by induction. Let us assume that the posterior at time $t - 1$ is already factored:

$$(13.7) \quad p(m \mid x_{1:t-1}, c_{1:t-1}, z_{1:t-1}) = \prod_{n=1}^{N} p(m_n \mid x_{1:t-1}, c_{1:t-1}, z_{1:t-1})$$

13.2 Factoring the SLAM Posterior

This statement is trivially true for $t = 1$, since in the beginning of time the robot has no knowledge about any feature; hence all estimates must be independent. At time t, the posterior is of the following form:

$$
\begin{aligned}
(13.8) \quad p(m \mid x_{1:t}, c_{1:t}, z_{1:t}) &= \frac{p(z_t \mid m, x_{1:t}, c_{1:t}, z_{1:t-1}) \, p(m \mid x_{1:t}, c_{1:t}, z_{1:t-1})}{p(z_t \mid x_{1:t}, c_{1:t}, z_{1:t-1})} \\
&= \frac{p(z_t \mid x_t, m_{c_t}, c_t) \, p(m \mid x_{1:t-1}, c_{1:t-1}, z_{1:t-1})}{p(z_t \mid x_{1:t}, c_{1:t}, z_{1:t-1})}
\end{aligned}
$$

Plugging in our inductive hypothesis (13.7) gives us:

$$
\begin{aligned}
(13.9) \quad & p(m \mid x_{1:t}, c_{1:t}, z_{1:t}) \\
&= \frac{p(z_t \mid x_t, m_{c_t}, c_t)}{p(z_t \mid x_{1:t}, c_{1:t}, z_{1:t-1})} \prod_{n=1}^{N} p(m_n \mid x_{1:t-1}, c_{1:t-1}, z_{1:t-1}) \\
&= \frac{p(z_t \mid x_t, m_{c_t}, c_t)}{p(z_t \mid x_{1:t}, c_{1:t}, z_{1:t-1})} \underbrace{p(m_{c_t} \mid x_{1:t-1}, c_{1:t-1}, z_{1:t-1})}_{Eq.\ (13.6)} \\
& \qquad\qquad\qquad\qquad\qquad\qquad \underbrace{\prod_{n \ne c_t} p(m_n \mid x_{1:t-1}, c_{1:t-1}, z_{1:t-1})}_{Eq.\ (13.4)} \\
&= p(m_{c_t} \mid x_{1:t}, c_{1:t}, z_{1:t}) \prod_{n \ne c_t} p(m_n \mid x_{1:t}, c_{1:t}, z_{1:t}) \\
&= \prod_{n=1}^{N} p(m_n \mid x_{1:t}, c_{1:t}, z_{1:t})
\end{aligned}
$$

Notice that we have substituted Equations (13.4) and (13.6) as indicated. This shows the correctness of Equation (13.3). The correctness of the main form (13.1) follows now directly from this result and the following generic transformation:

$$
\begin{aligned}
(13.10) \quad p(y_{1:t} \mid z_{1:t}, u_{1:t}, c_{1:t}) &= p(x_{1:t} \mid z_{1:t}, u_{1:t}, c_{1:t}) \, p(m \mid x_{1:t}, z_{1:t}, u_{1:t}, c_{1:t}) \\
&= p(x_{1:t} \mid z_{1:t}, u_{1:t}, c_{1:t}) \, p(m \mid x_{1:t}, c_{1:t}, z_{1:t}) \\
&= p(x_{1:t} \mid z_{1:t}, u_{1:t}, c_{1:t}) \prod_{n=1}^{N} p(m_n \mid x_{1:t}, c_{1:t}, z_{1:t})
\end{aligned}
$$

We note that conditioning on the entire path $x_{1:t}$ is indeed essential for this result. Conditioning on the most recent pose x_t would be insufficient as conditioning variable, as dependencies may arise through previous poses.

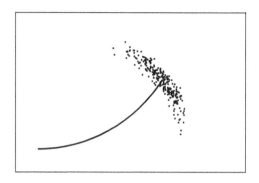

Figure 13.4 Samples drawn from the probabilistic motion model.

13.3 FastSLAM with Known Data Association

The factorial nature of the posterior provides us with significant computational advantages over SLAM algorithms that estimate an unstructured posterior distribution. FastSLAM exploits the factored representation by maintaining $MN + 1$ filters, M for each factor in (13.1). By doing so, all $MN + 1$ filters are low-dimensional.

As noted, FastSLAM estimates the path posterior using a particle filter. The map feature locations are estimated using EKFs. Because of our factorization, FastSLAM can maintain a separate EKF for each feature—which makes the update more efficient than in EKF SLAM. Each individual EKF is conditioned on a robot path. Hence, each particle possesses its own set of EKFs. In total there are NM EKFs, one for each feature in the map and one for each particle in the particle filter.

Let us begin with the FastSLAM algorithm in the case of known data association. *Particles* in FastSLAM will be denoted

$$(13.11) \quad Y_t^{[k]} = \left\langle x_t^{[k]}, \mu_{1,t}^{[k]}, \Sigma_{1,t}^{[k]}, \ldots, \mu_{N,t}^{[k]}, \Sigma_{N,t}^{[k]} \right\rangle$$

As usual, the bracketed notation $[k]$ indicates the index of the particle; $x_t^{[k]}$ is the path estimate of the robot, and $\mu_{n,t}^{[k]}$ and $\Sigma_{n,t}^{[k]}$ are the mean and variance of the Gaussian representing the n-th feature location, relative to the k-th particle. Together, all these quantities form the k-th particle $Y_t^{[k]}$, of which there are a total of M in the FastSLAM posterior.

Filtering, or calculating the posterior at time t from the one at time $t-1$, involves generating a new particle set Y_t from Y_{t-1}, the particle set one time step earlier. This new particle set incorporates a new control u_t and a measurement z_t with associated correspondence c_t. This update is performed in the following steps:

1. **Extending the path posterior by sampling new poses.** FastSLAM 1.0 uses the control u_t to sample new robot pose x_t for each particle in Y_{t-1}. More specifically, consider the k-the particle $Y_t^{[k]}$. FastSLAM 1.0 samples the pose x_t in accordance with this k-th particle, by drawing a sample according to the motion posterior

$$x_t^{[k]} \sim p(x_t \mid x_{t-1}^{[k]}, u_t) \tag{13.12}$$

Here $x_{t-1}^{[k]}$ is the posterior estimate for the robot location at time $t-1$, residing in the k-th particle. The resulting sample $x_t^{[k]}$ is then added to a temporary set of particles, along with the path of previous poses, $x_{1:t-1}^{[k]}$. The sampling step is graphically depicted in Figure 13.4, which illustrates a set of pose particles drawn from a single initial pose.

2. **Updating the observed feature estimate.** Next, FastSLAM 1.0 updates the posterior over the feature estimates, represented by the mean $\mu_{n,t-1}^{[k]}$ and the covariance $\Sigma_{n,t-1}^{[k]}$. The updated values are then added to the temporary particle set, along with the new pose.

The exact update equation depends on whether or not a feature m_n was observed at time t. For $n \neq c_t$ we did *not* observe feature n, we already established in Equation (13.4) that the posterior over the feature remains unchanged. This implies the simple update:

$$\left\langle \mu_{n,t}^{[k]}, \Sigma_{n,t}^{[k]} \right\rangle = \left\langle \mu_{n,t-1}^{[k]}, \Sigma_{n,t-1}^{[k]} \right\rangle \tag{13.13}$$

For the observed feature $n = c_t$, the update is specified through Equation (13.5), restated here with the normalizer denoted by η:

$$\begin{aligned}
p(m_{c_t} \mid x_{1:t}, z_{1:t}, c_{1:t}) \\
= \eta\, p(z_t \mid x_t, m_{c_t}, c_t)\, p(m_{c_t} \mid x_{1:t-1}, z_{1:t-1}, c_{1:t-1})
\end{aligned} \tag{13.14}$$

The probability $p(m_{c_t} \mid x_{1:t-1}, c_{1:t-1}, z_{1:t-1})$ at time $t-1$ is represented by a Gaussian with mean $\mu_{n,t-1}^{[k]}$ and covariance $\Sigma_{n,t-1}^{[k]}$. For the new estimate at time t to be Gaussian as well, FastSLAM linearizes the perceptual model $p(z_t \mid x_t, m_{c_t}, c_t)$ in the same way as EKF SLAM. As usual, we approximate the measurement function h by Taylor expansion:

$$(13.15) \quad h(m_{c_t}, x_t^{[k]}) \approx \underbrace{h(\mu_{c_t,t-1}^{[k]}, x_t^{[k]})}_{=:\, \hat{z}_t^{[k]}} + \underbrace{h'(x_t^{[k]}, \mu_{c_t,t-1}^{[k]})}_{=:\, H_t^{[k]}}(m_{c_t} - \mu_{c_t,t-1}^{[k]})$$

$$= \hat{z}_t^{[k]} + H_t^{[k]}(m_{c_t} - \mu_{c_t,t-1}^{[k]})$$

Here the derivative h' is taken with respect to the feature coordinates m_{c_t}. This linear approximation is tangent to h at $x_t^{[k]}$ and $\mu_{c_t,t-1}^{[k]}$. Under this approximation, the posterior for the location of feature c_t is indeed Gaussian. The new mean and covariance are obtained using the standard EKF measurement update:

$$(13.16) \quad K_t^{[k]} = \Sigma_{c_t,t-1}^{[k]} H_t^{[k]T} (H_t^{[k]} \Sigma_{c_t,t-1}^{[k]} H_t^{[k]T} + Q_t)^{-1}$$

$$(13.17) \quad \mu_{c_t,t}^{[k]} = \mu_{c_t,t-1}^{[k]} + K_t^{[k]}(z_t - \hat{z}_t^{[k]})$$

$$(13.18) \quad \Sigma_{c_t,t}^{[k]} = (I - K_t^{[k]} H_t^{[k]}) \Sigma_{c_t,t-1}^{[k]}$$

Steps 1 and 2 are repeated M times, resulting in a temporary set of M particles.

3. **Resampling.** In a final step, FastSLAM resamples this set of particles. We already encountered resampling in a number of algorithms. FastSLAM draws from its temporary set M particles (with replacement) according to a yet-to-be-defined importance weight. The resulting set of M particles then forms the new and final particle set, Y_t. The necessity to resample arises from the fact that the particles in the temporary set are not distributed according to the desired posterior: Step 1 generates poses x_t only in accordance with the most recent control u_t, paying no attention to the measurement z_t. As the reader should know well by now, resampling is the common technique in particle filtering to correct for such mismatches.

This situation is once again illustrated in Figure 13.5, for a 1-D example. Here the dashed line symbolizes the *proposal distribution*, which is the distribution at which particles are generated, and the solid line is the target distribution. In FastSLAM, the proposal distribution does not depend on z_t, but the target distribution does. By weighing particles as shown in the bottom of this figure, and resampling according to those weights, the resulting particle set indeed approximates the target distribution.

13.3 FastSLAM with Known Data Association

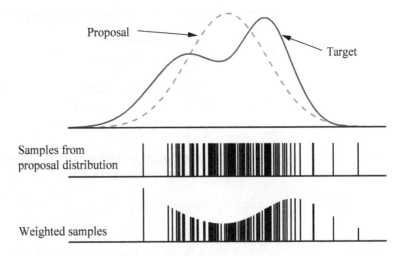

Figure 13.5 Samples cannot be drawn conveniently from the target distribution (shown as a solid line). Instead, the importance sampler draws samples from the proposal distribution (dashed line), which has a simpler form. Below, samples drawn from the proposal distribution are drawn with lengths proportional to their importance weights.

To determine the importance factor, it will prove useful to calculate the actual proposal distribution of the path particles in the temporary set. Under the assumption that the set of path particles in Y_{t-1} is distributed according to $p(x_{1:t-1} \mid z_{1:t-1}, u_{1:t-1}, c_{1:t-1})$ (which is an asymptotically correct approximation), path particles in the temporary set are distributed according to:

$$(13.19) \quad p(x_{1:t}^{[k]} \mid z_{1:t-1}, u_{1:t}, c_{1:t-1}) = p(x_t^{[k]} \mid x_{t-1}^{[k]}, u_t)\, p(x_{1:t-1}^{[k]} \mid z_{1:t-1}, u_{1:t-1}, c_{1:t-1})$$

The factor $p(x_t^{[k]} \mid x_{t-1}^{[k]}, u_t)$ is the sampling distribution used in Equation (13.12).

The *target distribution* takes into account the measurement at time z_t, along with the correspondence c_t:

$$(13.20) \quad p(x_{1:t}^{[k]} \mid z_{1:t}, u_{1:t}, c_{1:t})$$

The resampling process accounts for the difference of the target and the proposal distribution. As usual, the *importance factor* for resampling is given by the quotient of the target and the proposal distribution:

$$(13.21) \quad w_t^{[k]} = \frac{\text{target distribution}}{\text{proposal distribution}}$$

$$= \frac{p(x_{1:t}^{[k]} \mid z_{1:t}, u_{1:t}, c_{1:t})}{p(x_{1:t}^{[k]} \mid z_{1:t-1}, u_{1:t}, c_{1:t-1})}$$

$$= \eta \, p(z_t \mid x_t^{[k]}, c_t)$$

The last transformation is a direct consequence of the following transformation of the enumerator in (13.21):

$$(13.22) \quad p(x_{1:t}^{[k]} \mid z_{1:t}, u_{1:t}, c_{1:t})$$

$$= \eta \, p(z_t \mid x_{1:t}^{[k]}, z_{1:t-1}, u_{1:t}, c_{1:t}) \, p(x_{1:t}^{[k]} \mid z_{1:t-1}, u_{1:t}, c_{1:t})$$

$$= \eta \, p(z_t \mid x_t^{[k]}, c_t) \, p(x_{1:t}^{[k]} \mid z_{1:t-1}, u_{1:t}, c_{1:t-1})$$

To calculate the probability $p(z_t \mid x_t^{[k]}, c_t)$ in (13.21), it will be necessary to transform it further. In particular, this probability is equivalent to the following integration, where we once again omit variables irrelevant to the prediction of sensor measurements:

$$(13.23) \quad w_t^{[k]} = \eta \int p(z_t \mid m_{c_t}, x_t^{[k]}, c_t) \, p(m_{c_t} \mid x_t^{[k]}, c_t) \, dm_{c_t}$$

$$= \eta \int p(z_t \mid m_{c_t}, x_t^{[k]}, c_t) \, \underbrace{p(m_{c_t} \mid x_{1:t-1}^{[k]}, z_{1:t-1}, c_{1:t-1})}_{\sim \mathcal{N}(\mu_{c_t,t-1}^{[k]}, \Sigma_{c_t,t-1}^{[k]})} \, dm_{c_t}$$

Here $\mathcal{N}(x; \mu, \Sigma)$ denotes a Gaussian distribution over the variable x with mean μ and covariance Σ.

The integration in (13.23) involves the estimate of the observed feature location at time t and the measurement model. To calculate (13.23) in closed form, FastSLAM employs the very same linear approximation used in the measurement update in Step 2. In particular, the importance factor is given by

$$(13.24) \quad w_t^{[k]} \approx \eta \, |2\pi Q_t^{[k]}|^{-\frac{1}{2}} \exp\left\{-\tfrac{1}{2}(z_t - \hat{z}_t^{[k]}) Q_t^{[k]-1}(z_t - \hat{z}_t^{[k]})\right\}$$

with the covariance

$$(13.25) \quad Q_t^{[k]} = H_t^{[k]T} \Sigma_{n,t-1}^{[k]} H_t^{[k]} + Q_t$$

13.3 FastSLAM with Known Data Association

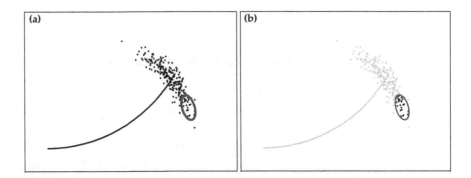

Figure 13.6 Mismatch between proposal and posterior distributions: (a) illustrates the forward samples generated by FastSLAM 1.0, and the posterior induced by the measurement (ellipse). Diagram (b) shows the sample set after the resampling step.

This expression is the probability of the actual measurement z_t under the Gaussian. It results from the convolution of the distributions in (13.23), exploiting our linear approximation of h. The resulting importance weights are used to draw with replacement M new samples from the temporary sample set. Through this resampling process, particles survive in proportion of their measurement probability.

These three steps together constitute the update rule of the FastSLAM 1.0 algorithm for SLAM problems with known data association. We note that the execution time of the update does not depend on the total path length t. In fact, only the most recent pose $x_{t-1}^{[k]}$ is used in the process of generating a new particle at time t. Consequently, past poses can safely be discarded. This has the pleasing consequence that neither the time requirements nor the memory requirements of FastSLAM depend on the total number of time steps spent on data acquisition.

A summary of the FastSLAM 1.0 algorithm with known data association is provided in Table 13.1. For simplicity, this implementation assumes that only a single feature is measured at each point in time. This algorithm implements the various update steps in a straightforward manner. Its implementation is relatively straightforward; in fact, FastSLAM 1.0 happens to be one of the easiest SLAM algorithms to implement!

1: **Algorithm FastSLAM 1.0_known_correspondence**(z_t, c_t, u_t, Y_{t-1}):
2:     for $k = 1$ to M do   // loop over all particles
3:         retrieve $\left\langle x_{t-1}^{[k]}, \left\langle \mu_{1,t-1}^{[k]}, \Sigma_{1,t-1}^{[k]} \right\rangle, \ldots, \left\langle \mu_{N,t-1}^{[k]}, \Sigma_{N,t-1}^{[k]} \right\rangle \right\rangle$ from Y_{t-1}
4:         $x_t^{[k]} \sim p(x_t \mid x_{t-1}^{[k]}, u_t)$ // sample pose
5:         $j = c_t$ // observed feature
6:         if feature j never seen before
7:             $\mu_{j,t}^{[k]} = h^{-1}(z_t, x_t^{[k]})$ // initialize mean
8:             $H = h'(x_t^{[k]}, \mu_{j,t}^{[k]})$ // calculate Jacobian
9:             $\Sigma_{j,t}^{[k]} = H^{-1} Q_t (H^{-1})^T$ // initialize covariance
10:           $w^{[k]} = p_0$ // default importance weight
11:         else
12:             $\hat{z} = h(\mu_{j,t-1}^{[k]}, x_t^{[k]})$ // measurement prediction
13:             $H = h'(x_t^{[k]}, \mu_{j,t-1}^{[k]})$ // calculate Jacobian
14:             $Q = H \, \Sigma_{j,t-1}^{[k]} \, H^T + Q_t$ // measurement covariance
15:             $K = \Sigma_{j,t-1}^{[k]} H^T Q^{-1}$ // calculate Kalman gain
16:             $\mu_{j,t}^{[k]} = \mu_{j,t-1}^{[k]} + K(z_t - \hat{z})$ // update mean
17:             $\Sigma_{j,t}^{[k]} = (I - K\,H)\Sigma_{j,t-1}^{[k]}$ // update covariance
18:             $w^{[k]} = |2\pi Q|^{-\frac{1}{2}} \exp\left\{-\frac{1}{2}(z_t - \hat{z}_n)^T \right.$
                                             $\left. Q^{-1}(z_t - \hat{z}_n) \right\}$ // importance factor
19:         endif
20:         for all other features $j' \neq j$ do   // unobserved features
21:             $\mu_{j',t}^{[k]} = \mu_{j',t-1}^{[k]}$ // leave unchanged
22:             $\Sigma_{j',t}^{[k]} = \Sigma_{j',t-1}^{[k]}$
23:         endfor
24:     endfor
25:     $Y_t = \emptyset$ // initialize new particle set
26:     do M times   // resample M particles
27:         draw random k with probability $\propto w^{[k]}$ // resample
28:         add $\left\langle x_t^{[k]}, \left\langle \mu_{1,t}^{[k]}, \Sigma_{1,t}^{[k]} \right\rangle, \ldots, \left\langle \mu_N^{[k]}, \Sigma_N^{[k]} \right\rangle \right\rangle$ to Y_t
29:     endfor
30:     return Y_t

Table 13.1 FastSLAM 1.0 with known correspondence.

13.4 Improving the Proposal Distribution

FASTSLAM 2.0   *FastSLAM 2.0* is largely equivalent to FastSLAM 1.0, with one important exception: Its proposal distribution takes the measurement z_t into account, when sampling the pose x_t. By doing so it can overcome a key limitation of FastSLAM 1.0.

On the surface, the difference looks rather marginal: The reader may recall that FastSLAM 1.0 samples poses based on the control u_t only, and then uses the measurement z_t to calculate importance weights. This is problematic when the accuracy of control is low relative to the accuracy of the robot's sensors. Such a situation is illustrated in Figure 13.6. Here the proposal generates a large spectrum of samples shown in Figure 13.6a, but only a small subset of these samples have high likelihood, as indicated by the ellipsoid. After resampling, only particles within the ellipsoid "survive" with reasonably high likelihood. FastSLAM 2.0 avoids this problem by sampling poses based on the measurement z_t in addition to the control u_t. Thus, as a result, FastSLAM 2.0 is more efficient than FastSLAM 1.0. On the downside, FastSLAM 2.0 is more difficult to implement than FastSLAM 1.0, and its mathematical derivation is more involved.

13.4.1 Extending the Path Posterior by Sampling a New Pose

In FastSLAM 2.0, the pose $x_t^{[k]}$ is drawn from the posterior

$$(13.26) \quad x_t^{[k]} \sim p(x_t \mid x_{1:t-1}^{[k]}, u_{1:t}, z_{1:t}, c_{1:t})$$

This distribution differs from the proposal distribution provided in (13.12) in that (13.26) takes the measurement z_t into consideration, along with the correspondence c_t. Specifically, the expression in (13.26) conditions on $z_{1:t}$, whereas the pose sampler in FastSLAM 1.0 conditions on $z_{1:t-1}$.

Unfortunately, it also comes with more complex math. In particular, the mechanism for sampling from (13.26) requires further analysis. First, we rewrite (13.26) in terms of the "known" distributions, such as the measurement and motion models, and the Gaussian feature estimates in the k-th particle.

$$(13.27) \quad p(x_t \mid x_{1:t-1}^{[k]}, u_{1:t}, z_{1:t}, c_{1:t})$$

$$\underset{\text{Bayes}}{=} \frac{p(z_t \mid x_t, x_{1:t-1}^{[k]}, u_{1:t}, z_{1:t-1}, c_{1:t}) \, p(x_t \mid x_{1:t-1}^{[k]}, u_{1:t}, z_{1:t-1}, c_{1:t})}{p(z_t \mid x_{1:t-1}^{[k]}, u_{1:t}, z_{1:t-1}, c_{1:t})}$$

$$= \eta^{[k]} \, p(z_t \mid x_t, x_{1:t-1}^{[k]}, u_{1:t}, z_{1:t-1}, c_{1:t}) \, p(x_t \mid x_{1:t-1}^{[k]}, u_{1:t}, z_{1:t-1}, c_{1:t})$$

$$\stackrel{\text{Markov}}{=} \eta^{[k]} \; p(z_t \mid x_t, x_{1:t-1}^{[k]}, u_{1:t}, z_{1:t-1}, c_{1:t}); p(x_t \mid x_{t-1}^{[k]}, u_t)$$

$$= \eta^{[k]} \int p(z_t \mid m_{c_t}, x_t, x_{1:t-1}^{[k]}, u_{1:t}, z_{1:t-1}, c_{1:t})$$

$$\qquad p(m_{c_t} \mid x_t, x_{1:t-1}^{[k]}, u_{1:t}, z_{1:t-1}, c_{1:t}) \, dm_{c_t} \; p(x_t \mid x_{t-1}^{[k]}, u_t)$$

$$\stackrel{\text{Markov}}{=} \eta^{[k]} \int \underbrace{p(z_t \mid m_{c_t}, x_t, c_t)}_{\sim \mathcal{N}(z_t; h(m_{c_t}, x_t), Q_t)} \underbrace{p(m_{c_t} \mid x_{1:t-1}^{[k]}, z_{1:t-1}, c_{1:t-1})}_{\sim \mathcal{N}(m_{c_t}; \mu_{c_t, t-1}^{[k]}, \Sigma_{c_t, t-1}^{[k]})} \, dm_{c_t}$$

$$\underbrace{p(x_t \mid x_{t-1}^{[k]}, u_t)}_{\sim \mathcal{N}(x_t; g(x_{t-1}^{[k]}, u_t), R_t)}$$

This expression makes apparent that our sampling distribution is truly the convolution of two Gaussians multiplied by a third. In the general SLAM case, the sampling distribution possesses no closed form from which we could easily sample. The culprit is the function h: If it were linear, this probability would be Gaussian, a fact that shall become more obvious below. In fact, not even the integral in (13.27) possesses a closed form solution. For this reason, sampling from the probability (13.27) is difficult.

This observation motivates the replacement of h by a linear approximation. As common in this book, this approximation is obtained through a first order Taylor expansion, given by the following linear function:

(13.28) $\quad h(m_{c_t}, x_t) \approx \hat{z}_t^{[k]} + H_m (m_{c_t} - \mu_{c_t, t-1}^{[k]}) + H_x (x_t - \hat{x}_t^{[k]})$

Here we use the following abbreviations:

(13.29) $\quad \hat{z}_t^{[k]} = h(\mu_{c_t, t-1}^{[k]}, \hat{x}_t^{[k]})$

(13.30) $\quad \hat{x}_t^{[k]} = g(x_{t-1}^{[k]}, u_t)$

The matrices H_m and H_x are the Jacobians of h. They are the derivatives of h with respect to m_{c_t} and x_t, respectively, evaluated at the expected values of their arguments:

(13.31) $\quad H_m = \nabla_{m_{c_t}} h(m_{c_t}, x_t) \big|_{x_t = \hat{x}_t^{[k]}; m_{c_t} = \mu_{c_t, t-1}^{[k]}}$

(13.32) $\quad H_x = \nabla_{x_t} h(m_{c_t}, x_t) \big|_{x_t = \hat{x}_t^{[k]}; m_{c_t} = \mu_{c_t, t-1}^{[k]}}$

Under this approximation, the desired sampling distribution (13.27) is a Gaussian with the following parameters:

(13.33) $\quad \Sigma_{x_t}^{[k]} = \left[H_x^T Q_t^{[k]-1} H_x + R_t^{-1} \right]^{-1}$

(13.34) $\quad \mu_{x_t}^{[k]} = \Sigma_{x_t}^{[k]} H_x^T Q_t^{[k]-1} (z_t - \hat{z}_t^{[k]}) + \hat{x}_t^{[k]}$

13.4 Improving the Proposal Distribution

where the matrix $Q_t^{[k]}$ is defined as follows:

$$(13.35) \quad Q_t^{[k]} = Q_t + H_m \Sigma_{c_t,t-1}^{[k]} H_m^T$$

To see, we note that under out linear approximation the convolution theorem provides us with a closed form for the integral term in (13.27):

$$(13.36) \quad \mathcal{N}(z_t; \hat{z}_t^{[k]} + H_x x_t - H_x \hat{x}_t^{[k]}, Q_t^{[k]})$$

The sampling distribution (13.27) is now given by the product of this normal distribution and the rightmost term in (13.27), the normal $\mathcal{N}(x_t; \hat{x}_t^{[k]}, R_t)$. Written in Gaussian form, we have

$$(13.37) \quad p(x_t \mid x_{1:t-1}^{[k]}, u_{1:t}, z_{1:t}, c_{1:t}) = \eta \exp\left\{-P_t^{[k]}\right\}$$

with

$$(13.38) \quad P_t^{[k]} = \tfrac{1}{2}\Big[(z_t - \hat{z}_t^{[k]} - H_x x_t + H_x \hat{x}_t^{[k]})^T Q_t^{[k]-1} (z_t - \hat{z}_t^{[k]} - H_x x_t + H_x \hat{x}_t^{[k]})$$
$$+ (x_t - \hat{x}_t^{[k]})^T R_t^{-1} (x_t - \hat{x}_t^{[k]})\Big]$$

This expression is obviously quadratic in our target variable x_t, hence $p(x_t \mid x_{1:t-1}^{[k]}, u_{1:t}, z_{1:t}, c_{1:t})$ is Gaussian. The mean and covariance of this Gaussian are equivalent to the minimum of $P_t^{[k]}$ and its curvature. Those are identified by calculating the first and second derivatives of $P_t^{[k]}$ with respect to x_t:

$$(13.39) \quad \frac{\partial P_t^{[k]}}{\partial x_t} = -H_x^T Q_t^{[k]-1}(z_t - \hat{z}_t^{[k]} - H_x x_t + H_x \hat{x}_t^{[k]}) + R_t^{-1}(x_t - \hat{x}_t^{[k]})$$
$$= (H_x^T Q_t^{[k]-1} H_x + R_t^{-1}) x_t - H_x^T Q_t^{[k]-1}(z_t - \hat{z}_t^{[k]} + H_x \hat{x}_t^{[k]}) - R_t^{-1} \hat{x}_t^{[k]}$$

$$(13.40) \quad \frac{\partial^2 P_t^{[k]}}{\partial x_t^2} = H_x^T Q_t^{[k]-1} H_x + R_t^{-1}$$

The covariance $\Sigma_{x_t}^{[k]}$ of the sampling distribution is now obtained by the inverse of the second derivative

$$(13.41) \quad \Sigma_{x_t}^{[k]} = \left[H_x^T Q_t^{[k]-1} H_x + R_t^{-1}\right]^{-1}$$

The mean $\mu_{x_t}^{[k]}$ of the sample distribution is obtained by setting the first derivative (13.39) to zero. This gives us:

$$(13.42) \quad \mu_{x_t}^{[k]} = \Sigma_{x_t}^{[k]} \left[H_x^T Q_t^{[k]-1}(z_t - \hat{z}_t^{[k]} + H_x \hat{x}_t^{[k]}) + R_t^{-1} \hat{x}_t^{[k]}\right]$$
$$= \Sigma_{x_t}^{[k]} H_x^T Q_t^{[k]-1}(z_t - \hat{z}_t^{[k]}) + \Sigma_{x_t}^{[k]} \left[H_x^T Q_t^{[k]-1} H_x + R_t^{-1}\right] \hat{x}_t^{[k]}$$
$$= \Sigma_{x_t}^{[k]} H_x^T Q_t^{[k]-1}(z_t - \hat{z}_t^{[k]}) + \hat{x}_t^{[k]}$$

This Gaussian is the approximation of the desired sampling distribution (13.26) in FastSLAM 2.0. Obviously, this proposal distribution is quite a bit more involved than the much simpler one for FastSLAM 1.0 in Equation (13.12).

13.4.2 Updating the Observed Feature Estimate

Just like our first version of the FastSLAM algorithm, FastSLAM 2.0 updates the posterior over the feature estimates based on the measurement z_t and the sampled pose $x_t^{[k]}$. The estimates at time $t-1$ are once again represented by the mean $\mu_{j,t-1}^{[k]}$ and the covariance $\Sigma_{j,t-1}^{[k]}$. The updated estimates are $\mu_{j,t}^{[k]}$ and $\Sigma_{j,t}^{[k]}$. The nature of the update depends on whether or not a feature j was observed at time t. For $j \neq c_t$, we already established in Equation (13.4) that the posterior over the feature remains unchanged. This implies that instead of updating the estimated, we merely have to copy it.

For the observed feature $j = c_t$, the situation is more intricate. Equation (13.5) already stated the posterior over observed features. Here we repeat it with the particle index k:

$$(13.43) \quad p(m_{c_t} \mid x_t^{[k]}, c_{1:t}, z_{1:t}) \\ = \eta \underbrace{p(z_t \mid m_{c_t}, x_t^{[k]}, c_t)}_{\sim \mathcal{N}(z_t; h(m_{c_t}, x_t^{[k]}), Q_t)} \underbrace{p(m_{c_t} \mid x_{1:t-1}^{[k]}, z_{1:t-1}, c_{1:t-1})}_{\sim \mathcal{N}(m_{c_t}; \mu_{c_t,t-1}^{[k]}, \Sigma_{c_t,t-1}^{[k]})}$$

As in (13.27), the nonlinearity of h causes this posterior to be non-Gaussian, which is at odds with FastSLAM 2.0's Gaussian representation for feature estimates. Luckily, the exact same linearization as above provides the solution:

$$(13.44) \quad h(m_{c_t}, x_t) \approx \hat{z}_t^{[k]} + H_m(m_{c_t} - \mu_{c_t,t-1}^{[k]})$$

Notice that x_t is not a free variable here, hence we can omit the third term in (13.28). This approximation renders the probability (13.43) Gaussian in the target variable m_{c_t}:

$$(13.45) \quad p(m_{c_t} \mid x_t^{[k]}, c_{1:t}, z_{1:t}) \\ = \eta \, \exp\Big\{ -\tfrac{1}{2}(z_t - \hat{z}_t^{[k]} - H_m(m_{c_t} - \mu_{c_t,t-1}^{[k]})) \, Q_t^{-1} \\ (z_t - \hat{z}_t^{[k]} - H_m(m_{c_t} - \mu_{c_t,t-1}^{[k]})) \\ - \tfrac{1}{2}(m_{c_t} - \mu_{c_t,t-1}^{[k]}) \, \Sigma_{c_t,t-1}^{[k]-1} (m_{c_t} - \mu_{c_t,t-1}^{[k]}) \Big\}$$

13.4 Improving the Proposal Distribution

The new mean and covariance are obtained using the standard EKF measurement update equations:

$$K_t^{[k]} = \Sigma_{c_t,t-1}^{[k]} H_m^T Q_t^{[k]-1} \tag{13.46}$$

$$\mu_{c_t,t}^{[k]} = \mu_{c_t,t-1}^{[k]} + K_t^{[k]}(z_t - \hat{z}_t^{[k]}) \tag{13.47}$$

$$\Sigma_{c_t,t}^{[k]} = (I - K_t^{[k]} H_m)\Sigma_{c_t,t-1}^{[k]} \tag{13.48}$$

We notice this is quite a bit more complicated than the update in FastSLAM 1.0, but the additional effort in implementing this often pays out, in terms of improved accuracy.

13.4.3 Calculating Importance Factors

The particles generated thus far do not yet match the desired posterior. In FastSLAM 2.0, the culprit is the normalizer $\eta^{[k]}$ in (13.27), which is usually different for each particle k. These differences are not yet accounted for in the resampling process. As in FastSLAM 1.0, the importance factor is given by the following quotient.

$$w_t^{[k]} = \frac{\text{target distribution}}{\text{proposal distribution}} \tag{13.49}$$

Once again, the target distribution that we would like our particles to assume is given by the path posterior, $p(x_t^{[k]} \mid z_{1:t}, u_{1:t}, c_{1:t})$. Under the asymptotically correct assumptions that paths in $x_{1:t-1}^{[k]}$ have been generated according to the target distribution one time step earlier, $p(x_{1:t-1}^{[k]} \mid z_{1:t-1}, u_{1:t-1}, c_{1:t-1})$, we note that the proposal distribution is now given by the product

$$p(x_{1:t-1}^{[k]} \mid z_{1:t-1}, u_{1:t-1}, c_{1:t-1})\, p(x_t^{[k]} \mid x_{1:t-1}^{[k]}, u_{1:t}, z_{1:t}, c_{1:t}) \tag{13.50}$$

The second term in this product is the pose sampling distribution (13.27). The importance weight is obtained as follows:

$$\begin{aligned}
w_t^{[k]} &= \frac{p(x_t^{[k]} \mid u_{1:t}, z_{1:t}, c_{1:t})}{p(x_t^{[k]} \mid x_{1:t-1}^{[k]}, u_{1:t}, z_{1:t}, c_{1:t})\, p(x_{1:t-1}^{[k]} \mid u_{1:t-1}, z_{1:t-1}, c_{1:t-1})} \\
&= \frac{p(x_t^{[k]} \mid x_{1:t-1}^{[k]}, u_{1:t}, z_{1:t}, c_{1:t})\, p(x_{1:t-1}^{[k]} \mid u_{1:t}, z_{1:t}, c_{1:t})}{p(x_t^{[k]} \mid x_{1:t-1}^{[k]}, u_{1:t}, z_{1:t}, c_{1:t})\, p(x_{1:t-1}^{[k]} \mid u_{1:t-1}, z_{1:t-1}, c_{1:t-1})} \\
&= \frac{p(x_{1:t-1}^{[k]} \mid u_{1:t}, z_{1:t}, c_{1:t})}{p(x_{1:t-1}^{[k]} \mid u_{1:t-1}, z_{1:t-1}, c_{1:t-1})}
\end{aligned} \tag{13.51}$$

$$\overset{\text{Bayes}}{=} \eta \, \frac{p(z_t \mid x_{1:t-1}^{[k]}, u_{1:t}, z_{1:t-1}, c_{1:t}) \, p(x_{1:t-1}^{[k]} \mid u_{1:t}, z_{1:t-1}, c_{1:t})}{p(x_{1:t-1}^{[k]} \mid u_{1:t-1}, z_{1:t-1}, c_{1:t-1})}$$

$$\overset{\text{Markov}}{=} \eta \, \frac{p(z_t \mid x_{1:t-1}^{[k]}, u_{1:t}, z_{1:t-1}, c_{1:t}) \, p(x_{1:t-1}^{[k]} \mid u_{1:t-1}, z_{1:t-1}, c_{1:t-1})}{p(x_{1:t-1}^{[k]} \mid u_{1:t-1}, z_{1:t-1}, c_{1:t-1})}$$

$$= \eta \, p(z_t \mid x_{1:t-1}^{[k]}, u_{1:t}, z_{1:t-1}, c_{1:t})$$

The reader may notice that this expression is the inverse of our normalization constant $\eta^{[k]}$ in (13.27). Further transformations give us the following form:

$$(13.52) \quad w_t^{[k]} = \eta \int p(z_t \mid x_t, x_{1:t-1}^{[k]}, u_{1:t}, z_{1:t-1}, c_{1:t})$$
$$p(x_t \mid x_{1:t-1}^{[k]}, u_{1:t}, z_{1:t-1}, c_{1:t}) \, dx_t$$

$$\overset{\text{Markov}}{=} \eta \int p(z_t \mid x_t, x_{1:t-1}^{[k]}, u_{1:t}, z_{1:t-1}, c_{1:t}) \, p(x_t \mid x_{t-1}^{[k]}, u_t) \, dx_t$$

$$= \eta \int \int p(z_t \mid m_{c_t}, x_t, x_{1:t-1}^{[k]}, u_{1:t}, z_{1:t-1}, c_{1:t})$$
$$p(m_{c_t} \mid x_t, x_{1:t-1}^{[k]}, u_{1:t}, z_{1:t-1}, c_{1:t}) \, dm_{c_t} \, p(x_t \mid x_{t-1}^{[k]}, u_t) \, dx_t$$

$$\overset{\text{Markov}}{=} \eta \int \underbrace{p(x_t \mid x_{t-1}^{[k]}, u_t)}_{\sim \mathcal{N}(x_t; g(\hat{x}_{t-1}^{[k]}, u_t), R_t)} \int \underbrace{p(z_t \mid m_{c_t}, x_t, c_t)}_{\sim \mathcal{N}(z_t; h(m_{c_t}, x_t), Q_t)}$$
$$\underbrace{p(m_{c_t} \mid x_{1:t-1}^{[k]}, u_{1:t-1}, z_{1:t-1}, c_{1:t-1})}_{\sim \mathcal{N}(m_{c_t}; \mu_{c_t,t-1}^{[k]}, \Sigma_{c_t,t-1}^{[k]})} \, dm_{c_t} \, dx_t$$

We find that this expression can once again be approximated by a Gaussian over measurements z_t by linearizing h. As it is easily shown, the mean of the resulting Gaussian is $\hat{z}_t$, and its covariance is

$$(13.53) \quad L_t^{[t]} = H_x^T Q_t H_x + H_m \Sigma_{c_t,t-1}^{[k]} H_m^T + R_t$$

Put differently, the (non-normalized) importance factor of the k-th particle is given by the following expression:

$$(13.54) \quad w_t^{[k]} = |2\pi L_t^{[t]}|^{-\frac{1}{2}} \exp\left\{-\tfrac{1}{2}(z_t - \hat{z}_t) \, L_t^{[t]-1} \, (z_t - \hat{z}_t)\right\}$$

As in FastSLAM 1.0, particles generated in Steps 1 and 2, along with their importance factor calculated in Step 3, are collected in a temporary particle set.

The final step of the FastSLAM 2.0 update is a resampling step. Just as in FastSLAM 1.0, FastSLAM 2.0 draws (with replacement) M particles from

13.5 Unknown Data Association

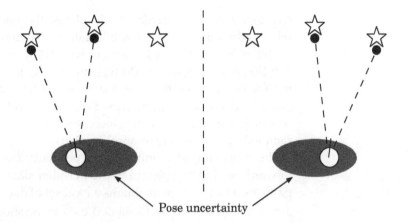

Figure 13.7 The data association problem in SLAM. This figure illustrates that the best data association may vary even within regions of high likelihood for the pose of the robot.

the temporary particle set. Each particle is drawn with a probability proportional to its importance factor $w_t^{[k]}$. The resulting particle set represent (asymptotically) the desired factored posterior at time t.

13.5 Unknown Data Association

This section extends both variants of the FastSLAM algorithm to cases where the correspondence variables $c_{1:t}$ are unknown. A key advantage of using particle filters for SLAM is that each particle can rely on its own, local data association decisions.

We remind the reader that the data association problem at time t is the problem of determining the variable c_t based on the available data. This problem is illustrated in Figure 13.7: Here a robot observes two features in the world. Depending on its actual pose relative to these features, these measurements correspond to different features in the map (depicted as stars in Figure 13.7).

So far, we discussed a number of data association technique using arguments such as maximum likelihood. Those techniques had in common that there is only a single data association per measurement, for the entire filter. FastSLAM, by virtue of using multiple particles, can determine the correspondence on a per-particle basis. Thus, the filter not only samples over robot paths, but also over possible data association decisions along the way.

This is one of the key features of FastSLAM, which sets it aside from the rich body of Gaussian SLAM algorithms. As long as a small subset of the particles is based on the correct data association, data association errors are not as fatal as in EKF approaches. Particles subject to such errors tend to possess inconsistent maps, which increases the probability that they are simply sampled away in future resampling steps.

The mathematical definition of the per-particle data association is straightforward, in that it generalizes the per-filter data association to individual particles. Each particle maintains a local set of data association variables, denoted $\hat{c}_t^{[k]}$. In maximum likelihood data association, each $\hat{c}_t^{[k]}$ is determined by maximizing the likelihood of the measurement z_t:

$$(13.55) \quad \hat{c}_t^{[k]} = \operatorname*{argmax}_{c_t} p(z_t \mid c_t, \hat{c}_{1:t-1}^{[k]}, x_{1:t}^{[k]}, z_{1:t-1}, u_{1:t})$$

An alternative is the data association sampler (DAS), which samples the data association variable according to their likelihood

$$(13.56) \quad \hat{c}_t \sim \eta\, p(z_t \mid c_t, \hat{c}_{1:t-1}^{[k]}, x_{1:t}^{[k]}, z_{1:t-1}, u_{1:t})$$

Both techniques, ML and DAS, make it possible to estimate the number of features in the map. SLAM techniques using ML create new features in the map if the likelihood falls below a threshold p_0 for all known features in the map. DAS associates an observed measurement with a new, previously unobserved feature stochastically. They do so with probability proportional to ηp_0, where η is a normalizer defined in (13.56).

$$(13.57) \quad \hat{c}_t^{[k]} \sim \eta\, p(z_t \mid c_t, \hat{c}_{1:t-1}^{[k]}, x_{1:t}^{[k]}, z_{1:t-1}, u_{1:t})$$

For both techniques, the likelihood is calculated as follows:

$$(13.58) \quad p(z_t \mid c_t, \hat{c}_{1:t-1}^{[k]}, x_{1:t}^{[k]}, z_{1:t-1}, u_{1:t})$$
$$= \int p(z_t \mid m_{c_t}, c_t, \hat{c}_{1:t-1}^{[k]}, x_{1:t}^{[k]}, z_{1:t-1}, u_{1:t})$$
$$\qquad p(m_{c_t} \mid c_t, \hat{c}_{1:t-1}^{[k]}, x_{1:t}^{[k]}, z_{1:t-1}, u_{1:t})\, dm_{c_t}$$
$$= \int \underbrace{p(z_t \mid m_{c_t}, c_t, x_t^{[k]})}_{\sim \mathcal{N}(z_t; h(m_{c_t}, x_t^{[k]}), Q_t)} \underbrace{p(m_{c_t} \mid \hat{c}_{1:t-1}^{[k]}, x_{1:t-1}^{[k]}, z_{1:t-1})}_{\sim \mathcal{N}(\mu_{c_t, t-1}^{[k]}, \Sigma_{c_t, t-1}^{[k]})}\, dm_{c_t}$$

Linearization of h enables us to obtain this in closed form:

$$(13.59) \quad p(z_t \mid c_t, \hat{c}_{1:t-1}^{[k]}, x_t^{[k]}, z_{1:t-1}, u_{1:t})$$

$$= |2\pi Q_t^{[k]}|^{-\frac{1}{2}} \exp\left\{-\tfrac{1}{2}(z_t - h(\mu_{c_t,t-1}^{[k]}, x_t^{[k]}))^T Q_t^{[k]-1}(z_t - h(\mu_{c_t,t-1}^{[k]}, x_t^{[k]}))\right\}$$

The variable $Q_t^{[k]}$ was defined in Equation (13.35), as a function of the data association variable c_t. New features are added to the map in exactly the same way as outlined above. In the ML approach, a new feature is added when the probability $p(z_t \mid c_t, \hat{c}_{1:t-1}^{[k]}, x_t^{[k]}, z_{1:t-1}, u_{1:t})$ falls beyond a threshold p_0. The DAS includes the hypothesis that an observation corresponds to a previously unobserved feature in its set of hypotheses, and samples it with probability ηp_0.

13.6 Map Management

Map management in FastSLAM is largely equivalent to EKF SLAM, with a few particulars arising from the fact that data association is handled on a per-particle level in FastSLAM.

As in the alternative SLAM algorithms, any newly added feature requires the initialization of a new Kalman filter. In many SLAM problems the measurement function h is *invertible*. This is the case, for example, for robots measuring range and bearing to features in the plane, in which a single measurement suffices to produce a (non-degenerate) estimate on the feature location. The initialization of the EKF is then straightforward:

$$\begin{align}
(13.60) \quad & x_t^{[k]} \sim p(x_t \mid x_{t-1}^{[k]}, u_t) \\
(13.61) \quad & \mu_{n,t}^{[k]} = h^{-1}(z_t, x_t^{[k]}) \\
(13.62) \quad & \Sigma_{n,t}^{[k]} = (H_{\hat{c}}^{[k]T} Q_t^{-1} H_{\hat{c}}^{[k]})^{-1} \quad \text{with} \quad H_{\hat{c}}^{[k]} = h'(\mu_{n,t}^{[k]}, x_t^{[k]}) \\
(13.63) \quad & w_t^{[k]} = p_0
\end{align}$$

Notice that for newly observed features, the pose $x_t^{[k]}$ is sampled according to the motion model $p(x_t \mid x_{t-1}^{[k]}, u_t)$. This distribution is equivalent to the FastSLAM sampling distribution (13.26) in situations where no previous location estimate for the observed feature is available.

Initialization techniques for situations in which h is not invertible are discussed in Deans and Hebert (2002). In general, such situations require the accumulation of multiple measurements, to obtain a good estimate for the linearization of h.

To accommodate features introduced erroneously into the map, FastSLAM features a mechanism for eliminating features that are not supported by sufficient evidence. Just as in EKF SLAM, FastSLAM does so by keeping track of the log odds of the actual existence of individual features in the map.

Specifically, when a feature is observed, its log odds for existence is incremented by a fixed amount, which is calculated using the standard Bayes filter formula. Similarly, when a feature is not observed even though it should have, such negative information results in a decrement of the feature existence variable by a fixed amount. Features whose variable sinks below a threshold value are then simply removed from the list of particles. It is also possible to implement a provisional feature list in FastSLAM. Technically this is trivial, since each feature possesses its own particle.

13.7 The FastSLAM Algorithms

Tables 13.2 and 13.3 summarize both FastSLAM variants with unknown data association. In both algorithms, particles are of the form

$$(13.64) \quad Y_t^{[k]} = \left\langle x_t^{[k]}, N_t^{[k]}, \left\langle \mu_{1,t}^{[k]}, \Sigma_{1,t}^{[k]}, \tau_1^{[k]} \right\rangle, \ldots, \left\langle \mu_{N_t^{[k]},t}^{[k]}, \Sigma_{N_t^{[k]},t}^{[k]}, \tau_{N_t^{[k]}}^{[k]} \right\rangle \right\rangle$$

In addition to the pose $x_t^{[k]}$ and the feature estimates $\mu_{n,t}^{[k]}$ and $\Sigma_{n,t}^{[k]}$, each particle maintains the number of features $N_t^{[k]}$ in its local map, and each feature carries a probabilistic estimate of its existence $\tau_n^{[k]}$. Iterating the filter requires time linear in the maximum number of features $\max_k N_t^{[k]}$ in each map, and it is also linear in the number of particles M. Further below, we will discuss advanced data structure that yield more efficient implementations.

We note that both versions of FastSLAM, as described here, consider a single measurement at a time. As before, this choice is made for notational convenience, and many of the techniques discussed in previous SLAM chapters can be brought to bear.

13.8 Efficient Implementation

At first glance, it may appear that each update in FastSLAM requires $O(MN)$ time, where M is the number of particles M and N the number of features in the map. The linear complexity in M is unavoidable, given that we have to process M particles with each update. The linear complexity in N is the result of the resampling process. Whenever a particle is drawn more than once in the resampling process, a "naive" implementation might duplicate the entire map attached to the particle. Such a duplication process is linear in the

13.7 The FastSLAM Algorithms

1: **Algorithm FastSLAM 1.0**(z_t, u_t, Y_{t-1}):

2:      for $k = 1$ to M do   // loop over all particles

3:          retrieve $\left\langle x_{t-1}^{[k]}, N_{t-1}^{[k]}, \left\langle \mu_{1,t-1}^{[k]}, \Sigma_{1,t-1}^{[k]}, i_1^{[k]} \right\rangle, \ldots, \right.$

                       $\left. \left\langle \mu_{N_{t-1}^{[k]},t-1}^{[k]}, \Sigma_{N_{t-1}^{[k]},t-1}^{[k]}, i_{N_{t-1}^{[k]},t-1}^{[k]} \right\rangle \right\rangle$ from Y_{t-1}

4:          $x_t^{[k]} \sim p(x_t \mid x_{t-1}^{[k]}, u_t)$ // sample new pose

5:          for $j = 1$ to $N_{t-1}^{[k]}$ do   // measurement likelihoods

6:             $\hat{z}_j = h(\mu_{j,t-1}^{[k]}, x_t^{[k]})$ // measurement prediction

7:             $H_j = h'(\mu_{j,t-1}^{[k]}, x_t^{[k]})$ // calculate Jacobian

8:             $Q_j = H_j \, \Sigma_{j,t-1}^{[k]} \, H_j^T + Q_t$ // measurement covariance

9:             $w_j = |2\pi Q_j|^{-\frac{1}{2}} \exp\left\{-\frac{1}{2}(z_t - \hat{z}_j)^T \right.$

                         $\left. Q_j^{-1}(z_t - \hat{z}_j)\right\}$ // likelihood of correspondence

10:        endfor

11:        $w_{1+N_{t-1}^{[k]}} = p_0$ // importance factor, new feature

12:        $w^{[k]} = \max w_j$ // max likelihood correspondence

13:        $\hat{c} = \operatorname{argmax} w_j$ // index of ML feature

14:        $N_t^{[k]} = \max\{N_{t-1}^{[k]}, \hat{c}\}$ // new number of features in map

15:        for $j = 1$ to $N_t^{[k]}$ do   // update Kalman filters

16:            if $j = \hat{c} = 1 + N_{t-1}^{[k]}$ then   // is new feature?

17:               $\mu_{j,t}^{[k]} = h^{-1}(z_t, x_t^{[k]})$ // initialize mean

18:               $H_j = h'(\mu_{j,t}^{[k]}, x_t^{[k]}); \Sigma_{j,t}^{[k]} = (H_j^{-1})^T Q_t H_j^{-1}$ // initialize covar.

19:               $i_{j,t}^{[k]} = 1$ // initialize counter

20:            else if $j = \hat{c} \leq N_{t-1}^{[k]}$ then   // is observed feature?

21:               $K = \Sigma_{j,t-1}^{[k]} H_j^T Q_{\hat{c}}^{-1}$ // calculate Kalman gain

22:               $\mu_{j,t}^{[k]} = \mu_{j,t-1}^{[k]} + K(z_t - \hat{z}_{\hat{c}})$ // update mean

23:               $\Sigma_{j,t}^{[k]} = (I - K H_j) \Sigma_{j,t-1}^{[k]}$ // update covariance

24:               $i_{j,t}^{[k]} = i_{j,t-1}^{[k]} + 1$ // increment counter

see next page for continuation

continued from the previous page

25:	else	// all other features
26:	$\mu_{j,t}^{[k]} = \mu_{j,t-1}^{[k]}$	// copy old mean
27:	$\Sigma_{j,t}^{[k]} = \Sigma_{j,t-1}^{[k]}$	// copy old covariance
28:	if $\mu_{j,t-1}^{[k]}$ outside perceptual range of $x_t^{[k]}$ then	// should feature have been seen?
29:	$i_{j,t}^{[k]} = i_{j,t-1}^{[k]}$	// no, do not change
30:	else	
31:	$i_{j,t}^{[k]} = i_{j,t-1}^{[k]} - 1$	// yes, decrement counter
32:	if $i_{j,t-1}^{[k]} < 0$ then	
33:	discard feature j	// discard dubious features
34:	endif	
35:	endif	
36:	endif	
37:	endfor	
38:	add $\left\langle x_t^{[k]}, N_t^{[k]}, \left\langle \mu_{1,t}^{[k]}, \Sigma_{1,t}^{[k]}, i_1^{[k]} \right\rangle, \ldots, \left\langle \mu_{N_t^{[k]},t}^{[k]}, \Sigma_{N_t^{[k]},t}^{[k]}, i_{N_t^{[k]}}^{[k]} \right\rangle \right\rangle$ to Y_{aux}	
39:	endfor	
40:	$Y_t = \emptyset$	// construct new particle set
41:	do M times	// resample M particles
42:	draw random index k with probability $\propto w^{[k]}$	// resample
43:	add $\left\langle x_t^{[k]}, N_t^{[k]}, \left\langle \mu_{1,t}^{[k]}, \Sigma_{1,t}^{[k]}, i_1^{[k]} \right\rangle, \ldots, \left\langle \mu_{N_t^{[k]},t}^{[k]}, \Sigma_{N_t^{[k]},t}^{[k]}, i_{N_t^{[k]}}^{[k]} \right\rangle \right\rangle$ to Y_t	
44:	enddo	
45:	return Y_t	

Table 13.2 The algorithm FastSLAM 1.0 with unknown data association. This version does not implement any of the efficient tree representations discussed in the chapter.

13.7 The FastSLAM Algorithms

1: **Algorithm FastSLAM 2.0**(z_t, u_t, Y_{t-1}):
2:     for $k = 1$ to M do   // loop over all particles
3:        retrieve $\left\langle x_{t-1}^{[k]}, N_{t-1}^{[k]}, \left\langle \mu_{1,t-1}^{[k]}, \Sigma_{1,t-1}^{[k]}, i_1^{[k]} \right\rangle, \ldots, \right.$
                 $\left. \left\langle \mu_{N_{t-1}^{[k]},t-1}^{[k]}, \Sigma_{N_{t-1}^{[k]},t-1}^{[k]}, i_{N_{t-1}^{[k]}}^{[k]} \right\rangle \right\rangle$ from Y_{t-1}
4:        for $j = 1$ to $N_{t-1}^{[k]}$ do   // calculate sampling distribution
5:           $\hat{x}_{j,t} = g(x_{t-1}^{[k]}, u_t)$ // predict pose
6:           $\bar{z}_j = h(\mu_{j,t-1}^{[k]}, \hat{x}_{j,t})$ // predict measurement
7:           $H_{x,j} = \nabla_{x_t} h(\mu_{j,t-1}^{[k]}, \hat{x}_{j,t})$ // Jacobian wrt pose
8:           $H_{m,j} = \nabla_{m_j} h(\mu_{j,t-1}^{[k]}, \hat{x}_{j,t})$ // Jacobian wrt map feature
9:           $Q_j = Q_t + H_{m,j} \Sigma_{j,t-1}^{[k]} H_{m,j}^T$ // measurement information
10:          $\Sigma_{x,j} = \left[H_{x,j}^T Q_j^{-1} H_{x,j} + R_t^{-1} \right]^{-1}$ // Cov of proposal distribution
11:          $\mu_{x_t,j} = \Sigma_{x,j} H_{x,j}^T Q_j^{-1}$
                     $(z_t - \bar{z}_j) + \hat{x}_{j,t}$ // mean of proposal distribution
12:           $x_{t,j}^{[k]} \sim \mathcal{N}(\mu_{x_t,j}, \Sigma_{x,j})$ // sample pose
13:           $\hat{z}_j = h(\mu_{j,t-1}^{[k]}, x_t^{[k]})$ // measurement prediction
14:           $\pi_j = |2\pi Q_j|^{-\frac{1}{2}} \exp\{-\frac{1}{2}$
                     $(z_t - \hat{z}_j)^T Q_j^{-1} (z_t - \hat{z}_j)\}$ // correspondence likelihood
15:        endfor
16:        $\pi_{1+N_{t-1}^{[k]}} = p_0$ // likelihood of new feature
17:        $\hat{c} = \operatorname{argmax} \pi_j$ // ML correspondence
18:        $N_t^{[k]} = \max\{N_{t-1}^{[k]}, \hat{c}\}$ // new number of features
19:        for $j = 1$ to $N_t^{[k]}$ do   // update Kalman filters
20:           if $j = \hat{c} = 1 + N_{t-1}^{[k]}$ then   // is new feature?
21:              $x_t^{[k]} \sim p(x_t \mid x_{t-1}^{[k]}, u_t)$ // sample pose
22:              $\mu_{j,t}^{[k]} = h^{-1}(z_t, x_t^{[k]})$ // initialize mean
23:              $H_{m,j} = \nabla_{m_j} h(\mu_{j,t}^{[k]}, x_t^{[k]})$ // Jacobian wrt map feature
24:              $\Sigma_{j,t}^{[k]} = (H_{m,j}^{-1})^T Q_t H_{m,j}^{-1}$ // initialize covariance
25:              $i_{j,t}^{[k]} = 1$ // initialize counter
26:              $w^{[k]} = p_0$ // importance weight
27:           else if $j = \hat{c} \leq N_{t-1}^{[k]}$ then   // is observed feature?
28:              $x_t^{[k]} = x_{t,j}^{[k]}$
29:              $K = \Sigma_{j,t-1}^{[k]} H_{m,j}^T Q_j^{-1}$ // calculate Kalman gain

see next page for continuation

continued from the previous page

30:	$\mu_{j,t}^{[k]} = \mu_{j,t-1}^{[k]} + K(z_t - \hat{z}_j)$	// update mean
31:	$\Sigma_{j,t}^{[k]} = (I - K\, H_{m,j})\, \Sigma_{j,t-1}^{[k]}$	// update covariance
32:	$i_{j,t}^{[k]} = i_{j,t-1}^{[k]} + 1$	// increment counter
33:	$L = H_{x,j}\, R_t\, H_{x,j}^T + H_{m,j}\, \Sigma_{j,t-1}^{[k]}\, H_{m,j}^T + Q_t$	
34:	$w^{[k]} = \lvert 2\pi L \rvert^{-\frac{1}{2}} \exp\left\{-\frac{1}{2}(z_t - \hat{z}_j)^T L^{-1} (z_t - \hat{z}_j)\right\}$	// importance weight
35:	else	// all other features
36:	$\mu_{j,t}^{[k]} = \mu_{j,t-1}^{[k]}$	// copy old mean
37:	$\Sigma_{j,t}^{[k]} = \Sigma_{j,t-1}^{[k]}$	// copy old covariance
38:	if $\mu_{j,t-1}^{[k]}$ outside perceptual range of $x_t^{[k]}$ then	// should feature have been seen?
39:	$i_{j,t}^{[k]} = i_{j,t-1}^{[k]}$	// no, do not change
40:	else	
41:	$i_{j,t}^{[k]} = i_{j,t-1}^{[k]} - 1$	// yes, decrement counter
42:	if $i_{j,t-1}^{[k]} < 0$ then	
43:	discard feature j	// discard dubious features
44:	endif	
45:	endif	
46:	endif	
47:	endfor	
48:	add $\left\langle x_t^{[k]}, N_t^{[k]}, \left\langle \mu_{1,t}^{[k]}, \Sigma_{1,t}^{[k]}, i_1^{[k]} \right\rangle, \ldots, \left\langle \mu_{N_t^{[k]},t}^{[k]}, \Sigma_{N_t^{[k]},t}^{[k]}, i_{N_t^{[k]}}^{[k]} \right\rangle \right\rangle$ to Y_{aux}	
49:	endfor	
50:	$Y_t = \emptyset$	// construct new particle set
51:	do M times	// resample M particles
52:	draw random index k with probability $\propto w^{[k]}$	// resample
53:	add $\left\langle x_t^{[k]}, N_t^{[k]}, \left\langle \mu_{1,t}^{[k]}, \Sigma_{1,t}^{[k]}, i_1^{[k]} \right\rangle, \ldots, \left\langle \mu_{N_t^{[k]},t}^{[k]}, \Sigma_{N_t^{[k]},t}^{[k]}, i_{N_t^{[k]}}^{[k]} \right\rangle \right\rangle$ to Y_t	
54:	enddo	
55:	return Y_t	

Table 13.3 The FastSLAM 2.0 Algorithm, stated here unknown data association.

size of the map N. Furthermore, a naive implementation of data association may result in evaluating the measurement likelihood for each of the N features in the map, resulting again in linear complexity in N. We note that a poor implementation of the sampling process might easily add another $\log N$ to the update complexity.

Efficient implementations of FastSLAM require only $O(M \log N)$ update time. This is logarithmic in the size of the map N. First, consider the situation with known data association. Linear copying costs can be avoided by introducing a data structure for representing particles that allow for more selective updates. The basic idea is to organize the map as a *balanced binary tree*. Figure 13.8a shows such a tree for a single particle, in the case of $M = 8$ features. Notice that all Gaussian parameters $\mu_j^{[k]}$ and $\Sigma_j^{[k]}$ for all j are located at the leaves of the tree. Assuming that the tree is approximately balanced, accessing a leaf requires time logarithmic in N.

Suppose FastSLAM incorporates a new control u_t and a new measurement z_t. Each new particle in Y_t will differ from the corresponding one in Y_{t-1} in two ways: First, it will possess a different pose estimate obtained via (13.26), and second, the observed feature's Gaussian will have been updated, as specified in Equations (13.47) and (13.48). All other Gaussian feature estimates, however, will be equivalent to the generating particle. When copying the particle, thus, only a single path has to be modified in the tree representing all Gaussians, leading to the logarithmic update time.

An illustration of this "trick" is shown in Figure 13.8b: Here we assume $c_t^i = 3$, hence only the Gaussian parameters $\mu_3^{[k]}$ and $\Sigma_3^{[k]}$ are updated. Instead of generating an entire new tree, only a single path is created, leading to the Gaussian $c_t^i = 3$. This path is an incomplete tree. The tree is completed by copying the missing pointers from the tree of the generating particle. Thus, branches that leave the path will point to the same (unmodified) subtree as that of the generating tree. Clearly, generating this tree takes only time logarithmic in N. Moreover, accessing a Gaussian also takes time logarithmic in N, since the number of steps required to navigate to a leaf of the tree is equivalent to the length of the path (which is by definition logarithmic). Thus, both generating and accessing a partial tree can be done in time $O(\log N)$. Since in each updating step M new particles are created, an entire update requires time in $O(M \log N)$.

Organizing particles in trees raises the question as to when to deallocate memory. Memory deallocation can equally be implemented in amortized logarithmic time. The idea is to assign a variable to each node—internal

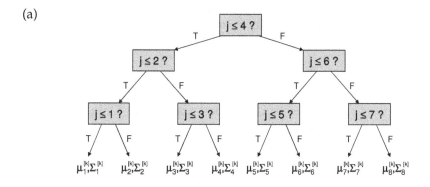

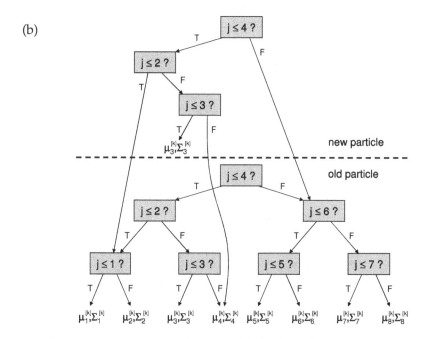

Figure 13.8 (a) A tree representing $N = 8$ feature estimates within a single particle. (b) Generating a new particle from an old one, while modifying only a single Gaussian. The new particle receives only a partial tree, consisting of a path to the modified Gaussian. All other pointers are copied from the generating tree. This can be done in time logarithmic in N.

13.8 Efficient Implementation

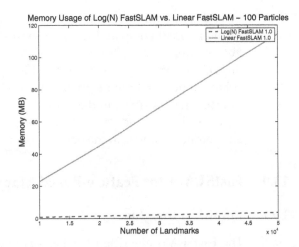

Figure 13.9 Memory requirements for linear and log(N) version of FastSLAM 1.0.

or leaf—that counts the number of pointers pointing to it. The counter of a newly created node will be initialized by 1. It will be incremented as pointers to a node are created in other particles. Decrements occur when pointers are removed (e.g., pointers of pose particles that fail to survive the resampling process). When a counter reaches zero, its children's counters are decremented and the memory of the corresponding node is deallocated. The process is then applied recursively to all children of the node whose counter may have reached zero. This recursive process will require $O(M \log N)$ time on average.

The tree also induces substantial memory savings. Figure 13.9 shows the effect of the efficient tree technique on the memory consumed by FastSLAM, measured empirically. This graph is the result of an actual implementation of FastSLAM 1.0 with $M = 100$ particles for acquiring feature-based maps. The graph shows nearly two orders of magnitude savings for a map with 50,000 features. The relative savings in update time are similar in value.

Obtaining logarithmic time complexity for FastSLAM with unknown data association is more difficult. Specifically, we need a technique that restricts data association search to the local neighborhood of a feature, to avoid calculating the data association probability for all N features in the map. Further, the tree has to remain approximately balanced.

There indeed exists variants of kernel density trees, or kd-trees, that can meet these assumptions, assuming that the variance in the sensor measurements is small compared to the overall size of the map. For example, the *bkd-tree* proposed by Procopiuc et al. (2003) maintains a sequence of trees of growing complexity. By carefully shifting items across those trees, a logarithmic time recall can be guaranteed under amortized logarithmic time for inserting new features in the map. Another is the DP-SLAM algorithm proposed by Eliazar and Parr (2003), which uses *history trees* for efficient storage and retrieval, similar to the one described here.

13.9 FastSLAM for Feature-Based Maps

13.9.1 Empirical Insights

The FastSLAM algorithm has been applied to a number of map representation and sensor data. The most basic application concerns feature-based maps, assuming that the robot is equipped with a sensor for detecting range and bearing to landmarks. One such data set is the Victoria Park dataset, which we already discussed in Chapter 12. Figure 13.10a shows the path of the vehicle obtained by integrating the estimated controls. Controls are a poor predictor of location for this vehicle; after 30 minutes of driving, the estimated position of the vehicle is well over 100 meters away from its GPS position.

The remaining three panels of Figure 13.10 show the output of FastSLAM 1.0. In all these diagrams, the path estimated by GPS is shown as a dashed line, and the output of FastSLAM is shown as a solid line. The RMS error of the resulting path is just over 4 meters over the 4 km traverse. This experiment was run with $M = 100$ particles. This error is indistinguishable from the error of other state-of-the-art SLAM algorithms, such as the ones discussed in previous chapters. The robustness of FastSLAM becomes apparent in Figure 13.10d, which plots the result for an experiment in which we simply ignored the motion information. Instead, the odometry-based motion model was replaced by a Brownian motion model. The average error of FastSLAM is statistically indistinguishable from the error obtained before.

When implementing FastSLAM in feature-based maps, it is important to consider negative information. When negative information is used to estimate the existence of each feature, as described in Chapter 13.6, many spurious features can be removed from the map. Figure 13.11 shows the Victoria Park map built with and without considering negative evidence. Here the use of negative information results in 44% percent fewer features in the re-

13.9   FastSLAM for Feature-Based Maps

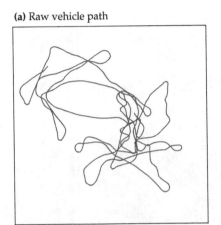

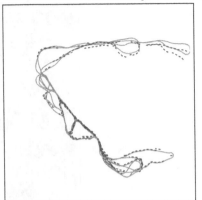

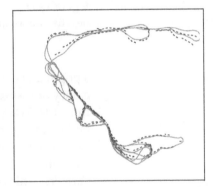

Figure 13.10   (a) Vehicle path predicted by the odometry; (b) True path (dashed line) and FastSLAM 1.0 path (solid line); (c) Victoria Park results overlayed on aerial imagery with the GPS path in blue (dashed), average FastSLAM 1.0 path in yellow (solid), and estimated features as yellow circles. (d) Victoria Park Map created without odometry information. Data and aerial image courtesy of José Guivant and Eduardo Nebot, Australian Centre for Field Robotics.

sulting map. While the *correct* number of features is not available, visual inspection of the maps suggests that many of the spurious features have been eliminated.

It makes sense to compare FastSLAM to EKF SLAM, which continues to be

Figure 13.11 FastSLAM 1.0 (a) without and (b) with feature elimination based on negative information.

a popular benchmark algorithm. For example, Figure 13.12 compares the accuracy of FastSLAM 1.0 with that of the EKF, for various numbers of particles from 1 to 5,000. The error of the EKF SLAM is shown as a dashed horizontal line in Figure 13.12 for comparison. The accuracy of FastSLAM 1.0 approaches the accuracy of the EKF as the number of particles is increased, and it is statistically indistinguishable from that of the EKF past approximately 10 particles. This is interesting because FastSLAM 1.0 with 10 particles and 100 features requires an order of magnitude fewer parameters than EKF SLAM in order to achieve this level of accuracy.

In practice, FastSLAM 2.0 yields superior results to FastSLAM 1.0, though the improvement is of significance only under certain circumstances. As a rule of thumb, both algorithms produce comparable results when the number of particles M is large, and when the measurement noise is large compared to the motion uncertainty. This is illustrated in Figure 13.13, which graphs the accuracy of either FastSLAM variant as a function of the measurement noise, using $M = 100$ particles. The most important finding here is FastSLAM 1.0's relatively poor performance in low-noise simulations. One way to test whether a FastSLAM 1.0 implementation suffers from this pathology is to artificially inflate the measurement noise in the probabilistic model $p(z \mid x)$. If, as a result of this inflation, the overall map error goes down—not

13.9 FastSLAM for Feature-Based Maps

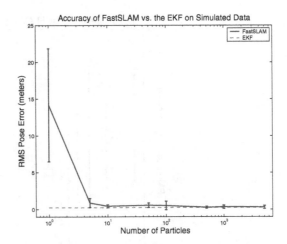

Figure 13.12 A comparison of the accuracy of FastSLAM 1.0 and the EKF on simulated data.

up—it is time to switch to FastSLAM 2.0.

13.9.2 Loop Closure

No algorithm is perfect. There exists problems in which FastSLAM is inferior to its Gaussian counterparts. One such problem involves *loop closure*. In loop closure, a robot moves through unknown terrain and at some point encounters features not seen for a long time. It is here that maintaining the correlations in a SLAM algorithm is particularly important, so that the information acquired when closing a loop can be propagated through the entire map. EKFs and GraphSLAM maintain such correlations directly, whereas FastSLAM maintains them through its diversity in the particle sets. Thus, the ability to close loops, depends on the number of particles M. Better diversity in the sample set results in better loop closing performance, because new observations can affect the pose of the vehicle further back in the past.

Unfortunately, by pruning away improbable trajectories of the vehicle, resampling eventually causes all of the FastSLAM particles to share a common history at some point in the past. New observations cannot affect the positions of features observed prior to this point. This common history point can be pushed back in time by increasing the number of particles M. This process of throwing away correlation data over time enables FastSLAM's effi-

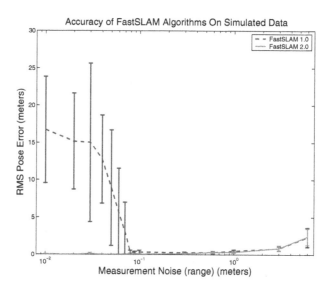

Figure 13.13 FastSLAM 1.0 and 2.0 with varying levels of measurement noise: As to be expected, FastSLAM 2.0 is uniformly superior to FastSLAM 1.0. The difference is particularly obvious for small particle sets, where the improved proposal distribution focuses the particles much better.

cient sensor updates. This efficiency comes at the cost of slower convergence speed. Throwing away correlation information means that more observations will be required to achieve a given level of accuracy. Clearly, FastSLAM 2.0's improved proposal distribution ensures that fewer particles are eliminated in resampling compared to FastSLAM 1.0, but it does not alleviate this problem.

In practice, diversity is important, and it is worthwhile to optimize the implementation so as to maintain maximum diversity. Examples of loop closure are shown in Figure 13.15. These figures show the histories of all M particles. In Figure 13.15a, the FastSLAM 1.0 particles share a common history part of the way around the loop. New observations can not affect the positions of features observed before this threshold. In this case of FastSLAM 2.0, the algorithm is able to maintain diversity that extends back to the beginning of the loop. This is crucial for reliable loop closing and fast convergence.

Figure 13.16a shows the result of an experiment comparing the loop closing performance of FastSLAM 1.0 and 2.0. As the size of the loop increases, the error of both algorithms increases. However, FastSLAM 2.0 consistently

13.9 FastSLAM for Feature-Based Maps

Figure 13.14 Map of Victoria Park by FastSLAM 2.0 with $M = 1$ particle.

outperforms FastSLAM 1.0. Alternately, this result can be rephrased in terms of particles. FastSLAM 2.0 requires fewer particles to close a given loop than FastSLAM 1.0.

Figure 13.16b shows the results of an experiment comparing the convergence speed of FastSLAM 2.0 and the EKF. FastSLAM 2.0 (with 1, 10, and 100 particles) and the EKF were each run 10 times around a large simulated loop of features, similar to the ones shown in Figure 13.16a&b. Different random seeds were used for each run, causing different controls and observations to be generated for each loop. The RMS position error in the map at every time step was averaged over the 10 runs for each algorithm.

As the vehicle goes around the loop, error should gradually build up in the map. When the vehicle closes the loop at iteration 150, revisiting old features should affect the positions of features all around the loop, causing the overall error in the map to decrease. This clearly happens in the EKF. FastSLAM 2.0 with a single particle has no way to affect the positions of past features so there is no drop in the feature error. As more particles are added to FastSLAM 2.0, the filter is able to apply observations to feature positions further back in time, gradually approaching the convergence speed of the EKF. Clearly, the number of particles necessary to achieve convergence time close to the EKF will increase with the size of the loop. The lack of long-range correlations in the FastSLAM representation is arguably the most important

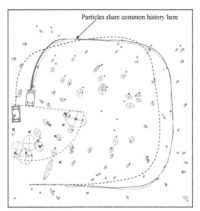 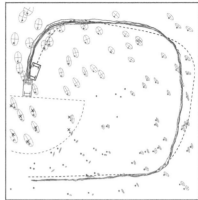

Figure 13.15 FastSLAM 2.0 can close larger loops than FastSLAM 1.0 given a constant number of particles.

weakness of FastSLAM algorithm over Gaussian SLAM techniques.

13.10 Grid-based FastSLAM

13.10.1 The Algorithm

In Chapter 9 we studied *occupancy grid maps* as a volumetric representation of robot environments. The advantage of such a representation is that it does not require any predefined definition of landmarks. Instead, in can model arbitrary types of environments. In the remainder of this chapter, we will therefore extend the FastSLAM algorithm to such representations.

To adapt the FastSLAM algorithm to occupancy grid maps, we need three functions that we already defined in previous sections. First, we have to sample from the motion posterior $p(x_t \mid x_{t-1}^{[k]}, u_t)$ as in Equation 13.12. Hence, we need such a sampling technique. Second, we need a technique for estimating the map of each particle. It turns out that we can rely on occupancy grid mapping, as described in Chapter 9. Finally, we need to compute the importance weights of the individual particles. That is, we require an approach to compute the likelihood $p(z_t \mid x_t^k, m^{[k]})$ of the observation z_t, conditioned on the pose x_t^k, the map $m^{[k]}$, and the most recent measurement z_t.

As it turns out, the extension of FastSLAM to occupancy grid maps is quite straightforward. Table 13.4 describes FastSLAM with occupancy grid maps. Not surprisingly, this algorithm borrows parts of Monte Carlo Localization

13.10 Grid-based FastSLAM

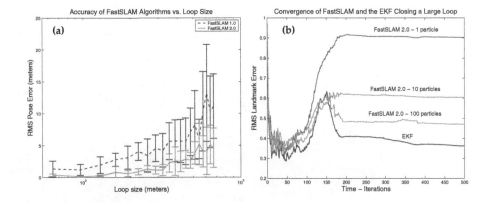

Figure 13.16 (a) Accuracy as a function of loop size: FastSLAM 2.0 can close larger loops than FastSLAM 1.0 given a fixed number of particles. (b) Comparison of the convergence speed of FastSLAM 2.0 and the EKF.

(see Table 8.2) and occupancy grid mapping (see Table 9.1). The individual functions used in this algorithm are variants of those used in the localization and mapping algorithms.

In particular, the function **measurement_model_map**$(z_t, x_t^{[k]}, m_{t-1}^{[k]})$ computes the likelihood of the measurement z_t given the pose $x_t^{[k]}$ represented by the k-th particle and given the map $m_{t-1}^{[k]}$ computed based on the previous measurement and the trajectory represented by this particle. Furthermore, the function **updated_occupancy_grid**$(z_t, x_t^{[k]}, m_{t-1}^{[k]})$ computes a new occupancy grid map, given the current pose $x_t^{[k]}$ of the k-th particle, the map $m_{t-1}^{[k]}$ associated to it, and the measurement z_t.

13.10.2 Empirical Insights

Figure 13.17 shows a typical situation of the application of the grid-based FastSLAM algorithm. Shown there are three particles together with their associated maps. Each particle represents a potential trajectory of the robot, which explains why each occupancy grid map looks different. The center map is the best in terms of global consistency.

A typical map acquired with the FastSLAM algorithm is depicted in Figure 13.19. The size of this environment is 28m × 28m. The length of the robot's trajectory is 491m and the average speed was 0.19m/s. The reso-

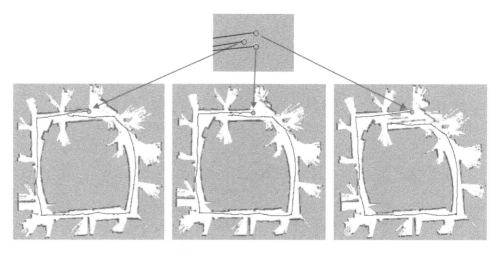

Figure 13.17 Application of the grid-based variant of the FastSLAM algorithm. Each particle carries its own map and the importance weights of the particles are computed based on the likelihood of the measurements given the particle's own map.

Figure 13.18 Occupancy grid map generated from laser range data and based on pure odometry. All images courtesy of Dirk Hähnel, University of Freiburg.

13.10 Grid-based FastSLAM

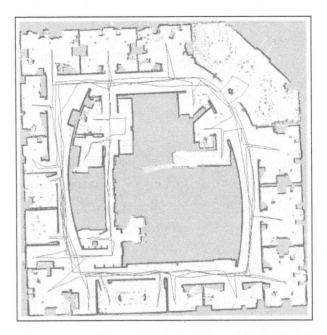

Figure 13.19 Occupancy grid map corresponding to the particle with the highest accumulated importance weight obtained by the algorithm listed in Table 13.4 from the data depicted in Figure 13.18. The number of particles to create this experiment was 500. Also depicted in this image is the path represented by the particle with the maximum accumulated importance weight.

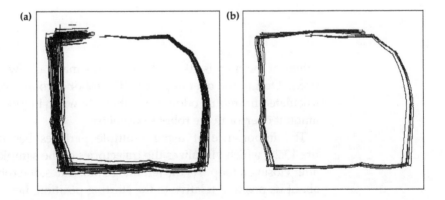

Figure 13.20 Trajectories of all samples shortly before (left) and after (right) closing the outer loop of the environment depicted in Figure 13.19. Images courtesy of Dirk Hähnel, University of Freiburg.

1:     **Algorithm FastSLAM_occupancy_grids**($\mathcal{X}_{t-1}, u_t, z_t$):
2:     $\bar{\mathcal{X}}_t = \mathcal{X}_t = \emptyset$
3:     for $k = 1$ to M do
4:         $x_t^{[k]} = $ **sample_motion_model**($u_t, x_{t-1}^{[k]}$)
5:         $w_t^{[k]} = $ **measurement_model_map**($z_t, x_t^{[k]}, m_{t-1}^{[k]}$)
5:         $m_t^{[k]} = $ **updated_occupancy_grid**($z_t, x_t^{[k]}, m_{t-1}^{[k]}$)
6:         $\bar{\mathcal{X}}_t = \bar{\mathcal{X}}_t + \langle x_t^{[k]}, m_t^{[k]}, w_t^{[k]} \rangle$
7:     endfor
8:     for $k = 1$ to M do
9:         draw i with probability $\propto w_t^{[i]}$
10:        add $\langle x_t^{[i]}, m_t^{[i]} \rangle$ to $\mathcal{X}_t$
11:    endfor
12:    return $\mathcal{X}_t$

Table 13.4   The FastSLAM algorithm for learning occupancy grid maps.

lution of the map is 10cm. To learn this map, as few as 500 particles were used. During the overall process the robot encountered two loops. A map calculated from pure odometry data is shown in Figure 13.18, illustrating the amount of error in the robot's odometry.

The importance of using multiple particles becomes evident in Figure 13.20, which visualizes the trajectories of the samples shortly before and after closing a loop. As the left image illustrates, the robot is quite uncertain about its position relative to the starting position, hence the wide spread of particles at the time of loop closure. However, a few resampling steps after the robot re-enters known terrain suffice to reduce the uncertainty drastically (right image).

13.11 Summary

This chapter has presented the particle filter approach to the SLAM problem, known as the FastSLAM algorithm.

- The basic idea of FastSLAM is to maintain a set of particles. Each particle contains a sampled robot path. It also contains a map, but here each feature in the map is represented by its own, local Gaussian. The resulting representation requires space linear in the size of the map, and linear in the number of particles.

- The "trick" to represent a map as a set of separate Gaussians—instead of one big joint Gaussian as was the case in the EKF SLAM algorithm—is possible because of the factorial structure of the SLAM problem. Specifically, we noticed that the map features are conditionally independent given the path. By factoring out the path (one per particle), we can then simply treat each map feature as independent, thereby avoiding the costly step of maintaining the correlations between them that plagued the EKF approach.

- The update in FastSLAM follows directly that of the conventional particle filter: Sample a new pose, then update the observed features. This update can be performed online, and FastSLAM is a solution to the online SLAM problem.

- Consequently, we noted that FastSLAM solves both SLAM problems: The offline SLAM problem and the online SLAM problem. Our derivation treated FastSLAM as an offline technique, in which particles represented samples in path space, not just in the momentary pose space. However, it so turned out that none of the update steps require knowledge of any pose other than the most recent one, so one can safely discard past pose estimates. This makes it possible to run FastSLAM as a filter. It also avoids that the size of particles grow linearly with time.

- We encountered two different instantiations of FastSLAM, with version numbers 1.0 and 2.0. FastSLAM 2.0 is an advanced version of FastSLAM. It differs from the base version in one key idea: FastSLAM 2.0 incorporates the measurement when sampling a new pose. The math for doing so was somewhat more involved, but FastSLAM 2.0 is found to be superior in that it requires fewer particles than FastSLAM 1.0.

- The idea of using particle filters makes it possible to estimate data association variables on a per-particle basis. Each particle can be based on a different data association. This provides FastSLAM with a straightforward and powerful mechanism for solving data association problems in SLAM. Previous algorithms, specifically the EKF, GraphSLAM, and the SEIF, are forced to adopt a single data association decision for the entire filter at any point in time, and therefore require more care in selecting the value of the data association.

- For efficiently updating the particles over time, we discussed tree representations of the map. These representations make it possible to reduce the complexity of FastSLAM updates from linear to logarithmic, enabling particles to share parts of the maps that are identical. The ideas of such trees is important in practice, as it enables FastSLAM to scale to 10^9 or more features in the map.

- We also discussed techniques for utilizing negative information. One pertains to the removal of features from the map that are not supported by sufficient evidence. Here FastSLAM adopts an evidence integration approach familiar from the occupancy grid map chapter. Another pertains to the weighting of particles themselves: When failing to observe a feature in the map of a particle, such a particle can be devalued by multiplying its importance weight accordingly.

- We discussed a number of practical properties of the two FastSLAM algorithms. Experiments show that both algorithms perform well in practice, both in feature-based maps and in volumetric occupancy grid-style maps. From a practical perspective, FastSLAM is among the best probabilistic SLAM techniques presently available. Its scaling properties are only rivaled by some of the information filter algorithms described in the two previous chapters.

- We extended FastSLAM 2.0 to different map representations. In one representation, the map was composed of points detected by a laser range finder. In this case, we were able to abandon the idea of modeling the uncertainty in the features through Gaussians, and rely on scan matching techniques to implement the forward sampling process of FastSLAM 2.0. The use of a particle filter leads to a robust loop closure technique.

Possibly the biggest limitation of FastSLAM is the fact that maintains dependencies in the estimates of feature locations only implicitly, through the diversity of its particle set. In certain environments this can negatively affect the convergence speed when compared to Gaussian SLAM techniques. When using FastSLAM care has to be taken to reduce the damaging effects of the particle deprivation problem in FastSLAM.

13.12 Bibliographical Remarks

The idea of computing distributions over sets of variables by combining samples with a parametric density function is due to Rao (1945) and Blackwell (1947). Today, this idea has become a common tool in the statistics literature (Gilks et al. 1996; Doucet et al. 2001). A first mapping algorithm that uses particle filters for closing loops can be found in Thrun et al. (2000b). The formal introduction of Rao-Blackwellized particle filters into the field of SLAM is due to Murphy (2000a); Murphy and Russell (2001), who developed this idea in the context of occupancy grid maps.

The FastSLAM algorithm was first developed by Montemerlo et al. (2002a), who also developed a tree representation for efficiently maintaining multiple maps. The extension of this algorithm to high-resolution maps is due to Eliazar and Parr, whose algorithm *DP-SLAM* has generated maps from laser range scans with unprecedented accuracy and detail. Their central data structure is called *ancestry trees*, which extends FastSLAM's trees to update problems in occupancy-style maps. A more efficient version is known as DP-SLAM 2.0 (Eliazar and Parr 2004). The FastSLAM 2.0 algorithm was developed by Montemerlo et al. (2003b). It is based on prior work by van der Merwe et al. (2001), who pioneered the idea of using the measurement as part of the proposal distribution in particle filter theory. The grid-based FastSLAM algorithm in this chapter is due to Hähnel et al. (2003b), who integrated the improved proposal distribution idea with Rao-Blackwellized filters applied to grid-based maps. A Rao-Blackwellized filter for tracking the status of doors in a dynamic office environment is described in Avots et al. (2002).

One of FastSLAM's most important contributions lies in the area of data association, so a word is on order on the literature of data association in SLAM. The original SLAM work (Smith et al. 1990; Moutarlier and Chatila 1989a) resorted to maximum likelihood data association, as derived in detail by Dissanayake et al. (2001). A key limitation of these data association techniques was the inability to enforce *mutual exclusivity*: Two different features seen within a single sensor measurement (or in short temporal succession) cannot possibly correspond to the same physical feature in the world. Realizing this, Neira and Tardós developed techniques for testing correspondence for sets of features, which greatly reduced the number of data association errors. To accommodate the huge number of potential associations (exponential in the number of features considered at each point in time), Neira et al. (2003) proposed random sampling techniques in the data association space. All of these techniques, however, maintained a single mode in the SLAM posterior. Feder et al. (1999) applied the greedy data association idea to sonar data, but implemented a delayed decision to resolve ambiguities.

The idea of maintaining multi-modal posterior in SLAM goes back to Durrant-Whyte et al. (2001), whose algorithm *sum of Gaussians* uses a Gaussian mixture for representing the posterior. Each mixture component corresponded to a different trace through the history of all data association decisions. FastSLAM follows this idea, but using particles instead of Gaussian mixture components. The idea of lazy data association can be traced back to other fields, such as the popular RANSAC algorithm (Fischler and Bolles 1981) in computer vision. The tree algorithm presented in the previous chapter is due to Hähnel et al. (2003a). As mentioned there, it parallels work by Kuipers et al. (2004). An entirely different approach for data association is described in Shatkay and Kaelbling (1997); Thrun et al. (1998b), which use the *expectation maximization* algorithm for resolving correspondence problems (see (Dempster et al. 1977)). The *EM algorithm* iterates a phase of data association for all features with a phase of map building, thereby performing search simultaneously in the space of numerical map parameters, and in the space of discrete correspondences. Araneda (2003) successfully uses MCMC techniques for data association in offline SLAM.

The data association problem arises naturally in the context of multi-robot map integration. A number of papers have developed algorithms for localizing one robot relative to another under the assumption that they both operate in the same environment, and their maps overlap (Gutmann and Konolige 2000; Thrun et al. 2000b). Roy and Dudek (2001) developed a technique by which robots had to rendezvous for integrating their information. For the general case, however, a data association technique must consider the possibility that maps might not even overlap. Stewart et al. (2003) developed a particle filter algorithm that explicitly models the possibility that maps do not overlap. Their algorithm contains a Bayesian estimator for calculating this probability, which takes into consideration how "common" specific local maps are in the environment. The idea of matching sets of features for data association in multi-robot mapping is due to Dedeoglu and Sukhatme (2000); see also Thrun and Liu (2003). A new non-planar representation of maps was proposed by Howard (2004), for circumvented inconsistency in incomplete maps. His approach led to remarkably accurate multi-robot mapping results (Howard et al. 2004).

13.13 Exercises

1. Name three key, distinct advantages for each of the following SLAM algorithms: EKF, GraphSLAM, and FastSLAM.

2. Describe a set of circumstances under which FastSLAM 1.0 will fail to converge and FastSLAM 2.0 will converge to a correct map (with probability 1).

3. On page 443, we stated that *conditioning on the most recent pose x_t instead of the entire path $x_{1:t}$ is insufficient, as dependencies may arise through previous poses*. Prove this assertion. You might prove it with an example.

4. FastSLAM generates many different maps, one for each particle. This chapter left open how to combine these maps into a single posterior estimate. Suggest two such methods, one for FastSLAM with known correspondence, and one for FastSLAM with per-particle data association.

5. The improvement of FastSLAM 2.0 over FastSLAM 1.0 lies in the nature of the proposal distribution. Develop the same for Monte Carlo Localization: Devise a similar proposal distribution for MCL in feature-based maps, and state the resulting algorithm "MCL 2.0." For this exercise you might assume known correspondence.

13.13 Exercises

6. Chapter 13.8 describes an efficient tree implementation, but it does not provide pseudo-code. In this exercise, you are asked to provide the corresponding data structures and update equations for the tree, assuming that the number of features is known in advance, and that we do not face a correspondence problem.

7. In this question, you are asked to verify empirically that FastSLAM indeed maintains the correlations between feature estimates and the robot pose estimate. Specifically, you are asked to implement a simple FastSLAM 1.0 algorithm for linear Gaussian SLAM. Recall from previous exercises that the motion and measurement equations in linear Gaussian SLAM are linear with additive Gaussian noise:

$$x_t \sim \mathcal{N}(x_{t-1} + u_t, R)$$
$$z_t = \mathcal{N}(m_j - x_t, Q)$$

Run FastSLAM 1.0 in simulation. After t steps, fit a Gaussian over the joint space of feature locations and the robot pose. Compute the correlation matrix from this Gaussian, and characterize the strength of the correlations as a function of t. What are your observations?

8. As mentioned in the text, FastSLAM is a Rao-Blackwellized particle filter. In this question you are asked to design a Rao-Blackwellized filter for a different problem: Localization with a robot that systematically drifts. Systematic drift is a common phenomena in odometry; just inspect Figures 9.1 and 10.7 for two particularly strong instances of drift. Suppose you are given a map of the environment. Can you design a Rao-Blackwellized filter that simultaneously estimates the drift parameters of the robot and the global location of the robot in the environment? Your filter should combine particle filters with Kalman filters.

Part IV

Planning and Control

14 *Markov Decision Processes*

14.1 Motivation

This is the first chapter on probabilistic *planning* and *control* in this book. Thus far, the book has focused exclusively on robot perception. We have discussed a range of probabilistic algorithms that estimate quantities of interest from sensor data. However, the ultimate goal of any robot software is to choose the right actions. This and the following chapters will discuss probabilistic algorithms for action selection.

To motivate the study of probabilistic planning algorithms, consider the following examples.

1. A robotic manipulator grasps and assembles parts arriving in random configuration on a conveyor belt. The configuration of a part is unknown at the time it arrives, yet the optimal manipulation strategy requires knowledge of the configuration. How can a robot manipulate such pieces? Will it be necessary to sense? If so, are all sensing strategies equally good? May there exist manipulation strategies that result in a well-defined configuration without sensing?

2. An underwater vehicle shall travel from Canada to the Caspian Sea. Shall it take the shortest route through the North Pole, running risk of loosing orientation under the Ice? Or should it take the longer route through open waters, where it can regularly relocalize using GPS, the satellite-based global positioning system? To what extent do such decisions depend on the accuracy of the submarine's inertial sensors?

3. A team of robots explores an unknown planet, seeking to acquire a joint map. Shall the robots seek each other to determine their relative location to each other? Or shall they instead avoid each other so that they can

cover more unknown terrain in shorter amounts of time? How does the optimal exploration strategy change when the relative starting locations of the robots are unknown?

These examples illustrate that action selection in many robotics tasks is closely tied to the notion of uncertainty. In some tasks, such as robot exploration, reducing uncertainty is the direct goal of action selection. Such problems are known as *information gathering tasks*. They will be studied in Chapter 17. In other cases, reducing uncertainty is merely a means to achieving some other goal, such as reliably arriving at a target location. These tasks will be studied in this and the next chapters.

From an algorithm design perspective, it is convenient to distinguish two types of uncertainty: uncertainty in action effects, and uncertainty in perception.

First, we distinguish *deterministic* from stochastic action effects. Many theoretical results in robotics are based on the assumption that the effects of control actions are deterministic. In practice, however, actions cause uncertainty, as outcomes of actions are non-deterministic. The uncertainty arising from the stochastic nature of the robot and its environments mandates that the robot senses at execution time, and reacts to unanticipated situations—even if the environment state is fully observable. It is insufficient to plan a single sequence of actions and blindly execute it at run-time.

Second, we distinguish *fully observable* from *partially observable systems*. Classical robotics often assumes that sensors can measure the full state of the environment. If this was always the case, then we would not have written this book! In fact, the contrary appears to be the case. In nearly all interesting real-world robotics problems, sensor limitations are a key factor.

Obviously, robots should consider their *current uncertainty* when determining what to do. When selecting a control action, at a minimum a robot should consider the various outcomes (which might include catastrophic failures), and weigh them by the probability that such outcomes might actually occur. However, robot control must also cope with future, *anticipated uncertainty*. An example of the latter was given above, where we discussed a robot that has the choice between a shorter path through a GPS-denied environment, with a longer one that reduces the danger of getting lost. Minimizing anticipated uncertainty is essential for many robotic applications.

Throughout this chapter, we will take a very liberal view and make virtually no distinction between planning and control. Fundamentally, both planning and control address the same problem: to select actions. They differ in the time constraints under which actions have to be selected, and in the role of sensing during execution. The algorithms described in this chapter are all similar in that they require an off-line optimization, or planning, phase. The result of this planning phase is a control policy, which prescribes a control action for any reasonable situation. In other words, the control policy is effectively a controller, in the sense that it can be used to determine robot actions with minimum computing time. By no means is the choice of algorithms meant to suggest that this is the only way to cope with uncertainty in robotics. However, it reflects the style of algorithms that are currently in use in the field of probabilistic robotics.

The majority of algorithms discussed in this chapter assume finite state and action spaces. Continuous spaces are approximated using grid-style representations.

The following four chapters are organized as follows.

- This chapter discusses in depth the role of the two types of uncertainty and lays out their implications on algorithm design. As a first solution to a restricted class of problems, we introduce *value iteration*, a popular planning algorithm for probabilistic systems. Our discussion in this chapter addresses only the first type of uncertainty: the uncertainty in robot motion. It rests on the assumption that the state is fully observable. The underlying mathematical framework is known as *Markov decision processes* (MDP).

- Chapter 15 generalizes the value iteration technique to both types of uncertainty, in action effects and perception. This algorithm applies value iteration to a belief state representation. The framework underlying this algorithm is called partially observable Markov decision processes (POMDPs). POMDP algorithms anticipate uncertainty, actively gather information, and explore optimally in pursuit of an arbitrary performance goal. Chapter 15 also discusses a number of efficient approximations which can calculate control policies more efficiently.

- Chapter 16 introduces a number of approximate value iteration algorithms for POMDPs. These algorithms have in common that they approximate the probabilistic planning process, so as to gain computational efficiency. One of these algorithms will shortcut probabilistic planning by assuming that at some point in the future, the state becomes fully observable. Another compresses the belief state into a lower dimensional representation, and plans using this representation. A third algorithm uses

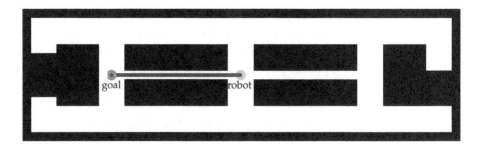

Figure 14.1 Near-symmetric environment with narrow and wide corridors. The robot starts at the center with unknown orientation. Its task is to move to the goal location on the left.

particle filters and a machine learning approach to condense the problem space. All three of these algorithms yield considerable computational improvements while still performing well in practical robotic applications.

- Chapter 17 addresses the specialized problem of robot exploration. Here the robot's goal is to accumulate information about its environment. While exploration techniques address the problem of sensor uncertainty, the problem is significantly easier than the full POMDP problem, and hence can be solved much more efficiently. Probabilistic exploration techniques are popular in robotics, as robots are frequently used for acquiring information about unknown spaces.

14.2 Uncertainty in Action Selection

Figure 14.1 shows a toy-like environment that we will use to illustrate the different types of uncertainty a robot may encounter. Shown in this figure is a mobile robot in a corridor-like environment. The environment is highly symmetric; the only distinguishing feature are its far ends, which are shaped differently. The robot starts at the location indicated, and it seeks to reach the location labeled goal. We notice there exist multiple paths to the goal, one that is short but narrow, and two others that are longer but wider.

In the classical robot planning paradigm, there is no uncertainty. The robot would simply know its initial pose and the location of the goal. Furthermore, actions when executed in the physical world have predictable effects, and such effects can be pre-planned. In such a situation, there is no need to sense. It suffices to plan off-line a single sequence of actions, which can

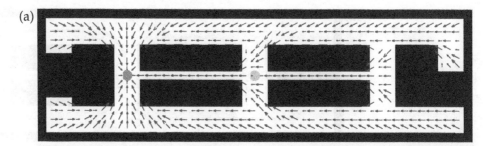
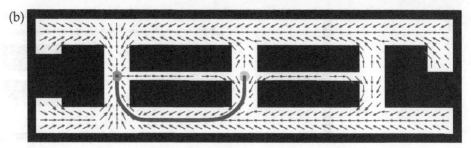

Figure 14.2 The value function and control policy for an MDP with (a) deterministic and (b) nondeterministic action effects. Under the deterministic model, the robot is perfectly fine to navigate through the narrow path; it prefers the longer path when action outcomes are uncertain, to reduce the risk of colliding with a wall. Panel (b) also shows a path.

then be executed at run-time. Figure 14.1 shows an example of such a plan. Obviously, in the absence of errors in the robot's motion, the narrow shorter path is superior to any of the longer, wider ones. Hence, a "classical" planner would choose the former path over the latter.

In practice, such plans tend to fail, for more than one reason. A robot blindly following the narrow hallway runs danger of colliding with the walls. Furthermore, a blindly executing robot might miss the goal location because of the error it accrued during plan execution. In practice, thus, planning algorithms of this type are often combined with a sensor-based, reactive control module that consults sensor readings to adjust the plan so as to avoid collisions. Such a module might prevent the robot from colliding in the narrow corridor. However, in order to do so it may have to slow down the robot, making the narrow path inferior to the wider, longer path.

A paradigm that encompasses uncertainty in robot motion is known as *Markov decision processes*, or *MDPs*. MDPs assume that the state of the en-

MARKOV DECISION PROCESS

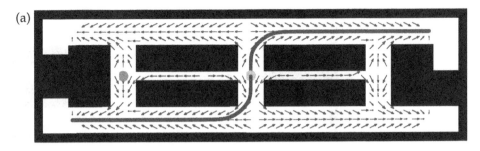
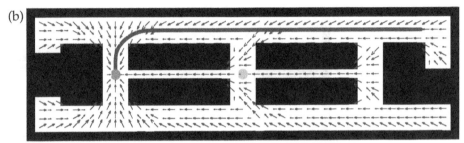
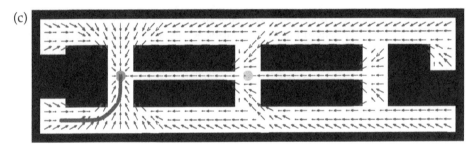

Figure 14.3 Knowledge gathering actions in POMDPs: To reach its goal with more than 50% chance, the belief space planner first navigates to a location where the global orientation can be determined. Panel (a) shows the corresponding policy, and a possible path the robot may take. Based on its location, the robot will then find itself in panel (b) or (c), from where it can safely navigate to the goal.

vironment can be fully sensed at all times. In other words, the perceptual model $p(z \mid x)$ is deterministic and bijective. However, the MDP framework allows for stochastic action effects. The action model $p(x' \mid u, x)$ may be non-deterministic. As a consequence, it is insufficient to plan a single sequence of actions. Instead, the planner has to generate actions for a whole range of situations that the robot might find itself in, because of its actions or

14.2 Uncertainty in Action Selection

other unpredictable environment dynamics. Our way to cope with the resulting uncertainty shall be to generate a *policy for action selection* defined for all states that the robot might encounter. Such mappings from states to actions are known under many names, such as *control policy, universal plans,* and *navigation functions.* An example of a policy is shown in Figure 14.2. Instead of a single sequence of actions, the robot calculates a mapping from states to actions indicated by the arrows. Once such a mapping is computed, the robot can accommodate non-determinism by sensing the state of the world, and acting accordingly. The panel in Figure 14.2a shows a policy for a robot with nearly no motion uncertainty, in which cases the narrow path is indeed acceptable. Figure 14.2b depicts the same situation for an increased randomness in robot motion. Here the narrow path makes a collision more likely, and the detour becomes preferable. This example illustrates two things: the importance of incorporating uncertainty in the motion planning process; and the ability of finding a good control policy using the algorithm described in this chapter.

CONTROL POLICY

Let us now return to the most general, fully probabilistic case, by dropping the assumption that the state is fully observable. This case is known as *partially observable Markov decision processes,* or *POMDPs.* In most if not all robotics applications, measurements z are noisy projections of the state x. Hence, the state can only be estimated up to a certain degree. To illustrate this, consider once again our example, but under different assumptions. Assume that the robot knows its initial location but does not know whether it is oriented towards the left or to the right. Further, it has no sensor to sense whether it has arrived at the goal location.

PARTIALLY OBSERVABLE MARKOV DECISION PROCESS

Clearly, the symmetry of the environment makes it difficult to disambiguate the orientation. By moving directly towards the projected goal state the robot faces a 50% chance of missing the goal, and instead moving to the symmetric location on the right size of the environment. The optimal plan, thus, is to move to any of the corners of the environment, which is perceptually sufficiently rich to disambiguate its orientation. A policy for moving to these locations is shown in Figure 14.3a. Based on its initial orientation, the robot may execute any of the two paths shown there. As it reaches a corner, its sensors now reveal its orientation, and hence its actual location relative to the environment. The robot might now find itself in any of the two situations depicted in Figures 14.3b&c, from where it can safely navigate to the goal location as shown.

This example illustrates one of the key aspects of probabilistic robotics. The robot has to actively gather information, and in order to do so, might be suffering a detour relative to a robot that knows its state with absolute certainty. This problem is paramount in robotics. For nearly any robotics task, the robot's sensors are characterized by intrinsic limitations as to what the robot can know, and where information can be acquired. Similar situations occur in tasks as diverse as locate-and-retrieve, planetary exploration, urban search-and-rescue, and so on.

The question now arises as to how can one devise an algorithm for action selection that can cope with this type of uncertainty. As we will learn, this is not a trivial question. One might be tempted to solve the problem of what to do by analyzing each possible situation that might be the case under the current state of knowledge. In our example, there are two such cases: the case where the goal is on the upper left relative to the initial robot heading, and the case where the goal is on the lower right. In both these cases, however, the optimal policy does not bring the agent to a location where it would be able to disambiguate its pose. That is, the planning problem in a partially observable environment cannot be solved by considering all possible environments and averaging the solution.

Instead, the key idea is to generate plans in the *belief space*, a synonym for *information space*. The belief space comprises the space of all posterior beliefs that the robot might hold about the world. The belief space for our simple example corresponds to the three panels in Figure 14.3. The top panel shows such a belief space policy. It displays the initial policy while the robot is unaware of its orientation. Under this policy, the robot navigates to one of the corners of the environment where it can localize. Once localized, it can safely navigate to the target location, as illustrated in the two lower panels of Figure 14.3. Since the a priori chance of each orientation is the same, the robot will experience a random transition with a 50% chance of ending up in either of the two bottom diagrams.

In our toy example, the number of different belief states happens to be finite: either the robot knows or it does not have a clue. In practical applications, this is usually not the case. In worlds with finitely many states the belief space is usually continuous, but of finite dimensionality. In fact, the number of dimensions of the belief space is of the same order as the number of states in the underlying state space. If the state space is continuous, the belief space possesses infinitely many dimensions.

This example illustrates a fundamental property that arises from the robot's inability to perfectly sense the state of the world—one whose importance for robotics has often been under-appreciated. In uncertain worlds a robot planning algorithm must consider the state of its knowledge when making control decisions. In general it does not suffice to consider the most likely state only. By conditioning the action on the belief state—as opposed to the most likely actual state—the robot can actively pursue information gathering. In fact, the optimal plan in belief state gathers information "optimally," in that it only seeks new information to the extent that it is actually beneficial to the expected utility of the robot's action. The ability to devise optimal control policies is a key advantage of the probabilistic approach to robotics over the classical deterministic, omniscient approach. However, as we shall see soon, it comes at the price of an increased complexity of the planning problem.

14.3 Value Iteration

Our first algorithm for finding control policies is called *value iteration*. Value iteration recursively calculates the utility of each action relative to a payoff function. Our discussion in this chapter will be restricted to the first type of uncertainty: stochasticity of the robot and the physical world. We defer our treatment of uncertainty arising from sensor limitations to the subsequent chapter. Thus, we will assume that the state of the world is fully observable at any point in time.

14.3.1 Goals and Payoff

GOAL

Before describing a concrete algorithm, let us first define the problem in more concise terms. In general, robotic action selection is driven by *goals*. Goals might correspond to specific configurations (e.g., a part has successfully been picked up and placed by a robot manipulator), or they might express conditions over longer periods of time (e.g., a robot balances a pole). In robotics, one is sometimes concerned with reaching a specific goal configuration, while simultaneously optimizing other variables, often thought of as *cost*. For example, one might be interested in moving the end-effector of a manipulator to a specific location, while simultaneously minimizing time, energy consumption, or the number of collisions with obstacles.

COST

At first glance, one might be tempted to express these desires by two quantities, one that is being maximized (e.g., the binary flag that indicates whether or not a robot reached its goal location), and the other one being minimized (e.g., the total energy consumed by the robot). However, both can be expressed using a single function called the *payoff function*.

PAYOFF FUNCTION

The payoff, denoted r, is a function of the state and the robot control. For example, a simple payoff function may be the following:

(14.1) $$r(x, u) = \begin{cases} +100 & \text{if } u \text{ leads to a goal configuration or state} \\ -1 & \text{otherwise} \end{cases}$$

This payoff function "rewards" the robot with $+100$ if a goal configuration is attained, while it "penalizes" the robot by -1 for each time step where it has not reached that configuration. Such a payoff function will yield maximum cumulative return if the robot reaches its goal configuration in the minimum possible time.

Why use a single payoff variable to express both goal achieval and costs? This is primarily because of two reasons: First, the notation avoids clutter in the formulae yet to come, as it shall unify our treatment of costs and goal achieval throughout this book. Second, and more importantly, it pays tribute to the fundamental trade-off between goal achievement and costs along the way. Since robots are inherently uncertain, they cannot know with certainty as to whether a goal configuration has been achieved; instead, all one can hope for is to maximize the chances of reaching a goal. This trade-off between goal achieval and cost is characterized by questions like *Is increasing the probability of reaching a goal worth the extra effort (e.g., in terms of energy, time)?* Treating both goal achieval and costs as a single numerical factor enables us to trade off one against the other, hence providing a consistent framework for selecting actions under uncertainty.

We are interested in devising programs that generate actions so as to optimize future payoff in expectation. Such a program is usually referred to as a

POLICY *control policy*, or simply *policy*. It will be denoted as follows:

(14.2) $\pi : z_{1:t-1}, u_{1:t-1} \longrightarrow u_t$

In the case of full observability, we assume the much simpler case:

(14.3) $\pi : x_t \longrightarrow u_t$

Thus, a policy π is a function that maps past data into controls, or states into controls when the state is observable.

So far, our definition of a control policy makes no statements about its computational properties. It might be a fast, reactive algorithm that bases its decision on the most recent data item only, or an elaborate planning algorithm. In practice, however, computational considerations are essential, in that any delay in calculating a control may negatively affect a robot's performance. Our definition of a policy π also makes no commitment as to whether it is deterministic or non-deterministic.

14.3 Value Iteration

PLANNING HORIZON

An interesting concept in the context of creating control policies is the *planning horizon*. Sometimes, it suffices to choose a control action so as to maximize the immediate next payoff value. Most of the time, however, an action might not pay off immediately. For example, a robot moving to a goal location will receive the final payoff for reaching its goal only after the very last action. Thus, payoff might be delayed. An appropriate objective is then to choose actions so that the sum of all future payoff is maximal. We will

CUMULATIVE PAYOFF

call this sum the *cumulative payoff*. Since the world is non-deterministic, the best one can optimize is the *expected cumulative payoff*, which is conveniently written as

$$(14.4) \quad R_T \;=\; E\left[\sum_{\tau=1}^{T} \gamma^{\tau} r_{t+\tau}\right]$$

Here the expectation $E[\,]$ is taken over future momentary payoff values $r_{t+\tau}$ which the robot might accrue between time t and time $t+T$. The individual

DISCOUNT FACTOR

payoffs $r_{t+\tau}$ are multiplied by a factor γ^{τ}, called the *discount factor*. The value of γ is a problem-specific parameter, and it is constrained to lie in the interval $[0;1]$. If $\gamma = 1$, we have $\gamma^{\tau} = 1$ for arbitrary values of τ, and hence the factor can be omitted in Equation (14.4). Smaller values of γ discount future payoff exponentially, making earlier payoffs exponentially more important than later ones. This discount factor, whose importance will be discussed later, bears resemblance to the value of money, which also loses value over time exponentially due to inflation.

PLANNING HORIZON

We notice that R_T is a sum of T time steps. T is called the *planning horizon*, or simply: *horizon*. We distinguish three important cases:

GREEDY CASE

1. $T = 1$. This is called the *greedy case*, where the robot only seeks to minimize the immediate next payoff. While this approach is degenerate in that it does not capture the effect of actions beyond the immediate next time step, it nevertheless plays an important role in practice. The reason for its importance stems from the fact that greedy optimization is much simpler than multi-step optimization. In many robotics problems, greedy algorithms are currently the best known solutions that can be computed in polynomial time. Obviously, greedy optimization is invariant with respect to the discount factor γ, but it requires that $\gamma > 0$.

FINITE-HORIZON CASE

2. T larger than 1 but finite. This case is known as the *finite-horizon case*. Typically, the payoff is not discounted over time, thus $\gamma = 1$. One might argue that the finite-horizon case is the only one that matters, since for all practical purposes time is finite. However, finite-horizon optimality is often

harder to achieve than optimality in the discounted infinite-horizon case. Why is this so? A first insight stems from the observation that the optimal control action is a function of time horizon. Near the far end of the time horizon, for example, the optimal policy might differ substantially from the optimal choice earlier in time, even under otherwise identical conditions (e.g., same state, same belief). As a result, planning algorithms with finite horizon are forced to maintain different plans for different horizons, which can add undesired complexity.

INFINITE-HORIZON CASE

3. T is infinite. This case is known as the *infinite-horizon case*. This case does not suffer the same problem as the finite horizon case, as the number of remaining time steps is the same for any point in time (it is infinite!). However, here the discount factor γ is essential. To see why, let us consider the case where we have two robot control programs, one that earns us $1 per hour, and another one that makes us $100 per hour. In the finite horizon case, the latter is clearly preferable to the former. No matter what the value of the horizon is, the expected cumulative payoff of the second program exceeds that of the first by a factor of a hundred. Not so in the infinite horizon case. Without discounting, both programs will earn us an infinite money, rendering the expected cumulative payoff R_T insufficient to select the better program.

Under the assumption that each individual payoff r is bounded in magnitude (that is, $|r| < r_{\max}$ for some value $r_{\max}$), discounting guarantees that R_∞ is finite—despite the fact that the sum has infinitely many terms. Specifically, we have

$$(14.5) \qquad R_\infty \;\leq\; r_{\max} + \gamma r_{\max} + \gamma^2 r_{\max} + \gamma^3 r_{\max} + \ldots \;=\; \frac{r_{\max}}{1-\gamma}$$

This shows that R_∞ is finite as long as γ is smaller than 1. As an aside, we note that an alternative to discounting involves maximizing the *average* payoff instead of the total payoff. Algorithms for maximizing average payoff will not be studied in this book.

Sometimes we will refer to the cumulative payoff R_T conditioned on a state x_t. This will be written as follows:

$$(14.6) \qquad R_T(x_t) \;=\; E\left[\sum_{\tau=1}^{T} \gamma^\tau r_{t+\tau} \;\big|\; x_t\right]$$

The cumulative payoff R_T is a function of the robot's policy for action selection. Sometimes, it is beneficial to make this dependence explicit:

$$(14.7) \quad R_T^\pi(x_t) = E\left[\sum_{\tau=1}^T \gamma^\tau r_{t+\tau} \mid u_{t+\tau} = \pi(z_{1:t+\tau-1}, u_{1:t+\tau-1})\right]$$

This notation enables us to compare two control policies π and π', and determine which one is better. Simply compare R_T^π to $R_T^{\pi'}$ and pick the algorithm with higher expected discounted future payoff!

14.3.2 Finding Optimal Control Policies for the Fully Observable Case

This chapter shall be concluded with a concreted value iteration algorithm, for calculating control policies in fully observable domains. At first glance, such algorithms depart from the basic assumption of probabilistic robotics, namely that state is not observable. However, in certain applications one can safely assume that the posterior $p(x_t \mid z_{1:t}, u_{1:t})$ is well-represented by its mean $E[p(x_t \mid z_{1:t}, u_{1:t})]$.

The fully observable case has some merit. The algorithm discussed in turn also prepares us for the more general case of partial observability.

We already noted that the framework of stochastic environments with fully observable state is known as Markov decision processes. Policies, in MDPs, are mappings from state to control actions:

$$(14.8) \quad \pi : x \longrightarrow u$$

The fact that the state is sufficient for determining the optimal control is a direct consequence of our Markov assumption that was discussed in length in Chapter 2.4.4. The goal of planning in the MDP framework is to identify the policy π that maximizes the future cumulative payoff.

Let us begin with defining the optimal policy for a planning horizon of $T = 1$, hence we are only interested in a policy that maximizes the immediate next payoff. This policy shall be denoted $\pi_1(x)$ and is obtained by maximizing the expected 1-step payoff over all controls:

$$(14.9) \quad \pi_1(x) = \operatorname*{argmax}_u r(x, u)$$

Thus, an optimal action is one that maximizes the immediate next payoff in expectation. The policy that chooses such an action is optimal in expectation.

VALUE FUNCTION

Every policy has an associated *value function*, which measures the expected value (cumulative discounted future payoff) of this specific policy. For π_1, the value function is simply the expected immediate payoff, discounted by the factor γ:

(14.10) $\quad V_1(x) \;=\; \gamma \, \max_u \, r(x, u)$

This value for longer planning horizons is now defined recursively. The optimal policy for horizon $T = 2$ selects the control that maximizes the sum of the one-step optimal value $V_1(x)$ and the immediate 1-step payoff:

(14.11) $\quad \pi_2(x) \;=\; \operatorname*{argmax}_u \left[r(x, u) + \int V_1(x') \, p(x' \mid u, x) \, dx' \right]$

It should be immediately obvious why this policy is optimal. The value of this policy conditioned on the state x is given by the following discounted expression:

(14.12) $\quad V_2(x) \;=\; \gamma \, \max_u \left[r(x, u) + \int V_1(x') \, p(x' \mid u, x) \, dx' \right]$

The optimal policy and its value function for $T = 2$ was constructed recursively, from the optimal value function for $T = 1$. This observation suggests that for any finite horizon T the optimal policy, and its associated value function, can be obtained recursively from the optimal policy and value function $T - 1$.

(14.13) $\quad \pi_T(x) \;=\; \operatorname*{argmax}_u \left[r(x, u) + \int V_{T-1}(x') \, p(x' \mid u, x) \, dx' \right]$

The resulting policy $\pi_T(x)$ is optimal for the planning horizon T. The associated value function is defined through the following recursion:

(14.14) $\quad V_T(x) \;=\; \gamma \, \max_u \left[r(x, u) + \int V_{T-1}(x') \, p(x' \mid u, x) \, dx' \right]$

In the infinite horizon case, the optimal value function tends to reach the equilibrium (with the exception of some rare deterministic systems, in which no such equilibrium exists):

(14.15) $\quad V_\infty(x) \;=\; \gamma \, \max_u \left[r(x, u) + \int V_\infty(x') \, p(x' \mid u, x) \, dx' \right]$

BELLMAN EQUATION   This invariance is known as *Bellman equation*. Without proof, we notice that every value function V that satisfies the condition (14.15) is both necessary and sufficient for the induced policy to be *optimal*.

14.3.3 Computing the Value Function

This consideration leads to the definition of a practical algorithm for calculating the optimal policy in stochastic systems with full state observability. Value iteration does this by successively approximating the optimal value functions, as defined in (14.15).

In detail, let us denote out value function approximation by $\hat{V}$. Initially, $\hat{V}$ is set to $r_{\min}$, the minimum possible immediate payoff:

(14.16) $$\hat{V} \longleftarrow r_{\min}$$

Value iteration then successively updates the approximation via the following recursive rule, which computes the value function for increasing horizons:

(14.17) $$\hat{V}(x) \longleftarrow \gamma \max_u \left[r(x,u) + \int \hat{V}(x') \, p(x' \mid u, x) \, dx' \right]$$

BACKUP STEP

Since each update propagates information in reverse temporal order through the value function, it is usually referred to as the *backup step*.

The value iteration rule bears close resemblance to the calculation of the horizon-T optimal policy above. Value iteration converges if $\gamma < 1$ and, in some special cases, even for $\gamma = 1$. The order in which states are updated in value iteration is irrelevant, as long as each state is updated infinitely often. In practice, convergence is usually observed after a much smaller number of iterations.

At any point in time, the value function $\hat{V}(x)$ defines a policy:

(14.18) $$\pi(x) = \operatorname*{argmax}_u \left[r(x,u) + \int \hat{V}(x') \, p(x' \mid u, x) \, dx' \right]$$

After convergence of value iteration, the policy that is greedy with respect to the final value function is optimal.

We note that all of these equations have been formulated for general state spaces. For finite state spaces, the integral in each of these equations can be implemented as a finite sum over all states. This sum can often be calculated efficiently, since $p(x' \mid u, x)$ will usually be non-zero for a relatively few states x and x'. This leads to an efficient family of algorithms for calculating value functions.

Table 14.1 shows three algorithms: The general value iteration algorithm **MDP_value_iteration** for arbitrary state and action spaces; its discrete variant for finite state spaces **MDP_discrete_value_iteration**, and the algorithm for retrieving the optimal control action from the value function, **policy_MDP**.

1: **Algorithm MDP_value_iteration( ):**
2:   for all x do
3:    $\hat{V}(x) = r_{\min}$
4:   endfor
5:   repeat until convergence
6:    for all x
7:     $\hat{V}(x) = \gamma \max_u \left[r(x,u) + \int \hat{V}(x') \, p(x' \mid u, x) \, dx' \right]$
8:    endfor
9:   endrepeat
10: return $\hat{V}$

1: **Algorithm MDP_discrete_value_iteration( ):**
2:   for $i = 1$ to N do
3:    $\hat{V}(x_i) = r_{\min}$
4:   endfor
5:   repeat until convergence
6:    for $i = 1$ to N do
7:     $\hat{V}(x_i) = \gamma \max_u \left[r(x_i, u) + \sum_{j=1}^{N} \hat{V}(x_j) \, p(x_j \mid u, x_i) \right]$
8:    endfor
9:   endrepeat
10: return $\hat{V}$

1: **Algorithm policy_MDP($x, \hat{V}$):**
2:   return $\operatorname*{argmax}_u \left[r(x,u) + \sum_{j=1}^{N} \hat{V}(x_j) \, p(x_j \mid u, x_i) \right]$

Table 14.1 The value iteration algorithm for MDPs, stated here in its most general form and for MDPs with finite state and control spaces. The bottom algorithm computes the best control action.

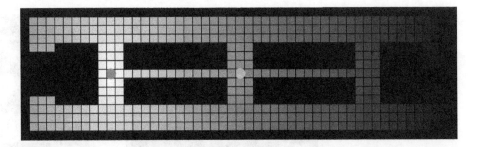

Figure 14.4 An example of an infinite-horizon value function T_∞, assuming that the goal state is an "absorbing state." This value function induces the policy shown in Figure 14.2a.

The first algorithm **MDP_value_iteration** initializes the value function in line 3. Line 5 through 9 implement the recursive calculation of the value function. Once value iteration converges, the resulting value function $\hat{V}$ induces the optimal policy. If the state space is finite, the integral is replaced by a finite sum, as shown in **MDP_discrete_value_iteration**. The factor γ is the discount factor. The algorithm **policy_MDP** processes the optimal value function along with a state x, and returns the control u that maximizes the expected value.

Figure 14.4 depicts an example value function, for our example discussed above. Here the shading of each grid cell corresponds to its value, with white being $V = 100$ and black being $V = 0$. Hill climbing in this value function using Equation 14.18 leads to the policy shown in Figure 14.2a.

14.4 Application to Robot Control

The simple value iteration algorithm is applicable to low-dimensional robot motion planning and control problems. To do so, we have to introduce two approximations.

First, the algorithm in Table 14.5 defines a value function over a continuous space, and requires maximization and integration over a continuous space. In practice, it is common to approximate the state space by a discrete decomposition, similar to our histogram representations in Chapter 4.1. Similarly, it is common to discretize the control space. The function $\hat{V}$ is then easily implemented as a look-up table. However, such a decomposition works only

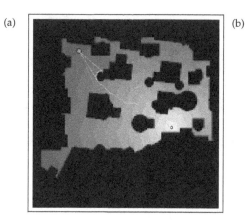 

Figure 14.5 Example of value iteration over state spaces in robot motion. Obstacles are shown in black. The value function is indicated by the gray shaded area. Greedy action selection with respect to the value function lead to optimal control, assuming that the robot's pose is observable. Also shown in the diagrams are example paths obtained by following the greedy policy.

for low-dimensional state and control spaces, due to the curse of dimensionality. In higher dimensional situations, it is common to introduce learning algorithms to represent the value function.

Second, we need a state! As noted above, it might be viable to replace the posterior by its mode

$$\hat{x}_t = E[p(x_t \mid z_{1:t}, u_{1:t})] \tag{14.19}$$

In the context of robot localization, for example, such an approximation works well if we can guarantee that the robot is always approximately localized, and the residual uncertainty in the posterior is local. It ceases to work when the robot performs global localization, or if it is being kidnapped.

Figure 14.5 illustrates value iteration in the context of a robotic path planning problem. Shown there is a two-dimensional projection of a configuration space of a circular robot. The configuration space is the space of all $\langle x, y, \theta \rangle$ coordinates that the robot can physically attain. For circular robots, the configuration space is obtained by "growing" the obstacles in the map by the radius of the robot. These increased obstacles shown in black in Figure 14.5.

The value function is shown in gray, where the brighter a location, the higher its value. The path obtained by following the optimal policy leads to

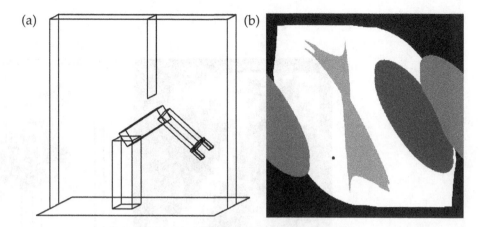

Figure 14.6 (a) 2-DOF robot arm in an environment with obstacles. (b) The *configuration space* of this arm: the horizontal axis corresponds to the shoulder joint, and the vertical axis to its elbow joint configuration. Obstacles are shown in gray. The small dot in this diagram corresponds to the configuration on the left.

the respective goal location, as indicated in Figure 14.5. The key observation is that the value function is defined over the entire state space, enabling the robot to select an action no matter where it is. This is important in non-deterministic worlds, where actions have stochastic effects on the robot's state.

The path planner that generated Figure 14.5 makes specific assumptions in order to keep the computational load manageable. For circular robots that can turn on the spot, it is common to compute the value function in the two-dimensional Cartesian coordinates only, ignoring the cost of rotation. We also ignore state variables such as the robot's velocity, despite the fact that velocity clearly constrains where a robot can move at any given point in time. To turn such a control policy into actual robot controls, it is therefore common practice to combine such path planners with fast, reactive collision avoidance modules that generate motor velocities while obeying dynamic constraints. A path planner that considers the full robot state would have to plan in at least five dimensions, comprising the full pose (three dimensions), the translational and the rotational velocity of the robot. In two dimensions, calculating the value function for environment like the one above takes only a fraction of a second on a low-end PC.

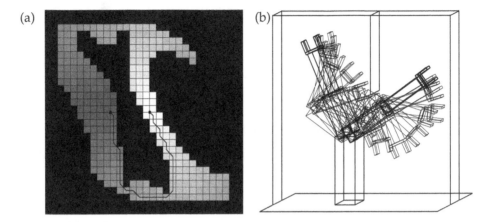

Figure 14.7 (a) Value iteration applied to a coarse discretization of the configuration space. (b) Path in workspace coordinates. The robot indeed avoids the vertical obstacle.

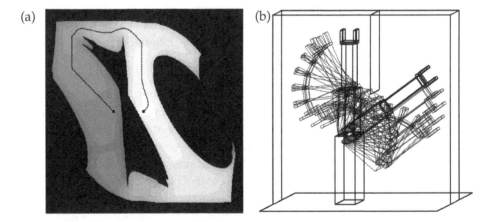

Figure 14.8 (a) Probabilistic value iteration, here over a fine-grained grid. (b) The corresponding path.

A second example is illustrated in Figure 14.6a. Shown there is a robot arm model with two rotational degrees of freedom (DOFs), a shoulder and an elbow joint. Determining the exact configuration of these joints is usually possible, through shaft encoders that are attached to the joints. Hence, the approximation in (14.19) is a valid one. However, robot arm motion is usually subject to noise, and as such the control noise should be taken into account during planning. This makes arm control a primary application for probabilistic MDP-style algorithms.

CONFIGURATION SPACE

Robot arm motion is usually tackled in the *configuration space*. The configuration space for the specific arm is shown in Figure 14.6b. Here the horizontal axis graphs the orientation of the shoulder, and the vertical orientation that of the elbow. Each point in this diagram, thus, corresponds to a specific configuration. In fact, the small dot in Figure 14.6b corresponds to the configuration shown in Figure 14.6a.

It is common to decompose the configuration space into areas in which the robot can move, and areas where it would collide. This is shown in Figure 14.6b. The white area in this figure corresponds to the collision-free configuration space, commonly called freespace. The black boundary in the configuration space is the constraint imposed by the table and the enclosing case. The vertical obstacle, protruding into the robot's workspace from above in Figure 14.6a, corresponds to the light gray obstacle in the center of Figure 14.6b. This figure is not at all obvious, and the reader may take a minute to visualize configurations in which the robot collides with this obstacle.

Figure 14.7a shows the result of value iteration using a coarse discretization of the configurations space. Here the value is propagated using a deterministic motion model and the resulting path is also shown. When executed, this policy leads to a motion shown in Figure 14.7b. Figure 14.8 shows a result for a probabilistic motion model, with the resulting motion of the arm. Again, this is the result of applying value iteration under the assumption that the configuration of the robot arm is fully observable—a rare instance in which this assumption is valid!

14.5 Summary

This chapter introduced the basic framework of probabilistic control.

- We identified the two basic types of uncertainty a robot may face: uncertainty with regards to controls, and uncertainty in perception. The former makes it difficult to determine what lies ahead, whereas the latter makes it hard to determine what is. Uncertainty from unpredictable events in the environment was silently subsumed into this taxonomy.

- We defined the control objective through a *payoff function*, which maps states and controls to a goodness value. The payoff enables us to express performance goals as well as costs of robot operation. The overall control objective is the maximization of all payoff, immediate and at future points in time. To avoid possibly infinite sums, we introduced a so-called discount factor that exponentially discounts future payoffs.

- We discussed an approach to solving probabilistic control problems by devising a control policy. A control policy defines the control action that is to be chosen, as a function of the robot's information about the world. A policy is optimal if it maximizes the sum of all future cumulative payoffs. The policy is computed in a planning phase that precedes robot operation. Once computed, it specifies the optimal control action for any possible situation a robot may encounter.

- We devised a concrete algorithm for finding the optimal control policy for the restricted case of fully observable domains, in which the state is fully observable. Those domains are known as Markov decision processes. The algorithm involved the calculation of a value function, which measures the expected cumulative payoff. A value function defines a policy of greedily selecting the control that maximizes value. If the value function is optimal, so is the policy. The value iteration algorithm successively improved the value function by updating it recursively.

- We discussed applications of MDP value iteration to probabilistic robotics problems. For that we extracted the mode of the belief as the state, and approximated the value function by a low-dimensional grid. The result was a motion planning algorithm for stochastic environments, which enables a robot to navigate even if its action effects are nondeterministic.

The material in this chapter lays the groundwork for the next chapter, in which we will tackle the more general problem of control under measurement uncertainty, also known as the partially observable MDP problem. We already intuitively discussed why this problem is much harder than the MDP problem. Nevertheless, some of the intuitions and basic algorithms carry over to this case.

We close this chapter by remarking that there exist a number of alternative techniques for probabilistic planning and control under uncertainty. Our choice of value iteration as the basic method is due to its popularity; further, value iteration techniques are among the best-understood techniques for the more general POMDP case.

Value iteration is by no means the most effective algorithm for generating control. Common planning algorithms include the A* algorithm, which uses a heuristic in the computation of the value function, or direct policy search techniques that identify a locally optimal policy through gradient descent. However, value iteration plays a pivotal role in the next chapter, when we address the much harder case of optimal control under sensor uncertainty.

14.6 Bibliographical Remarks

The idea of dynamic programming goes back to Bellman (1957) and Howard (1960). Bellman (1957) identified the equilibrium equation for value iteration, which has henceforth been called the Bellman equation. Markov decision processes (MDPs) with incomplete state estimation were first discussed by Astrom (1965); see also Mine and Osaki (1970) for early work on MDPs. Since then, dynamic programming for control has been a vast field, as a recent book on this topic attests (Bertsekas and Tsitsiklis 1996). Recent improvements to the basic paradigm include techniques for real-time value iteration (Korf 1988), value iteration guided through the interaction with an environment (Barto et al. 1991), model free value iteration (Watkins 1989), and value iteration with parametric representation of the value function (Roy and Tsitsiklis 1996; Gordon 1995), or using trees (Moore 1991) (see also (Mahadevan and Kaelbling 1996)). Hierarchical value iteration techniques have been developed by Parr and Russell (1998) and Dietterich (2000), and Boutilier et al. (1998) improved the efficiency of MDP value iteration by reachability analysis. There exists also a rich literature on applications of value iteration, for example work by Barniv (1990) on moving target detection. In the light of this rich literature, the material in this chapter is a basic exposition of the most simple value iteration techniques, in preparation of the techniques described in the chapters that follow.

Within robotics, the issue of robot motion planning has typically been investigated in a non-probabilistic framework. As noted, the assumption is usually that the robot and its environment is perfectly known, and controls have deterministic effects. Complications arise from the fact that the state space is continuous and high-dimensional. The standard text in this field is Latombe (1991). It is predated by seminal work on a number of basic motion planning techniques, visibility graphs (Wesley and Lozano-Perez 1979), potential field control (Khatib 1986), and Canny's (1987) famous silhouette algorithm. Rowat (1979) introduced the idea of Voronoi graphs into the field of robot control, and Guibas et al. (1992) showed how to compute them efficiently. Choset (1996) developed this paradigm into a family of efficient online exploration and mapping techniques. Another set of techniques used randomized (but not probabilistic!) techniques for searching the space of possible robot paths (Kavraki and Latombe 1994). A contemporary book on this topic is due to Choset et al. (2004).

APPROXIMATE CELL DECOMPOSITION

In the language of the robotic motion planning field, the methods discussed in this chapter are *approximate cell decompositions* which, in the deterministic case, provide no guarantee of completeness. Decompositions of continuous spaces into finite graphs have been studied in robotics for decades. Reif (1979) developed a number of techniques for decomposing continuous spaces into finitely many cells that retained completeness in motion planning. The idea of configuration spaces, necessary to check collision with the techniques described in this chapter, was originally proposed by Lozano-Perez (1983). Recursive cell decomposition techniques for configuration space planning can be found in Brooks and Lozano-Perez (1985), all under the non-probabilistic assumptions of perfect world models and perfect robots. Acting under partial knowledge has been addressed by Goldberg (1993), who in his Ph.D. thesis developed algorithms for orienting parts in the absence of sensors. Sharma (1992) developed robot path planning techniques with stochastic obstacles.

Policy functions that assign to every possible robot state a control action are known as *controllers*. An algorithm for devising a control policy is often called *optimal controller* (Bryson and Yu-Chi 1975). Control policies are also known as *navigation function* (Koditschek 1987; Rimon and Koditschek 1992). Within AI, they are known as *universal plans* (Schoppers 1987), and a number of symbolic planning algorithms have addressed the problem of finding such universal plans (Dean et al. 1995; Kushmerick et al. 1995; Hoey et al. 1999). Some robotic work has been devoted to bridging the gap between universal plans and open-loop action sequences, such as in Nourbakhsh's (1987) Ph.D. work.

We also remark that the notion of "control" in this book is somewhat narrowly defined. The chapter deliberately did not address standard techniques in the rich field of control, such as PID control and other popular techniques often found in introductory textbooks (Dorf and Bishop 2001). Clearly, such techniques are both necessary and applicable in many real-world robot systems. Our choice to omit them is based on space constraints, and on the fact that most of these techniques do not rely on explicit representations of uncertainty.

14.7 Exercises

1. The dynamic programming algorithm uses the most likely state to determine its action. Can you draw a robot environment in which conditioning actions on the most likely state is fundamentally the wrong choice? Can you give a concise reason why this might sometimes be a poor choice?

2. Suppose we run value iteration to completion, for a fixed cost function. Then the cost function changes. We would now like to adjust the value function by further iterations of the algorithm, using the previous value function as a starting point.

   (a) Is this a good or a bad idea? Does your answer depend on whether the cost increased or decreased?

   (b) Can you flesh out an algorithm that would be more efficient than simply continuing value iteration after the cost changes? If you can, argue why your algorithm is more efficient. If not, argue why no such algorithm may exist.

3. **Heaven or Hell?** In this exercise, you are asked to extend dynamic programming to an environment with a single hidden state variable. The

environment is a maze with a designated start marked "S," and two possible goal states, both marked "H."

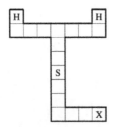

What the agent does not know is which of the two goal states provides a positive reward. One will give $+100$, whereas the other will give -100. There is a .5 probability that either of those situations is true. The cost of moving is -1; the agent can only move into the four directions north, south, east, and west. Once a state labeled "H" has been reached, the play is over.

(a) Implement a value iteration algorithm for this scenario (and ignore the label "X" in the figure). Have your implementation compute the value of the starting state. What is the optimal policy?

(b) Modify your value algorithm to accommodate a probabilistic motion model: with .9 chance the agent moves as desired; with .1 chance it will select any of the three other directions at random. Run your value iteration algorithm again, and compute both the value of the starting state, and the optimal policy.

(c) Now suppose the location labeled "X" contains a sign that informs the agent of the correct assignment of rewards to the two states labeled "H." How does this affect the optimal policy?

(d) How can you modify your value iteration algorithm to find the optimal policy? Be concise. State any modifications to the space over which the value function is defined.

(e) Implement your modification, and compute both the value of the starting state and the optimal policy.

15 Partially Observable Markov Decision Processes

15.1 Motivation

This chapter discusses algorithms for the partially observable robot control problem. These algorithms address both the uncertainty in measurement and the uncertainty in control effects. They generalize the value iteration algorithm discussed in the previous chapter, which was restricted to action effect uncertainty. The framework studied here is known as *partially observable Markov decision processes*, or *POMDPs*. This name was coined in the operations research literature. The term *partial* indicates that the state of the world cannot be sensed directly. Instead, the measurements received by the robot are incomplete and usually noisy projections of this state.

As has been discussed in so many chapters of this book, partial observability implies that the robot has to estimate a posterior distribution over possible world states. Algorithms for finding the optimal control policy exist for finite worlds, where the state space, the action space, the space of observations, and the planning horizon T are all finite. Unfortunately, these exact methods are computationally involved. For the more interesting continuous case, the best known algorithms are approximate.

All algorithms studied in this chapter build on the value iteration approach discussed previously. Let us restate Equation (14.14), which is the central update equation in value iteration in MDPs:

$$(15.1) \quad V_T(x) = \gamma \max_u \left[r(x,u) + \int V_{T-1}(x') \, p(x' \mid u, x) \, dx' \right]$$

with $V_1(x) = \gamma \max_u r(x,u)$. In POMDPs, we apply the very same idea. However, the state x it not observable. The robot has to make its decision in the belief state, which is the space of posterior distributions over states.

Throughout this and the next chapters, we will abbreviate a belief by the symbol b, instead of the more elaborate bel used in previous chapters.

POMDPs compute a value function over belief space:

$$(15.2) \quad V_T(b) = \gamma \max_u \left[r(b,u) + \int V_{T-1}(b') \, p(b' \mid u, b) \, db' \right]$$

with $V_1(b) = \gamma \max_u E_x[r(x,u)]$. The induced control policy is as follows:

$$(15.3) \quad \pi_T(b) = \operatorname*{argmax}_u \left[r(b,u) + \int V_{T-1}(b') \, p(b' \mid u, b) \, db' \right]$$

A belief is a probability distribution; thus, each value in a POMDP is a function of an entire probability distribution. This is problematic. If the state space is finite, the belief space is continuous, since it is the space of all distributions over the state space. Thus, there is a continuum of different values; whereas there was only a finite number of different values in the MDP case. The situation is even more delicate for continuous state spaces, where the belief space is an infinitely-dimensional continuum.

An additional complication arises from the computational properties of the value function calculation. Equations (15.2) and (15.3) integrate over all beliefs b'. Given the complex nature of the belief space, it is not at all obvious that the integration can be carried out exactly, or that effective approximations can be found. It should therefore come at no surprise that calculating the value function V_T is more complicated in belief space than it is in state space.

Luckily, an exact solution exists for the interesting special case of finite worlds, in which the state space, the action space, the space of observations, and the planning horizon are all finite. This solution represents value functions by *piecewise linear functions* over the belief space. As we shall see, the linearity of this representation arises directly from the fact that the expectation is a linear operator. The piecewise nature is the result of the fact that the robot has the ability to select controls, and in different parts of the belief space it will select different controls. All these statements will be proven in this chapter.

PIECEWISE LINEAR FUNCTION

This chapter discusses the general POMDP algorithm for calculating policies defined over the space of all belief distributions. This algorithm is computationally cumbersome but correct for finite POMDPs; although a variant will be discussed that is highly tractable. The subsequent chapter will discuss a number of more efficient POMDP algorithms, which are approximate but scale to actual robotics problems.

15.2 An Illustrative Example

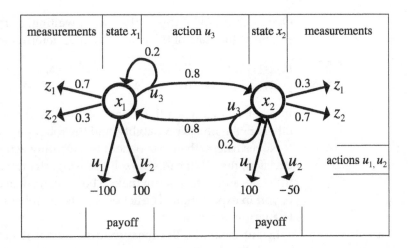

Figure 15.1 The two-state environment used to illustrate value iteration in belief space.

15.2 An Illustrative Example

15.2.1 Setup

We illustrate value iteration in belief spaces through a numerical example. This example is simplistic, but by discussing it we identify all major elements of value iteration in belief spaces.

Our example is the two-state world in Figure 15.1. The states are labeled x_1 and x_2. The robot can choose among three different control actions, u_1, u_2, and u_3. Actions u_1 and u_2 are terminal: When executed, they result in the following immediate payoff:

(15.4) $\quad r(x_1, u_1) = -100 \qquad r(x_2, u_1) = +100$

(15.5) $\quad r(x_1, u_2) = +100 \qquad r(x_2, u_2) = -50$

The dilemma is that both actions provide opposite payoffs in each of the states. Specifically, when in state x_1, u_2 is the optimal action, whereas it is u_1 in state x_2. Thus, knowledge of the state translates directly into payoff when selecting the optimal action.

To acquire such knowledge, the robot is provided with a third control action, u_3. Executing this control comes at a mild cost of -1:

(15.6) $\quad r(x_1, u_3) = r(x_2, u_3) = -1$

One might think of this as the cost of waiting, or the cost of sensing. Action u_3 affects the state of the world in a non-deterministic manner:

(15.7) $\quad p(x_1'|x_1, u_3) = 0.2 \qquad p(x_2'|x_1, u_3) = 0.8$

(15.8) $\quad p(x_1'|x_2, u_3) = 0.8 \qquad p(x_2'|x_2, u_3) = 0.2$

In other words, when the robot executes u_3, the state flips to the respective other state with 0.8 probability, and the robot pays a unit cost.

Nevertheless, there is a benefit to executing action u_3. Before each control decision, the robot can sense. By sensing, the robot gains knowledge about the state, and in turn it can make a better control decision that leads to higher payoff in expectation. The action u_3 lets the robot sense without committing to a terminal action.

In our example, the measurement model is governed by the following probability distribution:

(15.9) $\quad p(z_1|x_1) = 0.7 \qquad p(z_2|x_1) = 0.3$

(15.10) $\quad p(z_1|x_2) = 0.3 \qquad p(z_2|x_2) = 0.7$

Put differently, if the robot measures z_1 its confidence increases for being in x_1, and the same is the case for z_2 relative to x_2.

The reason for selecting a two-state example is that it makes it easy to graph functions over the belief space. In particular, a belief state b is characterized by $p_1 = b(x_1)$ and $p_2 = b(x_2)$. However, we know $p_2 = 1 - p_1$, hence it suffices to graph p_1. The corresponding control policy π is a function that maps the unit interval $[0; 1]$ to the space of all actions:

(15.11) $\quad \pi : [0; 1] \longrightarrow u$

15.2.2 Control Choice

In determining when to execute what control, let us start our consideration with the immediate payoff for each of the three control choices, u_1, u_2, and u_3. In the previous chapter, payoff was considered a function of state and actions. Since we do not know the state, we have to generalize the notion of a payoff to accommodate belief state. Specifically, for any given belief $b = (p_1, p_2)$, the expected payoff under this belief is given by the following expectation:

(15.12) $\quad r(b, u) = E_x[r(x, u)] = p_1\, r(x_1, u) + p_2\, r(x_2, u)$

PAYOFF IN POMDPs   The function $r(b, u)$ defines the *payoff in POMDPs*.

15.2 An Illustrative Example

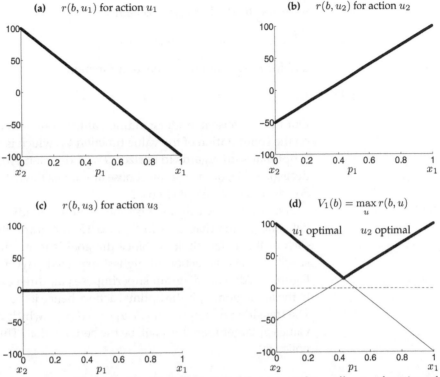

Figure 15.2 Diagrams (a), (b), and (c) depict the expected payoff r as a function of the belief state parameter $p_1 = b(x_1)$, for each of the three actions $u_1, u_2,$ and u_3. (d) The value function at horizon $T = 1$ corresponds to the maximum of these three linear functions.

Figure 15.2a graphs the expected payoff $r(b, u_1)$ for control choice u_1, parameterized by the parameter p_1. On the leftmost end of this diagram, we have $p_1 = 0$, hence the robot believes the world to be in state x_2 with absolute confidence. Executing action u_1 hence yields $r(x_2, u_1) = 100$, as specified in Equation (15.4). On the rightmost end, we have $p_1 = 1$, hence the state is x_1. Consequently, control choice u_1 will result in $r(x_1, u_1) = -100$. In between, the expectation provides a linear combination of these two values:

(15.13) $\quad r(b, u_1) \quad = \quad -100\, p_1 + 100\, p_2 \quad = \quad -100\, p_1 + 100\, (1 - p_1)$

This function is graphed in Figure 15.2a.

Figures Figure 15.2b&c show the corresponding functions for action u_2 and

u_3, respectively. For u_2, we obtain

(15.14) $$r(b, u_2) = 100\, p_1 - 50\, (1 - p_1)$$

and for u_3 we obtain the constant function

(15.15) $$r(b, u_3) = -1\, p_1 - 1\, (1 - p_1) = -1$$

Our first exercise in understanding value iteration in belief spaces will focus on the computation of the value function V_1, which is the value function that is optimal with regards to horizon $T = 1$ decision processes. Within a single decision cycle, our robot can choose among its three different control choices. So which one should it choose?

The answer is easily read off the diagrams studied thus far. For any belief state p_1, the diagrams in Figures 15.2a-c graph the expected payoff for each of the action choices. Since the goal is to maximize payoff, the robot simply selects the action of highest expected payoff. This is visualized in Figure 15.2d: This diagram superimposes all three expected payoff graphs. In the left region, u_1 is the optimal action, hence its value function dominates. The transition occurs when $r(b, u_1) = r(b, u_2)$, which resolves to $p_1 = \frac{3}{7}$. For values p_1 larger than $\frac{3}{7}$, u_2 will be the better action. Thus the $(T = 1)$-optimal policy is

(15.16) $$\pi_1(b) = \begin{cases} u_1 & \text{if } p_1 \leq \frac{3}{7} \\ u_2 & \text{if } p_1 > \frac{3}{7} \end{cases}$$

The corresponding value is then the thick upper graph in Figure 15.2d. This graph is a piecewise linear, convex function. It is the maximum of the individual payoff functions in Figures 15.2a-c. Thus, we can write it as a maximum over 3 functions:

(15.17) $$\begin{aligned} V_1(b) &= \max_u r(b, u) \\ &= \max \begin{Bmatrix} -100\, p_1 & +100\, (1 - p_1) \\ 100\, p_1 & -50\, (1 - p_1) \\ -1 & \end{Bmatrix} \begin{matrix} (*) \\ (*) \\ \end{matrix} \end{aligned}$$

Obviously, only the linear functions marked $(*)$ in (15.17) contribute. The remaining linear function can safely be pruned away:

(15.18) $$V_1(b) = \max \begin{Bmatrix} -100\, p_1 & +100\, (1 - p_1) \\ 100\, p_1 & -50\, (1 - p_1) \end{Bmatrix}$$

We will use this pruning trick repeatedly in our example. Prunable linear constraints are shown as dashed lines in Figure 15.2d and many graphs to follow.

15.2.3 Sensing

The next step in our reasoning involves perception. What if the robot can sense before it chooses its control? How does this affect the optimal value function? Obviously, sensing provides information about the state, hence should enable the robot to choose a better control action. Specifically, for the worst possible belief thus far, $p_1 = \frac{3}{7}$, the expected payoff in our example was $\frac{100}{7} \approx 14.3$, which is the value at the kink in Figure 15.2d. Clearly, if we can sense first, we find ourselves in a different belief after sensing. The value of this belief will be better than 14.3, but by how much?

The answer is surprising. Suppose we sense z_1. Figure 15.3a shows the belief after sensing z_1 as a function of the belief before sensing. Let us dissect this function. If our pre-sensing belief is $p_1 = 0$, our post-sensing belief is also $p_1 = 0$, regardless of the measurement. Similarly for $p_1 = 1$. Hence, at the extreme ends, this function is the identity. In between, we are uncertain as to what the state of the world is, and measuring z_1 does shift our belief. The amount it shifts is governed by Bayes rule:

$$(15.19) \quad p'_1 = p(x_1 \mid z)$$
$$= \frac{p(z_1 \mid x_1)\, p(x_1)}{p(z_1)}$$
$$= \frac{0.7\, p_1}{p(z_1)}$$

and

$$(15.20) \quad p'_2 = \frac{0.3\,(1 - p_1)}{p(z_1)}$$

The normalizer $p(z_1)$ adds the non-linearity in Figure 15.3a. In our example, it resolves to

$$(15.21) \quad p(z_1) = 0.7\, p_1 + 0.3\,(1 - p_1) = 0.4\, p_1 + 0.3$$

and hence $p'_1 = \frac{0.7\, p_1}{0.4\, p_1 + 0.3}$. However, as we shall see below, this normalizer nicely cancels out. More on this in a minute.

Let us first study the effect of this non-linear transfer function on the value function V_1. Suppose we know that we observed z_1, and then have

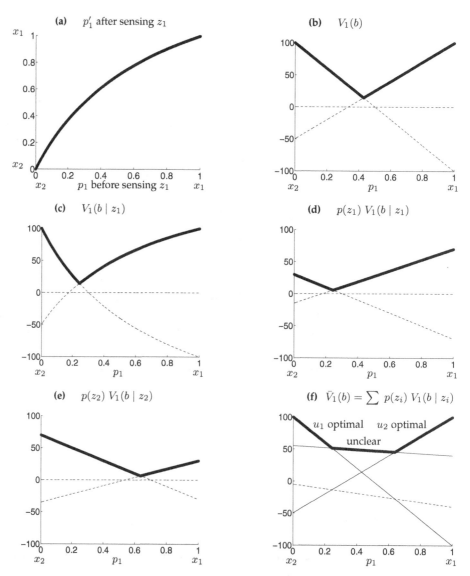

Figure 15.3 The effect of sensing on the value function: (a) The belief after sensing z_1 as a function of the belief before sensing z_1. Sensing z_1 makes the robot more confident that the state is x_1. Projecting the value function in (b) through this nonlinear function results in the non-linear value function in (c). (d) Dividing this value function by the probability of observing z_1 results in a piecewise linear function. (e) The same piecewise linear function for measurement z_2. (f) The expected value function after sensing.

15.2 An Illustrative Example

to make an action choice. What would that choice be, and what would the corresponding value function look like? The answer is given graphically in Figure 15.3c. This figure depicts the piecewise linear value function in Figure 15.3b, mapped through the nonlinear measurement function discussed above (and shown in Figure 15.3a). The reader may take a moment to get oriented here: Take a belief p_1, map it to a corresponding belief p'_1 according to our non-linear function, and then read off its value in Figure 15.3b. This procedure, for all $p_1 \in [0; 1]$, leads to the graph in Figure 15.3c.

Mathematically, this graph is given by

$$(15.22) \quad V_1(b \mid z_1) = \max \left\{ \begin{array}{ll} -100 \cdot \frac{0.7\, p_1}{p(z_1)} & +100 \cdot \frac{0.3\,(1-p_1)}{p(z_1)} \\ 100 \cdot \frac{0.7\, p_1}{p(z_1)} & -50 \cdot \frac{0.3\,(1-p_1)}{p(z_1)} \end{array} \right\}$$

$$= \frac{1}{p(z_1)} \max \left\{ \begin{array}{ll} -70\, p_1 & +30\,(1-p_1) \\ 70\, p_1 & -15\,(1-p_1) \end{array} \right\}$$

which is simply the result of replacing p_1 by p'_1 in the value function V_1 specified in (15.18). We note in Figure 15.3c that the belief of "worst" value has shifted to the left. Now the worst belief is the one that, after sensing z_1, makes us believe with $\frac{3}{7}$ probability we are in state x_1.

However, this is the consideration for one of the two measurements only, the value before sensing has to take both measurements into account. Specifically, the value before sensing, denoted $\bar{V}_1$, is given by the following expectation:

$$(15.23) \quad \bar{V}_1(b) = E_z[V_1(b \mid z)] = \sum_{i=1}^{2} p(z_i)\, V_1(b \mid z_i)$$

We immediately notice that in this expectation, each contributing value function $V_1(b \mid z_i)$ is multiplied by the probability $p(z_i)$, which was the cause of the nonlinearity in the pre-measurement value function. Plugging (15.19) into this expression gives us

$$(15.24) \quad \bar{V}_1(b) = \sum_{i=1}^{2} p(z_i)\, V_1\left(\frac{p(z_i \mid x_1)\, p_1}{p(z_i)}\right)$$

$$= \sum_{i=1}^{2} p(z_i)\, \frac{1}{p(z_i)}\, V_1(p(z_i \mid x_1)\, p_1)$$

$$= \sum_{i=1}^{2} V_1(p(z_i \mid x_1)\, p_1)$$

This transformation is true because each element in V_1 is linear in $1/p(z_i)$, as illustrated by example in (15.22). There we were able to move the factor $1/p(z_i)$ out of the maximization, since each term in the maximization is a product of this factor. After restoring the terms accordingly, the term $p(z_i)$ simply cancels out!

In our example, we have two measurements, hence we can compute the expectation $p(z_i) \, V_1(b \mid z_i)$ for each of these measurements. The reader may recall that these terms are added in the expectation (15.23). For z_1, we already computed $V_1(b \mid z_i)$ in (15.22), hence

$$(15.25) \quad p(z_1) \, V_1(b \mid z_1) \;=\; \max \left\{ \begin{array}{ll} -70 \, p_1 & +30 \, (1 - p_1) \\ 70 \, p_1 & -15 \, (1 - p_1) \end{array} \right\}$$

This function is shown in Figure 15.3d: It is indeed the maximum of two linear functions. Similarly, for z_2 we obtain

$$(15.26) \quad p(z_2) \, V_1(b \mid z_2) \;=\; \max \left\{ \begin{array}{ll} -30 \, p_1 & +70 \, (1 - p_1) \\ 30 \, p_1 & -35 \, (1 - p_1) \end{array} \right\}$$

This function is depicted in Figure 15.3e.

The desired value function before sensing is then obtained by adding those two terms, according to Equation (15.23):

$$(15.27) \quad \bar{V}_1 \;=\; \max \left\{ \begin{array}{ll} -70 \, p_1 & +30 \, (1 - p_1) \\ 70 \, p_1 & -15 \, (1 - p_1) \end{array} \right\} + \max \left\{ \begin{array}{ll} -30 \, p_1 & +70 \, (1 - p_1) \\ 30 \, p_1 & -35 \, (1 - p_1) \end{array} \right\}$$

This sum is shown in Figure 15.3f. It has a remarkable shape: Instead of a single kink, it possesses two different kinks, separating the value function into three different linear segments. For the left segment, u_1 is the optimal action, no matter what additional information the robot may reap through future sensing. Similarly for the right segment, u_2 is the optimal control action no matter what. In the center region, however, sensing matters. The optimal action is determined by what the robot senses. In doing so, the center segment defines a value that is significantly higher than the corresponding value without sensing, shown in Figure 15.2d. Essentially, the ability to sense lifted an entire region in the value function to a higher level, in the region where the robot was least certain about the state of the world. This remarkable finding shows that value iteration in belief space indeed values sensing, but only to the extent that it matters for future control choices.

Let us return to computing this value function, since it may appear easier than it is. Equation (15.27) requires us to compute the sum of two maxima over linear functions. Bringing this into our canonical form—which is the

15.2 An Illustrative Example

maximum over linear functions without the sum—requires some thought. Specifically, our new value function $\bar{V}_1$ will be bounded below by *any* sum that adds a linear function from the first max-expression to a linear function from the second max-expression. This leaves us with four possible combinations:

$$(15.28) \quad \bar{V}_1(b) = \max \begin{cases} -70\,p_1 & +30\,(1-p_1) & -30\,p_1 & +70\,(1-p_1) \\ -70\,p_1 & +30\,(1-p_1) & +30\,p_1 & -35\,(1-p_1) \\ 70\,p_1 & -15\,(1-p_1) & -30\,p_1 & +70\,(1-p_1) \\ 70\,p_1 & -15\,(1-p_1) & +30\,p_1 & -35\,(1-p_1) \end{cases}$$

$$= \max \begin{cases} -100\,p_1 & +100\,(1-p_1) \\ -40\,p_1 & -5\,(1-p_1) \\ 40\,p_1 & +55\,(1-p_1) \\ 100\,p_1 & -50\,(1-p_1) \end{cases} \begin{matrix} (*) \\ \\ (*) \\ (*) \end{matrix}$$

$$= \max \begin{cases} -100\,p_1 & +100\,(1-p_1) \\ 40\,p_1 & +55\,(1-p_1) \\ 100\,p_1 & -50\,(1-p_1) \end{cases}$$

Once again, we use $(*)$ to denote constraints that actually contribute to the definition of the value function. As shown in Figure 15.3f, only three of these four linear functions are required, and the fourth can safely be pruned away.

15.2.4 Prediction

Our final step concerns state transitions. When the robot selects an action, its state changes. To plan at a horizon larger than $T = 1$, we have to take this into consideration and project our value function accordingly. In our example, u_1 and u_2 are both terminal actions. Thus, we only have to consider the effect of action u_3.

Luckily, state transitions are not anywhere as intricate as measurements in POMDPs. Figure 15.4a shows mapping of the belief upon executing u_3. Specifically, suppose we start out in state x_1 with absolute certainty, hence $p_1 = 1$. Then according to our transition probability model in Equation (15.7), we have $p_1' = p(x_1'|x_1, u_3) = 0.2$. Similarly, for $p_1 = 0$ we get $p_1' = p(x_1'|x_2, u_3) = 0.8$. In between the expectation is linear:

$$(15.29) \quad p_1' = E_x[p(x_1'|x, u_3)]$$
$$= \sum_{i=1}^{2} p(x_1'|x_i, u_3)\, p_i$$

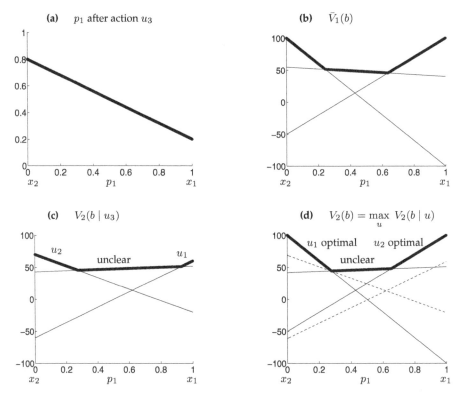

Figure 15.4 (a) The belief state parameter p'_1 after executing action u_3, as a function of the parameter p_1 before the action. Propagating the belief shown in (b) through the inverse of this mapping results in the belief shown in (c). (d) The value function V_2 obtained by maximizing the propagated belief function, and the payoff of the two remaining actions, u_1 and u_2.

$$= 0.2\, p_1 + 0.8\, (1 - p_1) = 0.8 - 0.6\, p_1$$

This is the function graphed in Figure 15.4a. If we now back-project the value function in Figure 15.4b—which is equivalent to the one shown in Figure 15.3f—we obtain the value function in Figure 15.4c. This value function is flatter than the one before the projection step, reflecting the loss of information through the state transition. It is also mirrored, since in expectation the state changes when executing u_3.

Mathematically, this value function is computed by projecting (15.28)

15.2 An Illustrative Example

through (15.29).

$$(15.30) \quad \bar{V}_1(b \mid u_3) = \max \left\{ \begin{array}{ll} -100\,(0.8 - 0.6\,p_1) & +100\,(1 - (0.8 - 0.6\,p_1)) \\ 40\,(0.8 - 0.6\,p_1) & +55\,(1 - (0.8 - 0.6\,p_1)) \\ 100\,(0.8 - 0.6\,p_1) & -50\,(1 - (0.8 - 0.6\,p_1)) \end{array} \right\}$$

$$= \max \left\{ \begin{array}{ll} -100\,(0.8 - 0.6\,p_1) & +100\,(0.2 + 0.6\,p_1) \\ 40\,(0.8 - 0.6\,p_1) & +55\,(0.2 + 0.6\,p_1) \\ 100\,(0.8 - 0.6\,p_1) & -50\,(0.2 + 0.6\,p_1) \end{array} \right\}$$

$$= \max \left\{ \begin{array}{ll} 60\,p_1 & -60\,(1 - p_1) \\ 52\,p_1 & +43\,(1 - p_1) \\ -20\,p_1 & +70\,(1 - p_1) \end{array} \right\}$$

These transformations are easily checked by hand. Figure 15.4c shows this function, along with the optimal control actions.

We have now almost completed the vale function V_2 with a planning horizon of $T = 2$. Once again, the robot is given a choice whether to execute the control u_3, or to directly engage in any of the terminal actions u_1 or u_2. As before, this choice is implemented by adding two new options to our consideration, in the form of the two linear functions $r(b, u_1)$ and $r(b, u_2)$. We also must subtract the cost of executing action u_3 from the value function.

This leads to the diagram in Figure 15.4d, which is of the form

$$(15.31) \quad \bar{V}_2(b) = \max \left\{ \begin{array}{ll} -100\,p_1 & +100\,(1 - p_1) \\ 100\,p_1 & -50\,(1 - p_1) \\ 59\,p_1 & -61\,(1 - p_1) \\ 51\,p_1 & +42\,(1 - p_1) \\ -21\,p_1 & +69\,(1 - p_1) \end{array} \right. \begin{array}{l} (*) \\ (*) \\ \\ (*) \\ \end{array} $$

Notice that we simply added the two options (lines one and two), and subtracted the uniform cost of u_3 from all other linear constraints (lines three through five). Once again, only three of those constraints are needed, as indicated by the $(*)$'s. The resulting value can thus be rewritten as

$$(15.32) \quad \bar{V}_2(b) = \max \left\{ \begin{array}{ll} -100\,p_1 & +100\,(1 - p_1) \\ 100\,p_1 & -50\,(1 - p_1) \\ 51\,p_1 & +42\,(1 - p_1) \end{array} \right\}$$

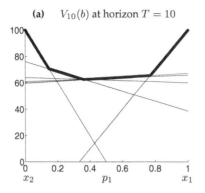 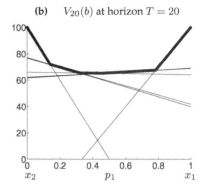

Figure 15.5 The value function V for horizons $T = 10$ and $T = 20$. Note that the vertical axis in these plots differs in scale from previous depictions of value functions.

15.2.5 Deep Horizons and Pruning

BACKUP STEP IN BELIEF SPACE

We have now executed a full *backup step in belief space*. This algorithm is easily recursed. Figure 15.5 shows the value function at horizon $T = 10$ and $T = 20$, respectively. Both of these value functions are seemingly similar. With appropriate pruning, V_{20} has only 13 components

$$(15.33) \quad \bar{V}_{20}(b) = \max \begin{Bmatrix} -100\, p_1 & +100\, (1-p_1) \\ 100\, p_1 & -50\, (1-p_1) \\ 64.1512\, p_1 & +65.9454\, (1-p_1) \\ 64.1513\, p_1 & +65.9454\, (1-p_1) \\ 64.1531\, p_1 & +65.9442\, (1-p_1) \\ 68.7968\, p_1 & +62.0658\, (1-p_1) \\ 68.7968\, p_1 & +62.0658\, (1-p_1) \\ 69.0914\, p_1 & +61.5714\, (1-p_1) \\ 68.8167\, p_1 & +62.0439\, (1-p_1) \\ 69.0369\, p_1 & +61.6779\, (1-p_1) \\ 41.7249\, p_1 & +76.5944\, (1-p_1) \\ 39.8427\, p_1 & +77.1759\, (1-p_1) \\ 39.8334\, p_1 & +77.1786\, (1-p_1) \end{Bmatrix}$$

We recognize the two familiar linear functions on the top; all others correspond to specific sequences of measurements and action choices.

As simple consideration shows that pruning is of essence. Without pruning, each update brings two new linear constraints (action choice), and then squares the number of constraints (measurement). Thus, an unpruned value function for $T = 20$ is defined over $10^{547,864}$ linear functions; at $T = 30$ we have $10^{561,012,337}$ linear constraints. The pruned value function, in comparison, contains only 13 such constraints.

This enormous explosion of linear pieces is a key reason why plain POMDPs are impractical. Figure 15.6 compares side-by-side the steps that led to the value function V_2. The left column shows our pruned functions, whereas the right row maintains all linear functions without pruning. While we only have a single measurement update in this calculation, the number of unused functions is already enormous. We will return to this point later, when we will devise efficient approximate POMDP algorithms.

A final observation of our analysis is that the optimal value function for any finite horizon is continuous, piecewise linear, and convex. Each linear piece corresponds to a different action choice at some point in the future. The convexity of the value function indicates the rather intuitive observation, namely that knowing is always superior to not knowing. Given two belief states b and b', the mixed value of the belief states is larger or equal to the value of the mixed belief state, for some mixing parameter β with $0 \leq \beta \leq 1$:

$$\text{(15.34)} \quad \beta V(b) + (1-\beta)V(b') \geq V(\beta b + (1-\beta)b')$$

This characterization only applies to the finite horizon case. Under infinite horizon, the value function can be discontinuous and nonlinear.

15.3 The Finite World POMDP Algorithm

The previous section showed, by example, how to calculate value functions in finite worlds. Here we briefly discuss a general algorithm for calculating a value function, before deriving it from first principles.

The algorithm **POMDP** is listed in Table 15.1. This algorithm accepts as an input just a single parameter: T, the planning horizon for the POMDP. It returns a set of parameter vectors, each of the form

$$\text{(15.35)} \quad (v_1, \ldots, v_N)$$

Each of these parameters specifies a linear function over the belief space of the form

$$\text{(15.36)} \quad \sum_i v_i \, p_i$$

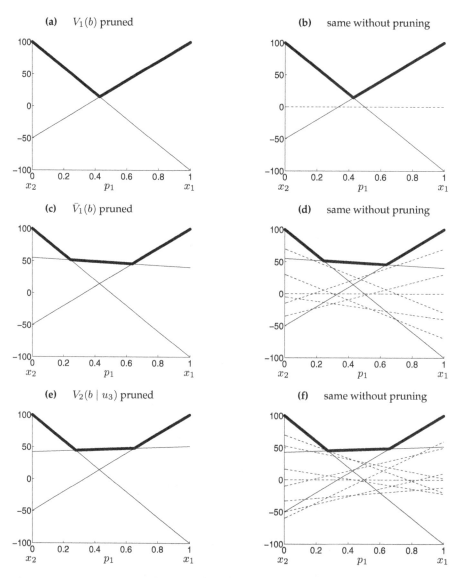

Figure 15.6 Comparison of an exact pruning algorithm (left row) versus a non-pruning POMDP algorithm (right row), for the first few steps of the POMDP planning algorithm. Obviously, the number of linear constraints increases dramatically without pruning. At $T = 20$, the unpruned value function is defined over $10^{547,864}$ linear functions, whereas the pruned one only uses 13 such functions.

1:      **Algorithm POMDP(T):**
2:         $\Upsilon = (0; 0, \ldots, 0)$
3:         for $\tau = 1$ to T do
4:            $\Upsilon' = \emptyset$
5:            for all $(u'; v_1^k, \ldots, v_N^k)$ in Υ do
6:               for all control actions u do
7:                   for all measurements z do
8:                       for $j = 1$ to N do
9:                          $v_{u,z,j}^k = \sum_{i=1}^{N} v_i^k \, p(z \mid x_i) \, p(x_i \mid u, x_j)$
10:                    endfor
11:                endfor
12:            endfor
13:         endfor
14:            for all control actions u do
15:                for all $k(1), \ldots, k(M) = (1, \ldots, 1)$ to $(|\Upsilon|, \ldots, |\Upsilon|)$ do
16:                   for $i = 1$ to N do
17:                       $v_i' = \gamma \left[r(x_i, u) + \sum_z v_{u,z,i}^{k(z)} \right]$
18:                  endfor
19:                add $(u; v_1', \ldots, v_N')$ to Υ'
20:                endfor
21:            endfor
22:            optional: prune Υ'
23:            $\Upsilon = \Upsilon'$
24:         endfor
25:         return Υ

Table 15.1 The POMDP algorithm for discrete worlds. This algorithm represents the optimal value function by a set of linear constraints, which are calculated recursively.

```
1:        Algorithm policy_POMDP(ϒ, b = (p_1, ..., p_N)):

2:              û =    argmax      Σ_{i=1}^{N} v_i^k p_i
                  (u;v_1^k,...,v_N^k)∈ϒ

3:              return û
```

Table 15.2 The algorithm for determining the optimal action for a policy represented by the set of linear functions Υ.

MAXIMUM OF LINEAR FUNCTIONS

The actual value is governed by the *maximum of all these linear functions*:

(15.37)
$$\max_{(p_1,\ldots,p_N)} \sum_i v_i \, p_i$$

The algorithm **POMDP** computes this value function recursively. An initial set for the pseudo-horizon $T = 0$ is set in line 2 of Table 15.1. The algorithm **POMDP** then recursively computes a new set in the nested loop of lines 3-24. A key computational step takes place in line 9: Here, the coefficients $v_{u,z,j}^k$ of the linear functions needed to compute the next set of linear constraints are computed. Each linear function results from executing control u, followed by observing measurement z, and then executing control u'. The linear constraint corresponding to u' was calculated in the previous iteration for a smaller planning horizon (taken in line 5). Thus, upon reaching line 14, the algorithm has generated one linear function for each combination of control action, measurement, and linear constraint of the previous value function.

The linear constraints of the new value function result by taking the expectations over measurements, as done in lines 14-21. For each control action, the algorithm generates K^M such linear constraints in line 15. This large number is due to the fact that each expectation is taken over the M possible measurements, each of which can be "combined" with any of the K constraints contained in the previous value function. Line 17 computes the expectation for each such combination. The resulting constraint is added to the new set of constraints in line 19.

The algorithm for finding the optimal control action is shown in Table 15.2. The input to this algorithm is a belief state, parameterized by

$b = (p_1, \ldots, p_N)$, along with the set of linear functions Υ. The optimal action is determined by search through all linear functions, and identifying the one that maximizes its value for b. This value is returned in line 3 of the algorithm **policy_POMDP:**, Table 15.2.

15.4 Mathematical Derivation of POMDPs

15.4.1 Value Iteration in Belief Space

The general update for the value function implements (15.2), restated here for convenience.

$$(15.38) \quad V_T(b) \;=\; \gamma \max_u \left[r(b,u) + \int V_{T-1}(b')\, p(b' \mid u, b)\, db' \right]$$

We begin by transforming this equation into a more practical form, one that avoids integration over the space of all possible beliefs.

A key factor in this update is the conditional probability $p(b' \mid u, b)$. This probability specifies a distribution over probability distributions. Given a belief b and a control action u, the outcome is indeed a distribution over distributions. This is because the concrete belief b' is also based on the next measurement, the measurement itself is generated stochastically. Dealing with distributions of distributions adds an element of complexity that is undesirable.

If we fix the measurement, the posterior b' is unique and $p(b' \mid u, b)$ degenerate to a point-mass distribution. Why is this so? The answer is provided by the Bayes filter. From the belief b before action execution, the action u, and the subsequent observation z, the Bayes filter calculates a single, posterior belief b' which is the single, correct belief. Thus, we conclude that if only we knew z, the integration over all beliefs in (15.38) would be obsolete.

This insight can be exploited by re-expressing

$$(15.39) \quad p(b' \mid u, b) \;=\; \int p(b' \mid u, b, z)\, p(z \mid u, b)\, dz$$

where $p(b' \mid u, b, z)$ is a point-mass distribution focused on the single belief calculated by the Bayes filter. Plugging this integral into Equation (15.38) gives us

$$(15.40) \quad V_T(b) \;=\; \gamma \max_u \left[r(b,u) + \int \left[\int V_{T-1}(b')\, p(b' \mid u, b, z)\, db' \right] p(z \mid u, b)\, dz \right]$$

The inner integral

$$(15.41) \quad \int V_{T-1}(b')\, p(b' \mid u, b, z)\, db'$$

contains only one non-zero term. This is the term where b' is the distribution calculated from b, u, and z using the Bayes filter. Let us call this distribution $B(b, u, z)$:

$$(15.42) \quad \begin{aligned} B(b, u, z)(x') &= p(x' \mid z, u, b) \\ &= \frac{p(z \mid x', u, b)\, p(x' \mid u, b)}{p(z \mid u, b)} \\ &= \frac{1}{p(z \mid u, b)}\, p(z \mid x') \int p(x' \mid u, b, s)\, p(x \mid u, b)\, dx \\ &= \frac{1}{p(z \mid u, b)}\, p(z \mid x') \int p(x' \mid u, x)\, b(x)\, dx \end{aligned}$$

The reader should recognize the familiar Bayes filter derivation that was extensively discussed in Chapter 2, this time with the normalizer made explicit.

We can now rewrite (15.40) as follows. Note that this expression no longer integrates over b'.

$$(15.43) \quad V_T(b) = \gamma \max_u \left[r(b, u) + \int V_{T-1}(B(b, u, z))\, p(z \mid u, b)\, dz \right]$$

This form is more convenient than the original one in (15.38), since it only requires integration over all possible measurements z, instead of all possible belief distributions b'. This transformation was used implicitly in the example above, where a new value function was obtained by mixing together finitely many piecewise linear functions.

Below, it will be convenient to split the maximization over actions from the integration. Hence, we notice that (15.43) can be rewritten as the following two equations:

$$(15.44) \quad V_T(b, u) = \gamma \left[r(b, u) + \int V_{T-1}(B(b, u, z))\, p(z \mid u, b)\, dz \right]$$

$$(15.45) \quad V_T(b) = \max_u V_T(b, u)$$

Here $V_T(b, u)$ is the horizon T-value function over the belief b, assuming that the immediate next action is u.

15.4.2 Value Function Representation

As in our example, we represent the value function by a maximum of a set of linear functions. We already discussed that any linear function over the belief simplex can be represented by the set of coefficients $v_1, \ldots, v_N$:

$$(15.46) \quad V(b) = \sum_{i=1}^{N} v_i\, p_i$$

where, as usual, $p_1, \ldots, p_N$ are the parameters of the belief distribution b. As in our example, a piecewise linear and convex value function $V_T(b)$ can be represented by the maximum of a finite set of linear functions

$$(15.47) \quad V(b) = \max_{k} \sum_{i=1}^{N} v_i^k\, p_i$$

where $v_1^k, \ldots, v_N^k$ denotes the parameters of the k-th linear function. The reader should quickly convince herself that the maximum of a finite set of linear functions is indeed a convex, continuous, and piecewise linear function.

15.4.3 Calculating the Value Function

We will now derive a recursive equation for calculating the value function $V_T(b)$. We assume by induction that $V_{T-1}(b)$, the value function for horizon $T-1$, is represented by a piecewise linear function as specified above. As part of the derivation, we will show that under the assumption that $V_{T-1}(b)$ is piecewise linear and convex, $V_T(b)$ is also piecewise linear and convex. Induction over the planning horizon T then proves that all value functions with finite horizon are indeed piecewise linear and convex.

We begin with Equations (15.44) and (15.45). If the measurement space is finite, we can replace the integration over z by a finite sum.

$$(15.48) \quad V_T(b, u) = \gamma \left[r(b, u) + \sum_{z} V_{T-1}(B(b, u, z))\, p(z \mid u, b) \right]$$

$$(15.49) \quad V_T(b) = \max_{u} V_T(b, u)$$

The belief $B(b, u, z)$ is obtained using the following expression, derived from Equation (15.42) by replacing the integral with a finite sum.

$$(15.50) \quad B(b, u, z)(x') = \frac{1}{p(z \mid u, b)}\, p(z \mid x') \sum_{x} p(x' \mid u, x)\, b(x)$$

If the belief b is represented by the parameters $\{p_1, \ldots, p_N\}$, and the belief $B(b, u, z)$ by $\{p_1', \ldots, p_N'\}$, it follows that the j-th parameter of the belief b' is computed as follows:

$$(15.51) \quad p'_j = \frac{1}{p(z \mid u, b)} p(z \mid x_j) \sum_{i=1}^{N} p(x_j \mid u, x_i) p_i$$

To compute the value function update (15.48), let us now find more practical expressions for the term $V_{T-1}(B(b,u,z))$, using the finite sums described above. Our derivation starts with the definition of V_{T-1} and substitutes the p'_j according to Equation (15.51):

$$(15.52) \quad V_{T-1}(B(b,u,z)) = \max_k \sum_{j=1}^{N} v_j^k p'_j$$

$$= \max_k \sum_{j=1}^{N} v_j^k \frac{1}{p(z \mid u, b)} p(z \mid x_j) \sum_{i=1}^{N} p(x_j \mid u, x_i) p_i$$

$$= \frac{1}{p(z \mid u, b)} \max_k \sum_{j=1}^{N} v_j^k p(z \mid x_j) \sum_{i=1}^{N} p(x_j \mid u, x_i) p_i$$

$$= \frac{1}{p(z \mid u, b)} \max_k \underbrace{\sum_{i=1}^{N} p_i \overbrace{\sum_{j=1}^{N} v_j^k p(z \mid x_j) p(x_j \mid u, x_i)}^{(**)}}_{(*)}$$

The term marked $(*)$ is independent of the belief. Hence, the function labeled $(**)$ is a linear function in the parameters of the belief space, $p_1, \ldots, p_N$. The term $1/p(z \mid u, b)$ is both nonlinear and difficult to compute, since it contains an entire belief b as conditioning variable. However, the beauty of POMDPs is that this expression cancels out. In particular, substituting this expression back into (15.48) yields the following update equation:

$$(15.53) \quad V_T(b, u) = \gamma \left[r(b, u) + \sum_z \max_k \sum_{i=1}^{N} p_i \sum_{j=1}^{N} v_j^k p(z \mid x_j) p(x_j \mid u, x_i) \right]$$

Hence, despite the non-linearity arising from the measurement update, $V_T(b, u)$ is once again piecewise linear.

Finally, we note that $r(b, u)$ is given by the expectation

$$(15.54) \quad r(b, u) = E_x[r(x, u)] = \sum_{i=1}^{N} p_i \, r(x_i, u)$$

15.4 Mathematical Derivation of POMDPs

Here we assumed that the belief b is represented by the parameters $\{p_1, \ldots, p_N\}$.

The desired value function V_T is now obtained by maximizing $V_T(b, u)$ over all actions u, as stated in (15.49):

$$
\begin{aligned}
(15.55) \quad V_T(b) &= \max_u V_T(b, u) \\
&= \gamma \max_u \left(\left[\sum_{i=1}^N p_i\, r(x_i, u) \right] + \sum_z \max_k \underbrace{\sum_{i=1}^N p_i \sum_{j=1}^N v_j^k\, p(z \mid x_j)\, p(x_j \mid u, x_i)}_{=:\, v_{u,z,i}^k} \right) \\
&= \gamma \max_u \left(\left[\sum_{i=1}^N p_i\, r(x_i, u) \right] + \underbrace{\sum_z \max_k \sum_{i=1}^N p_i\, v_{u,z,i}^k}_{(*)} \right)
\end{aligned}
$$

with

$$
(15.56) \quad v_{u,z,i}^k = \sum_{j=1}^N v_j^k\, p(z \mid x_j)\, p(x_j \mid u, x_i)
$$

as indicated. This expression is not yet in the form of a maximum of linear functions. In particular, we now need to change the sum-max-sum expression labeled $(*)$ in (15.55) into a max-sum-sum expression, which is the familiar form of a maximum over a set of linear functions.

We utilize the same transformation as in our example, Chapter 15.2.3. Specifically, suppose we would like to compute the maximum

$$
(15.57) \quad \max\{a_1(x), \ldots, a_n(x)\} + \max\{b_1(x), \ldots, b_n(x)\}
$$

for some functions $a_1(x), \ldots, a_n(x)$ and $b_1(x), \ldots, b_n(x)$ over a variable x. This maximum is attained at

$$
(15.58) \quad \max_i \max_j\, [a_i(x) + b_j(x)]
$$

This follows from the fact that each $a_i + b_j$ is indeed a lower bound. Further for any x there must exist an i and j such that $a_i(x) + b_j(x)$ defines the maximum. By including all such potential pairs in (15.58) we obtain a tight lower bound, i.e., the solution.

This is now easily generalized into arbitrary sums over max expressions:

(15.59) $$\sum_{j=1}^{m} \max_{i=1}^{N} a_{i,j}(x) = \max_{i(1)=1}^{N} \max_{i(2)=1}^{N} \cdots \max_{i(m)=1}^{N} \sum_{j=1}^{m} a_{i(j),j}$$

We apply now this "trick" to our POMDP value function calculation and obtain for the expression $(*)$ in (15.55). Let M be the total number of measurements.

(15.60) $$\sum_{z} \max_{k} \sum_{i=1}^{N} p_i\, v_{u,z,i}^{k} = \max_{k(1)} \max_{k(2)} \cdots \max_{k(M)} \sum_{z} \sum_{i=1}^{N} p_i\, v_{u,z,i}^{k(z)}$$
$$= \max_{k(1)} \max_{k(2)} \cdots \max_{k(M)} \sum_{i=1}^{N} p_i \sum_{z} v_{u,z,i}^{k(z)}$$

Here each $k(\ )$ is a separate variable, each of which takes on the values of the variable k on the left hand side. There are as many such variables as there are measurements. As a result, the desired value function is now obtained as follows:

(15.61) $$V_T(b) = \gamma \max_{u}\left[\sum_{i=1}^{N} p_i\, r(x_i, u)\right] + \max_{k(1)} \max_{k(2)} \cdots \max_{k(M)} \sum_{i=1}^{N} p_i \sum_{z} v_{u,z,i}^{k(z)}$$
$$= \gamma \max_{u} \max_{k(1)} \max_{k(2)} \cdots \max_{k(M)} \sum_{i=1}^{N} p_i \left[r(x_i, u) + \sum_{z} v_{u,z,i}^{k(z)}\right]$$

In other words, each combination

$$\left(\left[r(x_1, u) + \sum_{z} v_{u,z,1}^{k(z)}\right]\left[r(x_2, u) + \sum_{z} v_{u,z,2}^{k(z)}\right] \cdots \left[r(x_N, u) + \sum_{z} v_{u,z,N}^{k(z)}\right]\right)$$

makes for a new linear constraint in the value function V_T.

There will be one such constraint for each unique joint setting of the variables $k(1), k(2), \ldots, k(M)$. Obviously, the maximum of these linear functions is once again piecewise linear and convex, which proves that this representation indeed is sufficient to represent the correct value function over the underlying continuous belief space. Further, the number of linear pieces will be doubly exponential in the size of the measurement space, at least for our naive implementation that retains all such constraints.

15.5 Practical Considerations

The value iteration algorithm discussed thus far is far from practical. For any reasonable number of distinct states, measurements, and controls, the

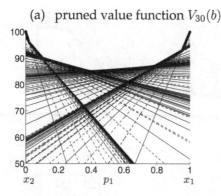 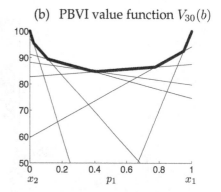

Figure 15.7 The benefit of point-based value iteration over general value iteration: Shown in (a) is the exact value function at horizon $T = 30$ for a different example, which consists of 120 constraints, after pruning. On the right is the result of the PBVI algorithm retaining only 11 linear functions. Both functions yield virtually indistinguishable results when applied to control.

complexity of the value function is prohibitive, even for relatively beginning planning horizons.

There exists a number of opportunities to implement more efficient algorithms. One was already discussed in our example: The number of linear constraints rapidly grows prohibitively large. Fortunately, a good number of linear constraints can safely be ignored, since they do not participate in the definition of the maximum.

Another related shortcoming of the value iteration algorithm is that it computes value functions for *all* belief states, not just for the relevant ones. When a robot starts at a well-defined belief state, the set of reachable belief states is often much smaller. For example, if the robot seeks to move through two doors for which it is uncertain as to whether they are open or closed, it surely knows the state of the first when reaching the second. Thus, a belief state in which the second door's state is known but the first one is not is physically unattainable. In many domains, huge subspaces of the belief space are unattainable.

Even for the attainable beliefs, some might only be attained with small probability; others may be plainly undesirable so that the robot will generally avoid them. Value iteration makes no such distinctions. In fact, the time and resources invested in value computations are independent of the odds that a belief state will actually be relevant.

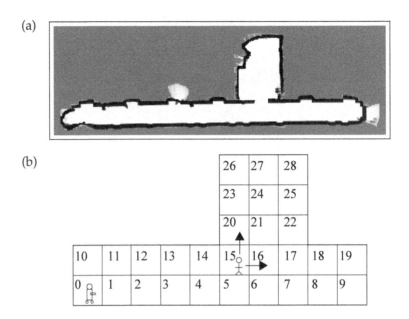

Figure 15.8 Indoor environment, in which we seek a control policy for finding a moving intruder. (a) Occupancy grid map, and (b) discrete state set used by the POMDP. The robot tracks its own pose sufficiently well that the pose uncertainty can be ignored. The remaining uncertainty pertains to the location of the person. Courtesy of Joelle Pineau, McGill University.

POINT-BASED VALUE ITERATION

There exists a flurry of algorithms that are more selective with regards to the subspace of the belief state for which a value function is computed. One of them is *point-based value iteration*, or *PBVI*. It is based on the idea of maintaining a set of exemplary belief states, and restricting the value function to constraints that maximize the value function for at least one of these belief states. More specifically, imagine we are given a set $B = \{b_1, b_2, \ldots\}$ of belief states, called *belief points*. Then the *reduced value function* V with respect to B is the set of constraints $v \in V$ for which we can find at least one $b_i \in B$ such that $v(b_i) = V(b_i)$. In other words, linear segments that do not coincide with any of the discrete belief points in B are discarded. The original PBVI algorithm calculates the value function efficiently by not even generating constraints that are not supported by any of the points; however, the same idea can also be implemented by pruning away all line segments after generating them in a standard POMDP backup.

15.5 Practical Considerations 539

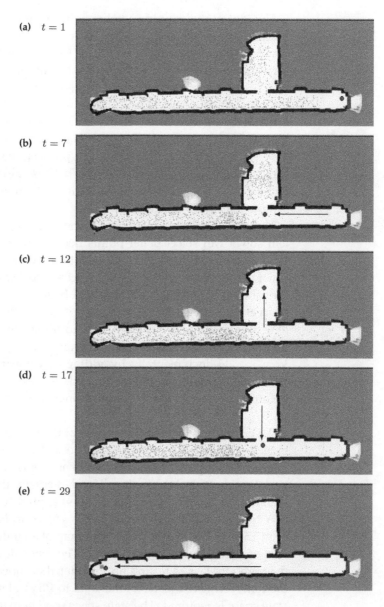

(a) $t = 1$

(b) $t = 7$

(c) $t = 12$

(d) $t = 17$

(e) $t = 29$

Figure 15.9 A successful search policy. Here the tracking of the intruder is implemented via a particle filter, which is then projected into a histogram representation suitable for the POMDP. The robot first clears the room on the top, then proceeds down the hallway. Courtesy of Joelle Pineau, McGill University.

The idea of maintaining a belief point set B can make value iteration significantly more efficient. Figure 15.7a shows the value function for a problem that differs from our example in Chapter 15.2 by only one aspect: The state transition function is deterministic (simply replace 0.8 by 1.0 in (15.7) and (15.8)). The value function in Figure 15.7a is optimal with respect to the horizon $T = 30$. Careful pruning along the way reduced it to 120 constraints, instead of the $10^{561,012,337}$ that a non-pruning implementation would give us—assuming the necessary patience. With a simple point set $B = \{p_1 = 0.0, p_1 = 0.1, p_1 = 0.2, \ldots, p_1 = 1\}$, we obtain the value function shown on the right side of Figure 15.7b. This value function is approximate, and it consists of only 11 linear functions. More importantly, its calculation is more than 1,000 times faster.

The use of belief points has a second important implication: The problem solver can select belief points deemed relevant for the planning process. There exists a number of heuristics to determine a set of belief points. Chief among them are to identify reachable beliefs (e.g., through simulating the robot in the POMDP), and to find beliefs that are spaced reasonably far apart from each other. By doing so it is usually possible to get many orders of magnitude faster POMDP algorithms. In fact, it is possible to grow the set B incrementally, and to therefore build up the set of value functions $V_1, v_2, \ldots, V_T$ incrementally, by adding new linear constraints to all of them whenever a new belief point is added. In this way, the planning algorithm becomes *anytime*, in that it produces increasingly better results as time goes on.

An emerging view in robotics is that the number of plausible belief states exceeds that of the number of states only by a constant factor. As a consequence, techniques that actively select appropriate regions in belief space for updating during planning have fundamentally different scaling properties than the flat, unselective value iteration approach.

A typical robotics result of PBVI is shown in Figures 15.8 and 15.9. Figure 15.8a depicts an occupancy grid map of an indoor environment that consists of a long corridor and a room. The robot starts on the right side of the diagram. Its task is to find an intruder that moves according to Brownian motion. To make this task amenable to PBVI planning, a low-dimensional state space is required. The state space used here is shown in Figure 15.8b. It tessellates the grid map into 22 discrete regions. The granularity of this representation is sufficient to solve this task, while it makes computing the PBVI value function computationally feasible. The task of finding such an intruder is inherently probabilistic. Any control policy has to be aware of the uncertainty in the environment, and seek its reduction. Further, it is inherently dynamic. Just moving to spaces not covered yet is generally insufficient. Figure 15.9 shows a typical result of POMDP planning. Here the robot has determined a control sequence that first explores the relatively small room, then progresses down the corridor. This control policy exploits the fact that while the robot clears the room, the intruder has insufficient time to pass through the corridor. Hence, this policy succeeds with high probability.

This example is paradigmatic of applying POMDP value iteration to actual robot control problem. Even when using aggressive pruning as in PBVI, the resulting value functions are still limited to a few dozen states. However, if such a low-dimensional state representation can be found, POMDP techniques yield excellent results through accommodating the inherent uncertainty in robotics.

15.6 Summary

In this section, we introduced the basic value iteration algorithm for robot control under uncertainty.

- POMDPs are characterized by multiple types of uncertainty: Uncertainty in the control effects, uncertainty in perception, and uncertainty with regards to the environment dynamics. However, POMDPs assume that we are given a probabilistic model of action and perception.

- The value function in POMDPs is defined over the space of all beliefs robots might have about the state of the world. For worlds with N states, this belief is defined over the $(N-1)$-dimensional belief simplex, characterized by the probability assigned to each of the N states.

- For finite horizons, the value function is piecewise linear in the belief space parameters. It is also continuous and convex. Thus, it can be represented as a maximum of a collection of finitely many linear functions. Further, these linear constraints are easily calculated.

- The POMDP planning algorithm computes a sequence of value functions, for increasing planning horizons. Each such calculation is recursive: Given the optimal value function at horizon $T-1$, the algorithm proceeds to computing the optimal value function at horizon T.

- Each recursive iteration combines a number of elements: The action choice is implemented by maximizing over sets of linear constraints, where each action carries its own set. The anticipated measurement is incorporated by combining sets of linear constraints, one for each measurement. The prediction is then implemented by linearly manipulating the set of linear constraints. Payoff is generalized into the belief space by calculating its expectation, which once again is linear in the belief space parameters. The result is a value backup routine that manipulates linear constraints.

- We find that the basic update produces intractably many linear constraints. Specifically, in each individual backup the measurement step increases the number of constraints by a factor that is exponential in the number of possible measurements. Most of these constraints are usually passive, and omitting them does not change the value function at all.

- Point-based value iteration (PBVI) is an approximate algorithm that maintains only constraints that are needed to support a finite set of representative belief states. In doing so, the number of constraints remains constant instead of growing doubly exponentially (in the worst case). Empirically PBVI provides good results when points are chosen to be representative and well-separated in the belief space.

In many ways, the material presented in this chapter is of theoretical interest. The value iteration algorithm defines the basic update mechanism that underlies a great number of efficient decision making algorithms. However, it in itself is not computationally tractable. Efficient implementations therefore resort to approximations, such as the PBVI technique we just discussed.

15.7 Bibliographical Remarks

EXPERIMENTAL DESIGN

The topic of decision making under uncertainty has been studied extensively in statistics, where it is known as *experimental design*. Key textbooks in this area include those by Winer et al. (1971) and Kirk and Kirk (1995); more recent work can be found in Cohn (1994).

The value iteration algorithm described in this paper goes back to Sondik (1971) and Smallwood and Sondik (1973), who were among the first to study the POMDP problem. Other early work can be found in Monahan (1982), with an early grid-based approximation in Lovejoy (1991). Finding policies for POMDPs was long deemed infeasible due to the enormous computational complexity involved. The problem was introduced into the field of Artificial Intelligence by Kaelbling et al. (1998). The pruning algorithms in Cassandra et al. (1997) and Littman et al. (1995) led to significant improvements over previous algorithms. Paired with remarkable increase of computer speed and memory available, their work enabled POMDPs to grow into a tool for solving small AI problems. Hauskrecht (1997) provided bounds on the complexity of POMDP problem solving.

The most significant wave of progress came with the advent of approximate techniques—some of which will be discussed in the next chapter. An improved grid approximation of POMDP belief spaces was devised by Hauskrecht (2000); variable resolution grids were introduced by Brafman (1997). Reachability analysis began to play a role in computing policies. Poon (2001) and Zhang and Zhang (2001) developed point-based POMDP techniques, in which the set of belief states were limited. Unlike Hauskrecht's (2000) work, these techniques relied on piecewise linear functions for representing value functions. This work culminated in the definition of the point based value iteration algorithm by Pineau et al. (2003b), who developed new any-time techniques for finding relevant belief space for solving POMDPs. Their work was later extended using tree-based representations (Pineau et al. 2003a).

Geffner and Bonet (1998) solved a number of challenge problems using dynamic programming applied to a discrete version of the belief space. This work was extended by Likhachev et al. (2004), who applied the A* algorithm (Nilsson 1982) to a restricted type of POMDP. Ferguson et al. (2004) extended this to D* planning for dynamic environments (Stentz 1995).

Another family of techniques used particles to compute policies, paired with nearest neighbor in particle set space to define approximations to the value function (Thrun 2000a). Particles were also used for POMDP monitoring by Poupart et al. (2001). Poupart and Boutilier (2000) devised an algorithm for approximating the value function using a technique sensitive to the value itself, which led to state-of-the-art results. A technique by Dearden and Boutilier (1994) gained efficiency through interleaving planning and execution of partial policies; see Smith and Simmons (2004) for additional research on interleaving heuristic search-type planning and execution. Exploiting domain knowledge was discussed in Pineau et al. (2003c), and Washington (1997) provided incremental techniques with bounds. Additional work on approximate POMDP solving is discussed in Aberdeen (2002); Murphy (2000b). One of the few fielded systems controlled by POMDP value iteration is the CMU Nursebot, whose high-level controller and dialog manager is a POMDP (Pineau et al. 2003d; Roy et al. 2000).

An alternative approach to finding POMDP control policies is to search directly in the space of policies, without computing a value function. This idea goes back to Williams (1992), who developed the idea of policy gradient search in the context of MDPs. Contemporary techniques for policy gradient search is described in Baxter et al. (2001) and Ng and Jordan (2000). Bagnell and Schneider (2001) and Ng et al. (2003) successfully applied this approach to the control of hovering an autonomous helicopter; in fact, Ng et al. (2003) reports that it took only 11 days to design such a controller using POMDP techniques, using a learned model. In more recent work, Ng et al. (2004) used these techniques to identify a controller capable of sustained inverted helicopter flight, a previously open problem. Roy and Thrun (2002) applied policy search techniques to mobile robot navigation, and discuss the combination of policy search and value iteration techniques.

Relatively little progress has been made on *learning* POMDP models. Early attempts to learn the model of a POMDP from interaction with an environment essentially failed (Lin and Mitchell 1992; Chrisman 1992), due to the hardness of the problem. Some more recent work on learning hierarchical models shows more promise (Theocharous et al. 2001). Recent work has moved away from learning HMM-style models, into alternative representations. Techniques for representing and learning the structure of partially observable stochastic environments can be found in Jaeger (2000); Littman et al. (2001); James and Singh. (2004); Rosencrantz et al. (2004). While none of these papers fully solve the POMDP problem, they nevertheless are intellectually relevant and promise new insights into the largely open problem of probabilistic robot control.

15.8 Exercises

TIGER PROBLEM

1. This problem is known as the *tiger problem* and is due to Cassandra, Littman and Kaelbling (Cassandra et al. 1994). A person faces two doors. Behind one is a tiger, behind the other a reward of $+10$. The person can either listen or open one of the doors. When opening the door with a tiger, the person will be eaten, which has an associated cost of -20. Listening costs -1. When listening, the person will hear a roaring noise that indicates the presence of the tiger, but only with 0.85 probability will the person be able to localize the noise correctly. With 0.15 probability, the noise will appear as if it came from the door hiding the reward.

   Your questions:

   (a) Provide the formal model of the POMDP, in which you define the state, action, and measurement spaces, the cost function, and the associated probability functions.

   (b) What is the expected cumulative payoff/cost of the open-loop action sequence: "Listen, listen, open door 1"? Explain your calculation.

   (c) What is the expected cumulative payoff/cost of the open-loop action sequence: "Listen, then open the door for which we did not hear a noise"? Again, explain your calculation.

   (d) Manually perform the one-step backup operation of the POMDP. Plot the resulting linear functions in a diagram just like the ones in Chapter 15.2. Provide diagrams of all intermediate steps, and don't forget to add units to your diagrams.

   (e) Manually perform the second backup, and provide all diagrams and explanations.

   (f) Implement the problem, and compute the solution for the planning horizons $T = 1, 2, \ldots, 8$. Make sure you prune the space of all linear functions. For what sequences of measurements would a person still choose to listen, even after 8 consecutive listening actions?

2. Show the correctness of Equation (15.26).

3. What is the worst-case computational complexity of a single POMDP value function backup? Provide your answer using the $O(\ )$ notation, where arguments may include the number of linear functions before a backup, and the number of states, actions, and measurements in a discrete POMDP.

15.8 Exercises

4. The POMDP literature often introduces a *discount factor*, which is analogous to the discount factor discussed in the previous section. Show that even with a discount factor, the resulting value functions are still piecewise linear.

5. Consider POMDP problems with finite state, action, and measurement space, but for which the horizon $T \uparrow \infty$.

    (a) Will the value function still be piecewise linear?

    (b) Will the value function still be continuous?

    (c) Will the value function still be convex?

    For all three questions, argue why the answer is positive, or provide a counterexample in case it is negative.

6. On page 28, we provided an example of a robot sensing and opening a door. In this exercise, you are asked to implement a POMDP algorithm for an optimal control policy. Most information can be found in the example on page 28. To turn this into a control task, let us assume that the robot has a third action: **go**. When it goes, it receives $+10$ payoff if the door is open, and -100 if it is closed. The action **go** terminates the episode. The action **do_nothing** costs the robot -1, and **push** costs the robot -5. Plot value functions for different time horizons up to $T = 10$, and explain the optimal policy.

16 Approximate POMDP Techniques

16.1 Motivation

In previous chapters, we have studied two main frameworks for action selection under uncertainty: MDPs and POMDPs. Both frameworks address non-deterministic action outcomes, but they differ in their ability to accommodate sensor limitations. Only the POMDP algorithms can cope with uncertainty in perception, whereas MDP algorithms assume that the state is fully observable. However, the computational expense of exact planning in POMDPs renders exact methods inapplicable to just about any practical problem in robotics.

This chapter describes POMDP algorithms that scale. As we shall see in this chapter, both MDP and POMDP are extreme ends of a spectrum of possible probabilistic planning and control algorithms. This chapter reviews a number of approximate POMDP techniques that fall in between MDPs and POMDPs. The algorithms discussed here share with POMDPs the use of value iteration in belief space. However, they approximate the value function in a number of ways. By doing so, they gain immense speed-ups over the full POMDP solution.

The techniques surveyed in this chapter have been chosen because they characterize different styles of approximation. In particular, we will discuss the following three algorithms:

- *QMDP* is a hybrid between MDPs and POMDPs. This algorithm generalizes the MDP-optimal value function defined over states, into a POMDP-style value function over beliefs. QMDP would be exact under the—usually false—assumption that after one step of control, the state becomes fully observable. Value iteration in QMDPs is of the same complexity as in MDPs.

1:     **Algorithm QMDP**($b = (p_1, \ldots, p_N)$):
2:     $\hat{V} = $ **MDP_discrete_value_iteration**() // see page 502
3:     *for all control actions u do*
4:     $$Q(x_i, u) = r(x_i, u) + \sum_{j=1}^{N} \hat{V}(x_j)\, p(x_j \mid u, x_i)$$
5:     *endfor*
6:     *return* $\displaystyle\operatorname*{argmax}_{u} \sum_{i=1}^{N} p_i\, Q(x_i, u)$

Table 16.1 The QMDP algorithm computes the expected return for each control action u, and then selects the action u that yields the highest value. The value function used here is MDP-optimal, hence dismisses the state uncertainty in the POMDP.

- The *augmented MDP*, or *AMDP*. This algorithm projects the belief state into a low-dimensional sufficient statistic and performs value iteration in this lower-dimensional space. The most basic implementation involves a representation that combines the most likely state and the degree of uncertainty, measured by entropy. The planning is therefore only marginally less efficient than planning in MDPs, but the result can be quite an improvement!

- *Monte Carlo POMDP*, or *MC-POMDP*. This is the particle filter version of the POMDP algorithm, where beliefs are approximated using particles. By constructing a belief point set dynamically—just like the PBVI algorithm described towards the end of the previous chapter—MC-POMDPs can maintain a relatively small set of beliefs. MC-POMDPs are applicable to continuous-valued states, actions, and measurements, but they are subject to the same approximations that we have encountered in all particle filter applications in this book, plus some additional ones that are unique to MC-POMDPs.

These algorithms cover some of the primary techniques for approximating value functions in the emerging literature on probabilistic planning and control.

16.2 QMDPs

QMDPs are an attempt to combine the best of MDPs and POMDPs. Value functions are easier to compute for MDPs than for POMDPs, but MDPs rely on the assumption that the state is fully observable. A QMDP is computationally just about as efficient as an MDP, but it returns a policy that is defined over the belief state.

The mathematical "trick" is relatively straightforward. The MDP algorithm discussed in Chapter 14 provides us with a state-based value function that is optimal under the assumption that the state is fully observable. The resulting value function $\hat{V}$ is defined over world states. The QMDP generalizes this value to the belief space through the mathematical expectation:

$$(16.1) \quad \hat{V}(b) = E_x[\hat{V}(x)] = \sum_{i=1}^{N} p_i \, \hat{V}(x_i)$$

Here we use our familiar notation $p_i = b(x_i)$. Thus, this value function is linear, with the parameters

$$(16.2) \quad u_i = \hat{V}(x_i)$$

This linear function is exactly of the form used by the POMDP value iteration algorithm. Hence the value function over the belief space is given by the following linear equation:

$$(16.3) \quad \hat{V}(b) = \sum_{i=1}^{N} p_i \, u_i$$

The MDP value function provides a *single* linear constraint in belief space. This enables us to apply the algorithm **policy_POMDP** in Table 15.2, with a single linear constraint.

The most basic version of this idea leads to the algorithm **QMDP** shown in Table 16.1. Here we use a slightly different notation than in Table 15.2: Instead of caching away one linear function for each action u and letting **policy_POMDP** determine the action, our formulation of **QMDP** directly computes the optimal value function through a function Q. The value of $Q(x_i, u)$, as calculated in line 4 in Table 16.1, is the MDP-value of the control u in state x_i. The generalization to belief states then follows in line 6, where the expectation is taken over the belief state. Line 6 also maximizes over all actions, and returns the control action with the highest expected value.

The insight that the MDP-optimal value function can be generalized to belief space enables us to arbitrarily combine MDP and POMDP backups.

In particular, the MDP-optimal value function $\hat{V}$ can be used as input to the POMDP algorithm in Table 15.1 (page 529). With T further POMDP backups, the resulting policy can actively engage in information gathering—as long as the information shows utility within the next T time steps. Even for very small values of T, we usually obtain a robust probabilistic control algorithm that is computationally vastly superior to the full POMDP solution.

16.3 Augmented Markov Decision Processes

16.3.1 The Augmented State Space

The augmented MDP, or *AMDP*, is an alternative to the QMDP algorithm. It too approximates the full POMDP value function. However, instead of ignoring state uncertainty beyond a small time horizon T, the AMDP compresses the belief state into a more compact representation, and then performs full POMDP-style probabilistic backups.

The fundamental assumption in AMDPs is that the belief space can be summarized by a lower-dimensional "sufficient" statistic f, which maps belief distributions into a lower dimensional space. Values and actions are calculated from this statistic $f(b)$ instead of the original belief b. The more compact the statistic, the more efficient the resulting value iteration algorithm.

In many situations a good choice of the statistic is the tuple

$$(16.4) \quad \bar{b} = \begin{pmatrix} \operatorname*{argmax}_{x} b(x) \\ H_b(x) \end{pmatrix}$$

Here $\operatorname{argmax}_x b(x)$ is the most likely state under the belief distribution b, and

$$(16.5) \quad H_b(x) = -\int b(x) \, \log b(x) \, dx$$

AUGMENTED STATE SPACE

is the *entropy* of the belief. This space will be called the *augmented state space*, since it augments the state space by a single value, the entropy of the belief distribution. Calculating a value function over the augmented state space, instead of the belief space, makes for a huge change in complexity. The augmented state avoids the high dimensionality of the belief space, which leads to enormous savings when computing a value function (from worst case doubly exponential to low-degree polynomial).

The augmented state representation is mathematically justified if $f(b)$ is a sufficient statistic of b with regards to the estimation of value:

(16.6) $$V(b) = V(f(b))$$

for all beliefs b the robot may encounter. In practice, this assumption will rarely hold true. However, the resulting value function might still be good enough for a sensible choice of control.

Alternatively, one might consider different statistics, such as the moments of the belief distribution (mean, variance, ...), the eigenvalues and vectors of the covariance, and so on.

16.3.2 The AMDP Algorithm

The AMDP algorithm performs value iteration in the augmented state space. To do so, we have to overcome two obstacles. First, the exact value update is non-linear for our augmented state representation. This is because the entropy is a non-linear function of the belief parameters. It therefore becomes necessary to approximate the value backup. AMDPs discretize the augmented state, representing the value function $\hat{V}$ by a look-up table. We already encountered such an approximation when discussing MDPs. In AMDPs, this table is one dimension larger than the state space table used by MDPs.

The second obstacle pertains to the transition probabilities and the payoff function in the augmented state space. We are normally given probabilities such as the motion model $p(x' \mid u, x)$, the measurement model $p(z \mid x)$, and the payoff function $r(x, u)$. But for value iteration in the augmented state space we need to define similar functions over the augmented state space.

AMDPs use a "trick" for constructing the necessary functions. The trick is to learn transition probabilities and payoffs from simulations. The learning algorithm is based on a frequency statistic, which counts how often an augmented belief $\bar{b}$ transitions to another belief $\bar{b}'$ under a control u, and what average payoff this transition induces.

Table 16.2 states the basic algorithm **AMDP_value_iteration**. The algorithm breaks down into two phases. In a first phase (lines 2–19), it constructs a transition probability table $\hat{\mathcal{P}}$ for the transition from an augmented state $\bar{b}$ and a control action u to a possible subsequent augmented state $\bar{b}'$. It also constructs a payoff function $\hat{\mathcal{R}}$ which measures the expected immediate payoff r when u is chosen in the augmented state $\bar{b}$.

These functions are estimated through a sampling procedure, in which we generate n samples for each combination of $\bar{b}$ and u (line 8). For each of

```
1:   Algorithm AMDP_value_iteration( ):
2:      for all b̄ do                                        // learn model
3:         for all u do
4:            for all b̄ do                                  // initialize model
5:               P̂(b̄, u, b̄') = 0
6:            endfor
7:            R̂(b̄, u) = 0
8:            repeat n times                                // learn model
9:               generate b with f(b) = b̄
10:              sample x ~ b(x)                            // belief sampling
11:              sample x' ~ p(x' | u, x)                   // motion model
12:              sample z ~ p(z | x')                       // measurement model
13:              calculate b' = B(b, u, z)                  // Bayes filter
14:              calculate b̄' = f(b')                       // belief state statistic
15:              P̂(b̄, u, b̄') = P̂(b̄, u, b̄') + 1/n           // learn transitions prob's
16:              R̂(b̄, u) = R̂(b̄, u) + r(u,s)/n              // learn payoff model
17:           endrepeat
18:        endfor
19:     endfor
20:     for all b̄                                           // initialize value function
21:        V̂(b̄) = r_min
22:     endfor
23:     repeat until convergence                            // value iteration
24:        for all b̄ do
25:           V̂(b̄) = γ max_u [ R̂(u, b̄) + ∑_{b̄'} V̂(b̄') P̂(b̄, u, b̄') ]
26:        endfor
27:     return V̂, P̂, R̂                                     // return value fct & model
```

```
1:   Algorithm policy_AMDP(V̂, P̂, R̂, b):
2:      b̄ = f(b)
3:      return argmax_u [ R̂(u, b̄) + ∑_{b̄'} V̂(b̄') P̂(b̄, u, b̄') ]
```

Table 16.2 Top: The value iteration algorithm for augmented MDPs. Bottom: The algorithm for selecting a control action.

these Monte Carlo simulations, the algorithm first generates a belief b for which $f(b) = \bar{b}$. This step is tricky (in fact, it is ill-defined): In the original AMDP model, the creators simply choose to set b to a symmetric Gaussian with parameters chosen to match $\hat{b}$. Next, the AMDP algorithm samples a pose x, a successor pose x', and a measurement z, all in the obvious ways. It then applies the Bayes filter to generate a posterior belief $B(b, u, z)$, for which it calculates the augmented statistics (line 14). The tables $\hat{\mathcal{P}}$ and $\hat{\mathcal{R}}$ are then updated in lines 15 and 16, using simple frequency counts weighted (in the case of the payoff) with the actual payoff for this Monte Carlo sample.

Once the learning is completed, AMDP continues with value iteration. This is implemented in lines 20-26. As usual, the value function is initialized by a large negative value. Iteration of the backup equation in line 25 leads to a value function defined over the augmented state space.

When using AMDPs, the state tracking usually takes place over the original belief space. For example, when using AMDPs for robot motion, one might use MCL for tracking the belief over the robot's pose. The algorithm **policy_AMDP** in Table 16.2 shows how to extract a policy action from the AMDP value function. It extracts the augmented state representation from the full belief in line 2, and then simply chooses the control action that maximizes the expected value (line 3).

16.3.3 Mathematical Derivation of AMDPs

The derivation of AMDP is relatively straightforward, under the assumption that f is a sufficient statistic of the belief state b; i.e., the world is *Markov* relative to the state $f(b)$. We start with an appropriate modification of the standard POMDP-style backup in Equation (15.2). Let f be the function that extracts the statistic $\bar{b}$ from b, hence $\bar{b} = f(b)$ for arbitrary beliefs b. Assuming that f is a sufficient statistic, the POMDP value iteration equation (15.2) can be defined over the AMDP state space

$$(16.7) \quad V_T(\bar{b}) = \gamma \max_u \left[r(\bar{b}, u) + \int V_{T-1}(\bar{b}') \, p(\bar{b}' \mid u, \bar{b}) \, d\bar{b}' \right]$$

where $\bar{b}$ refers to the low-dimensional statistic of b defined in (16.4). Here $V_{T-1}(\bar{b}')$ and $V_T(\bar{b})$ are realized through look-up tables.

This equation contains the probability $p(\bar{b}' \mid u, \bar{b})$, which needs further explanation. Specifically, we have

$$(16.8) \quad p(\bar{b}' \mid u, \bar{b}) = \int p(\bar{b}' \mid u, b) \, p(b \mid \bar{b}) \, db$$

$$= \int\int p(\bar{b}' \mid z, u, b) \, p(z \mid b) \, p(b \mid \bar{b}) \, dz \, db$$

$$= \int\int p(\bar{b}' = f(B(b, u, z))) \, p(z \mid b) \, p(b \mid \bar{b}) \, dz \, db$$

$$= \int\int p(\bar{b}' = f(B(b, u, z))) \int p(z \mid x')$$

$$\int p(x' \mid u, x) \, b(x) \, dx \, dx' \, dz \, p(b \mid \bar{b}) \, db$$

This transformation exploited the fact that the posterior belief b' is uniquely determined once we know the prior belief b, the control u, and the measurement z. This same "trick" was already exploited in our derivation of POMDPs. It enabled us to replace the distribution over posterior beliefs by the Bayes filter result $B(b, u, z)$. In the augmented state space, we therefore could replace $p(\bar{b}' \mid z, u, b)$ by a point-mass distribution centered on $f(B(b, u, z))$.

Our learning algorithm in Table 16.2 approximates this equation through Monte Carlo sampling. It replaced each of the integrals by a sampler. The reader should take a moment to establish the correspondence: Each of the nested integrals in (16.8) maps directly to one of the sampling steps in Table 16.2.

Along the same lines, we can derive an expression for the expected payoff $r(\bar{b}, u)$:

(16.9) $$r(\bar{b}, u) = \int r(b, u) \, p(b \mid \bar{b}) \, db$$

$$= \int\int r(x, u) \, b(x) \, dx \, p(b \mid \bar{b}) \, db$$

Once again, the algorithm **AMDP_value_iteration** approximates this integral using Monte Carlo sampling. The resulting learned payoff function resides in the lookup $\hat{\mathcal{R}}$. The value iteration backup in lines 20–26 of **AMDP_value_iteration** is essentially identical to the derivation of MDPs.

As noted above, this Monte Carlo approximation is only legitimate when $\bar{b}$ is a sufficient statistics of b and the system is Markov with respect to $\bar{b}$. In practice, this is usually *not* the case, and proper sampling if augmented states would therefore have to be conditioned on past actions and measurements—a nightmare! The AMDP algorithm ignores this, by simply generating b's with $f(b) = \bar{b}$. Our example above involved a symmetric Gaussian whose parameters matched $\bar{b}$. An alternative—which deviates from the original AMDP algorithm but would be mathematically more sound—would involve

16.3 Augmented Markov Decision Processes

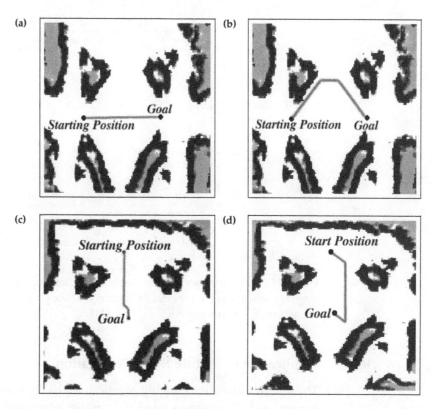

Figure 16.1 Examples of robot paths in a large, open environment, for two different configurations (top row and bottom row). The diagrams (a) and (c) show paths generated by a conventional dynamic programming path planner that ignores the robot's perceptual uncertainty. The diagrams (b) and (d) are obtained using the augmented MDP planner, which anticipates uncertainty and avoids regions where the robot is more likely to get lost. Courtesy of Nicholas Roy, MIT.

simulation of entire traces of belief states using the motion and measurement models, and using subsequent pairs of simulated belief states to learn $\hat{P}$ and $\hat{R}$. Below, when discussing MC-POMDPs, we will encounter such a technique. MC-POMDPs sidestep this issue by using simulation to generate plausible pairs of belief states.

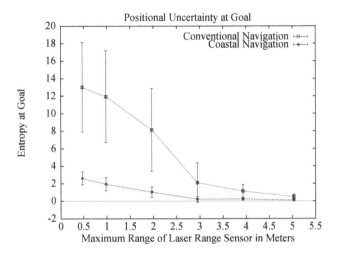

Figure 16.2 Performance comparison of MDP planning and Augmented MDP planning. Shown here is the uncertainty (entropy) at the goal location as a function of the sensor range. Courtesy of Nicholas Roy, MIT.

16.3.4 Application to Mobile Robot Navigation

The AMDP algorithm is highly practical. In the context of mobile robot navigation, AMDPs enable a robot to consider its general level of "confusion" in its action choice. This pertains not just to the momentary uncertainty, but also future expected uncertainties a robot may experience through the choice of its actions.

Our example involves a robot navigating in a known environment. It was already given as an example in the introduction to this book; see Figure 1.2 on page 7. Clearly, the level of confusion depends on where the robot navigates. A robot traversing a large featureless area is likely to gradually lose information as to where it is. This is reflected in the conditional probability $p(\bar{b}' \mid u, \bar{b})$, which with high likelihood increases the entropy of the belief in such areas. In areas populated with localization features, e.g., near the walls with distinctive features, the uncertainty is more likely to decrease. The AMDP anticipates such situations and generates policies that minimize the time of arrival while simultaneously maximizing the certainty at the time of arrival at a goal location. Since the uncertainty is an estimate of the true positioning error, it is a good measure for the chances of actually arriving at the desired location.

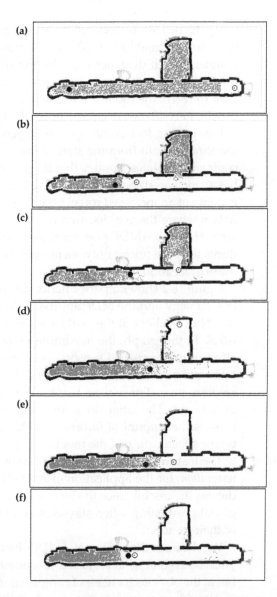

Figure 16.3 The policy computed using an advanced version of AMDP, with a learned state representation. The task is to find an intruder. The gray particles are drawn from the distribution of where the person might be, initially uniformly distributed in (a). The black dot is the true (unobservable) position of the person. The open circle is the observable position of the robot. This policy succeeds with high likelihood. Courtesy of Nicholas Roy, MIT, and Geoffrey Gordon, CMU.

Figure 16.1 shows example trajectories for two constellations (two different start and goal locations). The diagrams on the left correspond to a MDP planner, which does not consider the robot's uncertainty. The augmented MDP planner generates trajectories like the ones shown on the right. In Figures 16.1a&b, the robot is requested to move through a large open area, approximately 40 meters wide. The MDP algorithm, not aware of the increased risk of getting lost in the open area, generates a policy that corresponds to the shortest path from the start to the goal location. The AMDP planner, in contrast, generates a policy that stays close to the obstacles, where the robot has an increased chance of receiving informative sensor measurements at the expense of an increased travel time. Similarly, Figure 16.1c&d considers a situation where the goal location is close to the center of the featureless, open area. Here the AMDP planner recognizes that passing by known objects reduces the pose uncertainty, increasing the chances of successfully arriving at the goal location.

Figure 16.2 shows a performance comparison between the AMDP navigation strategy and the MDP approach. In particular, it depicts the entropy of the robot's belief b at the goal location, as a function of the sensor characteristics. In this graph, the maximum perceptual range is varied, to study the effect of impoverished sensors. As the graph suggests, the AMDP has significantly higher chances of success. The difference is largest if the sensors are very poor. For sensors that have a long range, the difference ultimately disappears. The latter does not come as a surprise, since with good range sensors the amount of information that can be perceived is less dependent on the specific pose of the robot.

COASTAL NAVIGATION

The feature to anticipate and avoid uncertainty has led to the name *coastal navigation*, for the application of AMDPs to robot navigation. This name indicates the resemblance to ships, which, before the advent of satellite-based global positioning, often stayed close to the coastline so as to not lose track of their location.

We close our discussion of AMDPs by noting that the choice of the statistic f has been somewhat arbitrary. It is possible to add more features as needed, but at the obvious increase of computational complexity. Recent work has led to algorithms that *learn* statistics f, using non-linear dimensionality reduction techniques. Figure 16.3 shows the result of such a learning algorithm, applied to the problem of clearing a building from a moving intruder. Here the learning algorithm identifies a 6-dimensional state representation, which captures the belief of the robot for any plausible pursuit strategy. The gray particles represent the robot's belief about the intruder's location. As this example illustrates, AMDPs with a learned state representation succeed in generating quite a sophisticated strategy: The robot first clears part of the corridor, but always stays close enough to the room on the top that an intruder cannot escape from there. It then clears the room, in time short enough to prevent an intruder to pass by the corridor undetected. The robot finally continues its pursuit in the corridor.

16.4 Monte Carlo POMDPs

16.4.1 Using Particle Sets

The final algorithm discussed in this chapter is the particle filter solution to POMDPs, called *MC-POMDP*, which is short for *Monte Carlo POMDP*. MC-POMDPs acquire a value function that is defined over particle sets. Let $\mathcal{X}$ be a particle set representing a belief b. Then the value function is represented as a function

(16.10) $\quad V : \mathcal{X} \longrightarrow \Re$

This representation has a number of advantages, but it also creates a number of difficulties. A key advantage is that we can represent value functions over arbitrary state spaces. In fact, among all the algorithms discussed thus far, MC-POMDPs are the only ones that do not require a finite state space. Further, MC-POMDPs use particle filters for belief tracking. We have already seen a number of successful particle filter applications. MC-POMDPs extend particle filters to planning and control problems.

The primary difficulty in using particle sets in POMDPs pertains to the representation of the value function. The space of all particle sets of any given size M is M-dimensional. Further, the probability that any particle set is ever observed twice is zero, due to the stochastic nature of particle generation. As a result, we need a representation for V that can be updated using some particle set, but then provides a value for other particle sets, which the MC-POMDP algorithm never saw before. In other words, we need a *learning algorithm*. MC-POMDP use a *nearest neighbor* algorithm using locally weighted interpolation when interpolating between different beliefs.

16.4.2 The MC-POMDP Algorithm

Table 16.3 lays out the basic MC-POMDP algorithm. The MC-POMDP algorithm required a number of nested loops. The innermost loop, in lines 6

1: **Algorithm MC-POMDP(b_0, V):**
2:     *repeat until convergence*
3:         sample $x \sim b(x)$ // initialization
4:         initialize $\mathcal{X}$ with M samples of $b(x)$
5:         *repeat until episode over*
6:             for all control actions u do   // update value function
7:                 $Q(u) = 0$
8:                 *repeat n times*
9:                     select random $x \in \mathcal{X}$
10:                    sample $x' \sim p(x' \mid u, x)$
11:                    sample $z \sim p(z \mid x')$
12:                    $\mathcal{X}' = \textbf{Particle\_filter}(\mathcal{X}, u, z)$
13:                    $Q(u) = Q(u) + \dfrac{1}{n}\gamma\ [r(x,u) + V(\mathcal{X}')]$
14:                 endrepeat
15:             endfor
16:             $V(\mathcal{X}) = \max_u Q(u)$ // update value function
17:             $u^* = \operatorname*{argmax}_u Q(u)$ // select greedy action
18:             sample $x' \sim p(x' \mid u, x)$ // simulate state transition
19:             sample $z \sim p(z \mid x')$
20:             $\mathcal{X}' = \textbf{Particle\_filter}(\mathcal{X}, u, z)$ // compute new belief
21:             set $x = x'$; $\mathcal{X} = \mathcal{X}'$ // update state and belief
22:         endrepeat
23:     endrepeat
24:     return V

Table 16.3 The MC-POMDP algorithm.

16.4 Monte Carlo POMDPs

through 16 in Table 16.3, updates the value function V for a specific belief $\mathcal{X}$. It does so by simulating for each applicable control action u, the set of possible successor beliefs. This simulation takes place in lines 9 through 12. From that, it gathers a local value for each of the applicable actions (line 13). The value function update takes place in line 16, in which V is simply set to the maximum of all Q_u's.

Following this local backup is a step in which MC-POMDPs simulate the physical system, to generate a new particle set $\mathcal{X}$. This simulation takes place in lines 17 through 21. In our example, the update always selects the greedy action (line 17); however, in practice it may be advantageous to occasionally select a random action. By transitioning to a new belief $\mathcal{X}$, the MC-POMDP value iteration performs the update for a different belief state. By iterating through entire episodes (outer loops in lines 2 through 5), the value function is eventually updated everywhere.

The key open question pertains to representation of the function V. MC-POMDP uses a local learning algorithm reminiscent of nearest neighbor. This algorithm grows a set of reference beliefs $\mathcal{X}_i$ with associated values V_i. When a query arrives with a previously unseen particle set $\mathcal{X}_{\text{query}}$, MC-POMDP identifies the K "nearest" particle sets in its memory. To define a suitable concept of nearness for particle sets requires additional assumptions. In the original implementation, MC-POMDP convolves each particle with a Gaussian with small, fixed covariance, and then measures the KL-divergence between the resulting mixtures of Gaussians. Leaving details aside, this step makes it possible to determine K nearest reference particle sets $\mathcal{X}_1, \ldots, \mathcal{X}_K$, with an associated measure of distance, denoted $d_1, \ldots, d_K$ (we note that KL-divergence is not a distance in the technical sense, since it is asymmetric). The value of the query set $\mathcal{X}_{\text{query}}$, is then obtained through the following formula

$$(16.11) \quad V(\mathcal{X}_{\text{query}}) = \eta \sum_{k=1}^{K} \frac{1}{d_k} V_k$$

SHEPARD'S INTERPOLATION

with $\eta = \left[\sum_k \frac{1}{d_k}\right]^{-1}$. Here $\mathcal{X}_k$ is the k-th reference belief in the set of K nearest neighbors, and d_k is the associated distance to the query set. This interpolation formula, known as *Shepard's interpolation*, explains how to calculate $V(\mathcal{X}')$ in line 13 of Table 16.3.

The update in line 16 involves an implicit case differentiation. If the reference set contains already K particle sets whose distance is below a user-defined threshold, the corresponding V-values are simply updated in proportion to their contribution in the interpolation:

$$(16.12) \quad V_k \longleftarrow V_k + \alpha \, \eta \, \frac{1}{d_k} \left(\max_u Q(u) - V_k \right)$$

where α is a learning rate. The expression $\max_u Q(u)$ is the "target" value for the function V, and $\eta \frac{1}{d_k}$ is the contribution of the k-th reference particle set under Shepard's interpolation formula.

If there are less than K particle sets whose distance falls below the threshold, the query particle set is simply added into the reference set, with the associated value $V = \max_u Q(u)$. In this way, the set of reference particle sets grow over time. The value of K and the user-specified distance threshold determine the smoothness of the MC-POMDP value function. In practice, selecting appropriate values will take some thought, since it is easy to exceed the memory of an ordinary PC with the reference set, when the threshold is chosen too tightly.

16.4.3 Mathematical Derivation of MC-POMDPs

The MC-POMDP algorithm relies on a number of approximations: the use of particle sets constitutes one such approximation. Another one is the local learning algorithm for representing V, which is clearly approximate. A third approximation is due to the Monte Carlo backup step of the value function. Each of these approximations jeopardizes convergence of the basic algorithm.

The mathematical justification for using particle filters was already provided in Chapter 4. The Monte Carlo update step follows from the general POMDP update Equation (15.43) on page 532, which is restated here:

$$(16.13) \quad V_T(b) = \gamma \max_u \left[r(b,u) + \int V_{T-1}(B(b,u,z)) \, p(z \mid u, b) \, dz \right]$$

The Monte Carlo approximation is now derived in a way entirely analogous to our AMDP derivation. We begin with the measurement probability $p(z \mid u, b)$, which resolves as follows:

$$(16.14) \quad p(z \mid u, b) = \int \int p(z \mid x') \, p(x' \mid u, x) \, b(x) \, dx \, dx'$$

Similarly, we obtain for $r(b, u)$:

$$(16.15) \quad r(b, u) = \int r(x, u) \, b(x) \, dx$$

This enables us to re-write (16.13) as follows:

$$
\begin{aligned}
(16.16) \quad V_T(b) &= \gamma \max_u \left[\int r(x,u)\, b(x)\, dx \right. \\
&\quad \left. + \int V_{T-1}(B(b,u,z)) \left[\int \int p(z \mid x')\, p(x' \mid u, x)\, b(x)\, dx\, dx' \right] dz \right] \\
&= \gamma \max_u \int \int \int [r(x,u) + V_{T-1}(B(b,u,z))\, p(z \mid x')\, p(x' \mid u, x)] \\
&\quad b(x)\, dx\, dx'\, dz
\end{aligned}
$$

The Monte Carlo approximation to this integral is now a multi-variable sampling algorithm, which requires us to sample $x \sim b(x)$, $x' \sim p(x' \mid u, x)$ and $z \sim p(z \mid x')$. Once we have x, x', and z, we can compute $B(b,u,z)$ via the Bayes filter. We then compute $V_{T-1}(B(b,u,z))$ using the local learning algorithm, and $r(x,u)$ by simply looking it up. We note that all these steps are implemented in 7 through 14 of Table 16.3, with the final maximization carried out in line 16.

The local learning algorithm, which plays such a central role in MC-POMDPs, may easily destroy any convergence we might otherwise obtain with the Monte Carlo algorithm. We will not attempt to characterize the conditions under which local learning may give accurate approximations, but instead simply state that care has to be taken setting its various parameters.

16.4.4 Practical Considerations

From the three POMDP approximations provided in this chapter, MC-POMDP is the least developed and potentially the least efficient one. Its approximation relies on a learning algorithm for representing the value function. Implementing an MC-POMDP algorithm can be tricky. A good understanding of the smoothness of the value function is required, as is the number of particles one seeks to employ.

The original implementation of the MC-POMDP algorithm led to the results shown in Figure 16.4. A robot, shown in Figure 16.4a, is placed near a graspable object located on the floor near the robot, which it can detect using a camera. However, initially the object is placed outside the robot's perceptual field. A successful policy will therefore exhibit three stages: A search stage, in which the robot rotates until it senses the object; a motion stage in which the robot centers itself relative to the object so that it can grasp it; and a final grasping action. The combination of active perception and goal-directed behavior make this a relatively challenging probabilistic control problem.

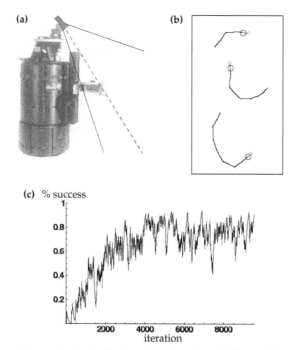

Figure 16.4 A robotic find and fetch task: (a) The mobile robot with gripper and camera, holding the target object. (b) 2-D trajectory of three successful policy executions, in which the robot rotates until it sees the object, and then initiates a successful grasp action (c) success rate as a function of number of planning steps, evaluated in simulation.

Figure 16.4b shows example episodes, in which the robot turned, moved, and grasped successfully. The trajectories shown there are projected motion directories in 2-D. Quantitative results are shown in Figure 16.4c, which plots the success rate as a function of update iterations of the MC-POMDP value iteration. 4,000 iterations of value backup require approximately 2 hours computation time, on a low-end PC, at which point the average performance levels off at 80%. The remaining 20% failures are largely due to configurations in which the robot fails to position itself to grasp the object—in part a consequence of the many approximations in MC-POMDPs.

16.5 Summary

In this section, we introduced three approximate probabilistic planning and control algorithms, with varying degrees of practical applicability. All three algorithms relied on approximations of the POMDP value function. However, they differed in the nature of their approximations.

- The QMDP framework considers uncertainty only for a single action choice: It is based on the assumption that after the immediate next control action, the state of the world suddenly becomes observable. The full observability made it possible to use the MDP-optimal value function. QMDP generalizes the MDP value function to belief spaces through the mathematical expectation operator. As a result, planning in QMDPs is as efficient as in MDPs, but the value function generally overestimates the true value of a belief state.

- Extensions of the QMDP algorithm combine the MDP-optimal value function with a sequence of POMDP backups. When combined with T POMDP backup steps, the resulting policy considers information-gathering actions within the horizon of T, and then relies on the QMDP assumption of a fully observable state. The larger the horizon T is, the closer the resulting policy to the full POMDP solution.

- The AMDP algorithm pursues a different approximation: It maps the belief into a lower-dimensional representation, over which it then performs exact value iteration. The "classical" representation consists of the most likely state under a belief, along with the belief entropy. With this representation, AMDPs are like MDPs with one added dimension in the state representation that measures the global degree of uncertainty of the robot.

- To implement an AMDP, it becomes necessary to *learn* the state transition and the reward function in the low-dimensional belief space. AMDPs achieve this by an initial phase, in which statistics are cached into look-up tables, representing the state transition and reward function. Thus, AMDPs operate over a learned model, and are only accurate to the degree that the learned model is accurate.

- AMDPs applied to mobile robot navigation in known environments is called *coastal navigation*. This navigation technique anticipates uncertainty, and selects motion that trades off overall path length with the uncertainty accrued along a path. The resulting trajectories differ significantly from any non-probabilistic solution: A "coastally" navigating robot stays away from areas in which chances of getting permanently lost are high. Being temporarily lost is acceptable, if the robot can later re-localize with sufficiently high probability.

- The MC-POMDP algorithm is the particle filter version of POMDPs. It calculates a value function defined over sets of particles. To implement such a value function, MC-POMDPs had to resort to a local learning technique, which used a locally weighted learning rule in combination with a proximity test based on KL-divergence. MC-POMDPs then apply Monte Carlo sampling to implement an approximate value backup. The resulting algorithm is a full-fledged POMDP algorithm whose computational complexity and accuracy are both functions of the parameters of the learning algorithm.

The key lesson to take away from this chapter is that there exists a number of approximations whose computational complexity is much closer to MDPs, but that still consider state uncertainty. No matter how crude the approximation, algorithms that consider state uncertainty tend to be significantly more robust than algorithms that entirely ignore state uncertainty. Even a single new element in the state vector—which measures global uncertainty in a one-dimensional way—can make a huge difference in the performance of a robot.

16.6 Bibliographical Remarks

The literature on approximate POMDP problem solving was already extensively discussed in the last chapter (15.7). The QMDP algorithm described in this chapter is due to Littman et al. (1995). The AMDP algorithm for a fixed augmented state representation was developed by Roy et al. (1999). Later, Roy et al. (2004) extended it to a learned state representation. Thrun (2000a) devised the Monte Carlo POMDP algorithm.

16.7 Exercises

1. In this question, you are asked to design an AMDP that solves a simple navigation problem. Consider the following environment with 12 discrete states.

16.7 Exercises

Initially, the robot is placed at a random location, chosen uniformly among all 12 states. Its goal is to advance to state 7. At any point in time, the robot goes north, east, west, or south. Its only sensor is a bumper: When it hits an obstacle, the bumper triggers and the robot does not change states. The robot cannot sense what state it is in, and it cannot sense the direction of its bumper. There is no noise in this problem, just the initial location uncertainty (which we will assume to be uniform).

(a) How many states will an AMDP minimally have to possess? Describe them all.

(b) How many of those states are reachable from the initial AMDP state? Describe them all.

(c) Now assume the robot starts at state 2 (but it still does not know, so its internal belief state will be different). Draw the state transition diagram between all AMDP states that can be reached within four actions.

(d) For this specific type problem (noise-free sensors and robot motion, finite state, action, and measurement space), can you think of a more compact representation than the one used by AMDPs, which is still sufficient to find the optimal solution?

(e) For this specific type problem (noise-free sensors and robot motion, finite state, action, and measurement space), can you craft a state space for which AMDPs will fail to find the optimal solution?

2. In the previous chapter, we learned about the *tiger problem* (Exercise 1 on page 544). What modification of this problem will enable QMDP to come up with an optimal solution? Hint: There are multiple possible answers.

3. In this question, we would like you to determine the *size* of the belief state space. Consider the following table:

problem number	number of states	sensors	state transition	initial state
#1	3	perfect	noise-free	known
#2	3	perfect	noisy	known
#3	3	noise-free	noise-free	unknown (uniform)
#4	3	noisy	noise-free	known
#5	3	noisy	noise-free	unknown (uniform)
#6	3	none	noise-free	unknown (uniform)
#7	3	none	noisy	known
#8	1-dim continuum	perfect	noisy	known
#9	1-dim continuum	noisy	noisy	known
#10	2-dim continuum	noisy	noisy	unknown (uniform)

A perfect sensor always provides the full state information. A noise-free sensor may provide partial state information, but it does so without any randomness. A noisy sensor may be partial and is also subject to noise. A noise-free state transition is deterministic, whereas a stochastic state transition is called noisy. Finally, we only distinguish two types of initial conditions, one in which the initial state is known with absolute certainty, and one in which it is entirely unknown and the prior over states is uniform.

Your question: What is the size of the reachable belief space for all 10 problems above? Hint: It may be finite or infinite, and in the infinite case you should be able to tell what dimension the belief state has.

4. We want you to brainstorm about the failure modes of an AMDP planner. In particular, the AMDP *learns* state transition and reward functions. Brainstorm what can go wrong with this when such learned models are used for value iteration. Identify at least three different types of problems, and discuss them in detail.

17 *Exploration*

17.1 Introduction

This chapter focuses on the robotic exploration problem. Exploration is the problem of controlling a robot so as to maximize its knowledge about the external world. In many robotics applications, the primary purpose of the robotic device is to provide us, the users, with information. Some environments may be plainly inaccessible to people. In others, it may be uneconomical to send people, and robots may be the most economical means of acquiring information. The exploration problem is paramount in robotics. Robots have explored abandoned mines, nuclear disaster sites, even Mars.

Exploration problems come in many different flavors. For example, a robot may seek to acquire a map of a static environment. If we represent the environment by an occupancy grid map, the exploration problem is the problem of maximizing the cumulative information we have about each grid cell. A more dynamic version of the problem may involve moving actors: For example, a robot might have the task to find a person in a known environment, as part of a *pursuit evasion problem*. The goal might be to maximize the information about the person's whereabouts, and finding a person might require exploring the environment. However, since the person might move, the exploration policy may have to explore areas multiple times. A third exploration problem arises when a robot seeks to determine its own pose during localization. This problem is commonly called *active localization*, and the goal is to maximize the information about the robot's own pose. In robotic manipulation, an exploration problem arises when a manipulator equipped with a sensor faces an unknown object. As this brief discussion suggests, exploration problems arise virtually everywhere in robotics.

At first glance, one might conclude that the exploration problem is already fully subsumed by the POMDP framework discussed in previous chapters. As we have shown, POMDPs naturally engage in information gathering. To turn a POMDP into an algorithm whose sole goal it is to maximize information, all we have to do is supply it with an appropriate payoff function. One appropriate choice is the information gain, which measures the reduction in entropy of a robot's belief as a function of its actions. With such a payoff function, POMDPs solve the exploration problem.

However, exploring using the POMDP framework is often not such a good idea. This is because in many exploration problems, the number of unknown state variables is huge, as is the number of possible observations. Consider, for example, the problem of exploring an unknown planet. The number of variables needed to describe the surface of a planet will be enormous. So will be the set of possible observations a robot may make. We already found that in the general POMDP framework the planning complexity grows doubly exponential in the number of observations (in the worst case); hence calculating a value function is plainly impossible. In fact, given the huge number of *possible* values for the unknown state variables in exploration, any algorithm that integrates over all possible such values will inevitably be inapplicable to high dimensional exploration problems, simply for computational reasons.

This chapter discusses a family of practical algorithms that can solve high-dimensional exploration problems. The techniques discussed here are all *greedy*. Put differently, their look-ahead is limited to one exploration action. However an exploration action can involve a sequence of control actions. For example, we will discuss algorithms that select a location anywhere in the map to explore; moving there is considered one *exploration action*. The algorithms discussed here also approximate the knowledge gain upon sensing, so as to reduce the computation involved.

EXPLORATION ACTION

This chapter is organized as follows

- We begin with the general definition of information gain in exploration, for the discrete and the continuous case. We define the basic greedy exploration algorithm that selects actions so as to maximize its information gain.

- We then analyze a first special case of robotic exploration: *active localization*. Active localization pertains to the choice of actions while globally localizing a robot. The application of our greedy exploration algorithm, under an appropriate definition of the action space yields a practical solution to this problem.

- We also consider the problem of exploration in occupancy grid mapping. We derive a popular technique called *frontier-based exploration*, in which a robot moves to its nearest frontier.

- Furthermore we describe an extension of our exploration algorithm to multi-robot systems and show how a team of mobile robots can be controlled so as to efficiently explore an unknown environment.

- Finally we consider the application of exploration techniques to the SLAM problem. We show how to control a robot so as to minimize the uncertainty, using FastSLAM as our example SLAM approach.

The exploration techniques described below have been widely used in the literature and in a variety of practical applications. They also span a number of different representations and robotics problems.

17.2 Basic Exploration Algorithms

17.2.1 Information Gain

The key to exploration is information. We already encountered a number of uses for information in probabilistic robotics. In the context of exploration, we define the *entropy* $H_p(x)$ of a probability distribution p as the *expected information* $E[-\log p]$

EXPECTED INFORMATION

$$(17.1) \quad H_p(x) = -\int p(x) \log p(x) \, dx \quad \text{or} \quad -\sum_x p(x) \log p(x)$$

The entropy was already briefly discussed in Chapter 2.2 of this book. $H_p(x)$ is at its maximum if p is a uniform distribution. It is at its minimum when p is a point-mass distribution; however, in certain continuous cases such as Gaussians we might never reach full confidence.

CONDITIONAL ENTROPY

Conditional entropy is defined as the entropy of a conditional distribution. In exploration, we seek to minimize the expected entropy of the belief after executing an action, hence it is natural to condition on the measurement z and the control u that define the belief state transition.

Holding with the notation introduced before, let us use $B(b, z, u)$ to denote the belief after executing control u and observing z under the belief b. The conditional entropy of the state x' after executing action u and measuring z is given by

$$(17.2) \quad H_b(x' \mid z, u) = -\int B(b, z, u)(x') \log B(b, z, u)(x') \, dx'$$

Here $B(b, z, u)$ is computed using the Bayes filter. In robotics, we only have a choice over the control action u, we cannot pick z. Consequently, we consider the conditional entropy of the control u, with the measurement integrated out:

$$(17.3) \quad H_b(x' \mid u) \approx E_z[H_b(x' \mid z, u)]$$
$$= \int \int H_b(x' \mid z, u) \, p(z \mid x') \, p(x' \mid u, x) \, b(x) \, dz \, dx' \, dx$$

INFORMATION GAIN

Notice that this is only an approximation, as the final expression inverts the oder of a summation and a logarithm. The *information gain* associated with action u in belief b is thus given by the difference

$$(17.4) \quad I_b(u) = H_p(x) - E_z[H_b(x' \mid z, u)]$$

17.2.2 Greedy Techniques

The expected information gain lets us phrase the exploration problem as a decision-theoretic problem of the type addressed in previous chapters. In particular, let $r(x, u)$ be the cost of applying control action u in state x; here we assume $r(x, u) < 0$ to keep our notation consistent. Then the optimal greedy exploration for the belief b maximizes the difference between the information gain and the costs, weighted by a factor α.

$$(17.5) \quad \pi(b) = \operatorname*{argmax}_u \alpha \underbrace{(H_p(x) - E_z[H_b(x' \mid z, u)])}_{\text{expected information gain}} + \underbrace{\int r(x, u) \, b(x) \, dx}_{\text{expected costs}}$$

The factor α relates information to the cost of executing u. It specifies the value a robot assigns to information, which measures the price it is willing to pay in terms of costs for obtaining information.

Equation (17.5) resolves to

$$(17.6) \quad \pi(b) = \operatorname*{argmax}_u -\alpha \, E_z[H_b(x' \mid z, u)] + \int r(x, u) \, b(x) \, dx$$
$$= \operatorname*{argmax}_u \int [r(x, u) - \alpha \int \int H_b(x' \mid z, u) \, p(z \mid x')$$
$$p(x' \mid u, x) \, dz \, dx'] \, b(x) \, dx$$

In short, to understand the utility of the control u, we need to compute the expected entropy after executing u and observing. This expected entropy is obtained by integrating over all possible measurements z that we might be

17.2 Basic Exploration Algorithms

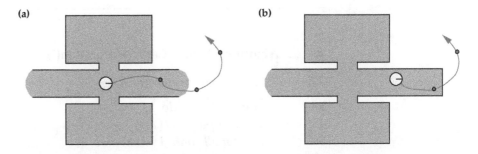

Figure 17.1 Unpredictability of the exploration problem: A robot in (a) might anticipate a sequence of three controls, but whether or not this sequence is executable depends on the things the robot finds out along the way. Any exploration policy has to be highly reactive.

receiving, times their probability. It is translated into utility via the constant α. We subsequently subtract the expected cost of executing action r.

Most exploration techniques employ this greedy policy, which is indeed optimal at horizon 1. The reason for relying on greedy techniques is because of the enormous branching factor in exploration, which renders multi-step planning impossible. The large branching factor is due to the very nature of the exploration problem. The goal of exploration is to acquire new information, but once such information has been acquired, the robot is in a new belief state, hence has to adjust its policy. Hence, measurements are inherently unpredictable.

Figure 17.1 illustrates such a situation. Here a robot has mapped two rooms and part of a corridor. At this point, the optimal exploration might involve exploring one corridor, and an appropriate sequence of actions might correspond to the one shown in Figure 17.1a. However, whether or not such an action sequence is executable is highly unpredictable. For example, the robot may find itself in a dead end as illustrated in Figure 17.1b where the anticipated action sequence is not applicable.

17.2.3 Monte Carlo Exploration

The algorithm **Monte_Carlo_exploration** is a simple probabilistic exploration algorithm. Table 17.1 depicts this Monte Carlo approximation of the greedy exploration rule in Equation (17.6). This algorithm simply replaces the integrals in the greedy approach by sampling. In line 4, it samples a

1:          **Algorithm Monte_Carlo_exploration(**b**):**
2:          set $\rho_u = 0$ for all actions u
3:          for $i = 1$ to N do
4:             sample $x \sim b(x)$
5:             for all control action u do
6:                 sample $x' \sim p(x' \mid u, x)$
7:                 sample $z \sim p(z \mid x')$
8:                 $b' =$ **Bayes_filter**(b, z, u)
9:                 $\rho_u = \rho_u + r(x, u) - \alpha \; H_{b'}(x')$
10:            endfor
11:          endfor
12:          return $\operatorname*{argmax}_{u} \rho_u$

Table 17.1 A Monte Carlo implementation of the greedy exploration algorithm, which chooses actions by maximizing the trade-off between information gain and cost.

state x from the momentary belief b, followed by sampling the next state x' and the corresponding observation z'. A new posterior belief is then calculated in line 8, and its entropy-cost trade-off is cached away in line 9. Line 12 then returns the action whose Monte Carlo information gain-cost value is the highest.

In general, the greedy Monte Carlo algorithm may still require substantial time, to a point that it becomes impractical. The main complication arises from the sampling of measurements z. When exploring an unknown environment during mapping, the number of possible measurements can be huge. For example, for a robot equipped with 24 ultrasound sensors that each report one byte of range data, the number of potential sonar scans obtained at a specific location is 256^{24}. Clearly, not all of those measurements are plausible, but the number of plausible measurements is at least as large as the number of plausible local maps. And for any realistic mapping problem, the number of possible maps is immense! Below, we will consider exploration techniques that side-step this integration through a closed-form analysis of the expected information gain.

17.2.4 Multi-Step Techniques

In situations where the measurement and the state spaces are confined, it may be possible to generalize the principle of information gathering to any finite horizon $T > 1$. Suppose we would like to optimize the information-cost trade-off at horizon T. This is achieved by defining the following exploration payoff function:

$$(17.7) \quad r_{\exp}(b_t, u_t) = \begin{cases} \int r(x_t, u_t)\, b(x_t)\, dx_t & \text{if } t < T \\ \alpha\, H_{b_t}(x_t) & \text{if } t = T \end{cases}$$

Under this payoff function, a POMDP planner then finds a control policy that minimizes the entropy of the final belief b_T minus the cost of achieving this belief, scaled appropriately. All of the previously discussed POMDP techniques are applicable.

The reader may notice the similarity to the augmented MDP discussed in the previous chapter. The difference here is that we only specify the payoff function, not the belief state representation. Because most exploration problems become computationally intractable under the general POMDP model, we will not pursue this approach any further in this book.

17.3 Active Localization

The simplest case of exploration occurs when estimating the state of a low-dimensional variable. *Active localization* is such a problem: Here we seek information about the robot pose x_t, but we are given a map of the environment. Active localization is particularly interesting during global localization, since here the control choice can have an enormous impact on the information gain. In many environments, wandering around aimlessly makes global localization hard, whereas moving to the right location can yield very fast localization.

Such an environment is depicted in Figure 17.2a. Here the robot is placed within a symmetric corridor, and no matter how long it explores, it cannot resolve its pose without leaving the corridor and moving through one of the open doors. Thus, any solution to the active localization problem must ultimately steer the robot out of the corridor and into one of the rooms.

Active localization can be solved greedily, along the lines of the algorithm just presented. The key insight pertains to the choice of action representation. Clearly, if actions are defined as low-level control actions as in much of

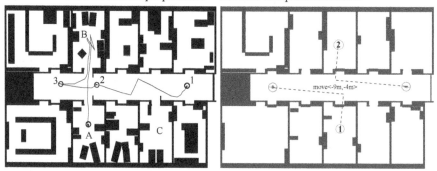

Figure 17.2 (a) Active localization in a symmetric environment: Shown here is an environment with a symmetric corridor, but an asymmetric arrangement of rooms, labeled A, B, and C. This figure also shows an exploration path. (b) An example of the exploration action "go backward 9 meters, go left 4 meters." If the robot's pose posterior possesses two distinct modes as shown here, the actual control in global world coordinates might lead to two different places.

this book, any viable exploration plan would have to involve a long chain of controls, before any pose ambiguity can be resolved. To solve the active localization problem through greedy exploration, we need a definition of actions through which the robot can gather pose information greedily.

One possible solution is to define an *exploration action* through target locations, expressed in the robot's local coordinate frame. For example, "*move to the point $\Delta x = -9m$ and $\Delta y = 4m$ relative to the robot's local coordinate frame*" can be considered an action, so long as we can devise a low level navigation module that can map this action back into low-level controls. Figure 17.2b visualizes the potential effect of this action in global world coordinates. In this example, the posterior possesses two modes, hence this action can carry the robot to two different locations.

The definition of relative motion actions makes it possible to solve the active localization problem through an algorithm that is essentially the same as the greedy exploration algorithm in Table 17.1. We will describe this algorithm through an example. Figure 17.3a shows the active localization path along with a number of labeled places. We begin in the middle of localization: Figure 17.3b shows the belief after moving from the location labeled "1" to the location labeled "2". This belief possesses six modes, each indicated by a circle in Figure 17.3b. For this belief, the expected occupancy in robot

17.3 Active Localization

(a) Path of the localizing robot

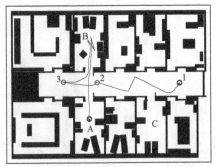

(b) Early belief distribution with six modes

(c) Occupancy probability in robot coordinates

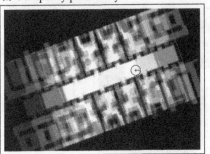

(d) Expected costs of motion

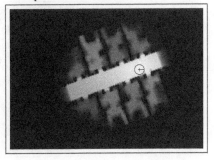

(e) Exp. information gain in robot coordinates

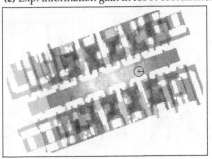

(f) Gain plus costs (the darker, the better)

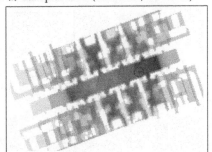

Figure 17.3 Illustration of active localization. This figure displays a number of auxiliary functions for computing the optimal action, for a multi-hypothesis pose distribution.

(a) Belief distribution

(b) Occupancy prob. in robot coordinates
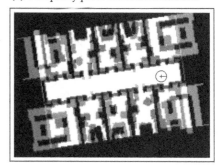

(c) Expected costs of motion
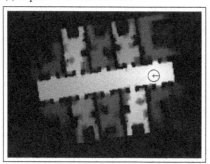

(d) Exp. information gain in robot coordinates

(e) Gain plus costs (the darker, the better)
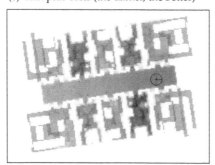

(f) Final belief after active localization

Figure 17.4 Illustration of active localization at some later point in time, for a belief with two distinct modes.

coordinates is shown in Figure 17.3c. This figure is simply obtained by superimposing the known occupancy grid map for each of the possible robot poses, weighted by the respective probability. Since the robot does not know its pose with certainty, it cannot know the occupancy of locations; hence the "fuzziness" of the expected cost map. However, with high likelihood the robot is in a corridor-shaped area that is traversable.

While Figure 17.3c depicts the cost of *being* at a target location, we need the cost of *moving* to such a target location. We have already encountered an algorithm for computing such motion costs, along with the optimal path: *value iteration* (see Chapter 14). Figure 17.3d shows the result of value iteration, applied to the map in Figure 17.3b as cost function. Here value is propagated from the robot outwards (as opposed to from the goal, as was the notion in Chapter 14). This makes it possible to calculate the cost of moving to *any* potential target location in this map.

As this figure illustrates, there exists a large region near the robot where motion is safe; in fact, this region corresponds to the corridor, no matter where the robot actually is located. Moving out of this region and into one of the rooms incurs higher expected costs, since the validity of such a motion hinges on the exact location of the robot, which is unknown.

For greedy exploration, we now need to determine the expected information gain. It can be approximated by placing the robot in a location, simulate possible range measurements, incorporate the result, and measure the information after the Bayesian update. Repetition of this evaluation step for each possible location yields an expected information gain map. Figure 17.3e shows the result: the darker a location in this map, the more information it provides relative to the robot's pose estimate. Obviously, any of the rooms will be most informative, as will be the far end of the corridors. Thus, from a pure information gathering perspective, the robot should seek to move into one of the rooms.

Adding this cost map to the expected information gain leads to the plot in Figure 17.3f: the darker a target location, the better. While the outside rooms fare still high in this combined function, their utility has been reduced by the relatively high costs of moving there. At this point, areas at the end of the corridor score the highest.

The robot now moves to the location of the highest combined value, which brings it to the outer area of the corridor that is still safe to travel. In Figure 17.3a, this corresponds to a transition from the location marked "2" to the one marked "3".

The greedy active exploration rule is now reiterated. The belief at location

"3" is depicted in Figure 17.4a. Obviously, the previous exploration action had the effect of reducing the modes in the posterior, from six down to two. Figure 17.4b shows the new occupancy map, in robot-centric coordinates; and Figure 17.4c depicts the corresponding value function. The expected information gain is now uniformly high only in the rooms, as illustrated in Figure 17.4d. Figure 17.4e shows the combined gain-cost map. At this point, the cost of moving into any of the symmetrically open rooms is the lowest, hence the robot moves there, at which the ambiguity is largely resolved. One time step later, after another round of deliberation, the final belief is the one shown in Figure 17.4f.

This greedy active localization algorithm is not without shortcomings. One shortcoming stems from its greediness: It cannot compose multiple exploration actions that together maximize the knowledge gain. Another shortcoming is the result of our action definition. The algorithm fails to consider the measurements that will be acquired while moving to a target location. Instead, it treats each such motion as an open-loop control, in which the robot is unable to react to its measurements. Clearly, when faced with a closed door, a real robot can abandon a target point even before reaching it; however, no such provision is considered during planning. This explains the sub-optimal choice in our example above, where the robot explores the room labeled "B" before it explores the room labeled "A." As a result, localization tends to take longer than theoretically necessary. Nevertheless, this algorithm performs well in practice.

17.4 Exploration for Learning Occupancy Grid Maps

17.4.1 Computing Information Gain

Greedy exploration can also be applied in robotic mapping. Mapping problems involve many more unknown variables than robot localization, hence we need a technique for calculating expected knowledge gain that scales to high-dimensional estimation problems. As we shall see, the "trick" for exploration in occupancy grid maps is the very same trick that led us to define an efficient update algorithm for occupancy grid maps: We treat information gain as *independent* between different grid cells.

Consider an occupancy grid map, such as the one shown in Figure 17.5a. Parts of this map are well-explored, such as the large free area in the center of the map, and the many walls and obstacles whose locations are well-known. Other parts remain unexplored, such as the large gray area outside the map.

17.4 Exploration for Learning Occupancy Grid Maps

(a) Occupancy grid map

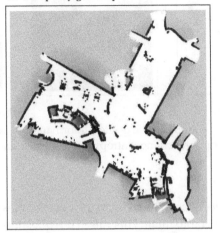

(b) Cell entropy

(c) Explored and unexplored space

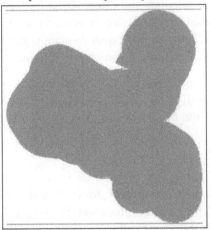

(d) Value function for exploration

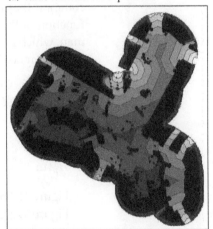

Figure 17.5 Example of the essential step in exploration for mapping. (a) shows a partial grid map; (b) depicts the map entropy; (c) shows the space for which we have zero information; and (d) displays the value function for optimal exploration.

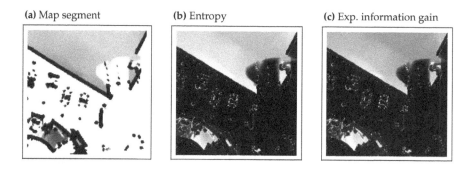

Figure 17.6 Map, entropy and expected information gain. This figure illustrates that with the appropriate scaling, entropy and expected information gain are nearly indistinguishable.

Greedy exploration steers the robot to the nearest unexplored area, where the information gain is maximal. This raises the question as to how to compute gain.

We will discuss three possible techniques. All three approaches have in common that they calculate the information gain *per grid cell*, and not as a function of the robot action. This conveniently leads to an information gain map, which is a 2-D map defined over the same grid as the original grid map. The difference between these techniques is the quality of the approximation.

- **Entropy.** Computing the entropy of each cell is straightforward. Let us denote the i-th cell by $\mathbf{m}_i$, and its occupancy probability $p_i = p(\mathbf{m}_i)$. Then the entropy of the binary occupancy variable is given by the following sum:

$$H_p(\mathbf{m}_i) \;=\; -p_i \, \log p_i - (1 - p_i) \, \log(1 - p_i) \tag{17.8}$$

Figure 17.5b depicts the entropy, for each of the cells in the map shown in Figure 17.5a. The brighter a location, the larger the entropy. Most of the central areas in the map exhibit low entropy, with the exception of a few areas near or inside obstacles. This matches our intuition, as most of the map is already well-explored. The outer areas possess high entropy, indicating that they might be good areas to explore. Thus, the entropy map indeed assigns high information values to places that remain unexplored.

- **Expected information gain.** Technically, the entropy only measures the current information. It does not specify the information a robot would gain through its sensors when in a vicinity of a cell. The calculation of the expected information gain is slightly more involved, and it requires additional assumptions on the nature of the information provided by the robot's sensors.

Suppose our sensor measures with probability p_{true} the correct occupancy, and it errs with probability $1 - p_{\text{true}}$. Then we would expect to measure "occupied" with the following probability:

$$p^+ = p_{\text{true}}\, p_i + (1 - p_{\text{true}})(1 - p_i) \tag{17.9}$$

The standard occupancy grid update will then yield the new probability, which follows directly from the occupancy grid mapping algorithm discussed in Chapter 9:

$$p'_i = \frac{p_{\text{true}}\, p_i}{p_{\text{true}}\, p_i + (1 - p_{\text{true}})(1 - p_i)} \tag{17.10}$$

The entropy of this posterior is now:

$$\begin{aligned} H^+_{p'}(\mathbf{m}_i) &= -\frac{p_{\text{true}}\, p_i}{p_{\text{true}}\, p_i + (1 - p_{\text{true}})(1 - p_i)} \log \frac{p_{\text{true}}\, p_i}{p_{\text{true}}\, p_i + (1 - p_{\text{true}})(1 - p_i)} \\ &\quad - \frac{(1 - p_{\text{true}})(1 - p_i)}{p_{\text{true}}\, p_i + (1 - p_{\text{true}})(1 - p_i)} \log \frac{(1 - p_{\text{true}})(1 - p_i)}{p_{\text{true}}\, p_i + (1 - p_{\text{true}})(1 - p_i)} \end{aligned} \tag{17.11}$$

By analogy, our sensor will sense "free" with probability

$$p^- = p_{\text{true}}(1 - p_i) + (1 - p_{\text{true}})\, p_i \tag{17.12}$$

in which case the posterior will become

$$p'_i = \frac{(1 - p_{\text{true}})\, p_i}{p_{\text{true}}(1 - p_i) + (1 - p_{\text{true}})\, p_i} \tag{17.13}$$

This posterior has the entropy

$$\begin{aligned} H^-_{p'}(\mathbf{m}_i) &= -\frac{(1 - p_{\text{true}})\, p_i}{p_{\text{true}}(1 - p_i) + (1 - p_{\text{true}})\, p_i} \log \frac{(1 - p_{\text{true}})\, p_i}{p_{\text{true}}(1 - p_i) + (1 - p_{\text{true}})\, p_i} \\ &\quad - \frac{p_{\text{true}}(1 - p_i)}{p_{\text{true}}(1 - p_i) + (1 - p_{\text{true}})\, p_i} \log \frac{p_{\text{true}}(1 - p_i)}{p_{\text{true}}(1 - p_i) + (1 - p_{\text{true}})\, p_i} \end{aligned} \tag{17.14}$$

Putting the previous equations together, we obtain the expected entropy after sensing:

$$
\begin{aligned}
E[H_{p'}(\mathbf{m}_i)] &= p^+ \, H^+_{p'}(\mathbf{m}_i) \; + \; p^- \, H^-_{p'}(\mathbf{m}_i) \\
&= -p_{\text{true}} \, p_i \, \log \frac{p_{\text{true}} \, p_i}{p_{\text{true}} \, p_i + (1 - p_{\text{true}})\,(1 - p_i)} \\
&\quad -(1 - p_{\text{true}})\,(1 - p_i) \, \log \frac{(1 - p_{\text{true}})\,(1 - p_i)}{p_{\text{true}} \, p_i + (1 - p_{\text{true}})\,(1 - p_i)} \\
&\quad -(1 - p_{\text{true}}) \, p_i \, \log \frac{(1 - p_{\text{true}}) \, p_i}{p_{\text{true}}\,(1 - p_i) + (1 - p_{\text{true}}) \, p_i} \\
&\quad -p_{\text{true}} \,(1 - p_i) \, \log \frac{(1 - p_{\text{true}}) \, p_i}{p_{\text{true}}\,(1 - p_i) + (1 - p_{\text{true}}) \, p_i}
\end{aligned}
\tag{17.15}
$$

Following our definition in Equation (17.4), the expected information gain upon sensing the grid cell $\mathbf{m}_i$ is simply the difference $H_p(\mathbf{m}_i) - E[H_{p'}(\mathbf{m}_i)]$.

So how much better is the expected information gain than just the entropy from which it was derived? The answer is: not much. Figure 17.6 plots the entropy in Panel (b), next to the expected information gain in Panel (c), all for the map segment shown in Panel (a). Visually, the entropy and the expected information gain are nearly indistinguishable, although the values are different. This "justifies" the common practice of using entropy—instead of the expected information gain—as a function to direct the exploration.

- **Binary gain.** The third approach is the simplest of all, and by far the most popular. A very crude approximation of the expected information gain is to mark cells that have been updated at least once as "explored," and all other cells 'unexplored.' Thus, the gain becomes a binary function.

  Figure 17.5c shows such a binary map: Only the outer white area promises any new information; the map interior is assumed to be fully explored. While this information gain map is clearly the crudest of all the approximations discussed here, it tends to work well in practice, in that it pushes an exploring robot continuously into unexplored terrain. This binary map is at the core of a popular exploration algorithm called *frontier-based exploration*, in which the robot continuously moves to the nearest unexplored frontier of the explored space.

FRONTIER-BASED EXPLORATION

17.4.2 Propagating Gain

The remaining question now pertains to the development of a greedy exploration technique that utilizes these maps. As in our active localization example, this requires the definition of an appropriate exploration action.

A simple yet effective definition of an *exploration action* involves moving to an x-y location along a minimum cost path, and then sensing all grid cells in a small circular diameter around the robot. Thus, each location in the map defines a potential exploration action.

The computation of the best greedy exploration action is now easily done using value iteration. Figure 17.5d shows the resulting value function for the map shown in Figure 17.5a, and the binary information gain map in Figure 17.5c. The value iteration approach assumes such a binary gain map: Gain can only be reaped at unexplored locations.

The central value update is implemented by the following recursion:

$$(17.16) \quad V_T(\mathbf{m}_i) = \begin{cases} \max_j r(\mathbf{m}_i, \mathbf{m}_j) + V_{T-1}(\mathbf{m}_j) & \text{if } I(\mathbf{m}_i) = 0 \\ I(\mathbf{m}_i) & \text{if } I(\mathbf{m}_i) > 0 \end{cases}$$

Here V is the value function, j indexes over all grid cells adjacent to $\mathbf{m}_i$, r measures the cost of moving there (usually a function of the occupancy grid map), and $I(\mathbf{m}_i)$ is the information we would gain in cell $\mathbf{m}_i$. The termination condition $I(\mathbf{m}_i) > 0$ is only true for unexplored grid cells in the binary information gain map.

Figure 17.5d shows such a value function after convergence. Here the value is highest near the open areas of the map, and lower in the interior of the map. The exploration technique now simply determines the optimal path by hill-climbing in this map. This path leads the robot directly to the nearest unexplored frontier.

Clearly, this exploration technique is just a crude approximation. It entirely ignores the information acquired as the robot moves to a target location, in that it falsely assumes that no sensing takes place along the way. However, it tends to work well in practice. Figure 17.7 shows a value function and a map of an exploring robot. This map is historical: it is taken from the 1994 *AAAI Mobile Robot Competition*, which involved the acquisition of an environment map at high speeds. The robot was equipped with an array of 24 sonar sensors, which accounts for the relatively low accuracy of the map. The most interesting aspect of this figure pertains to the robot path: As can be seen in Figure 17.7b, initially the exploration is highly effective, with the robot

(a) Exploring value function

(b) Exploration path

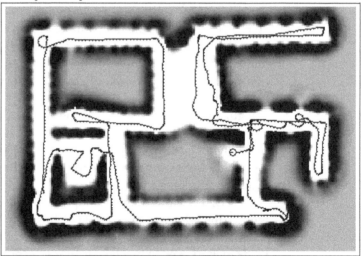

Figure 17.7 Illustration of autonomous exploration. (a) Exploration values V, computed by value iteration. White regions are completely unexplored. By following the gray-scale gradient, the robot moves to the next unexplored area on a minimum-cost path. The large black rectangle indicates the global wall orientation θ_{wall}. (b) Actual path traveled during autonomous exploration, along with the resulting metric map.

17.4 Exploration for Learning Occupancy Grid Maps

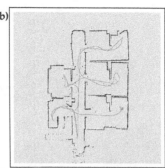

Figure 17.8 (a) Urban robot for indoor and outdoor exploration. The urban robot's odometry happens to be poor. (b) Exploration path of the autonomously exploring robot, using the exploration techniques described in the text.

roaming unexplored corridors. Later, however, the robot begins to alternate between different goal locations. Such alternate behavior is quite common for greedy exploration techniques, and most contemporary implementations provide additional mechanisms for avoiding such behavior.

A second example is shown in Figure 17.8. The path of the robot, shown on the right, illustrates the efficiency of the greedy exploration algorithm.

17.4.3 Extension to Multi-Robot Systems

The gain-seeking exploration rule has frequently been extended to multi-robot systems, in which the robot seeks to acquire a map through cooperative exploration. In general, the speed-up when using K robots is linear. It can be *super-unitary*: K robots tend to speed-up the time required for exploration by more than a factor of K when compared to one robot. Such super-unitary speed-up is due to the fact that a single robot might have to traverse many areas twice, once to go out and explore, and once to come back to explore some other part of the environment. With an appropriate number of robots, the return portion may become unnecessary, and the speed-up is closer to a factor of $2K$. This factor of $2K$ is an upper bound for robots that can navigate easily in all directions.

The key provision for multi-robot exploration pertains to coordination. In static exploration, this is easily accomplished through greedy task allocation techniques. Consider a situation in which K robots are placed in a partially explored map. The setup now is that there exists a number of frontier places

1:	**Algorithm multi_robot_exploration**($m, x_1, \ldots, x_K$):
2:	for each robot $k = 1$ to K do
3:	Let $\mathbf{m}_k$ be the grid cell that contains x_k
4:	$V_k(\mathbf{m}_i) = \begin{cases} \infty & \text{if } \mathbf{m}_i \neq \mathbf{m}_k \\ 0 & \text{if } \mathbf{m}_i = \mathbf{m}_k \end{cases}$
5:	do until V_k converged
6:	for all i do
7:	$V_k(\mathbf{m}_i) \longleftarrow \min_j \{V_k(\mathbf{m}_i), r(\mathbf{m}_i, \mathbf{m}_j) + V_k(\mathbf{m}_j)\}$
8:	endfor
9:	endwhile
10:	endfor
11:	compute the binary gain map $\bar{m}$ from the map m
12:	for each robot $k = 1$ to K do
13:	set $\text{goal}_k = \underset{i \text{ such that } \bar{m}_i = 1}{\operatorname{argmin}} V_k(\mathbf{m}_i)$
14:	for all cells $\bar{\mathbf{m}}_j$ in ε-neighborhood of goal_k
15:	set $\bar{\mathbf{m}}_j = 0$
16:	endfor
17:	endfor
18:	return $\{\text{goal}_1, \text{goal}_2, \ldots, \text{goal}_K\}$

Table 17.2 Multi-robot exploration algorithm.

to explore, and we need an algorithm that assigns such robots to places in a way that greedily maximizes the overall exploration effect.

The algorithm **multi_robot_exploration** in Table 17.2 is a very simple such algorithm. It computes for a set of K robots a set of K exploration goals, which are locations to which the robots move during coordinated exploration.

The algorithm first computes value function V_k, one for each of the robots (lines 2 through 10). However, these value functions are defined differently from the ones encountered thus far: The minimum value is attained at the robot pose. The further out a cell, the higher its value. Figure 17.9 and 17.10 show several examples of such value functions. In each case, the location of the robot possesses minimum value, with the value increasing throughout the reachable space.

It is easy to show that these value functions measure for any possible grid cell the cost of moving there. For each individual robot, the optimal frontier cell to explore is now computed via line 13: it is the minimum cost cell in V_k which is yet unexplored, according to a binary gain map computed in line 11. This cell is used as the goal point. However, to "discourage" other robots from using the same or a nearby goal location, our algorithm resets the gain map to zero in the vicinity of the chosen goal location. This takes place in lines 14 through 16.

The *coordination mechanism* in **multi_robot_exploration** can be summarized as follows: Each robot greedily picks the best available exploration goal point, and then prohibits other robots from selecting the same or a nearby point. Figures 17.9 and 17.10 illustrate the effect of this coordination. Both robots in Figure 17.9, although located at different places, identify the same frontier cell for exploration. Thus, when exploring without coordination, they would both aim for the same area to explore. This is different in Figure 17.10. Here the first robot makes its choice, out-ruling this location for the second robot. In turn, the second robot selects a superior location. The resulting joint exploration action avoids this conflict and, as a result, explores more efficiently.

Clearly, the coordination mechanism is quite simplistic, and it is easily trapped in a local minimum. What would have happened if in Figure 17.10 we would have let the second robot choose first? Then the first robot would have been forced to select a far-away target location, and both robots' paths would have crossed along the way. Obviously, a good criterion for a suboptimal assignment is that paths cross. However, the absence of such a path crossing does not guarantee an optimal assignment.

Improved coordination techniques take such conflicts into consideration, and enable robots to trade goal points with each other. A popular family of techniques enables robots to swap assignments of goal points with each other, if this reduces the overall exploration costs. Such algorithms are often phrased in terms of *auction mechanisms*. The resulting algorithms are often characterized as *market-based algorithms*.

AUCTION MECHANISM

Figure 17.11 shows an application of algorithm **multi_robot_exploration** in a real-world experiment. Here three robots are exploring an unknown environment. The leftmost image shows all the robots at their starting lo-

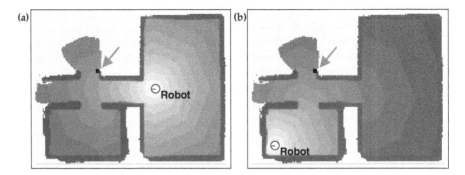

Figure 17.9 Two robots exploring an environment. Without any coordination both vehicles would decide to approach the same target location. Each image shows the robot, the map, and its value function. The black rectangle indicates the target points with minimum cost.

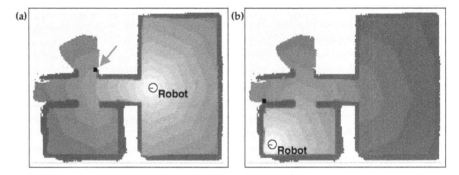

Figure 17.10 Target positions obtained using the coordination approach. In this case the target point for the second robot is to the left in the corridor.

cations. The other images depict different situations during the coordinated exploration. A map constructed by the same robots in an additional run is depicted in Figure 17.12. As can be seen, the robots in fact were nicely distributed over the environment.

Figure 17.13 depicts the performance of this algorithm compared to a team of robots from which all robots apply Algorithm **Monte_Carlo_exploration** without any coordination. Whereas the horizontal axis depicts the number of robots in the team, the vertical axis shows the number of time steps needed to complete the exploration task. In these experiments it was assumed that the robots always share their local maps. It was furthermore assumed that

17.4 Exploration for Learning Occupancy Grid Maps

Figure 17.11 Coordinated exploration by a team of mobile robots. The robots distribute themselves throughout the environment.

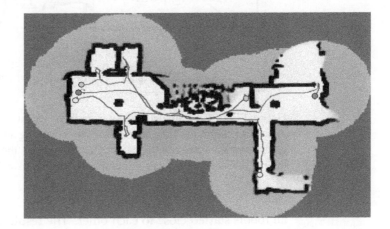

Figure 17.12 Map of a $62 \times 43m^2$ large environment learned by three robots in 8 minutes.

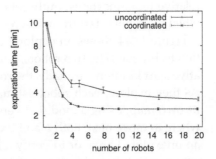

Figure 17.13 Exploration time obtained in a simulation experiment in which robot teams of different sizes explore the environment shown in the left image.

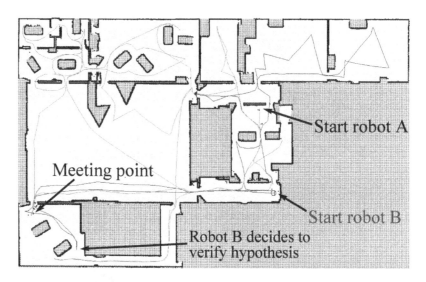

Figure 17.14 Coordinated exploration from unknown start locations. The robots establish a common frame of reference by estimating and verifying their relative locations using a rendezvous approach. Once they meet, they share a map and coordinate their exploration. Courtesy of Jonathan Ko and Benson Limketkai.

all robots started close to each other. The result is quite telling. Obviously, the uncoordinated robots are significantly less efficient than the coordinated team.

The coordination strategies discussed so far assume that the robots share a map and know their relative locations. By considering the uncertainty in the relative robot locations, multi robot exploration can be extended to the case when the robots start from different, unknown locations.

Figure 17.14 shows an exploration run using such an extended coordination technique. The two robots, A and B, have no knowledge about their relative start locations. Initially, the robots explore independently of each other. As they explore, each robot estimates the other robot's location relative to its own map, using a modified version of MCL localization. When deciding where to move next, both A and B consider whether it is "better" to move to an unexplored area, or to verify a hypothesis for the other robot's location. At one point, B decides to verify a hypothesis for A's location. It sends A a message to stop and moves to A's hypothesized location (marked as meeting point in Figure 17.14). Upon reaching this location, both robots check the presence of the other robot using their laser range-finders (robots are tagged with highly reflective tape). When they detect each other, their maps are merged and the robots share a common frame of reference. From then on, they explore the environment using algorithm **multi_robot_exploration**. Such an exploration technique from unknown start locations can be applied to scenarios involving more than two robots.

17.5 Exploration for SLAM

The final algorithm in this book applies the greedy exploration idea to a full SLAM algorithm. In the previous sections we always assumed that either the map or the pose of the robot was known. In SLAM, however, we do not know either. Accordingly, we have to consider the uncertainty in the map as well as the uncertainty in the robot's pose when choosing how to explore, and we can gain or lose knowledge about both. Obviously, without knowledge about the pose of the vehicle, the integration of sensor information into the map can lead to serious errors. On the other hand, a robot that solely focuses on reducing the uncertainty about its pose will simply not move, and thus it will never acquire any information about the environment beyond its initial sensor radius.

17.5.1 Entropy Decomposition in SLAM

The key insights for optimal exploration in SLAM is that the entropy of the SLAM posterior can be *decomposed* into two terms: One pertaining to the entropy in the pose posterior, and one pertaining to the expected entropy of the map. In this way, an exploring SLAM robot trades off the uncertainty in the robot pose with the uncertainty in the map. Control actions tend to reduce only one of both: when closing a loop, a robot will mainly reduce its pose uncertainty. When moving into open unexplored terrain, it will mainly reduce its map uncertainty. By considering both, whatever reduction is larger will win, and the robot may sometimes move into open terrain, sometimes re-localize by moving back into known terrain.

The decomposition of the entropy is in fact universal for the full SLAM problem. Consider the full SLAM posterior.

(17.17) $\quad p(x_{1:t}, m \mid z_{1:t}, u_{1:t})$

This posterior can be factored as follows:

(17.18) $\quad p(x_{1:t}, m \mid z_{1:t}, u_{1:t}) = p(x_{1:t} \mid z_{1:t}, u_{1:t}) \, p(m \mid x_{1:t}, z_{1:t}, u_{1:t})$

This is trivial, and it was already stated in Equation (13.2) on page 442. It is less obvious that this implies

$$
\begin{aligned}
(17.19)\quad & H[p(x_{1:t}, m \mid z_{1:t}, u_{1:t})] \\
& = H[p(x_{1:t} \mid z_{1:t}, u_{1:t})] + E\left[H[p(m \mid x_{1:t}, z_{1:t}, u_{1:t})]\right]
\end{aligned}
$$

where the expectation is taken over the posterior probability $p(x_{1:t} \mid z_{1:t}, u_{1:t})$.

Writing $p(x, m)$ as an abbreviation for the full posterior $p(x_{1:t}, m \mid z_{1:t}, u_{1:t})$, we can derive the additive decomposition from first principles:

$$
\begin{aligned}
(17.20)\quad H(x, m) & = E_{x,m}[-\log p(x, m)] \\
& = E_{x,m}[-\log(p(x)\, p(m \mid x))] \\
& = E_{x,m}[-\log p(x) - \log p(m \mid x)] \\
& = E_{x,m}[-\log p(x)] + E_{x,m}[-\log p(m \mid x)] \\
& = E_x[-\log p(x)] + \int_{x,m} -p(x, m) \log p(m \mid x)\, dx\, dm \\
& = E_x[-\log p(x)] + \int_{x,m} -p(m \mid x) p(x) \log p(m \mid x)\, dx\, dm \\
& = E_x[-\log p(x)] + \int_x p(x) \int_m -p(m \mid x) \log p(m \mid x)\, dx\, dm \\
& = H(x) + \int_x p(x)\, H(m \mid x)\, dx \\
& = H(x) + E_x[H(m \mid x)]
\end{aligned}
$$

This transformation directly implies the decomposition in (17.19). It proves that the SLAM entropy is the sum of the path entropy and the expected entropy of the map.

17.5.2 Exploration in FastSLAM

The entropy decomposition is now leveraged into an actual SLAM exploration algorithm. Our approach is based on the FastSLAM algorithm described in Chapter 13 of this book, and in particular the grid-based FastSLAM algorithm described in Chapter 13.10. Recall that FastSLAM represents the SLAM posterior by a set of particles. Each particle contains a robot path. In the case of the grid-based implementation, each particle also contains an occupancy grid map. This makes the entropy measures for occupancy grid maps applicable, as were just discussed in the previous section.

1:	**Algorithm FastSLAM_exploration(Y_t):**
2:	initialize $\hat{h} = \infty$
3:	repeat
4:	propose an exploration control sequence $u_{t+1:T}$
5:	select a random particle $y_t \in Y_t$
6:	for $\tau = t + 1$ to T
7:	draw $x_\tau \sim p(x_\tau \mid u_\tau, x_{\tau-1})$
8:	draw $z_\tau \sim p(z_\tau \mid x_\tau)$
9:	compute $Y_\tau =$ **FastSLAM**$(z_\tau, u_\tau, Y_{\tau-1})$
10:	endfor
11:	fit a Gaussian μ_x, Σ_x to all pose particles $\{x_T^{[k]}\} \in Y_T$
12:	$h = \frac{1}{2} \log \det(\Sigma_x)$
13:	for particles $k = 1$ to M do
14:	let m be the map $m_T^{[k]}$ from the k-th particle in Y_T
15:	update $h = h + \frac{1}{M} H[m]$
16:	endfor
17:	if $h < \hat{h}$ then
18:	set $\hat{h} = h$
19:	set $\hat{u}_{t+1:T} = u_{t+1:T}$
20:	endif
21:	until convergence
22:	return $\hat{u}_{t+1:T}$

Table 17.3 The exploration algorithm for the grid-based version of FastSLAM. It accepts as input a set of particles Y_t. Each particle $y_t^{[k]}$ contains a sampled robot path $x_{1:t}^{[k]}$ and an associated occupancy grid map $m^{[k]}$. It outputs an exploration path, expressed in relative motion commands.

An algorithm for determining an exploration action sequence is illustrated in Table 17.3. This algorithm leaves a number of important implementation questions open, hence it shall only serve as a schematic illustration. However, it characterizes all main steps of an actual implementation of the SLAM exploration ideas.

The FastSLAM exploration algorithm is essentially a test-and-evaluate algorithm. It proposes a course of action for exploration. It then evaluates these actions by measuring the residual entropy. Building on the fundamental insight discussed above, the entropy is computed by adding two terms; a term corresponding to the robot pose at the end of the proposed exploration sequence, and a term pertaining to the expected map entropy. The exploration algorithm then selects the controls that minimize the resulting entropy.

In detail, the algorithm **FastSLAM_exploration** accepts as input a set of particles and returns as output a proposed sequence of control for exploration. Line 4 of Table 17.3 proposes a potential control sequence. The evaluation of a control sequence takes place in lines 5 through 16. It is organized in three parts. First, the robot is simulated, based on a random particle in the particle set. This simulation uses the stochastic model of the robot and its environment. The result is a sequence of particle sets, all the way to the end of the control trajectory. This simulation takes place in lines 5 through 9.

Subsequently, the entropy of the final particle set is calculated. Through the mathematical decomposition stated in Equation 17.19, the entropy is factored into two terms: A term related to the entropy of the robot pose estimate at time T, and a term related to the expected map uncertainty. The first term is calculated in lines 11 and 12. Its correctness follows from the calculation of the entropy of a Gaussian, which is derived in Table 17.4.

The second entropy term is computed in lines 13 through 16. Notice that the calculation of the second term involves the entropy of a map m. For occupancy grid maps, this computation is analogous to the one discussed in the previous section. Lines 13 through 16 compute the average entropy of the map, where the average is taken over all particles at time T. The result is a value h which measures the expected entropy at time T, conditioned on the proposed control sequence. Lines 17 through 20 then select the action sequence that minimizes this expected entropy, which is ultimately returned in line 22.

Note that this algorithm uses an approximation to compute the entropy of the posterior about the trajectories in line 11. Instead of fitting a Gaussian

17.5 Exploration for SLAM

> **Lemma.** The *entropy of a multivariate Gaussian* of d dimensions and covariance Σ is given by
>
> $$H \;=\; \frac{d}{2}(1 + \log 2\pi) + \frac{1}{2}\log\det(\Sigma)$$
>
> **Proof.** With
>
> $$p(x) \;=\; (2\pi)^{-\frac{d}{2}}\,\det(\Sigma)^{-\frac{1}{2}}\,\exp\left\{-\frac{1}{2}\,x^T\,\Sigma^{-1}\,x + x^T\,\Sigma^{-1}\,\mu - \frac{1}{2}\,\mu^T\,\Sigma^{-1}\,\mu\right\}$$
>
> we get
>
> $$\begin{aligned}H_p[x] \;&=\; E[-\log p(x)] \\ &=\; \frac{1}{2}\left(d\log 2\pi + \log\det(\Sigma) + E[x^T\,\Sigma^{-1}\,x] - 2E[x^T]\,\Sigma^{-1}\,\mu + \mu^T\,\Sigma^{-1}\,\mu\right)\end{aligned}$$
>
> Here $E[x^T] = \mu^T$, and $E[x^T\,\Sigma^{-1}\,x]$ resolve as follows (where "$\cdot$" denotes the dot product)
>
> $$\begin{aligned}E[x^T\,\Sigma^{-1}\,x] \;&=\; E[x\,x^T \cdot \Sigma^{-1}] \\ &=\; E[x\,x^T] \cdot \Sigma^{-1} \\ &=\; \mu\,\mu^T \cdot \Sigma^{-1} + \Sigma \cdot \Sigma^{-1} \\ &=\; \mu^T\,\Sigma^{-1}\,\mu + d\end{aligned}$$
>
> It follows that
>
> $$\begin{aligned}H_p[x] \;&=\; \frac{1}{2}\left(d\log 2\pi + \log\det(\Sigma) + \mu^T\Sigma^{-1}\mu + d - 2\mu^T\Sigma^{-1}\mu + \mu^T\Sigma^{-1}\mu\right) \\ &=\; \frac{d}{2}(1 + \log 2\pi) + \frac{1}{2}\log\det(\Sigma)\end{aligned}$$

Table 17.4 The entropy of a multivariate Gaussian.

to all trajectory particles $y_T^{[k]}$, it only computes it based on the last poses $x_T^{[k]}$. This approximation works well in practice, and it bears resemblance to our notion of exploration action approximation.

In summary, the FastSLAM exploration algorithm is essentially an extension of the Monte Carlo exploration algorithm stated in Table 17.1, with two insights. First, it is applied to entire sequences of controls, not just a single control. Second—and more importantly—the FastSLAM exploration algorithm computes two types of entropies, one pertaining to the robot path, and one to the map.

17.5.3 Empirical Characterization

The exploration algorithm leads to appropriate exploratory behavior, specifically in cyclic environments. Figure 17.15 shows a situation in which a robot explores a *cyclic environment* containing a *loop*. The robot started in the lower right corner of the loop, labeled "start." At time step 35, the actions considered by the robot lead back to the start, and also the unknown area to the left of this marker. The anticipated knowledge gain for moving back to the start is higher than moving into unexplored terrain, since in addition to new map information the pose uncertainty will be reduced. Thus, the exploring robot actively decided to close the loop and move towards previously explored terrain.

Figure 17.16 investigates the cost-benefit trade off closer. Shown there are 8 different actions as indicated. The utility of action 1 happens to be the highest. It specifically outweighs the utility of action 4 (and any other non-loop-closing action), which is significantly lower.

In time step 45, the robot closes the loop, as indicated in Figure 17.15. At this point, the pose uncertainty is minimal and the map uncertainty begins to dominate. As a result, the loop-closing action becomes unattractive. At time 88, the robot chooses to explore the open area, to which it then advances as visualized in Figure 17.15.

Figure 17.17 depicts the evolution of the overall entropy during this experiment over time. Up to time step 30, the reduction of the map uncertainty compensates the increase of uncertainty about the robot's trajectory. Therefore, the entropy stays more or less constant. Whereas the execution of the loop-closing action reduces the entropy in the belief about the trajectory of the robot, the change in map uncertainty is relatively small. This leads to a reduction in the overall entropy. As soon as the robot integrates measurements covering so far unknown areas in the horizontal corridor, the changes in the map and pose uncertainties again compensate each other. The decline of the overall entropy around time step 90 is caused by observations in the wider parts of the horizontal corridor. This is due to the fact that in occupancy grid maps the reduction of the map uncertainty by incorporating a range scan generally depends linearly on the size of the unknown area covered by the scan.

The intricate interplay between path and map uncertainty, and the corresponding knowledge gain terms, play an essential role in the exploration approach. When appropriate, the robot sometimes prefers localization, and sometimes prefers moving into unmapped terrain.

17.5 Exploration for SLAM

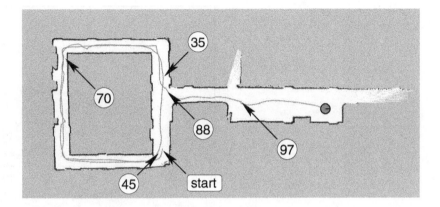

Figure 17.15 A mobile robot explores an environment with a loop. The robot starts in the lower right corner of the loop. After traversing it, it decides to follow the previous trajectory again in order to reduce its uncertainty. Then it continues to explore the corridor. Courtesy of Cyrill Stachniss, University of Freiburg.

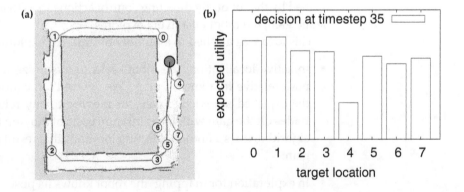

Figure 17.16 In this situation the robot determines the expected utility of possible actions. (a) the exploration actions considered by the robot; (b) the expected utility of each action. Action 1 is selected because it maximizes the expected utility. Courtesy of Cyrill Stachniss, University of Freiburg.

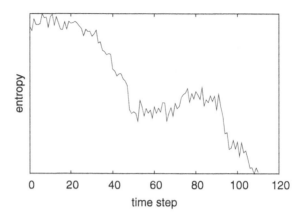

Figure 17.17 The evolution of the entropy during the exploration experiment shown in Figure 17.15. Courtesy of Cyrill Stachniss, University of Freiburg.

17.6 Summary

In this chapter, we have learned about robot exploration. All algorithms in this chapter are motivated through a single aim: to maximize the knowledge gained by the robot. By directing control actions so as to maximize the knowledge gain, the robot effectively explores.

This idea was applied to a number of different exploration problems:

- In active localization, the robot seeks to maximize the knowledge of its pose relative to a known map. We devised an algorithm that calculates the expected knowledge gain for moving to any relative robot pose. It trades off the gain with the minimum costs of moving there. The resulting algorithm does a fine job selecting locations that result in high information gain.

- In exploration for mapping, the robot knows its pose at all times. Instead, it has to gather information about its environment. Building on the occupancy grid mapping paradigm, we noted that the information gain can be computed separately for each grid cell in the map. We compared a number of different techniques for computing information gain and noted that simple techniques—such as the entropy—do quite well. We then devised a dynamic programming algorithm for moving to the nearby point that optimizes the trade-off between information gain and costs of moving there.

- We augmented the knowledge gain exploration technique to the case of multiple robots. This extension proved remarkably straightforward. This is because an easy modification of the dynamic programming paradigm makes it possible to compute the cost of moving to *any* location, and trade it off with the knowledge gain. By comparing costs and knowledge gain for different target locations, multiple robots can coordinate their exploration assignments so as to minimize the overall exploration time.

- Finally, we discussed exploration techniques for the full SLAM problem, in which both the robot pose and the environment map are unknown. Our approach observed a fundamental decomposition of the entropy into two terms, one pertaining to the path uncertainty, and one to the uncertainty in the map (averaged over all paths). This insight was leveraged into a generate-and-test algorithm for exploration, which generates control sequences, computes future estimates, and trades off path uncertainty versus map uncertainty when evaluating such control sequences. The result was an exploration technique that sometimes leads the robot into unknown terrain so as to improve the maps, and sometimes back to previously mapped terrain so as to improve the pose estimates.

Most of the exploration techniques in this chapter were greedy in the sense that the robot only considers a single action choice in its exploration decision. This greediness is the result of the enormous branching factor in most exploration problems, which renders multi-step planning intractable. However, the choice of the right exploration action required some thought.

The algorithms in this chapter consider moving to any point in the robot's local coordinate system as an exploration action. Thus, the basic action unit considered here goes significantly beyond the basic robot control action as defined in Chapter 5 of this book. It is this definition of action that makes seemingly simple exploration techniques applicable to complex multi-step robot exploration problems.

It was noted that exploration can also be formulated as a general POMDP problem, using a payoff function that rewards information gain. However, POMDPs are best when the branching factor is small, and the number of possible observations is limited. Exploration problems are characterized by huge state and observation spaces. Hence they are best solved through greedy techniques that directly maximize information gain.

17.7 Bibliographical Remarks

Exploration has been a primary application domain in robotics systems development, with applications as far reaching as volcano exploration (Bares and Wettergreen 1999; Wettergreen et al. 1999), planetary/lunar exploration (Gat et al. 1994; Höllerer et al. 1999; Krotkov et al. 1999; Matthies et al. 1995; Li et al. 2000), search and rescue (Murphy 2004), abandoned mine mapping (Madhavan et al. 1999; Thrun et al. 2004c; Baker et al. 2004), meteorite search in Antarctica (Urmson et al. 2001; Apostolopoulos et al. 2001), desert exploration (Bapna et al. 1998) and underwater exploration (Ballard 1987; Sandwell 1997; Smith and Dunn 1995; Whitcomb 2000).

The literature on algorithm design for robotic exploration draws its root in the various fields of information gathering and decision theory referenced in the two previous chapters. One of the early approaches to robot exploration is the algorithm described by Kuipers and Byun (1990) and Kuipers and Byun (1991), see also Pierce and Kuipers (1994). In this approach, the robot identifies so-called locally distinguishable places that allow it to distinguish between already visited and so far unexplored areas. A similar seminal paper is by Dudek et al. (1991), who developed an exploring strategy for exploring an unknown graph-like environment. Their algorithm does not consider distance metrics and is specifically designed for robots with very limited perceptual capabilities.

An early exploration technique for learning topological maps with mobile robots was proposed by Koenig and Simmons (1993). The idea of actively exploring for occupancy grid mapping using dynamic programming goes back to Moravec (1988) and Thrun (1993). Tailor and Kriegman (1993) described an approach to visit all landmarks in an environment for learning a feature-based map. In this system the robot maintains a list of all unvisited landmarks in the environment. The idea of information maximization for exploration using a statistical formulation can be found in Kaelbling et al. (1996). Yamauchi et al. (1999) introduced the frontier-based approach to mobile robot exploration, specifically seeking out the frontiers between the explored and the unexplored for directing a robot's actions. More recently, González-Baños and Latombe (2001) proposed an exploration strategy that considers the amount of unseen area that might be visible to the robot from possible view points, to determine the next action. Similar exploration strategies have also become popular in the area of 3-D object modeling. For example, Whaite and Ferrie (1997) study the problem of scanning objects and consider the uncertainty in the parameters of a model to determine the next best view-point.

The exploration approach has also been extended to teams of collaboratively exploring robots. Burgard et al. (2000), and Simmons et al. (2000a) extended the greedy exploration framework to robot teams who jointly explore and seek to maximize their map information; see also Burgard et al. (2004). This approach is similar to the incremental deployment technique introduced by Howard et al. (2002) as well as to the algorithm proposed by Stroupe (2004). Market-based techniques for coordinated exploration have been investigated by Zlot et al. (2002). Dias et al. (2004) analyzed potential failures during multi-robot coordination, and provided an improved algorithm. An approach to deal with heterogeneous teams has been presented by Singh and Fujimura (1993). An extension to coordinated exploration from multiple, unknown start locations was introduced by Ko et al. (2003) and tested thoroughly in Konolige et al. (2005). The approach uses a structural estimate of the environment along with a modified version of MCL localization to estimate the relative robot locations (Fox et al. 2005). In Rekleitis et al. (2001b), the authors propose an exploration technique in which one robot observes another exploring, thereby reducing its location uncertainty. Some of the multi-robot exploration experiments presented in this chapter were first published by Thrun (2001).

COVERAGE PROBLEM   In some papers, the map exploration problem has been studied as a *coverage problem,* which addresses the problem of algorithm design for exhaustively covering an unknown environment. A recent paper by Choset (2001) provides a comprehensive survey into this field. Newer techniques (Acar and Choset 2002; Acar et al. 2003) have approached this problem from a statistical technique, with algorithms not dissimilar from the ones discussed here.

Within the context of SLAM, several authors have devised exploration techniques that can jointly optimize for map coverage and active localization. Makarenko et al. (2002) describe an approach that determines the actions to be carried out based on expected information gain obtained by re-observing landmarks (to more accurately determine their location or the pose of the robot) and by exploring unknown terrain. In a similar vein, Newman et al. (2003) describe an exploration approach in the context of the Atlas (Bosse et al. 2003) framework for efficient SLAM. Here the robot builds a graph-structure to represent visited areas. Sim et al. (2004) specifically addresses the problem of trajectory planning in SLAM. He considers a parameterized class of spiral trajectory policies in the context of an EKF-based approach to the SLAM problem. The FastSLAM exploration technique described in this chapter is due to Stachniss and Burgard (2003, 2004). A technique for SLAM exploration whereby a robot drops markers to aid the localization problem is described in Batalin and Sukhatme (2003).

The performance analysis of robotic exploration strategies has also been the subject of considerable interest. Several authors provided mathematical or empirical analyses of the complexity of different exploration strategies (Albers and Henzinger 2000; Deng and Papadimitriou 1998; Koenig and Tovey 2003; Lumelsky et al. 1990; Rao et al. 1993). For example, Lee and Recce (1997) presented an experimental study in which they compare the performance of different exploration strategies for single robots.

Our technique for active localization in mobile robotics was first published in Burgard et al. (1997) and Fox et al. (1998). More recently, Jensfelt and Christensen (2001a) presented a system that uses a mixture of Gaussians to represent the posterior about the pose of the robot and describe how to perform active localization given this representation. The problem of active localization has also been studied theoretically, for example, by Kleinberg (1994) under the assumptions of perfect sensors.

Several authors have developed robotic exploration strategies for dynamic environments. Of specific interest have been pursuit evasion games, as discussed in the rich literature on differential games (Isaacs 1965; Bardi et al. 1999). Techniques for pursuit evasion in indoor mobile robotics are due to LaValle et al. (1997) and Guibas et al. (1999), and have recently been extended by Gerkey et al. (2004).

Finally, exploration has been intensively studied in automata theory. The sequential decision making paradigm in which a learner receives payoff as it experiments has originally been studied in the context of simple finite state automata known as bandits, see Feldman (1962); Berry and Fristedt (1985); Robbins (1952). Techniques for learning the structure of finite state automata go back to Rivest and Schapire (1987a,b) and Mozer and Bachrach (1989), who developed techniques for generating sequences of tests that distinguish different states in an FSA. State-based bounds on the complexity of exploring deterministic environments have been derived by Koenig and Simmons (1993) and Thrun (1992), which were later extended to stochastic environments by Kearns and Singh (2003).

17.8 Exercises

1. Consider a robot that operates in the triangular environment with three types of landmarks:

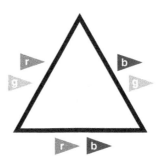

    Each location has two different landmarks, each with a different color. Let us assume that in every round the robot can only inquire about the presence of one landmark type: either the one labeled "r", the one labeled "g", or the one labeled "b". Suppose the robot first fires the detector for "b" landmarks and moves clockwise to the next arc. What would be the optimal landmark detector to use next? How would the answer change if it does not move or if it moved counterclockwise to the next arc?

2. Suppose you are given K omnidirectional robots, which for the sake of this question can move and sense in all directions at all times. In this question, we would like to see each visible location once; we do not care about the benefits of seeing locations more than once.

    The text noted that the use of multiple robots can speed-up exploration by more than just a unitary factor (meaning: K robots can be more than K times faster than 1 robot).

    (a) How much faster can a team of K robots be when compared to a single exploring robot?

    (b) Provide an example environment that maximizes the speed-up for $K = 4$ robots, and discuss the exploration strategy that leads to this speed-up.

3. Suppose you are chasing a moving intruder through a known, bounded environment. Can you draw an environment where K robots can succeed in finding the intruder in finite time, but $K - 1$ robots cannot? Draw such

an environment for $K = 2$, $K = 3$, and $K = 4$ robots. Notice that your result may make no assumption on the motion strategy of the intruder, other than: if she is in your field of view, you see her.

BANDIT PROBLEM

4. A very simplistic exploration problem is known as the *K-arm bandit problem*. In it you face a slot machine with K arms. Each arm provides a payoff of $1 with p_K probability, where p_K is fixed but unknown to you. Your task is to select arms to play such that your overall payoff is maximal.

   (a) Prove that the greedy exploration strategy can be suboptimal, where "greedy" is defined through the optimal choice of action relative to a maximum likelihood estimator of the probabilities p_k. (After n plays of arm k, the maximum likelihood estimate of p_k is given by n_k/n, where n_k is the number of times you received the $1 payoff.)

   (b) Prove that an optimal exploration strategy may never abandon any of the arms.

   (c) Implement the K-arm bandit for $K = 2$, with both probabilities p_1 and p_2 chosen uniformly from the interval $[0;1]$. Implement as good an exploration strategy as you can find. Your exploration strategy may only depend on the variables n_i for $i = 1, 2$. Explain your strategy, and measure your overall return for 1,000 games, each lasting 100 steps.

5. In Chapter 17.4, we empirically compared two different ways of calculating the information gain for a grid cell: entropy and expected entropy gain. Provide a mathematical bound for the error between these two quantities under the assumptions stated in the chapter. For which map occupancy value(s) will this error be maximal? For which value(s) will it be minimal?

6. In the text we encountered an expression for the entropy of a Gaussian. We would like you to compute the expected information gain for a simple Gaussian update. Suppose we are estimating an unknown state variable x, and our present maximum likelihood estimates are μ and Σ. Let us further assume our sensor can measure x, but the measurement will be corrupted by Gaussian noise with covariance Q. Provide an expression for the expected knowledge gain through taking a sensor measurement. *Hint: Focus on the covariance, not the mean.*

Bibliography

Aberdeen, D. 2002. A survey of approximate methods for solving partially observable Markov decision processes. Technical report, Australia National University.

Acar, E.U., and H. Choset. 2002. Sensor-based coverage of unknown environments. *International Journal of Robotic Research* 21:345–366.

Acar, E.U., H. Choset, Y. Zhang, and M.J. Schervish. 2003. Path planning for robotic demining: Robust sensor-based coverage of unstructured environments and probabilistic methods. *International Journal of Robotic Research* 22:441–466.

Albers, S., and M.R. Henzinger. 2000. Exploring unknown environments. *SIAM Journal on Computing* 29:1164–1188.

Anguelov, D., R. Biswas, D. Koller, B. Limketkai, S. Sanner, and S. Thrun. 2002. Learning hierarchical object maps of non-stationary environments with mobile robots. In *Proceedings of the 17th Annual Conference on Uncertainty in AI (UAI)*.

Anguelov, D., D. Koller, E. Parker, and S. Thrun. 2004. Detecting and modeling doors with mobile robots. In *Proceedings of the International Conference on Robotics and Automation (ICRA)*.

Apostolopoulos, D., L. Pedersen, B. Shamah, K. Shillcutt, M.D. Wagner, and W.R. Whittaker. 2001. Robotic antarctic meteorite search: Outcomes. In *Proceedings of the International Conference on Robotics and Automation (ICRA)*, pp. 4174–4179.

Araneda, A. 2003. Statistical inference in mapping and localization for a mobile robot. In J. M. Bernardo, M.J. Bayarri, J.O. Berger, A. P. Dawid, D. Heckerman, A.F.M. Smith, and M. West (eds.), *Bayesian Statistics 7*. Oxford, UK: Oxford University Press.

Arkin, R. 1998. *Behavior-Based Robotics*. Cambridge, MA: MIT Press.

Arras, K.O., and S.J Vestli. 1998. Hybrid, high-precision localisation for the mail distributing mobile robot system MOPS. In *Proceedings of the International Conference on Robotics and Automation (ICRA)*.

Astrom, K.J. 1965. Optimal control of Markov decision processes with incomplete state estimation. *Journal of Mathematical Analysis and Applications* 10:174–205.

Austin, D.J., and P. Jensfelt. 2000. Using multiple Gaussian hypotheses to represent probability-distributions for mobile robots. In *Proceedings of the IEEE International Conference on Robotics and Automation (ICRA)*.

Avots, D., E. Lim, R. Thibaux, and S. Thrun. 2002. A probabilistic technique for simultaneous localization and door state estimation with mobile robots in dynamic environments. In *Proceedings of the IEEE/RSJ Int. Conf. on Intelligent Robots and Systems (IROS)*.

B, Triggs, McLauchlan P, Hartley R, and Fitzgibbon A. 2000. Bundle adjustment – A modern synthesis. In W. Triggs, A. Zisserman, and R. Szeliski (eds.), *Vision Algorithms: Theory and Practice*, LNCS, pp. 298–375. Springer Verlag.

Bagnell, J., and J. Schneider. 2001. Autonomous helicopter control using reinforcement learning policy search methods. In *Proceedings of the International Conference on Robotics and Automation (ICRA)*.

Bailey, T. 2002. *Mobile Robot Localisation and Mapping in Extensive Outdoor Environments*. PhD thesis, University of Sydney, Sydney, NSW, Australia.

Baker, C., A. Morris, D. Ferguson, S. Thayer, C. Whittaker, Z. Omohundro, C. Reverte, W. Whittaker, D. Hähnel, and S. Thrun. 2004. A campaign in autonomous mine mapping. In *Proceedings of the International Conference on Robotics and Automation (ICRA)*.

Ballard, R.D. 1987. *The Discovery of the Titanic*. New York, NY: Warner/Madison Press.

Bapna, D., E. Rollins, J. Murphy, M. Maimone, W.L. Whittaker, and D. Wettergreen. 1998. The Atacama Desert trek: Outcomes. In *Proceedings of the International Conference on Robotics and Automation (ICRA)*, volume 1, pp. 597–604.

Bar-Shalom, Y., and T.E. Fortmann. 1988. *Tracking and Data Association*. Academic Press.

Bar-Shalom, Y., and X.-R. Li. 1998. *Estimation and Tracking: Principles, Techniques, and Software*. Danvers, MA: YBS.

Bardi, M., Parthasarathym T., and T.E.S. Raghavan. 1999. *Stochastic and Differential Games: Theory and Numerical Methods*. Boston: Birkhauser.

Bares, J., and D. Wettergreen. 1999. Dante II: Technical description, results and lessons learned. *International Journal of Robotics Research* 18:621–649.

Barniv, Y. 1990. Dynamic programming algorithm for detecting dim moving targets. In Y. Bar-Shalom (ed.), *Multitarget-Multisensor Tracking: Advanced Applications*, pp. 85–154. Boston: Artech House.

Barto, A.G., S.J. Bradtke, and S.P. Singh. 1991. Real-time learning and control using asynchronous dynamic programming. Technical Report COINS 91-57, Department of Computer Science, University of Massachusetts, MA.

Batalin, M., and G. Sukhatme. 2003. Efficient exploration without localization. In *Proceedings of the International Conference on Robotics and Automation (ICRA)*.

Baxter, J., L. Weaver, and P. Bartlett. 2001. Infinite-horizon gradient-based policy search: II. Gradient ascent algorithms and experiments. *Journal of Artificial Intelligence Research*. To appear.

Bekker, G. 1956. *Theory of Land Locomotion*. University of Michigan.

Bekker, G. 1969. *Introduction to Terrain-Vehicle Systems*. University of Michigan.

Bellman, R.E. 1957. *Dynamic Programming*. Princeton, NJ: Princeton University Press.

Berry, D., and B. Fristedt. 1985. *Bandit Problems: Sequential Allocation of Experiments*. Chapman and Hall.

Bertsekas, Dimitri P., and John N. Tsitsiklis. 1996. *Neuro-Dynamic Programming*. Belmont, MA: Athena Scientific.

Besl, P., and N. McKay. 1992. A method for registration of 3d shapes. *Transactions on Pattern Analysis and Machine Intelligence* 14:239–256.

Betgé-Brezetz, S., R. Chatila, and M. Devy. 1995. Object-based modelling and localization in natural environments. In *Proceedings of the International Conference on Robotics and Automation (ICRA)*.

Betgé-Brezetz, S., P. Hébert, R. Chatila, and M. Devy. 1996. Uncertain map making in natural environments. In *Proceedings of the IEEE International Conference on Robotics and Automation (ICRA)*, Minneapolis.

Betke, M., and K. Gurvits. 1994. Mobile robot localization using landmarks. In *Proceedings of the IEEE International Conference on Robotics and Automation (ICRA)*, pp. 135–4142.

Biswas, R., B. Limketkai, S. Sanner, and S. Thrun. 2002. Towards object mapping in dynamic environments with mobile robots. In *Proceedings of the IEEE/RSJ Int. Conf. on Intelligent Robots and Systems (IROS)*.

Blackwell, D. 1947. Conditional expectation and unbiased sequential estimation. *Annals of Mathematical Statistics* 18:105–110.

Blahut, R.E., W. Miller, and C.H. Wilcox. 1991. *Radar and Sonar: Parts I&II*. New York, NY: Springer-Verlag.

Borenstein, J., B. Everett, and L. Feng. 1996. *Navigating Mobile Robots: Systems and Techniques*. Wellesley, MA: A.K. Peters, Ltd.

Borenstein, J., and Y. Koren. 1991. The vector field histogram – fast obstacle avoidance for mobile robots. *IEEE Transactions on Robotics and Automation* 7:278–288.

Bosse, M., P. Newman, J. Leonard, M. Soika, W. Feiten, and S. Teller. 2004. Simultaneous localization and map building in large-scale cyclic environments using the atlas framework. *International Journal of Robotics Research* 23:1113–1139.

Bosse, M., P. Newman, M. Soika, W. Feiten, J. Leonard, and S. Teller. 2003. An atlas framework for scalable mapping. In *Proceedings of the International Conference on Robotics and Automation (ICRA)*.

Bouguet, J.-Y., and P. Perona. 1995. Visual navigation using a single camera. In *Proceedings of the International Conference on Computer Vision (ICCV)*, pp. 645–652.

Boutilier, C., R. Brafman, and C. Geib. 1998. Structured reachability analysis for Markov decision processes. In *Proceedings of the Conference on Uncertainty in AI (UAI)*, pp. 24–32.

Brafman, R.I. 1997. A heuristic variable grid solution method for POMDPs. In *Proceedings of the AAAI National Conference on Artificial Intelligence*.

Brooks, R.A. 1986. A robust layered control system for a mobile robot. *IEEE Journal of Robotics and Automation* 2:14–23.

Brooks, R.A. 1990. Elephants don't play chess. *Autonomous Robots* 6:3–15.

Brooks, R.A., and T. Lozano-Perez. 1985. A subdivision algorithm in configuration space for findpath with rotation. *IEEE Transactions on Systems, Man, and Cybernetics* 15:224–233.

Bryson, A.E., and H. Yu-Chi. 1975. *Applied Optimal Control*. Halsted Press, John Wiley & Sons.

Bulata, H., and M. Devy. 1996. Incremental construction of a landmark-based and topological model of indoor environments by a mobile robot. In *Proceedings of the International Conference on Robotics and Automation (ICRA)*, Minneapolis, USA.

Burgard, W., A.B. Cremers, D. Fox, D. Hähnel, G. Lakemeyer, D. Schulz, W. Steiner, and S. Thrun. 1999a. Experiences with an interactive museum tour-guide robot. *Artificial Intelligence* 114:3–55.

Burgard, W., A. Derr, D. Fox, and A.B. Cremers. 1998. Integrating global position estimation and position tracking for mobile robots: the Dynamic Markov Localization approach. In *Proceedings of the IEEE/RSJ Int. Conf. on Intelligent Robots and Systems (IROS)*.

Burgard, W., D. Fox, D. Hennig, and T. Schmidt. 1996. Estimating the absolute position of a mobile robot using position probability grids. In *Proceedings of the National Conference on Artificial Intelligence (AAAI)*.

Burgard, W., D. Fox, H. Jans, C. Matenar, and S. Thrun. 1999b. Sonar-based mapping of large-scale mobile robot environments using EM. In *Proceedings of the International Conference on Machine Learning*, Bled, Slovenia.

Burgard, W., D. Fox, M. Moors, R.G. Simmons, and S. Thrun. 2000. Collaborative multi-robot exploration. In *Proceedings of the International Conference on Robotics and Automation (ICRA)*.

Burgard, W., D. Fox, and S. Thrun. 1997. Active mobile robot localization. In *Proceedings of the Fourteenth International Joint Conference on Artificial Intelligence (IJCAI)*, San Mateo, CA. Morgan Kaufmann.

Burgard, W., M. Moors, C. Stachniss, and F. Schneider. 2004. Coordinated multi-robot exploration. *IEEE Transactions on Robotics and Automation*. To appear.

Canny, J. 1987. *The Complexity of Robot Motion Planning*. Cambridge, MA: MIT Press.

Casella, G.C., and R.L. Berger. 1990. *Statistical Inference*. Pacific Grove, CA: Wadsworth & Brooks.

Cassandra, A.R., L.P. Kaelbling, and M.L. Littman. 1994. Acting optimally in partially observable stochastic domains. In *Proceedings of the AAAI National Conference on Artificial Intelligence*, pp. 1023–1028.

Cassandra, A., M. Littman, and N. Zhang. 1997. Incremental pruning: A simple, fast, exact method for partially observable Markov decision processes. In *Proceedings of the Conference on Uncertainty in AI (UAI)*.

Castellanos, J.A., J.M.M. Montiel, J. Neira, and J.D. Tardós. 1999. The SPmap: A probabilistic framework for simultaneous localization and map building. *IEEE Transactions on Robotics and Automation* 15:948–953.

Castellanos, J.A., J. Neira, and J.D. Tardós. 2001. Multisensor fusion for simultaneous localization and map building. *IEEE Transactions on Robotics and Automation* 17: 908–914.

Castellanos, J.A., J. Neira, and J.D. Tardós. 2004. Limits to the consistency of the EKF-based SLAM. In M.I. Ribeiro and J. Santos-Victor (eds.), *Proceedings of Intelligent Autonomous Vehicles (IAV-2004)*, Lisboa, PT. IFAC/EURON and IFAC/Elsevier.

Chatila, R., and J.-P. Laumond. 1985. Position referencing and consistent world modeling for mobile robots. In *Proceedings of the International Conference on Robotics and Automation (ICRA)*, pp. 138–145.

Cheeseman, P., and P. Smith. 1986. On the representation and estimation of spatial uncertainty. *International Journal of Robotics* 5:56–68.

Choset, H. 1996. *Sensor Based Motion Planning: The Hierarchical Generalized Voronoi Graph*. PhD thesis, California Institute of Technology.

Choset, H. 2001. Coverage for robotics — a survey of recent results. *Annals of Mathematical Artificial Intelligence* 31:113–126.

Choset, H., K. Lynch, S. Hutchinson, G. Kantor, W. Burgard, L. Kavraki, and S. Thrun. 2004. *Principles of Robotic Motion: Theory, Algorithms, and Implementation*. Cambridge, MA: MIT Press.

Chown, E., S. Kaplan, and D. Kortenkamp. 1995. Prototypes, location, and associative networks (plan): Towards a unified theory of cognitive mapping. *Cognitive Science* 19:1–51.

Chrisman, L. 1992. Reinforcement learning with perceptual aliasing: The perceptual distinction approach. In *Proceedings of 1992 AAAI Conference*, Menlo Park, CA. AAAI Press / The MIT Press.

Cid, R.M., C. Parra, and M. Devy. 2002. Visual navigation in natural environments: from range and color data to a landmark-based model. *Autonomous Robots* 13:143–168.

Cohn, D. 1994. Queries and exploration using optimal experiment design. In J.D. Cowan, G. Tesauro, and J. Alspector (eds.), *Advances in Neural Information Processing Systems 6*, San Mateo, CA. Morgan Kaufmann.

Connell, J. 1990. *Minimalist Mobile Robotics*. Boston: Academic Press.

Coppersmith, D., and S. Winograd. 1990. Matrix multiplication via arithmetic progressions. *Journal of Symbolic Computation* 9:251–280.

Cover, T.M., and J.A. Thomas. 1991. *Elements of Information Theory*. Wiley.

Cowell, R.G., A.P. Dawid, S.L. Lauritzen, and D.J. Spiegelhalter. 1999. *Probabilistic Networks and Expert Systems*. Berlin, New York: Springer Verlag.

Cox, I.J. 1991. Blanche—an experiment in guidance and navigation of an autonomous robot vehicle. *IEEE Transactions on Robotics and Automation* 7:193–204.

Cox, I.J., and J.J. Leonard. 1994. Modeling a dynamic environment using a Bayesian multiple hypothesis approach. *Artificial Intelligence* 66:311–344.

Cox, I.J., and G.T. Wilfong (eds.). 1990. *Autonomous Robot Vehicles*. Springer Verlag.

Craig, J.J. 1989. *Introduction to Robotics: Mechanics and Control (2nd Edition)*. Reading, MA: Addison-Wesley Publishing, Inc. 3rd edition.

Crowley, J. 1989. World modeling and position estimation for a mobile robot using ultrasonic ranging. In *Proceedings of the International Conference on Robotics and Automation (ICRA)*, pp. 674–680.

Csorba, M. 1997. *Simultaneous Localisation and Map Building*. PhD thesis, University of Oxford.

Davison, A. 1998. *Mobile Robot Navigation Using Active Vision*. PhD thesis, University of Oxford, Oxford, UK.

Davison, A. 2003. Real time simultaneous localisation and mapping with a single camera. In *Proceedings of the International Conference on Computer Vision (ICCV)*, Nice, France.

Dean, L.P. Kaelbling, J. Kirman, and A. Nicholson. 1995. Planning under time constraints in stochastic domains. *Artificial Intelligence* 76:35–74.

Deans, M., and M. Hebert. 2000. Invariant filtering for simultaneous localization and mapping. In *Proceedings of the International Conference on Robotics and Automation (ICRA)*, pp. 1042–1047.

Deans, M.C., and M. Hebert. 2002. Experimental comparison of techniques for localization and mapping using a bearing-only sensor. In *Proceedings of the International Symposium on Experimental Robotics (ISER)*, Sant'Angelo d'Ischia, Italy.

Dearden, R., and C. Boutilier. 1994. Integrating planning and execution in stochastic domains. In *Proceedings of the AAAI Spring Symposium on Decision Theoretic Planning*, pp. 55–61, Stanford, CA.

Dedeoglu, G., and G. Sukhatme. 2000. Landmark-based matching algorithm for cooperative mapping by autonomous robots. In *Proceedings of the International Symposium on Distributed Autonomous Robotic Systems (DARS 2000)*, Knoxville, Tenneessee.

DeGroot, Morris H. 1975. *Probability and Statistics*. Reading, MA: Addison-Wesley.

Dellaert, F. 2005. Square root SAM. In S. Thrun, G. Sukhatme, and S. Schaal (eds.), *Proceedings of the Robotics Science and Systems Conference*. Cambridge, MA: MIT Press.

Dellaert, F., D. Fox, W. Burgard, and S. Thrun. 1999. Monte Carlo localization for mobile robots. In *Proceedings of the International Conference on Robotics and Automation (ICRA)*.

Dellaert, F., S.M. Seitz, C. Thorpe, and S. Thrun. 2003. EM, MCMC, and chain flipping for structure from motion with unknown correspondence. *Machine Learning* 50:45–71.

Dempster, A.P., A.N. Laird, and D.B. Rubin. 1977. Maximum likelihood from incomplete data via the EM algorithm. *Journal of the Royal Statistical Society, Series B* 39:1–38.

Deng, X., and C. Papadimitriou. 1998. How to learn in an unknown environment: The rectilinear case. *Journal of the ACM* 45:215–245.

Devroye, L., L. Györfi, and G. Lugosi. 1996. *A Probabilistic Theory of Pattern Recognition*. New York, NY: Springer-Verlag.

Devy, M., and H. Bulata. 1996. Multi-sensory perception and heterogeneous representations for the navigation of a mobile robot in a structured environment. In *Proceedings of the Symposium on Intelligent Robot Systems*, Lisboa.

Devy, M., and C. Parra. 1998. 3-d scene modelling and curve-based localization in natural environments. In *Proceedings of the International Conference on Robotics and Automation (ICRA)*.

Dias, M.B., M. Zinck, R. Zlot, and A. Stentz. 2004. Robust multirobot coordination in dynamic environments. In *Proceedings of the International Conference on Robotics and Automation (ICRA)*.

Dickmanns, E.D. 2002. Vision for ground vehicles: history and prospects. *International Journal of Vehicle Autonomous Systems* 1:1–44.

Dickmanns, E.D., and V. Graefe. 1988. Application of monocular machine vision. *Machine Vision and Applications* 1:241–261.

Diebel, J., K. Reuterswärd, J. Davis, and S. Thrun. 2004. Simultaneous localization and mapping with active stereo vision. In *Proceedings of the IEEE/RSJ Int. Conf. on Intelligent Robots and Systems (IROS)*.

Dietterich, T.G. 2000. Hierarchical reinforcement learning with the MAXQ value function decomposition. *Journal of Artificial Intelligence Research* 13:227–303.

Dissanayake, G., P. Newman, S. Clark, H.F. Durrant-Whyte, and M. Csorba. 2001. A solution to the simultaneous localisation and map building (SLAM) problem. *IEEE Transactions on Robotics and Automation* 17:229–241.

Dissanayake, G., S.B. Williams, H. Durrant-Whyte, and T. Bailey. 2002. Map management for efficient simultaneous localization and mapping (SLAM). *Autonomous Robots* 12:267–286.

Dorf, R.C., and R.H. Bishop. 2001. *Modern Control Systems (Ninth Edition)*. Englewood Cliffs, NJ: Prentice Hall.

Doucet, A. 1998. On sequential simulation-based methods for Bayesian filtering. Technical Report CUED/F-INFENG/TR 310, Cambridge University, Department of Engineering, Cambridge, UK.

Doucet, A., J.F.G. de Freitas, and N.J. Gordon (eds.). 2001. *Sequential Monte Carlo Methods In Practice*. New York: Springer Verlag.

Driankov, D., and A. Saffiotti (eds.). 2001. *Fuzzy Logic Techniques for Autonomous Vehicle Navigation*, volume 61 of *Studies in Fuzziness and Soft Computing*. Berlin, Germany: Springer-Verlag.

Duckett, T., S. Marsland, and J. Shapiro. 2000. Learning globally consistent maps by relaxation. In *Proceedings of the International Conference on Robotics and Automation (ICRA)*, pp. 3841–3846.

Duckett, T., S. Marsland, and J. Shapiro. 2002. Fast, on-line learning of globally consistent maps. *Autonomous Robots* 12:287 – 300.

Duckett, T., and U. Nehmzow. 2001. Mobile robot self-localisation using occupancy histograms and a mixture of Gaussian location hypotheses. *Robotics and Autonomous Systems* 34:119–130.

Duda, R.O., P.E. Hart, and D. Stork. 2000. *Pattern classification and scene analysis (2nd edition)*. New York: John Wiley and Sons.

Dudek, G., and D. Jegessur. 2000. Robust place recognition using local appearance based methods. In *Proceedings of the International Conference on Robotics and Automation (ICRA)*, pp. 466–474.

Dudek, G., and M. Jenkin. 2000. *Computational Principles of Mobile Robotics*. Cambridge CB2 2RU, UK: Cambridge University Press.

Dudek, G., M. Jenkin, E. Milios, and D. Wilkes. 1991. Robotic exploration as graph construction. *IEEE Transactions on Robotics and Automation* 7:859–865.

Durrant-Whyte, H.F. 1988. Uncertain geometry in robotics. *IEEE Transactions on Robotics and Automation* 4:23 – 31.

Durrant-Whyte, H.F. 1996. Autonomous guided vehicle for cargo handling applications. *International Journal of Robotics Research* 15.

Durrant-Whyte, H., S. Majumder, S. Thrun, M. de Battista, and S. Scheding. 2001. A Bayesian algorithm for simultaneous localization and map building. In *Proceedings of the 10th International Symposium of Robotics Research (ISRR'01)*, Lorne, Australia.

Elfes, A. 1987. Sonar-based real-world mapping and navigation. *IEEE Transactions on Robotics and Automation* pp. 249–265.

Eliazar, A., and R. Parr. 2003. DP-SLAM: Fast, robust simultaneous localization and mapping without predetermined landmarks. In *Proceedings of the Sixteenth International Joint Conference on Artificial Intelligence (IJCAI)*, Acapulco, Mexico. IJCAI.

Eliazar, A., and R. Parr. 2004. DP-SLAM 2.0. In *Proceedings of the International Conference on Robotics and Automation (ICRA)*, New Orleans, USA.

Elliott, R.J., L.Aggoun, and J.B. Moore. 1995. *Hidden Markov Models: Estimation and Control*. New York, NY: Springer-Verlag.

Engelson, S., and D. McDermott. 1992. Error correction in mobile robot map learning. In *Proceedings of the International Conference on Robotics and Automation (ICRA)*, pp. 2555–2560.

Etter, P.C. 1996. *Underwater Acoustic Modeling: Principles, Techniques and Applications*. Amsterdam: Elsevier.

Featherstone, R. 1987. *Robot Dynamics Algorithms*. Boston, MA: Kluwer Academic Publishers.

Feder, H.J.S., J.J. Leonard, and C.M. Smith. 1999. Adaptive mobile robot navigation and mapping. *International Journal of Robotics Research* 18:650–668.

Feldman, D. 1962. Contributions to the two-armed bandit problem. *Ann. Math. Statist* 33:847–856.

Feller, W. 1968. *An Introduction To Probability Theory And Its Applications (3rd edition)x*. Quinn-Woodbine.

Feng, L., J. Borenstein, and H.R. Everett. 1994. "Where am I?" Sensors and methods for autonomous mobile robot positioning. Technical Report UM-MEAM-94-12, University of Michigan, Ann Arbor, MI.

Fenwick, J., P. Newman, and J. Leonard. 2002. Collaborative concurrent mapping and localization. In *Proceedings of the International Conference on Robotics and Automation (ICRA)*.

Ferguson, D., T. Stentz, and S. Thrun. 2004. PAO* for planning with hidden state. In *Proceedings of the International Conference on Robotics and Automation (ICRA)*.

Fischler, M.A., and R.C. Bolles. 1981. Random sample consensus: A paradigm for model fitting with applications to image analysis and automated cartography. *Communications of the ACM* 24:381–395.

Folkesson, J., and H.I. Christensen. 2003. Outdoor exploration and SLAM using a compressed filter. In *Proceedings of the IEEE International Conference on Robotics and Automation (ICRA)*, pp. 419–427.

Folkesson, J., and H.I. Christensen. 2004a. Graphical SLAM: A self-correcting map. In *Proceedings of the International Conference on Robotics and Automation (ICRA)*.

Folkesson, J., and H.I. Christensen. 2004b. Robust SLAM. In *Proceedings of the International Symposium on Autonomous Vehicles*, Lisboa, PT.

Fox, D. 2003. Adapting the sample size in particle filters through KLD-sampling. *International Journal of Robotics Research* 22:985 – 1003.

Fox, D., W. Burgard, F. Dellaert, and S. Thrun. 1999a. Monte Carlo localization: Efficient position estimation for mobile robots. In *Proceedings of the National Conference on Artificial Intelligence (AAAI)*, Orlando, FL. AAAI.

Fox, D., W. Burgard, H. Kruppa, and S. Thrun. 2000. A probabilistic approach to collaborative multi-robot localization. *Autonomous Robots* 8.

Fox, D., W. Burgard, and S. Thrun. 1998. Active Markov localization for mobile robots. *Robotics and Autonomous Systems* 25:195–207.

Fox, D., W. Burgard, and S. Thrun. 1999b. Markov localization for mobile robots in dynamic environments. *Journal of Artificial Intelligence Research (JAIR)* 11:391–427.

Fox, D., W. Burgard, and S. Thrun. 1999c. Markov localization for mobile robots in dynamic environments. *Journal of Artificial Intelligence Research* 11:391–427.

Fox, D., J. Ko, K. Konolige, and B. Stewart. 2005. A hierarchical Bayesian approach to mobile robot map structure learning. In P. Dario and R. Chatila (eds.), *Robotics Research: The Eleventh International Symposium*, Springer Tracts in Advanced Robotics (STAR). Springer Verlag.

Freedman, D., and P. Diaconis. 1981. On this histogram as a density estimator: L_2 theory. *Zeitschrift für Wahrscheinlichkeitstheorie und verwandte Gebiete* 57:453–476.

Frese, U. 2004. *An $O(logn)$ Algorithm for Simultaneous Localization and Mapping of Mobile Robots in Indoor Environments*. PhD thesis, University of Erlangen-Nürnberg, Germany.

Frese, U., and G. Hirzinger. 2001. Simultaneous localization and mapping—a discussion. In *Proceedings of the IJCAI Workshop on Reasoning with Uncertainty in Robotics*, pp. 17–26, Seattle, WA.

Frese, U., P. Larsson, and T. Duckett. 2005. A multigrid algorithm for simultaneous localization and mapping. *IEEE Transactions on Robotics.* To appear.

Frueh, C., and A. Zakhor. 2003. Constructing 3d city models by merging ground-based and airborne views. In *Proceedings of the IEEE Computer Society Conference on Computer Vision and Pattern Recognition (CVPR)*, Madison, Wisconsin.

Gat, E. 1998. Three-layered architectures. In D. Kortenkamp, R.P. Bonasso, and R. Murphy (eds.), *AI-based Mobile Robots: Case Studies of Successful Robot Systems*, pp. 195–210. Cambridge, MA: MIT Press.

Gat, E., R. Desai, R. Ivlev, J. Loch, and D.P. Miller. 1994. Behavior control for robotic exploration of planetary surfaces. *IEEE Transactions on Robotics and Automation* 10: 490–503.

Gauss, K.F. 1809. *Theoria Motus Corporum Coelestium (Theory of the Motion of the Heavenly Bodies Moving about the Sun in Conic Sections).* Republished in 1857, and by Dover in 1963: Little, Brown, and Co.

Geffner, H., and B. Bonet. 1998. Solving large POMDPs by real time dynamic programming. In *Working Notes Fall AAAI Symposium on POMDPs*, Stanford, CA.

Gerkey, B., S. Thrun, and G. Gordon. 2004. Parallel stochastic hill-climbing with small teams. In L. Parker, F. Schneider, and A. Schultz (eds.), *Proceedings of the 3rd International Workshop on Multi-Robot Systems*, Amsterdam. NRL, Kluwer Publisher.

Gilks, W.R., S. Richardson, and D.J. Spiegelhalter (eds.). 1996. *Markov Chain Monte Carlo in Practice.* Chapman and Hall/CRC.

Goldberg, K. 1993. Orienting polygonal parts without sensors. *Algorithmica* 10:201–225.

Golfarelli, M., D. Maio, and S. Rizzi. 1998. Elastic correction of dead-reckoning errors in map building. In *Proceedings of the IEEE/RSJ Int. Conf. on Intelligent Robots and Systems (IROS)*, pp. 905–911.

Golub, G.H., and C.F. Van Loan. 1986. *Matrix Computations.* North Oxford Academic.

González-Baños, H.H., and J.C. Latombe. 2001. Navigation strategies for exploring indoor environments. *International Journal of Robotics Research.*

Gordon, G. J. 1995. Stable function approximation in dynamic programming. In A. Prieditis and S. Russell (eds.), *Proceedings of the Twelfth International Conference on Machine Learning.* Also appeared as Technical Report CMU-CS-95-103, Carnegie Mellon University, School of Computer Science, Pittsburgh, PA.

Greiner, R., and R. Isukapalli. 1994. Learning to select useful landmarks. In *Proceedings of 1994 AAAI Conference*, pp. 1251–1256, Menlo Park, CA. AAAI Press / The MIT Press.

Grunbaum, F.A., M.Bernfeld, and R.E. Blahut (eds.). 1992. *Radar and Sonar: Part II.* New York, NY: Springer-Verlag.

Guibas, L.J., D.E. Knuth, and M. Sharir. 1992. Randomized incremental construction of Delaunay and Voronoi diagrams. *Algorithmica* 7:381–413. See also *17th Int. Coll. on Automata, Languages and Programming*, 1990, pp. 414–431.

Guibas, L.J., J.-C. Latombe, S.M. LaValle, D. Lin, and R. Motwani. 1999. A visibility-based pursuit-evasion problem. *International Journal of Computational Geometry and Applications* 9:471–493.

Guivant, J., and E. Nebot. 2001. Optimization of the simultaneous localization and map building algorithm for real time implementation. *IEEE Transactions on Robotics and Automation* 17:242–257. In press.

Guivant, J., and E. Nebot. 2002. Improving computational and memory requirements of simultaneous localization and map building algorithms. In *Proceedings of the International Conference on Robotics and Automation (ICRA)*, pp. 2731–2736.

Guivant, J., E. Nebot, and S. Baiker. 2000. Autonomous navigation and map building using laser range sensors in outdoor applications. *Journal of Robotics Systems* 17: 565–583.

Guivant, J.E., E.M. Nebot, J. Nieto, and F. Masson. 2004. Navigation and mapping in large unstructured environments. *International Journal of Robotics Research* 23.

Gutmann, J.S., and D. Fox. 2002. An experimental comparison of localization methods continued. In *Proc. of the IEEE/RSJ International Conference on Intelligent Robots and Systems (IROS)*.

Gutmann, J.-S., W. Burgard, D. Fox, and K. Konolige. 1998. An experimental comparison of localization methods. In *Proceedings of the IEEE/RSJ Int. Conf. on Intelligent Robots and Systems (IROS)*.

Gutmann, J.-S., and K. Konolige. 2000. Incremental mapping of large cyclic environments. In *Proceedings of the IEEE International Symposium on Computational Intelligence in Robotics and Automation (CIRA)*.

Gutmann, J.-S., and B. Nebel. 1997. Navigation mobiler roboter mit laserscans. In *Autonome Mobile Systeme*. Berlin: Springer Verlag. In German.

Gutmann, J.-S., and C. Schlegel. 1996. AMOS: Comparison of scan matching approaches for self-localization in indoor environments. In *Proc. of the 1st Euromicro Workshop on Advanced Mobile Robots*. IEEE Computer Society Press.

Hähnel, D., W. Burgard, B. Wegbreit, and S. Thrun. 2003a. Towards lazy data association in SLAM. In *Proceedings of the 11th International Symposium of Robotics Research (ISRR'03)*, Sienna, Italy. Springer.

Hähnel, D., D. Fox, W. Burgard, and S. Thrun. 2003b. A highly efficient FastSLAM algorithm for generating cyclic maps of large-scale environments from raw laser range measurements. In *Proceedings of the IEEE/RSJ Int. Conf. on Intelligent Robots and Systems (IROS)*.

Hähnel, D., D. Schulz, and W. Burgard. 2003c. Mobile robot mapping in populated environments. *Autonomous Robots* 17:579–598.

Hartley, R., and A. Zisserman. 2000. *Multiple View Geometry in Computer Vision*. Cambridge University Press.

Hauskrecht, M. 1997. Incremental methods for computing bounds in partially observable Markov decision processes. In *Proceedings of the AAAI National Conference on Artificial Intelligence*, pp. 734–739, Providence, RI.

Hauskrecht, M. 2000. Value-function approximations for partially observable Markov decision processes. *Journal of Artificial Intelligence Research* 13:33–94.

Hayet, J.B., F. Lerasle, and M. Devy. 2002. A visual landmark framework for indoor mobile robot navigation. In *Proceedings of the International Conference on Robotics and Automation (ICRA)*, Washington, DC.

Hertzberg, J., and F. Kirchner. 1996. Landmark-based autonomous navigation in sewerage pipes. In *Proc. of the First Euromicro Workshop on Advanced Mobile Robots*.

Hinkel, R., and T. Knieriemen. 1988. Environment perception with a laser radar in a fast moving robot. In *Proceedings of Symposium on Robot Control*, pp. 68.1–68.7, Karlsruhe, Germany.

Hoey, J., R. St-Aubin, A. Hu, and C. Boutilier. 1999. SPUDD: Stochastic planning using decision diagrams. In *Proceedings of the Conference on Uncertainty in AI (UAI)*, pp. 279–288.

Höllerer, T., S. Feiner, T. Terauchi, G. Rashid, and D. Hallaway. 1999. Exploring MARS: Developing indoor and outdoor user interfaces to a mobile augmented reality system. *Computers and Graphics* 23:779–785.

Howard, A. 2004. Multi-robot mapping using manifold representations. In *Proceedings of the International Conference on Robotics and Automation (ICRA)*, pp. 4198–4203.

Howard, A., M.J. Matarić, and G.S. Sukhatme. 2002. An incremental deployment algorithm for mobile robot teams. In *Proceedings of the IEEE/RSJ Int. Conf. on Intelligent Robots and Systems (IROS)*.

Howard, A., M.J. Matarić, and G.S. Sukhatme. 2003. Cooperative relative localization for mobile robot teams: An ego-centric approach. In *Proceedings of the Naval Research Laboratory Workshop on Multi-Robot Systems*, Washington, D.C.

Howard, A., L.E. Parker, and G.S. Sukhatme. 2004. The SDR experience: Experiments with a large-scale heterogenous mobile robot team. In *Proceedings of the 9th International Symposium on Experimental Robotics 2004*, Singapore.

Howard, R.A. 1960. *Dynamic Programming and Markov Processes*. MIT Press and Wiley.

Iagnemma, K., and S. Dubowsky. 2004. *Mobile Robots in Rough Terrain: Estimation, Motion Planning, and Control with Application to Planetary Rovers*. Springer.

Ilon, B.E., 1975. Wheels for a course stable selfpropelling vehicle movable in any desired direction on the ground or some other base. United States Patent #3,876,255.

Iocchi, L., K. Konolige, and M. Bajracharya. 2000. Visually realistic mapping of a planar environment with stereo. In *Proceesings of the 2000 International Symposium on Experimental Robotics*, Waikiki, Hawaii.

IRobots Inc., 2004. Roomba robotic floor vac. On the Web at http://www.irobot.com/consumer/.

Isaacs, R. 1965. *Differential Games–A Mathematical Theory with Applications to Warfare and Pursuit, Control and Optimization*. John Wiley and Sons, Inc.

Isard, M., and A. Blake. 1998. CONDENSATION: conditional density propagation for visual tracking. *International Journal of Computer Vision* 29:5–28.

Jaeger, H. 2000. Observable operator processes and conditioned continuation representations. *Neural Computation* 12:1371–1398.

James, M., and S. Singh. 2004. Learning and discovery of predictive state representations in dynamical systems with reset. In *Proceedings of the Twenty-First International Conference on Machine Learning (ICML)*, pp. 417–424.

Jazwinsky, A.M. 1970. *Stochastic Processes and Filtering Theory*. New York: Academic.

Jensfelt, P., D. Austin, O. Wijk, and M. Andersson. 2000. Feature based condensation for mobile robot localization. In *Proceedings of the International Conference on Robotics and Automation (ICRA)*, pp. 2531–2537.

Jensfelt, P., and H.I. Christensen. 2001a. Active global localisation for a mobile robot using multiple hypothesis tracking. *IEEE Transactions on Robotics and Automation* 17:748–760.

Jensfelt, P., and H.I. Christensen. 2001b. Pose tracking using laser scanning and minimalistic environmental models. *IEEE Transactions on Robotics and Automation* 17:138–147.

Jensfelt, P., H.I. Christensen, and G. Zunino. 2002. Integrated systems for mapping and localization. In *Proceedings of the International Conference on Robotics and Automation (ICRA)*.

Julier, S., and J. Uhlmann. 1997. A new extension of the Kalman filter to nonlinear systems. In *International Symposium on Aerospace/Defense Sensing, Simulate and Controls*, Orlando, FL.

Julier, S.J., and J.K. Uhlmann. 2000. Building a million beacon map. In *Proceedings of the SPIE Sensor Fusion and Decentralized Control in Robotic Systems IV, Vol. #4571*.

Jung, I.K., and S. Lacroix. 2003. High resolution terrain mapping using low altitude aerial stereo imagery. In *Proceedings of the International Conference on Computer Vision (ICCV)*, Nice, France.

Kaelbling, L.P., A.R. Cassandra, and J.A. Kurien. 1996. Acting under uncertainty: Discrete Bayesian models for mobile-robot navigation. In *Proceedings of the IEEE/RSJ Int. Conf. on Intelligent Robots and Systems (IROS)*.

Kaelbling, L.P., M.L. Littman, and A.R. Cassandra. 1998. Planning and acting in partially observable stochastic domains. *Artificial Intelligence* 101:99–134.

Kaelbling, L. P., and S. J. Rosenschein. 1991. Action and planning in embedded agents. In *Designing Autonomous Agents*, pp. 35–48. Cambridge, MA: The MIT Press (and Elsevier).

Kalman, R.E. 1960. A new approach to linear filtering and prediction problems. *Trans. ASME, Journal of Basic Engineering* 82:35–45.

Kanazawa, K., D. Koller, and S.J. Russell. 1995. Stochastic simulation algorithms for dynamic probabilistic networks. In *Proceedings of the 11th Annual Conference on Uncertainty in AI*, Montreal, Canada.

Kavraki, L., and J.-C. Latombe. 1994. Randomized preprocessing of configuration space for fast path planning. In *Proceedings of the International Conference on Robotics and Automation (ICRA)*, pp. 2138–2145.

Kavraki, L., P. Svestka, J.-C. Latombe, and M. Overmars. 1996. Probabilistic roadmaps for path planning in high-dimensional configuration spaces. *IEEE Transactions on Robotics and Automation* 12:566–580.

Kearns, M., and S. Singh. 2003. Near-optimal reinforcement learning in polynomial time. *Machine Learning* 49:209–232.

Khatib, O. 1986. Real-time obstacle avoidance for robot manipulator and mobile robots. *The International Journal of Robotics Research* 5:90–98.

Kirk, R.E., and P. Kirk. 1995. *Experimental Design: Procedures for the Behavioral Sciences*. Pacific Grove, CA: Brooks/Cole.

Kitagawa, G. 1996. Monte Carlo filter and smoother for non-Gaussian nonlinear state space models. *Journal of Computational and Graphical Statistics* 5:1–25.

Kleinberg, J. 1994. The localization problem for mobile robots. In *Proc. of the 35th IEEE Symposium on Foundations of Computer Science*.

Ko, J., B. Stewart, D. Fox, K. Konolige, and B. Limketkai. 2003. A practical, decision-theoretic approach to multi-robot mapping and exploration. In *Proc. of the IEEE/RSJ International Conference on Intelligent Robots and Systems (IROS)*, pp. 3232–3238.

Koditschek, D.E. 1987. Exact robot navigation by means of potential functions: Some topological considerations. In *Proceedings of the International Conference on Robotics and Automation (ICRA)*, pp. 1–6.

Koenig, S., and R.G. Simmons. 1993. Exploration with and without a map. In *Proceedings of the AAAI Workshop on Learning Action Models at the Eleventh National Conference on Artificial Intelligence (AAAI)*, pp. 28–32. Also available as AAAI Technical Report WS-93-06.

Koenig, S., and R. Simmons. 1998. A robot navigation architecture based on partially observable Markov decision process models. In Kortenkamp et al. (1998).

Koenig, S., and C. Tovey. 2003. Improved analysis of greedy mapping. In *Proceedings of the IEEE/RSJ Int. Conf. on Intelligent Robots and Systems (IROS)*.

Konecny, G. 2002. *Geoinformation: Remote Sensing, Photogrammetry and Geographical Information Systems*. Taylor & Francis.

Konolige, K. 2004. Large-scale map-making. In *Proceedings of the AAAI National Conference on Artificial Intelligence*, pp. 457–463, San Jose, CA. AAAI.

Konolige, K., and K. Chou. 1999. Markov localization using correlation. In *Proceedings of the International Joint Conference on Artificial Intelligence (IJCAI)*.

Konolige, K., D Fox, C. Ortiz, A. Agno, M. Eriksen, B. Limketkai, J. Ko, B. Morisset, D. Schulz, B. Stewart, and R. Vincent. 2005. Centibots: Very large scale distributed robotic teams. In M. Ang and O. Khatib (eds.), *Experimental Robotics: The 9th International Symposium*, Springer Tracts in Advanced Robotics (STAR). Springer Verlag.

Konolige, K., J.-S. Gutmann, D. Guzzoni, R. Ficklin, and K. Nicewarner. 1999. A mobile robot sense net. In *Proceedings of SPIE 3839 Sensor Fusion and Decentralized Control in Robotic Systmes II*, Boston.

Korf, R.E. 1988. Real-time heuristic search: New results. In *Proceedings of the sixth National Conference on Artificial Intelligence (AAAI-88)*, pp. 139–143, Los Angeles, CA 90024. Computer Science Department, University of California, AAAI Press/MIT Press.

Kortenkamp, D., R.P. Bonasso, and R. Murphy (eds.). 1998. *Artificial Intelligence and Mobile Robots: Case Studies of Successful Robot Systems*. Cambridge, MA: MIT/AAAI Press.

Kortenkamp, D., and T. Weymouth. 1994. Topological mapping for mobile robots using a combination of sonar and vision sensing. In *Proceedings of the Twelfth National Conference on Artificial Intelligence*, pp. 979–984, Menlo Park. AAAI, AAAI Press/MIT Press.

Kröse, B., N. Vlassis, and R. Bunschoten. 2002. Omnidirectional vision for appearance-based robot localization. In G.D. Hagar, H.I. Cristensen, H. Bunke, and R. Klein (eds.), *Sensor Based Intelligent Robots (Lecture Notes in Computer Science #2238)*, pp. 39–50. Springer Verlag.

Krotkov, E., M. Hebert, L. Henriksen, P. Levin, M. Maimone, R.G. Simmons, and J. Teza. 1999. Evolution of a prototype lunar rover: Addition of laser-based hazard detection, and results from field trials in lunar analog terrain. *Autonomous Robots* 7:119–130.

Kuipers, B., and Y.-T. Byun. 1990. A robot exploration and mapping strategy based on a semantic hierarchy of spatial representations. Technical report, Department of Computer Science, University of Texas at Austin, Austin, Texas 78712.

Kuipers, B., and Y.-T. Byun. 1991. A robot exploration and mapping strategy based on a semantic hierarchy of spatial representations. *Robotics and Autonomous Systems* pp. 47–63.

Kuipers, B.J., and T.S. Levitt. 1988. Navigation and mapping in large-scale space. *AI Magazine*.

Kuipers, B., J. Modayil, P. Beeson, M. MacMahon, and F. Savelli. 2004. Local metrical and global topological maps in the hybrid spatial semantic hierarchy. In *Proceedings of the International Conference on Robotics and Automation (ICRA)*.

Kushmerick, N., S. Hanks, and D.S. Weld. 1995. An algorithm for probabilistic planning. *Artificial Intelligence* 76:239–286.

Kwok, C.T., D. Fox, and M. Meilă. 2004. Real-time particle filters. *Proceedings of the IEEE* 92:469–484. Special Issue on Sequential State Estimation.

Latombe, J.-C. 1991. *Robot Motion Planning*. Boston, MA: Kluwer Academic Publishers.

LaValle, S.M., H. Gonzalez-Banos, C. Becker, and J.C. Latombe. 1997. Motion strategies for maintaining visibility of a moving target. In *Proceedings of the International Conference on Robotics and Automation (ICRA)*.

Lawler, E.L., and D.E. Wood. 1966. Branch-and-bound methods: A survey. *Operations Research* 14:699–719.

Lebeltel, O., P. Bessière, J. Diard, and E. Mazer. 2004. Bayesian robot programming. *Autonomous Robots* 16:49–97.

Lee, D., and M. Recce. 1997. Quantitative evaluation of the exploration strategies of a mobile robot. *International Journal of Robotics Research* 16:413–447.

Lenser, S., and M. Veloso. 2000. Sensor resetting localization for poorly modelled mobile robots. In *Proceedings of the International Conference on Robotics and Automation (ICRA)*.

Leonard, J.J., and H.F. Durrant-Whyte. 1991. Mobile robot localization by tracking geometric beacons. *IEEE Transactions on Robotics and Automation* 7:376–382.

Leonard, J.J., and H.J.S. Feder. 1999. A computationally efficient method for large-scale concurrent mapping and localization. In J. Hollerbach and D. Koditschek (eds.), *Proceedings of the Ninth International Symposium on Robotics Research*, Salt Lake City, Utah.

Leonard, J.J., and H.J.S. Feder. 2001. Decoupled stochastic mapping. *IEEE Journal of Ocean Engineering* 26:561–571.

Leonard, J., and P. Newman. 2003. Consistent, convergent, and constant-time SLAM. In *Proceedings of the IJCAI Workshop on Reasoning with Uncertainty in Robot Navigation*, Acapulco, Mexico.

Leonard, J.J., R.J. Rikoski, P.M. Newman, and M. Bosse. 2002a. Mapping partially observable features from multiple uncertain vantage points. *International Journal of Robotics Research* 21:943–975.

Leonard, J., J.D. Tardós, S. Thrun, and H. Choset (eds.). 2002b. *Workshop Notes of the ICRA Workshop on Concurrent Mapping and Localization for Autonomous Mobile Robots (W4)*. Washington, DC: ICRA Conference.

Levenberg, K. 1944. A method for the solution of certain problems in least squares. *Quarterly Applied Mathematics* 2:164–168.

Li, R., F. Ma, F. Xu, L. Matthies, C. Olson, and Y. Xiong. 2000. Large scale mars mapping and rover localization using descent and rover imagery. In *Proceedings of the ISPRS 19th Congress, IAPRS Vol. XXXIII*, Amsterdam.

Likhachev, M., G. Gordon, and S. Thrun. 2004. Planning for Markov decision processes with sparse stochasticity. In L. Saul, Y. Weiss, and L. Bottou (eds.), *Proceedings of Conference on Neural Information Processing Systems (NIPS)*. Cambridge, MA: MIT Press.

Lin, L.-J., and T.M. Mitchell. 1992. Memory approaches to reinforcement learning in non-Markovian domains. Technical Report CMU-CS-92-138, Carnegie Mellon University, Pittsburgh, PA.

Littman, M.L., A.R. Cassandra, and L.P. Kaelbling. 1995. Learning policies for partially observable environments: Scaling up. In A. Prieditis and S. Russell (eds.), *Proceedings of the Twelfth International Conference on Machine Learning*.

Littman, M.L., R.S. Sutton, and S. Singh. 2001. Predictive representations of state. In *Advances in Neural Information Processing Systems 14*.

Liu, J., and R. Chen. 1998. Sequential Monte Carlo methods for dynamic systems. *Journal of the American Statistical Association* 93:1032–1044.

Liu, Y., and S. Thrun. 2003. Results for outdoor-SLAM using sparse extended information filters. In *Proceedings of the International Conference on Robotics and Automation (ICRA)*.

Lovejoy, W.S. 1991. A survey of algorithmic methods for partially observable Markov decision processes. *Annals of Operations Research* 28:47–65.

Lozano-Perez, T. 1983. Spatial planning: A configuration space approach. *IEEE Transactions on Computers* pp. 108–120.

Lu, F., and E. Milios. 1994. Robot pose estimation in unknown environments by matching 2d range scans. In *IEEE Computer Vision and Pattern Recognition Conference (CVPR)*.

Lu, F., and E. Milios. 1997. Globally consistent range scan alignment for environment mapping. *Autonomous Robots* 4:333–349.

Lu, F., and E. Milios. 1998. Robot pose estimation in unknown environments by matching 2d range scans. *Journal of Intelligent and Robotic Systems* 18:249–275.

Lumelsky, S., S. Mukhopadhyay, and K. Sun. 1990. Dynamic path planning in sensor-based terrain acquisition. *IEEE Transactions on Robotics and Automation* 6:462–472.

MacDonald, I.L., and W. Zucchini. 1997. *Hidden Markov and Other Models for Discrete-Valued Time Series*. London, UK: Chapman and Hall.

Madhavan, R., G. Dissanayake, H. Durrant-Whyte, J. Roberts, P. Corke, and J. Cunningham. 1999. Issues in autonomous navigation of underground vehicles. *Journal of Mineral Resources Engineering* 8:313–324.

Mahadevan, S., and L. Kaelbling. 1996. The NSF workshop on reinforcement learning: Summary and observations. *AI Magazine* Winter:89–97.

Mahadevan, S., and N. Khaleeli. 1999. Robust mobile robot navigation using partially-observable semi-Markov decision processes. Internal report.

Makarenko, A.A., S.B. Williams, F. Bourgoult, and F. Durrant-Whyte. 2002. An experiment in integrated exploration. In *Proceedings of the IEEE/RSJ Int. Conf. on Intelligent Robots and Systems (IROS)*.

Marquardt, D. 1963. An algorithm for least-squares estimation of nonlinear parameters. *SIAM Journal of Applied Mathematics* 11:431–441.

Mason, M.T. 2001. *Mechanics of Robotic Manipulation*. Cambridge, MA: MIT Press.

Matarić, M.J. 1990. A distributed model for mobile robot environment-learning and navigation. Master's thesis, MIT, Cambridge, MA. Also available as MIT Artificial Intelligence Laboratory Tech Report AITR-1228.

Matthies, L., E. Gat, R. Harrison, B. Wilcox, R. Volpe, and T. Litwin. 1995. Mars microrover navigation: Performance evaluation and enhancement. *Autonomous Robots* 2:291–311.

Maybeck, P.S. 1990. The Kalman filter: An introduction to concepts. In I.J. Cox and G.T. Wilfong (eds.), *Autonomous Robot Vehicles*. Springer Verlag.

Metropolis, N., and S. Ulam. 1949. The Monte Carlo method. *Journal of the American Statistical Association* 44:335–341.

Mikhail, E. M., J. S. Bethel, and J. C. McGlone. 2001. *Introduction to Modern Photogrammetry*. John Wiley and Sons, Inc.

Mine, H., and S. Osaki. 1970. *Markovian Decision Processes*. American Elsevier.

Minka, T. 2001. *A family of algorithms for approximate Bayesian inference*. PhD thesis, MIT Media Lab, Cambridge, MA.

Monahan, G.E. 1982. A survey of partially observable Markov decision processes: Theory, models, and algorithms. *Management Science* 28:1–16.

Montemerlo, M., N. Roy, and S. Thrun. 2003a. Perspectives on standardization in mobile robot programming: The Carnegie Mellon navigation (CARMEN) toolkit. In *Proceedings of the Conference on Intelligent Robots and Systems (IROS)*. Software package for download at www.cs.cmu.edu/∼carmen.

Montemerlo, M., and S. Thrun. 2004. Large-scale robotic 3-d mapping of urban structures. In *Proceedings of the International Symposium on Experimental Robotics (ISER)*, Singapore. Springer Tracts in Advanced Robotics (STAR).

Montemerlo, M., S. Thrun, D. Koller, and B. Wegbreit. 2002a. FastSLAM: A factored solution to the simultaneous localization and mapping problem. In *Proceedings of the AAAI National Conference on Artificial Intelligence*, Edmonton, Canada. AAAI.

Montemerlo, M., S. Thrun, D. Koller, and B. Wegbreit. 2003b. FastSLAM 2.0: An improved particle filtering algorithm for simultaneous localization and mapping that provably converges. In *Proceedings of the Sixteenth International Joint Conference on Artificial Intelligence (IJCAI)*, Acapulco, Mexico. IJCAI.

Montemerlo, M., W. Whittaker, and S. Thrun. 2002b. Conditional particle filters for simultaneous mobile robot localization and people-tracking. In *Proceedings of the International Conference on Robotics and Automation (ICRA)*.

Moore, A.W. 1991. Variable resolution dynamic programming: Efficiently learning action maps in multivariate real-valued state-spaces. In *Proceedings of the Eighth International Workshop on Machine Learning*, pp. 333–337.

Moravec, H.P. 1988. Sensor fusion in certainty grids for mobile robots. *AI Magazine* 9:61–74.

Moravec, H.P., and M.C. Martin, 1994. Robot navigation by 3D spatial evidence grids. Mobile Robot Laboratory, Robotics Institute, Carnegie Mellon University.

Moutarlier, P., and R. Chatila. 1989a. An experimental system for incremental environment modeling by an autonomous mobile robot. In *1st International Symposium on Experimental Robotics*, Montreal.

Moutarlier, P., and R. Chatila. 1989b. Stochastic multisensory data fusion for mobile robot location and environment modeling. In *5th Int. Symposium on Robotics Research*, Tokyo.

Mozer, M.C., and J.R. Bachrach. 1989. Discovering the structure of a reactive environment by exploration. Technical Report CU-CS-451-89, Dept. of Computer Science, University of Colorado, Boulder.

Murphy, K. 2000a. Bayesian map learning in dynamic environments. In *Advances in Neural Information Processing Systems (NIPS)*. Cambridge, MA: MIT Press.

Murphy, K. 2000b. A survey of POMDP solution techniques. Technical report, UC Berkeley, Berkeley, CA.

Murphy, K., and S. Russell. 2001. Rao-Blackwellized particle filtering for dynamic Bayesian networks. In A. Doucet, N. de Freitas, and N. Gordon (eds.), *Sequential Monte Carlo Methods in Practice*, pp. 499–516. Springer Verlag.

Murphy, R. 2000c. *Introduction to AI Robotics*. Cambridge, MA: MIT Press.

Murphy, R. 2004. Human-robot interaction in rescue robotics. *IEEE Systems, Man and Cybernetics Part C: Applications and Reviews* 34.

Narendra, P.M., and K. Fukunaga. 1977. A branch and bound algorithm for feature subset selection. *IEEE Transactions on Computers* 26:914–922.

Neira, J., M.I. Ribeiro, and J.D. Tardós. 1997. Mobile robot localisation and map building using monocular vision. In *Proceedings of the International Symposium On Intelligent Robotics Systems*, Stockholm, Sweden.

Neira, J., and J.D. Tardós. 2001. Data association in stochastic mapping using the joint compatibility test. *IEEE Transactions on Robotics and Automation* 17:890–897.

Neira, J., J.D. Tardós, and J.A. Castellanos. 2003. Linear time vehicle relocation in SLAM. In *Proceedings of the International Conference on Robotics and Automation (ICRA)*.

Nettleton, E.W., P.W. Gibbens, and H.F. Durrant-Whyte. 2000. Closed form solutions to the multiple platform simultaneous localisation and map building (slam) problem. In Bulur V. Dasarathy (ed.), *Sensor Fusion: Architectures, Algorithms, and Applications IV*, volume 4051, pp. 428–437, Bellingham.

Nettleton, E., S. Thrun, and H. Durrant-Whyte. 2003. Decentralised slam with low-bandwidth communication for teams of airborne vehicles. In *Proceedings of the International Conference on Field and Service Robotics*, Lake Yamanaka, Japan.

Newman, P. 2000. *On the Structure and Solution of the Simultaneous Localisation and Map Building Problem*. PhD thesis, Australian Centre for Field Robotics, University of Sydney, Sydney, Australia.

Newman, P., M. Bosse, and J. Leonard. 2003. Autonomous feature-based exploration. In *Proceedings of the International Conference on Robotics and Automation (ICRA)*.

Newman, P.M., and H.F. Durrant-Whyte. 2001. A new solution to the simultaneous and map building (SLAM) problem. In *Proceedings of SPIE*.

Newman, P., and J.L. Rikoski. 2003. Towards constant-time slam on an autonomous underwater vehicle using synthetic aperture sonar. In *Proceedings of the International Symposium of Robotics Research*, Sienna, Italy.

Neyman, J. 1934. On the two different aspects of the representative model: the method of stratified sampling and the method of purposive selection. *Journal of the Royal Statistical Society* 97:558–606.

Ng, A.Y., A. Coates, M. Diel, V. Ganapathi, J. Schulte, B. Tse, E. Berger, and E. Liang. 2004. Autonomous inverted helicopter flight via reinforcement learning. In *Proceedings of the International Symposium on Experimental Robotics (ISER)*, Singapore. Springer Tracts in Advanced Robotics (STAR).

Ng, A.Y., and M. Jordan. 2000. PEGASUS: a policy search method for large MDPs and POMDPs. In *Proceedings of Uncertainty in Artificial Intelligence*.

Ng, A.Y., J. Kim, M.I. Jordan, and S. Sastry. 2003. Autonomous helicopter flight via reinforcement learning. In S. Thrun, L. Saul, and B. Schölkopf (eds.), *Proceedings of Conference on Neural Information Processing Systems (NIPS)*. Cambridge, MA: MIT Press.

Nieto, J., J.E. Guivant, and E.M. Nebot. 2004. The hybrid metric maps (HYMMs): A novel map representation for dense SLAM. In *Proceedings of the International Conference on Robotics and Automation (ICRA)*.

Nilsson, N.J. 1982. *Principles of Artificial Intelligence*. Berlin, New York: Springer Publisher.

Nilsson, N. 1984. Shakey the robot. Technical Report 323, SRI International, Menlo Park, CA.

Nourbakhsh, I. 1987. *Interleaving Planning and Execution for Autonomous Robots*. Boston, MA: Kluwer Academic Publishers.

Nourbakhsh, I., R. Powers, and S. Birchfield. 1995. DERVISH an office-navigating robot. *AI Magazine* 16.

Nüchter, A., H. Surmann, K. Lingemann, J. Hertzberg, and S. Thrun. 2004. 6D SLAM with application in autonomous mine mapping. In *Proceedings of the International Conference on Robotics and Automation (ICRA)*.

Oore, S., G.E. Hinton, and G. Dudek. 1997. A mobile robot that learns its place. *Neural Computation* 9:683–699.

Ortin, D., J. Neira, and J.M. Montiel. 2004. Relocation using laser and vision. In *Proceedings of the International Conference on Robotics and Automation (ICRA)*, New Orleans.

Park, S., F. Pfenning, and S. Thrun. 2005. A probabilistic progamming language based upon sampling functions. In *Proceedings of the ACM Symposium on Principles of Programming Languages (POPL)*, Long Beach, CA. ACM SIGPLAN - SIGACT.

Parr, R., and S. Russell. 1998. Reinforcement learning with hierarchies of machines. In *Advances in Neural Information Processing Systems 10*. Cambridge, MA: MIT Press.

Paskin, M.A. 2003. Thin junction tree filters for simultaneous localization and mapping. In *Proceedings of the Sixteenth International Joint Conference on Artificial Intelligence (IJCAI)*, Acapulco, Mexico. IJCAI.

Paul, R.P. 1981. *Robot Manipulators: Mathematics, Programming, and Control*. Cambridge, MA: MIT Press.

Pearl, J. 1988. *Probabilistic reasoning in intelligent systems: networks of plausible inference*. San Mateo, CA: Morgan Kaufmann.

Pierce, D., and B. Kuipers. 1994. Learning to explore and build maps. In *Proceedings of the Twelfth National Conference on Artificial Intelligence*, pp. 1264–1271, Menlo Park. AAAI, AAAI Press/MIT Press.

Pineau, J., G. Gordon, and S. Thrun. 2003a. Applying metric trees to belief-point POMDPs. In S. Thrun, L. Saul, and B. Schölkopf (eds.), *Proceedings of Conference on Neural Information Processing Systems (NIPS)*. Cambridge, MA: MIT Press.

Pineau, J., G. Gordon, and S. Thrun. 2003b. Point-based value iteration: An anytime algorithm for POMDPs. In *Proceedings of the Sixteenth International Joint Conference on Artificial Intelligence (IJCAI)*, Acapulco, Mexico. IJCAI.

Pineau, J., G. Gordon, and S. Thrun. 2003c. Policy-contingent abstraction for robust robot control. In *Proceedings of the Conference on Uncertainty in AI (UAI)*, Acapulco, Mexico.

Pineau, J., M. Montemerlo, N. Roy, S. Thrun, and M. Pollack. 2003d. Towards robotic assistants in nursing homes: challenges and results. *Robotics and Autonomous Systems* 42:271–281.

Pitt, M., and N. Shephard. 1999. Filtering via simulation: auxiliary particle filter. *Journal of the American Statistical Association* 94:590–599.

Poon, K.-M. 2001. A fast heuristic algorithm for decision-theoretic planning. Master's thesis, The Hong Kong University of Science and Technology.

Poupart, P., and C. Boutilier. 2000. Value-directed belief state approximation for POMDPs. In *Proceedings of the Conference on Uncertainty in AI (UAI)*, pp. 279–288.

Poupart, P., L.E. Ortiz, and C. Boutilier. 2001. Value-directed sampling methods for monitoring POMDPs. In *Proceedings of the 17th Annual Conference on Uncertainty in AI (UAI)*.

Procopiuc, O., P.K. Agarwal, L. Arge, and J.S. Vitter. 2003. Bkd-tree: A dynamic scalable kd-tree. In T. Hadzilacos, Y. Manolopoulos, J.F. Roddick, and Y. Theodoridis (eds.), *Advances in Spatial and Temporal Databases*, Santorini Island, Greece. Springer Verlag.

Rabiner, L.R., and B.H. Juang. 1986. An introduction to hidden Markov models. *IEEE ASSP Magazine* 3:4–16.

Raibert, M.H. 1991. Trotting, pacing, and bounding by a quadruped robot. *Journal of Biomechanics* 23:79–98.

Raibert, M.H., M. Chepponis, and H.B. Brown Jr. 1986. Running on four legs as though they were one. *IEEE Transactions on Robotics and Automation* 2:70–82.

Rao, C.R. 1945. Information and accuracy obtainable in estimation of statistical parameters. *Bulletin of the Calcutta Mathematical Society* 37:81–91.

Rao, N., S. Hareti, W. Shi, and S. Iyengar. 1993. Robot navigation in unknown terrains: Introductory survey of non-heuristic algorithms. Technical Report ORNL/TM-12410, Oak Ridge National Laboratory.

Reed, M.K., and P.K. Allen. 1997. A robotic system for 3-d model acquisition from multiple range images. In *Proceedings of the International Conference on Robotics and Automation (ICRA)*.

Rees, W.G. 2001. *Physical Principles of Remote Sensing (Topics in Remote Sensing)*. Cambridge, UK: Cambridge University Press.

Reif, J.H. 1979. Complexity of the mover's problem and generalizations. In *Proceedings of the 20th IEEE Symposium on Foundations of Computer Science*, pp. 421–427.

Rekleitis, I.M., G. Dudek, and E.E. Milios. 2001a. Multi-robot collaboration for robust exploration. *Annals of Mathematics and Artificial Intelligence* 31:7–40.

Rekleitis, I., R. Sim, G. Dudek, and E. Milios. 2001b. Collaborative exploration for map construction. In *IEEE International Symposium on Computational Intelligence in Robotics and Automation*.

Rencken, W.D. 1993. Concurrent localisation and map building for mobile robots using ultrasonic sensors. In *Proceedings of the IEEE/RSJ Int. Conf. on Intelligent Robots and Systems (IROS)*, pp. 2129–2197.

Reuter, J. 2000. Mobile robot self-localization using PDAB. In *Proceedings of the International Conference on Robotics and Automation (ICRA)*.

Rikoski, R., J. Leonard, P. Newman, and H. Schmidt. 2004. Trajectory sonar perception in the ligurian sea. In *Proceedings of the International Symposium on Experimental Robotics (ISER)*, Singapore. Springer Tracts in Advanced Robotics (STAR).

Rimon, E., and D.E. Koditschek. 1992. Exact robot navigation using artificial potential functions. *IEEE Transactions on Robotics and Automation* 8:501–518.

Rivest, R.L., and R.E. Schapire. 1987a. Diversity-based inference of finite automata. In *Proceedings of Foundations of Computer Science*.

Rivest, R.L., and R.E. Schapire. 1987b. A new approach to unsupervised learning in deterministic environments. In P. Langley (ed.), *Proceedings of the Fourth International Workshop on Machine Learning*, pp. 364–375, Irvine, California.

Robbins, H. 1952. Some aspects of the sequential design of experiments. *Bulletin of the American Mathemtical Society* 58:529–532.

Rosencrantz, M., G. Gordon, and S. Thrun. 2004. Learning low dimensional predictive representations. In *Proceedings of the Twenty-First International Conference on Machine Learning*, Banff, Alberta, Canada.

Roumeliotis, S.I., and G.A. Bekey. 2000. Bayesian estimation and Kalman filtering: A unified framework for mobile robot localization. In *Proceedings of the International Conference on Robotics and Automation (ICRA)*, pp. 2985–2992.

Rowat, P.F. 1979. *Representing the Spatial Experience and Solving Spatial problems in a Simulated Robot Environment*. PhD thesis, University of British Columbia, Vancouver, BC, Canada.

Roy, B.V., and J.N. Tsitsiklis. 1996. Stable linear approximations to dynamic programming for stochastic control problems with local transitions. In D. Touretzky, M. Mozer, and M.E. Hasselmo (eds.), *Advances in Neural Information Processing Systems 8*. Cambridge, MA: MIT Press.

Roy, N., W. Burgard, D. Fox, and S. Thrun. 1999. Coastal navigation: Robot navigation under uncertainty in dynamic environments. In *Proceedings of the International Conference on Robotics and Automation (ICRA)*.

Roy, N., and G. Dudek. 2001. Collaborative exploration and rendezvous: Algorithms, performance bounds and observations. *Autonomous Robots* 11:117–136.

Roy, N., G. Gordon, and S. Thrun. 2004. Finding approximate POMDP solutions through belief compression. *Journal of Artificial Intelligence Research*. To appear.

Roy, N., J. Pineau, and S. Thrun. 2000. Spoken dialogue management using probabilistic reasoning. In *Proceedings of the 38th Annual Meeting of the Association for Computational Linguistics (ACL-2000)*, Hong Kong.

Roy, N., and S. Thrun. 2002. Motion planning through policy search. In *Proceedings of the IEEE/RSJ Int. Conf. on Intelligent Robots and Systems (IROS)*.

Rubin, D.B. 1988. Using the SIR algorithm to simulate posterior distributions. In M.H. Bernardo, K.M. an DeGroot, D.V. Lindley, and A.F.M. Smith (eds.), *Bayesian Statistics 3*. Oxford, UK: Oxford University Press.

Rubinstein, R.Y. 1981. *Simulation and the Monte Carlo Method*. John Wiley and Sons, Inc.

Russell, S., and P. Norvig. 2002. *Artificial Intelligence: A Modern Approach*. Englewood Cliffs, NJ: Prentice Hall.

Saffiotti, A. 1997. The uses of fuzzy logic in autonomous robot navigation. *Soft Computing* 1:180–197.

Sahin, E., P. Gaudiano, and R. Wagner. 1998. A comparison of visual looming and sonar as mobile robot range sensors. In *Proceedings of the Second International Conference on Cognitive And Neural Systems*, Boston, MA.

Salichs, M.A., J.M. Armingol, L. Moreno, and A. Escalera. 1999. Localization system for mobile robots in indoor environments. *Integrated Computer-Aided Engineering* 6: 303–318.

Salichs, M.A., and L. Moreno. 2000. Navigation of mobile robots: Open questions. *Robotica* 18:227–234.

Sandwell, D.T., 1997. Exploring the ocean basins with satellite altimeter data. http://julius.ngdc.noaa.gov/mgg/bathymetry/predicted/explore.HTML.

Saranli, U., and D.E. Koditschek. 2002. Back flips with a hexapedal robot. In *Proceedings of the International Conference on Robotics and Automation (ICRA)*, volume 3, pp. 128–134.

Schiele, B., and J.L. Crowley. 1994. A comparison of position estimation techniques using occupancy grids. In *Proceedings of the International Conference on Robotics and Automation (ICRA)*.

Schoppers, M.J. 1987. Universal plans for reactive robots in unpredictable environments. In J. McDermott (ed.), *Proceedings of the Tenth International Joint Conference on Artificial Intelligence (IJCAI-87)*, pp. 1039–1046, Milan, Italy. Morgan Kaufmann.

Schulz, D., W. Burgard, D. Fox, and A.B. Cremers. 2001a. Tracking multiple moving objects with a mobile robot. In *Proceedings of the IEEE Computer Society Conference on Computer Vision and Pattern Recognition (CVPR)*, Kauai, Hawaii.

Schulz, D., W. Burgard, D. Fox, and A.B. Cremers. 2001b. Tracking multiple moving targets with a mobile robot using particle filters and statistical data association. In *Proceedings of the International Conference on Robotics and Automation (ICRA)*.

Schulz, D., and D. Fox. 2004. Bayesian color estimation for adaptive vision-based robot localization. In *Proceedings of the IEEE/RSJ Int. Conf. on Intelligent Robots and Systems (IROS)*.

Schwartz, J.T., M. Scharir, and J. Hopcroft. 1987. *Planning, Geometry and Complexity of Robot Motion*. Norwood, NJ: Ablex Publishing Corporation.

Scott, D.W. 1992. *Multivariate density estimation: theory, practice, and visualization*. John Wiley and Sons, Inc.

Shaffer, G., J. Gonzalez, and A. Stentz. 1992. Comparison of two range-based estimators for a mobile robot. In *SPIE Conf. on Mobile Robots VII*, pp. 661–667.

Sharma, R. 1992. Locally efficient path planning in an uncertain, dynamic environment using a probabilistic model. *T-RA* 8:105–110.

Shatkay, H, and L. Kaelbling. 1997. Learning topological maps with weak local odometric information. In *Proceedings of IJCAI-97*. IJCAI, Inc.

Siegwart, R., K.O. Arras, S. Bouabdallah, D. Burnier, G. Froidevaux, X. Greppin, B. Jensen, A. Lorotte, L. Mayor, M. Meisser, R. Philippsen, R. Piguet, G. Ramel, G. Terrien, and N. Tomatis. 2003. A large scale installation of personal robots. Special issue on socially interactive robots. *Robotics and Autonomous Systems* 42: 203–222.

Siegwart, R., and I. Nourbakhsh. 2004. *Introduction to Autonomous Mobile Robots*. Cambridge, MA: MIT Press.

Sim, R., G. Dudek, and N. Roy. 2004. Online control policy optimization for minimizing map uncertainty during exploration. In *Proceedings of the International Conference on Robotics and Automation (ICRA)*.

Simmons, R.G., D. Apfelbaum, W. Burgard, D. Fox, M. Moors, S. Thrun, and H. Younes. 2000a. Coordination for multi-robot exploration and mapping. In *Proc. of the National Conference on Artificial Intelligence (AAAI)*.

Simmons, R.G., J. Fernandez, R. Goodwin, S. Koenig, and J. O'Sullivan. 2000b. Lessons learned from Xavier. *IEEE Robotics and Automation Magazine* 7:33–39.

Simmons, R.G., and S. Koenig. 1995. Probabilistic robot navigation in partially observable environments. In *Proceedings of the International Joint Conference on Artificial Intelligence (IJCAI)*.

Simmons, R.G., S. Thrun, C. Athanassiou, J. Cheng, L. Chrisman, R. Goodwin, G.-T. Hsu, and H. Wan. 1992. Odysseus: An autonomous mobile robot. *AI Magazine*. extended abstract.

Singh, K., and K. Fujimura. 1993. Map making by cooperating mobile robots. In *Proceedings of the International Conference on Robotics and Automation (ICRA)*, pp. 254–259.

Smallwood, R.W., and E.J. Sondik. 1973. The optimal control of partially observable Markov processes over a finite horizon. *Operations Research* 21:1071–1088.

Smith, A.F.M., and A.E. Gelfand. 1992. Bayesian statistics without tears: a sampling-resampling perspective. *American Statistician* 46:84–88.

Smith, R.C., and P. Cheeseman. 1986. On the representation and estimation of spatial uncertainty. *International Journal of Robotics Research* 5:56–68.

Smith, R., M. Self, and P. Cheeseman. 1990. Estimating uncertain spatial relationships in robotics. In I.J. Cox and G.T. Wilfong (eds.), *Autonomous Robot Vehicles*, pp. 167–193. Springer-Verlag.

Smith, S. M., and S. E. Dunn. 1995. The ocean explorer AUV: A modular platform for coastal sensor deployment. In *Proceedings of the Autonomous Vehicles in Mine Countermeasures Symposium*. Naval Postgraduate School.

Smith, T., and R.G. Simmons. 2004. Heuristic search value iteration for POMDPs. In *Proceedings of the 20th Annual Conference on Uncertainty in AI (UAI)*.

Soatto, S., and R. Brockett. 1998. Optimal structure from motion: Local ambiguities and global estimates. In *Proceedings of the Conference on Computer Vision and Pattern Recognition (CVPR)*, pp. 282–288.

Sondik, E. 1971. *The Optimal Control of Partially Observable Markov Processes*. PhD thesis, Stanford University.

Stachniss, C., and W. Burgard. 2003. Exploring unknown environments with mobile robots using coverage maps. In *Proceedings of the Sixteenth International Joint Conference on Artificial Intelligence (IJCAI)*, Acapulco, Mexico. IJCAI.

Stachniss, C., and W. Burgard. 2004. Exploration with active loop-closing for Fast-SLAM. In *Proceedings of the IEEE/RSJ Int. Conf. on Intelligent Robots and Systems (IROS)*.

Steels, L. 1991. Towards a theory of emergent functionality. In J-A. Meyer and R. Wilson (eds.), *Simulation of Adaptive Behavior*. Cambridge, MA: MIT Press.

Stentz, A. 1995. The focussed D* algorithm for real-time replanning. In *Proceedings of IJCAI-95*.

Stewart, B., J. Ko, D. Fox, and K. Konolige. 2003. The revisiting problem in mobile robot map building: A hierarchical Bayesian approach. In *Proceedings of the Conference on Uncertainty in AI (UAI)*, Acapulco, Mexico.

Strassen, V. 1969. Gaussian elimination is not optimal. *Numerische Mathematik* 13: 354–356.

Stroupe, A.W. 2004. Value-based action selection for exploration and mapping with robot teams. In *Proceedings of the International Conference on Robotics and Automation (ICRA)*.

Sturges, H. 1926. The choice of a class-interval. *Journal of the American Statistical Association* 21:65–66.

Subrahmaniam, K. 1979. *A Primer In Probability*. New York, NY: M. Dekker.

Swerling, P. 1958. A proposed stagewise differential correction procedure for satellite tracking and prediction. Technical Report P-1292, RAND Corporation.

Tailor, C.J., and D.J. Kriegman. 1993. Exloration strategies for mobile robots. In *Proceedings of the International Conference on Robotics and Automation (ICRA)*, pp. 248–253.

Tanner, M.A. 1996. *Tools for Statistical Inference*. New York: Springer Verlag. 3rd edition.

Tardós, J.D., J. Neira, P.M. Newman, and J.J. Leonard. 2002. Robust mapping and localization in indoor environments using sonar data. *International Journal of Robotics Research* 21:311–330.

Teller, S., M. Antone, Z. Bodnar, M. Bosse, S. Coorg, M. Jethwa, and N. Master. 2001. Calibrated, registered images of an extended urban area. In *Proceedings of the Conference on Computer Vision and Pattern Recognition (CVPR)*.

Theocharous, G., K. Rohanimanesh, and S. Mahadevan. 2001. Learning hierarchical partially observed Markov decision process models for robot navigation. In *Proceedings of the International Conference on Robotics and Automation (ICRA)*.

Thorp, E.O. 1966. *Elementary Probability*. R.E. Krieger.

Thrun, S. 1992. Efficient exploration in reinforcement learning. Technical Report CMU-CS-92-102, Carnegie Mellon University, Computer Science Department, Pittsburgh, PA.

Thrun, S. 1993. Exploration and model building in mobile robot domains. In E. Ruspini (ed.), *Proceedings of the IEEE International Conference on Neural Networks*, pp. 175–180, San Francisco, CA. IEEE Neural Network Council.

Thrun, S. 1998a. Bayesian landmark learning for mobile robot localization. *Machine Learning* 33.

Thrun, S. 1998b. Learning metric-topological maps for indoor mobile robot navigation. *Artificial Intelligence* 99:21–71.

Thrun, S. 2000a. Monte Carlo POMDPs. In S.A. Solla, T.K. Leen, and K.-R. Müller (eds.), *Advances in Neural Information Processing Systems 12*, pp. 1064–1070. Cambridge, MA: MIT Press.

Thrun, S. 2000b. Towards programming tools for robots that integrate probabilistic computation and learning. In *Proceedings of the IEEE International Conference on Robotics and Automation (ICRA)*, San Francisco, CA. IEEE.

Thrun, S. 2001. A probabilistic online mapping algorithm for teams of mobile robots. *International Journal of Robotics Research* 20:335–363.

Thrun, S. 2002. Robotic mapping: A survey. In G. Lakemeyer and B. Nebel (eds.), *Exploring Artificial Intelligence in the New Millenium*. Morgan Kaufmann.

Thrun, S. 2003. Learning occupancy grids with forward sensor models. *Autonomous Robots* 15:111–127.

Thrun, S., M. Beetz, M. Bennewitz, W. Burgard, A.B. Cremers, F. Dellaert, D. Fox, D. Hähnel, C. Rosenberg, N. Roy, J. Schulte, and D. Schulz. 2000a. Probabilistic algorithms and the interactive museum tour-guide robot minerva. *International Journal of Robotics Research* 19:972–999.

Thrun, S., A. Bücken, W. Burgard, D. Fox, T. Fröhlinghaus, D. Henning, T. Hofmann, M. Krell, and T. Schmidt. 1998a. Map learning and high-speed navigation in RHINO. In D. Kortenkamp, R.P. Bonasso, and R. Murphy (eds.), *AI-based Mobile Robots: Case Studies of Successful Robot Systems*, pp. 21–52. Cambridge, MA: MIT Press.

Thrun, S., W. Burgard, and D. Fox. 2000b. A real-time algorithm for mobile robot mapping with applications to multi-robot and 3D mapping. In *Proceedings of the International Conference on Robotics and Automation (ICRA)*.

Thrun, S., M. Diel, and D. Hähnel. 2003. Scan alignment and 3d surface modeling with a helicopter platform. In *Proceedings of the International Conference on Field and Service Robotics*, Lake Yamanaka, Japan.

Thrun, S., D. Fox, and W. Burgard. 1998b. A probabilistic approach to concurrent mapping and localization for mobile robots. *Machine Learning* 31:29–53. Also appeared in Autonomous Robots 5, 253–271 (joint issue).

Thrun, S., D. Fox, and W. Burgard. 2000c. Monte Carlo localization with mixture proposal distribution. In *Proceedings of the AAAI National Conference on Artificial Intelligence*, Austin, TX. AAAI.

Thrun, S., D. Koller, Z. Ghahramani, H. Durrant-Whyte, and A.Y. Ng. 2002. Simultaneous mapping and localization with sparse extended information filters. In J.-D. Boissonnat, J. Burdick, K. Goldberg, and S. Hutchinson (eds.), *Proceedings of the Fifth International Workshop on Algorithmic Foundations of Robotics*, Nice, France.

Thrun, S., and Y. Liu. 2003. Multi-robot SLAM with sparse extended information filers. In *Proceedings of the 11th International Symposium of Robotics Research (ISRR'03)*, Sienna, Italy. Springer.

Thrun, S., Y. Liu, D. Koller, A.Y. Ng, Z. Ghahramani, and H. Durrant-Whyte. 2004a. Simultaneous localization and mapping with sparse extended information filters. *International Journal of Robotics Research* 23.

Thrun, S., C. Martin, Y. Liu, D. Hähnel, R. Emery-Montemerlo, D. Chakrabarti, and W. Burgard. 2004b. A real-time expectation maximization algorithm for acquiring multi-planar maps of indoor environments with mobile robots. *IEEE Transactions on Robotics* 20:433–443.

Thrun, S., S. Thayer, W. Whittaker, C. Baker, W. Burgard, D. Ferguson, D. Hähnel, M. Montemerlo, A. Morris, Z. Omohundro, C. Reverte, and W. Whittaker. 2004c. Autonomous exploration and mapping of abandoned mines. *IEEE Robotics and Automation Magazine*. Forthcoming.

Tomasi, C., and T. Kanade. 1992. Shape and motion from image streams under orthography: A factorization method. *International Journal of Computer Vision* 9: 137–154.

Tomatis, N., I. Nourbakhsh, and R. Siegwart. 2002. Hybrid simultaneous localization and map building: closing the loop with multi-hypothesis tracking. In *Proceedings of the International Conference on Robotics and Automation (ICRA)*.

Uhlmann, J., M. Lanzagorta, and S. Julier. 1999. The NASA mars rover: A testbed for evaluating applications of covariance intersection. In *Proceedings of the SPIE 13th Annual Symposium in Aerospace/Defence Sensing, Simulation and Controls*.

United Nations, and International Federation of Robotics. 2004. *World Robotics 2004*. New York and Geneva: United Nations.

Urmson, C., B. Shamah, J. Teza, M.D. Wagner, D. Apostolopoulos, and W.R. Whittaker. 2001. A sensor arm for robotic antartic meteorite search. In *Proceedings of the 3rd International Conference on Field and Service Robotics*, Helsinki, Finland.

Vaganay, J., J. Leonard, J.A. Curcio, and J.S. Willcox. 2004. Experimental validation of the moving long base-line navigation concept. In *Proceedings of the IEEE Conference on Autonomous Underwater Vehicles*.

van der Merwe, R. 2004. *Sigma-Point Kalman Filters for Probabilistic Inference in Dynamic State-Space Models*. PhD thesis, OGI School of Science & Engineering.

van der Merwe, R., N. de Freitas, A. Doucet, and E. Wan. 2001. The unscented particle filter. In *Advances in Neural Information Processing Systems 13*.

Vlassis, N., B. Terwijn, and B. Kröse. 2002. Auxiliary particle filter robot localization from high-dimensional sensor observations. In *Proceedings of the International Conference on Robotics and Automation (ICRA)*.

Vukobratović, M. 1989. *Introduction to Robotics*. Berlin, New York: Springer Publisher.

Wang, C.-C., C. Thorpe, and S. Thrun. 2003. Online simultaneous localization and mapping with detection and tracking of moving objects: Theory and results from a ground vehicle in crowded urban areas. In *Proceedings of the International Conference on Robotics and Automation (ICRA)*.

Washington, R. 1997. BI-POMDP: Bounded, incremental, partially-observable Markov-model planning. In *Proceedings of the European Conference on Planning (ECP)*, Toulouse, France.

Watkins, C.J.C.H. 1989. *Learning from Delayed Rewards*. PhD thesis, King's College, Cambridge, England.

Weiss, G., C. Wetzler, and E. von Puttkamer. 1994. Keeping track of position and orientation of moving indoor systems by correlation of range-finder scans. In *Proceedings of the International Conference on Intelligent Robots and Systems*, pp. 595–601.

Wesley, M.A., and T. Lozano-Perez. 1979. An algorithm for planning collision-free paths among polyhedral objects. *Communications of the ACM* 22:560–570.

West, M., and P.J. Harrison. 1997. *Bayesian Forecasting and Dynamic Models*, 2nd edition. New York: Springer-Verlag.

Wettergreen, D., D. Bapna, M. Maimone, and H. Thomas. 1999. Developing Nomad for robotic exploration of the Atacama Desert. *Robotics and Autonomous Systems* 26:127–148.

Whaite, P., and F.P. Ferrie. 1997. Autonomous exploration: Driven by uncertainty. *IEEE Transactions on Pattern Analysis and Machine Intelligence* 19:193–205.

Whitcomb, L. 2000. Underwater robotics: out of the research laboratory and into the field. In *Proceedings of the International Conference on Robotics and Automation (ICRA)*, pp. 85–90.

Williams, R.J. 1992. Simple statistical gradient-following algorithms for connectionist reinforcement learning. *Machine Learning* 8:229–256.

Williams, S.B. 2001. *Efficient Solutions to Autonomous Mapping and Navigation Problems*. PhD thesis, ACFR, University of Sydney, Sydney, Australia.

Williams, S.B., G. Dissanayake, and H.F. Durrant-Whyte. 2001. Constrained initialization of the simultaneous localization and mapping algorithm. In *Proceedings of the Symposium on Field and Service Robotics*. Springer Verlag.

Williams, S.B., G. Dissanayake, and H. Durrant-Whyte. 2002. An efficient approach to the simultaneous localisation and mapping problem. In *Proceedings of the International Conference on Robotics and Automation (ICRA)*, pp. 406–411.

Winer, B.J., D.R. Brown, and K.M. Michels. 1971. *Statistical Principles in Experimental Design*. New York: Mc-Graw-Hill.

Winkler, G. 1995. *Image Analysis, Random Fields, and Dynamic Monte Carlo Methods*. Berlin: Springer Verlag.

Wolf, D.F., and G.S. Sukhatme. 2004. Mobile robot simultaneous localization and mapping in dynamic environments. *Autonomous Robots*.

Wolf, J., W. Burgard, and H. Burkhardt. 2005. Robust vision-based localization by combining an image retrieval system with Monte Carlo localization. *IEEE Transactions on Robotics and Automation*.

Wong, J. 1989. *Terramechanics and Off-Road Vehicles*. Elsevier.

Yamauchi, B., and P. Langley. 1997. Place recognition in dynamic environments. *Journal of Robotic Systems* 14:107–120.

Yamauchi, B., A. Schultz, and W. Adams. 1999. Integrating exploration and localization for mobile robots. *Adaptive Systems 7*.

Yoshikawa, T. 1990. *Foundations of Robotics: Analysis and Control*. Cambridge, MA: MIT Press.

Zhang, N.L., and W. Zhang. 2001. Speeding up the convergence of value iteration in partially observable Markov decision processes. *Journal of Artificial Intelligence Research* 14:29–51.

Zhao, H., and R. Shibasaki. 2001. A vehicle-borne system of generating textured CAD model of urban environment using laser range scanner and line cameras. In *Proc. International Workshop on Computer Vision Systems (ICVS)*, Vancouver, Canada.

Zlot, R., A.T. Stenz, M.B. Dias, and S. Thayer. 2002. Multi-robot exploration controlled by a market economy. In *Proceedings of the International Conference on Robotics and Automation (ICRA)*.

Index

action selection, 490
active feature, 387
active localization, 195, 569, 575
adaptive algorithm, 86
adjusting model parameters, 158
AMDP, 550
  algorithm, 551, 552
amortized map recovery, 402
anchoring constraint, 343
anticipated uncertainty, 488
anytime algorithm, 540
approximate cell decomposition, 509
assumed density filter, 65
Atlas, 376
auction mechanism, 589
augmented state
  in a POMDP, 550
  in localization, 269

backpropagation, 296
backup step, 501
  in belief space, 526
bandit problem, 605
Bayes filter, 8, 26
  algorithm, 26, 27
Bayes rule, 16
bearing of a robot, 119
bearing only SLAM, 320, 334
behavior-based robotics, 11
belief, 25
  in a POMDP, 494

Bellman equation, 500
bias of a sampler, 112
bicycle model, 146
binary Bayes filter, 94
  algorithm, 94, 95
  in mapping, 285
binary information gain, 584
binary tree, 465
bin in a histogram, 87
branch-and-bound data association, 415
bundle adjustment, 381

canonical parameterization, 40, 71
catastrophic failure in localization, 249
clustering, 105
CML, *see* SLAM
coastal navigation, 8, 558
coin flip, 14
combined state vector, 313
complete state, 21, 24
complete state assumption, 33
computer vision, 332
concurrent mapping and localization, *see* SLAM
condensation algorithm, 276
conditional entropy, 571
conditional independence, 17, 24
  in SLAM, 437
conditional of a Gaussian, 359

conditional probability, 16
conditioning lemma, 359
configuration of a mobile robot, 118
configuration space, 141, 507
conjugate gradient
    in GraphSLAM, 376
consistency in SLAM, 435
constant time SLAM, 395
control, 22, 487
control action, 22
control policy, 493, 496
coordinate descent, 403
coordination between robots, 589
correction, 26
    in an UKF, 225
    in localization, 207
correlation-based measurement
    model, 174
correspondence, 178
    between different maps, 429
    in EKF, 323
    in GraphSLAM, 362
    in localization, 201, 215
    in SEIF, 409
    likelihood test, 363
    through maximum likelihood, 215
cost, 495
covariance intersection, 434
covariance matrix, 15
coverage problem, 603
cumulative payoff, 497
cyclic environment, 282
    in exploration, 598
    in FastSLAM, 471
    in SLAM, 343

data association, 178
decomposition of entropy, 593
delayed motion update in
    localization, 244
De Morgan's law, 293
density tree, 92

    for a particle filter, 105
derivative-free filter, 70
discount factor, 497
discrete Bayes filter, 86
    algorithm, 86, 87
dynamic Bayes network, 25
dynamic environment, 194, 267
dynamics, 145
dynamic state, 20

EIF, 75
    algorithm, 75, 76
EKF, 54
    algorithm, 59
EKF localization, 201
    algorithm with known
        correspondences, 203, 204
    general algorithm, 216, 217
    illustration, 201, 210
EKF SLAM, 312
    algorithm with known
        correspondence, 313, 314
    fundamental dilemma, 330
    general algorithm, 321, 323
    numerical instability, 329
EM algorithm, 333
    for data association, 482
    for learning a sensor model, 165
    for outlier rejection, 269
emergent behavior, 12
entropy, 18
    decomposition, 593
    of a multivariate Gaussian, 597
entropy lemma, 597
environment, 19
equivalence constraints, 419
expectation maximization, *see* EM
    algorithm
expectation of a random variable, 18
expected cumulative payoff, 497
expected information
    in exploration, 571

experimental design, 542
exploration, 569
    exploration action, 570, 576, 585
    for SLAM, 593
    frontier-based, 584
    greedy, 572
    in occupancy grids, 580
    Monte Carlo, 573
    Monte Carlo algorithm, 573, 574
    multi-robot, 587
    multi-robot algorithm, 588
    multi step, 575
exponential decay, 307
extended information filter, *see* EIF
extended information form, *see* GraphSLAM
extended Kalman filter, *see* EKF
extended Kalman filter SLAM, *see* EKF SLAM

factorization
    in GraphSLAM, 344
    in SLAM, 439
FastSLAM, 437
    exploration algorithm, 595, 596
    grid-based, 474
    occupancy grid algorithm, 478
    v1.0, 439
    v1.0 algorithm with known correspondence, 449, 450
    v1.0 general algorithm, 460, 461
    v1.0 with known correspondence, 444
    v2.0, 451
    v2.0 general algorithm, 460, 463
    with unknown correspondence, 457
feature-based
    measurement model, 176
feature-based map, 152
    in SLAM, 312
feature in a map, 176

finite-horizon case, 497
frontier
    in exploration, 584
    of a search tree, 416
frontier-based exploration, 584
full SLAM problem, 310

Gaussian, 14
    as a posterior, 40
    canonical parameterization, 71
    conditioning lemma, 359
    for a particle filter, 104
    marginalization lemma, 358
    mixture, 63
    multivariate, 15
    noise, in SLAM, 312
Gaussian filter, 39
generative model, 17
global correspondence problem, 424
global localization, 5, 194, 198
goal, 495
gradient descent, 376
    in learning, 296
graphical SLAM, 382
GraphSLAM, 337
    algorithm, 346, 347, 365
    correspondence test algorithm, 364
    with unknown correspondence, 363
greedy case, 497
grid-based FastSLAM, 474
grid localization, 238
    algorithm, 238, 239
    illustration, 245

heading direction, 119
heaven-or-hell problem, 510
hidden Markov model, 25, 87
histogram filter, 86
    dynamic decomposition, 92
    for localization, 238
    selective updating, 93

static decomposition, 92
histogram in a particle filter, 105
holonomic robot, 147
horizon, 497
hybrid control, 12
hybrid state, 21

importance factor
   in a particle filter, 99
   in FastSLAM, 447, 455
importance sampling
   in a particle filter, 99
incomplete state, 21
independence, 15
   between sensor beams, 152
   conditional, 17
infinite-horizon case, 498
information filter, 71
   algorithm, 73
information gain, 572
information gathering, 488
information matrix, 71
information state, 26
   in a POMDP, 494
information vector, 71
injection of random particles, 256
innovation, 43, 214
inverse measurement model, 95, 294
inverse range sensor algorithm, 287, 288
inversion lemma, 50

Jacobian, 58
joint distribution, 15

k-means algorithm, 104
Kalman filter, 40
   algorithm, 42, 43
   multi-hypothesis, 63
   unscented, 65
Kalman gain, 43, 52
   in EKF SLAM, 315
kernel density estimation, 105

kidnapped robot problem, 194
kinematics, 118
kinematic state, 20
KLD-sampling, 263
   algorithm, 264
   comparison, 265
knowledge state, 26
Kullback-Leibler divergence, 263

landmark, 21, 177
landmark existence probability, 329
landmark model
   algorithm, 178, 179
   sampling algorithm, 179, 180
laser range scan, 150
lazy SLAM, 339, 385
learning
   in an AMDP, 558
   in an MC-POMDP, 559
   of a measurement model, 158
least squares
   in SLAM, 337
Levenberg Marquardt, 376
likelihood field
   algorithm, 171, 172
   for range finders, 169
likelihood test for correspondence, 363
linear Gaussian SLAM, 382
linear Gaussian system, 40
   measurement probability, 42
   state transition, 41
linearization, 56
   in GraphSLAM, 356
localization
   comparison chart, 273
   in a dynamic environment, 267
   Markov, 197
   of a mobile robot, 5
   taxonomy, 193
   with a grid, 238
   with an EKF, 201

Index

with MCL, 250
local submaps, 369
location
   of a robot, 119
location-based map, 152
logistic regression, 294
log odds ratio, 94
   in an occupancy grid map, 286
loop closure, 471
loopy belief propagation, 79
low variance sampling, 109
   algorithm, 109, 110

Mahalanobis distance, 72
map, 152
   feature-based, 152
   location-based, 152
   metric, 93
   of an environment, 152
   topological, 93
   volumetric, 152
map-based motion model, 140
map management
   in FastSLAM, 459
   in SLAM, 328
map matching, 174, 233, 275
MAP occupancy grid map, 299
MAP occupancy grid mapping
   algorithm, 301, 302
mapping with known poses, 283
marginalization lemma, 358
marginal of a Gaussian, 358
market-based algorithms, 589
Markov assumption, 33
Markov blanket, 419
Markov chain, 21
Markov localization, 197
   algorithm, 197
   illustration, 5, 200
Markov random field, 79
matrix inversion lemma, 50
max-range measurement, 156

maximum likelihood
   correspondence, 215
   estimator, 159
maximum likelihood correspondence
   in SLAM, 323
maximum of linear functions, 530
MC-POMDP, 559
   algorithm, 559, 560
MCL, 250
   algorithm, 252
   algorithm with adaptive sample
     set, 264
   augmented algorithm, 257, 258
   with adaptive sample set, 263
   with mixture proposal distribution,
     262
   with random particles, 256
MDP, 491
measurement, 22
measurement arcs, 337
measurement innovation, 214
measurement likelihood, 208
measurement model, 149
   correlation-based, 174
   feature-based, 176
   for a range finder, 153
measurement noise, 154, 169
measurement probability, 25
   in a linear Gaussian, 42
measurement update, 26
   in a Bayes filter, 27
   in an information filter, 75
metric map, 93, 241
Mixture MCL, 262
mixture of Gaussians, 63
mixture weight in an MHT, 219
model-based paradigm, 11
moments matching, 65
moments parameterization, 40
Monte Carlo exploration, 573
   algorithm, 573, 574
Monte Carlo localization, *see* MCL

Monte Carlo POMDP, *see* MC-POMDP
motion arcs, 337
motion model, 119
    algorithm with maps, 141
    map-based, 140
    nonlinear, 54
    odometry, 132
    odometry algorithm, 134
    velocity, 121
    velocity algorithm, 121, 123, 129
multi-hypothesis EKF, 63
multi-hypothesis tracking, 218
multi-robot
    coordination, 589
    exploration, 587
    exploration algorithm, 588
    localization, 196
    SLAM, 424
multi-sensor fusion, 293
multi-step exploration, 575
multivariate distribution, 15
mutual exclusion principle, 229, 369

natural parameterization, 40
navigation function, 11, 493
nearest neighbor, 559
negative information
    in a SEIF, 421
    in FastSLAM, 459
    in GraphSLAM, 370
    in localization, 231
neural network, 294, 296
nonlinear motion model, 54
nonlinear state transition, 54
nonparametric filter, 85
normal distribution, 14, 128
    algorithm for computing, 123
    algorithm for sampling, 124
normalizer in Bayes rule, 17
numerical instability of EKF SLAM, 329

occupancy grid
    in FastSLAM, 478
occupancy grid map, 94, 281
    algorithm, 286
    MAP technique, 299
odds of an event, 94
odometer, 23
odometry algorithm
    sampling algorithm, 136
odometry model
    algorithm, 134, 135
    sampling algorithm, 137
odometry motion model, 132
online SLAM problem, 309
outlier rejection
    in localization, 229
    in MCL, 269
    in SLAM, 328
overconfidence in perception, 183

partially observable Markov decision process, *see* POMDP
partial observability, 488
particle
    in a particle filter, 97
    in FastSLAM, 444
particle deprivation, 112
particle filter, 96
    algorithm, 98
    Gaussian approximation, 104
    histogram approximation, 105
    in localization, 250
    in SLAM, 437
    low variance sampling, 109
    particle deprivation, 112
    stratified sampling, 111
passive localization, 195
payoff function, 495
    in a POMDP, 516
PBVI, 538
photogrammetry, 332
piecewise linear function, 514

planning, 487
planning horizon, 497
point-based value iteration, 538
policy, 496
policy for action selection, 493
POMDP, 493, 513
    algorithm, 527, 529
    approximate algorithms, 547
    numerical example, 515
pose, 20
pose of a robot, 118
position estimation, 191
position tracking, 193, 197
positive information, 313
posterior probability, 16
potential field, 11
pre-caching of a measurement model, 167, 243
precision matrix, 71
prediction, 26
    in a Bayes filter, 27
    in an information filter, 74
    in an UKF, 223
    in localization, 205
prior probability, 16
    in a binary Bayes filter, 96
proactive SLAM, 339
probabilistic robotics, 4
probability density function, 14
proposal distribution
    in a particle filter, 102
    in FastSLAM, 446
provisional landmark list
    in EKF SLAM, 328
    in SEIF, 411
pruning, 220
pursuit evasion problem, 569

QMDP, 549
    algorithm, 548, 549

random particles in MCL, 256

random variables, 14
range and bearing sensor, 177
range finder algorithm, 158
Rao-Blackwellized particle filter, 437
ray casting, 154
relative motion information, 133
relaxation algorithm, 390
resampling, 99
resource-adaptive algorithms, 86
robot environment, 19
robot learning problem, 9
rotational velocity, 121

sampling
    from a normal distribution, 124
    from a probability density function, 122
    from a triangular distribution, 124
sampling algorithm
    landmark model, 179, 180
    model model, 141
    normal distribution, 124
    odometry model, 136, 137
    triangular distribution, 124
    velocity motion model, 122, 124
scan matching
    for localization, 234
Schur complement, 359
SEIF, 385
    algorithm with known correspondence, 391, 392
    correspondence test, 416
    correspondence test algorithm, 418
    map fusion algorithm, 425, 426
selective updating
    in localization, 244
    of a histogram filter, 93
semi Markov decision process, 276
sensor failure, 156, 171
sensor measurement, 22
sensor resetting in localization, 276

sensor subsampling in localization, 244
Shepard's interpolation, 561
Sherman/Morrison formula, 50
sigma point, 65
signature of a landmark, 177
significant place, 239
simultaneous localization and mapping, *see* SLAM
single-robot localization, 196
situated agent, 12
SLAM, 309
   exploration, 593
soft correspondence, 419
soft data association constraints, 415
sonar range scan, 149
sparse extended information filter, *see* SEIF
sparseness
   in SEIF, 387
   in the full SLAM problem, 337
sparsification, 390
   in SEIF, 400
Spatial Semantic hierarchy, 306
specular reflection, 150
state
   augmented, 550
   complete, 21, 24
   dynamic, 20
   hybrid, 21
   incomplete, 21
   kinematic, 20
   of an environment, 20
   of knowledge, 26
   static, 20
   transition probability, 25
state augmentation, 269
state transition
   in a linear Gaussian, 41
   nonlinear, 54
static environment, 194
static state, 20

stochastic action effects, 488
stratified sampling, 111
structure from motion, 334, 381
sufficient statistics, 181
supervised learning algorithm, 294

target distribution
   in a particle filter, 100
   in FastSLAM, 447
taxonomy of localization problems, 193
Taylor expansion, 57
textbooks on mobile robotics, 145
Theorem of total probability, 16
thin junction filter, 434
tiger problem, 544
topological map, 93, 239
track, 220
training example, 295
translational velocity, 121
triangular distribution, 128
   algorithm for computing, 123
   algorithm for sampling, 124

UKF
   algorithm, 70
UKF localization
   algorithm, 221
uncertainty ellipse, 82
undo a data association, 363, 369
unexplainable measurements, 157, 171
universal plan, 493
unscented Kalman filter, 65
unscented transform, 65
update, 27

value function, 499
value iteration
   algorithm for an MDP, 501, 502
   algorithm for a POMDP, 527, 529
   for exploration, 579
   in an MDP, 495

Index 647

    in a POMDP, 531
variable elimination algorithm, 344
variance of a sampler, 108
variance reduction, 109
velocity motion model, 121
    algorithm, 121, 123
velocity motion model algorithm, 129
volumetric map, 152